Superbases for Organic Synthesis

Superbases for Organic Synthesis: Guanidines, Amidines, Phosphazenes and Related Organocatalysts

Editor

PROFESSOR TSUTOMU ISHIKAWA

Graduate School of Pharmaceutical Sciences, Chiba University, Japan

A John Wiley and Sons, Ltd, Publication

This edition first published 2009
© 2009 John Wiley & Sons, Ltd

Registered office
John Wiley & Sons Ltd, The Atrium, Southern Gate, Chichester, West Sussex, PO19 8SQ, United Kingdom

For details of our global editorial offices, for customer services and for information about how to apply for permission to reuse the copyright material in this book please see our website at www.wiley.com.

The right of the author to be identified as the author of this work has been asserted in accordance with the Copyright, Designs and Patents Act 1988.

All rights reserved. No part of this publication may be reproduced, stored in a retrieval system, or transmitted, in any form or by any means, electronic, mechanical, photocopying, recording or otherwise, except as permitted by the UK Copyright, Designs and Patents Act 1988, without the prior permission of the publisher.

Wiley also publishes its books in a variety of electronic formats. Some content that appears in print may not be available in electronic books.

Designations used by companies to distinguish their products are often claimed as trademarks. All brand names and product names used in this book are trade names, service marks, trademarks or registered trademarks of their respective owners. The publisher is not associated with any product or vendor mentioned in this book. This publication is designed to provide accurate and authoritative information in regard to the subject matter covered. It is sold on the understanding that the publisher is not engaged in rendering professional services. If professional advice or other expert assistance is required, the services of a competent professional should be sought.

The publisher and the author make no representations or warranties with respect to the accuracy or completeness of the contents of this work and specifically disclaim all warranties, including without limitation any implied warranties of fitness for a particular purpose. This work is sold with the understanding that the publisher is not engaged in rendering professional services. The advice and strategies contained herein may not be suitable for every situation. In view of ongoing research, equipment modifications, changes in governmental regulations, and the constant flow of information relating to the use of experimental reagents, equipment, and devices, the reader is urged to review and evaluate the information provided in the package insert or instructions for each chemical, piece of equipment, reagent, or device for, among other things, any changes in the instructions or indication of usage and for added warnings and precautions. The fact that an organization or Website is referred to in this work as a citation and/or a potential source of further information does not mean that the author or the publisher endorses the information the organization or Website may provide or recommendations it may make. Further, readers should be aware that Internet Websites listed in this work may have changed or disappeared between when this work was written and when it is read. No warranty may be created or extended by any promotional statements for this work. Neither the publisher nor the author shall be liable for any damages arising herefrom.

Library of Congress Cataloging-in-Publication Data

Ishikawa, Tsutomu.
  Superbases for organic synthesis : guanidines, amidines and phosphazenes and related organocatalysts / Tsutomu Ishikawa.
     p. cm.
  Includes bibliographical references and index.
  ISBN 978-0-470-51800-7 (cloth : alk. paper)
  1. Amidines. 2. Guanidines. 3. Phosphazo compounds. 4. Organic bases. I. Title.
  QD341.A7I84 2008
  547′.2–dc22   2008044058

A catalogue record for this book is available from the British Library.

ISBN 978-0-470-51800-7 (H/B)

Typeset in 10/12pt Times by Thomson Digital, Noida, India.
Printed and bound in Great Britain by CPI Antony Rowe, Chippenham, Wiltshire.

To the late Professor Hisashi Ishii and Kimiko

Contents

Preface	xiii
Acknowledgements	xv
Contributors	xvii

1. General Aspects of Organosuperbases   1
Tsutomu Ishikawa
    References   6

2. Physico-Chemical Properties of Organosuperbases   9
Davor Margetic
    2.1   Introduction   9
    2.2   Proton Sponges   10
        2.2.1   'Classical' Proton Sponges   10
        2.2.2   Proton Sponges with Other Aromatic Backbones   12
        2.2.3   Polycyclic Proton Sponges   14
    2.3   Amidines   20
    2.4   Guanidines   24
    2.5   Phosphazenes   31
    2.6   Guanidinophosphazenes   35
    2.7   Other Phosphorus Containing Superbases: Verkade's Proazaphosphatranes   37
    2.8   Theoretical Methods   41
    2.9   Concluding Remarks   41
    References   42

3. Amidines in Organic Synthesis   49
Tsutomu Ishikawa and Takuya Kumamoto
    3.1   Introduction   49
    3.2   Preparation of Amidines   52
        3.2.1   Alkylation of Amidines   52
        3.2.2   Condensation of 1,2-Diamine   53
        3.2.3   Coupling of Imines (Isoamarine Synthesis)   53
        3.2.4   Modification of Amide Derivatives   54
        3.2.5   Multi-Component Reaction   59

	3.2.6	Oxidative Amidination	62
	3.2.7	Oxidative Cyclization to Bisamidine	63
	3.2.8	Ring Opening of Aziridine	63
3.3	Application of Amidines to Organic Synthesis		65
	3.3.1	Acetoxybromination	65
	3.3.2	Aldol-Like Reaction	66
	3.3.3	Azidation	67
	3.3.4	Aziridination	68
	3.3.5	Baylis–Hillman Reaction	68
	3.3.6	Cycloaddition	68
	3.3.7	Dehydrohalogenation	70
	3.3.8	Deprotection	70
	3.3.9	Deprotonation	71
	3.3.10	Displacement Reaction	72
	3.3.11	Horner–Wadsworth–Emmons Reaction	72
	3.3.12	Intramolecular Cyclization	72
	3.3.13	Isomerization	72
	3.3.14	Metal-Mediated Reaction	74
	3.3.15	Michael Reaction	77
	3.3.16	Nef Reaction	78
	3.3.17	Nucleophilic Epoxidation	79
	3.3.18	Oxidation	80
	3.3.19	Pudovik-phospha-Brook Rearrangement	80
	3.3.20	[1,4]-Silyl Transfer	80
	3.3.21	Tandem Reaction	81
3.4	Amidinium Salts: Design and Synthesis		82
	3.4.1	Catalyst	82
	3.4.2	Molecular Recognition	82
	3.4.3	Reagent Source	85
3.5	Concluding Remarks		86
References			86

4. Guanidines in Organic Synthesis — 93
Tsutomu Ishikawa

4.1.	Introduction		93
4.2.	Preparation of Chiral Guanidines		94
	4.2.1	Polysubstituted Acyclic and Monocyclic Guanidines	95
	4.2.2	Monosubstituted Guanidines (Guanidinylation)	95
	4.2.3	Bicyclic Guanidines	97
	4.2.4	Preparation Based on DMC Chemistry	98
4.3	Guanidines as Synthetic Tools		99
	4.3.1	Addition	99
	4.3.2	Substitution	112
	4.3.3	Others	117
4.4	Guanidinium Salt		125

		4.4.1	Guanidinium Ylide	125
		4.4.2	Ionic Liquid	128
		4.4.3	Tetramethylguanidinium Azide (TMGA)	131
	4.5	Concluding Remarks		136
	References			136

5. Phosphazene: Preparation, Reaction and Catalytic Role   145
Yoshinori Kondo

	5.1	Introduction		145
	5.2	Deprotonative Transformations Using Stoichiometric Phosphazenes		150
		5.2.1	Use of P1 Base	151
		5.2.2	Use of P2 Base	156
		5.2.3	Use of P4 Base	159
		5.2.4	Use of P5 Base	164
	5.3	Transformation Using Phosphazene Catalyst		164
		5.3.1	Addition of Nucleophiles to Alkyne	164
		5.3.2	Catalytic Activation of Silylated Nucleophiles	165
	5.4	Proazaphosphatrane Base (Verkade's Base)		176
		5.4.1	Properties of Proazaphosphatrane	176
		5.4.2	Synthesis Using Proazaphosphatrane	176
	5.5	Concluding Remarks		181
	References			181

6. Polymer-Supported Organosuperbases   187
Hiyoshizo Kotsuki

6.1	Introduction	187
6.2	Acylation Reactions	188
6.3	Alkylation Reactions	190
6.4	Heterocyclization	198
6.5	Miscellaneous	200
6.6	Concluding Remarks	205
References		205

7. Application of Organosuperbases to Total Synthesis   211
Kazuo Nagasawa

	7.1	Introduction		211
	7.2	Carbon–Carbon Bond Forming Reactions		211
		7.2.1	Aldol Reaction	211
		7.2.2	Michael Reaction	215
		7.2.3	Pericyclic Reaction	217
		7.2.4	Wittig Reaction	220
	7.3	Deprotection		225
	7.4	Elimination		225

7.5	Ether Synthesis	230
7.6	Heteroatom Conjugate Addition	233
7.7	Isomerization	237
7.8	Concluding Remarks	247
References		247

8. Related Organocatalysts (1): A Proton Sponge — 251
Kazuo Nagasawa

8.1	Introduction	251
8.2	Alkylation and Hetero Michael Reaction	252
	8.2.1 Amine Synthesis by *N*-Alkylation	252
	8.2.2 Ether Synthesis by *O*-Alkylation	252
8.3	Amide Formation	256
8.4	Carbon–Carbon Bond Forming Reaction	259
	8.4.1 Alkylation and Nitro Aldol Reaction	259
	8.4.2 Pericyclic Reaction	261
8.5	Palladium Catalyzed Reaction	264
8.6	Concluding Remarks	268
References		268

9. Related Organocatalysts (2): Urea Derivatives — 273
Waka Nakanishi

9.1	Introduction	273
9.2	Bisphenol as an Organoacid Catalyst	274
	9.2.1 Role of Phenol as Hydrogen Donor	274
	9.2.2 Bisphenol Catalysed Reaction	276
9.3	Urea and Thiourea as Achiral Catalysts	277
	9.3.1 Role of Urea and Thiourea as Hydrogen Donors	277
	9.3.2 Urea and Thiourea Catalysed Reactions	278
9.4	Urea and Thiourea as Chiral Catalysts	282
	9.4.1 Monothiourea Catalysts	284
	9.4.2 Bisthiourea Catalysts	289
	9.4.3 Urea-Sulfinimide Hybrid Catalyst	290
9.5	Concluding Remarks	291
References		292

10. Amidines and Guanidines in Natural Products and Medicines — 295
Takuya Kumamoto

10.1 Introduction	295
10.2 Natural Amidine Derivatives	295
10.2.1 Natural Amidines from Microorganisms and Fungi	296
10.2.2 Natural Amidines from Marine Invertebrates	298
10.2.3 Natural Amidines from Higher Plants	299
10.3 Natural Guanidine Derivatives	299
10.3.1 Natural Guanidines from Microorganisms	300

	10.3.2 Natural Guanidines from Marine Invertebrates	301
	10.3.3 Natural Guanidines from Higher Plant	302
10.4	Medicinal Amidine and Guanidine Derivatives	303
	10.4.1 Biguanides	305
	10.4.2 Cimetidine	305
	10.4.3 Imipenem	306
	10.4.4 NOS Inhibitors	307
	10.4.5 Pentamidine	307
References		308

11. Perspectives — **315**
Tsutomu Ishikawa and Davor Margetic
References — 319

Index — **321**

Preface

Science has developed progressively from the late nineteenth century to the twentieth century and, without stopping, has continuously been evolving through the twenty first century to explore new technologies in various fields. These new technologies give us their benefits with the result that our standard of living is improving. However, negative aspects behind their application to economic development, such as public pollution and global warming etc., have accompanied them and lead to the destruction of the natural environment in some cases. Reconstruction of environmental conditions at a global level is the biggest problem to be solved by us and, therefore, as scientists we should always keep in mind the mission for the next generation.

There are without exception the same pressures on the field of chemistry. Synthetic organic chemistry especially has grown quickly through the design of new intelligent reagents and the discovery of innovative and widely applicable reaction methods. Now, total synthesis of target compounds with an even more complex structure can be formally achieved by the skilful combination of conventional and new methods; there is strong competition between research groups throughout the world to do this.

In spite of the large contribution of chemistry to the improvement in our standard of living, the negative and dangerous image of chemistry has been spread in human society because of destructive damage, such as big explosions in chemical factories in some cases. Therefore, chemistry should progress with the creation of a comfortable world in harmony with nature.

High efficiency based on environmentally benign concepts is strongly required of synthetic organic chemistry in twenty first century. The efficiency involves not only a short reaction process and higher yield in each step, but also lower energy costs and reaction with less waste (high atom economy), and of course from the economical aspect the selection of cheap and easily available materials for the reaction sequence. An important mission of organic chemistry in twenty first century is the establishment of new sustainable chemistry and, in order to achieve the mission, the efficient and repeated use of limited resources is essential. Thus, various types of new recyclable catalysts with high potency have been extensively explored. Organosuperbases are one compound group of promising catalysts in organic chemistry because of their easy molecular modification, possible recyclability, and non or lower toxicity.

Recently, nitrogen-containing organobases, such as guanidines and amidines, have been attracting much attention in organic synthesis due to their potential functionality. It is known that nitrogen–phosphorus hybrid organobases such as phosphazenes show stronger basicity than the nitrogen bases. This book will review the multi-functional ability of these organosuperbases and related molecules in organic synthesis and will discuss their possible perspective as intelligent molecules. I am very happy if this book is able to give a hint in research activity to organic chemists who are interested in organobase catalysts.

I am responsible for this book as editor. Thus, please do not to hesitate to contact me benti@p.chiba-u.ac.jp with any questions.

Tsutomu Ishikawa
Chiba Japan, June 2008

Acknowledgements

This book was completed not only by through the hard work of my team mates, Drs Davor Margetic, Yoshinori Kondo, Hiyoshizo Kotsuki, Kazuo Nagasawa, Takuya Kumamoto and Waka Nakanishi, but also by editorial support from John Wiley, especially Richard Davies. Paul Deards gave me the chance to write this book. I express my sincere appreciation of their cooperation and I am proud to work together with them. My family and laboratory members of the Graduate School of Pharmaceutical Sciences, Chiba University, should be thanked also for their encouragement during the writing.

Contributors

Tsutomu Ishikawa, Graduate School of Pharmaceutical Sciences, Chiba University, 1-33 Yayoi, Inage, Chiba 263-8522, Japan

Yoshinori Kondo, Graduate School of Pharmaceutical Sciences, Tohoku University, Aramaki Aza Aoba 6-3, Aoba-ku, Sendai 980-8578, Japan

Hiyoshizo Kotsuki, Faculty of Science, Kochi University, 2-5-1 Akebono-cho, Kochi 780-8520, Japan

Takuya Kumamoto, Graduate School of Pharmaceutical Sciences, Chiba University, 1-33 Yayoi, Inage, Chiba 263-8522, Japan

Davor Margetic, Rudjer Bošković Institute, Bijenička c. 54, 10001 Zagreb, Croatia

Kazuo Nagasawa, Tokyo University of Agriculture and Technology, 2-24-16 Naka-cho, Koganei, Tokyo 184-8588, Japan

Waka Nakanishi, Graduate School of Pharmaceutical Sciences, Chiba University, 1-33 Yayoi, Inage, Chiba 263-8522, Japan

1
General Aspects of Organosuperbases

Tsutomu Ishikawa

Graduate School of Pharmaceutical Sciences, Chiba University, 1-33 Yayoi, Inage, Chiba 263-8522, Japan

In the field of organic chemistry, a base is generally defined as a reagent capable of abstracting proton to yield a carbanion species. At a basic textbook level, organobases are normally limited to amines, which are categorized as very weak bases according to the above definition. The introduction of an imine function (=NH) to the α-carbon of amines affords more basic amine species, amidines, which correspond structurally to amine equivalents of carboxylic esters (carboxylic acid imidates). Guanidines, which carry three nitrogen functions (one amine and two imines) and correspond to amine equivalents of ortho esters (carbonimidic diamides), show the strongest Brønsted basicity among these amine derivatives [1]. Thus, basicity is proportional to the number of the substituted nitrogen functions at the same carbon atom; representative examples are shown in Figure 1.1. The basicity of guanidine is comparable to the hydroxyl ion (OH^-) [2]. Basic amino acids, lysine and arginine, have amino and guanidine groups, respectively, at the side chains as additional functional groups and can act as base catalysts responsible for important biological actions, such as enzymatic reactions in living organisms, through hydrogen bonding networks caused by these basic characters [3]. On the other hand, histidine belongings to an acidic amino acid in spite of carrying an imidazole ring involving an amidine function as a partial structure [4].

The basicity of these amine derivatives is due to the construction of highly effective conjugation system after protonation under reversible conditions; primitively, it is a reflection of the number of canonical forms, especially isoelectronic forms, in the resonance system (Figure 1.2). This is one of the reasons why guanidines are stronger bases than amidines [5].

Superbases for Organic Synthesis: Guanidines, Amidines, Phosphazenes and Related Organocatalysts
Edited by Tsutomu Ishikawa
© 2009 John Wiley & Sons, Ltd

2   General Aspects of Organosuperbases

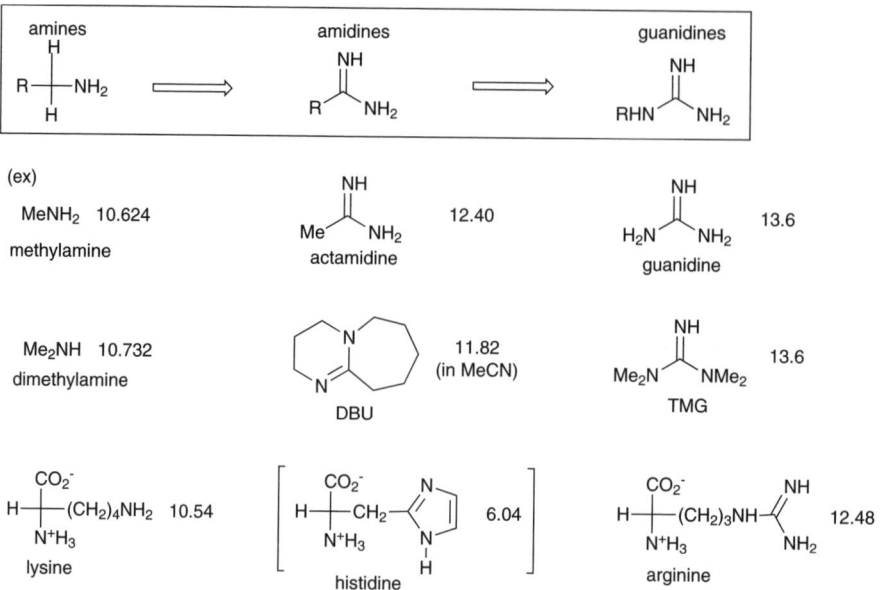

Figure 1.1 Structures of amine derivatives and their representative examples (pKa of the conjugated acids in H_2O): DBU = 1,5-diazabicyclo[5.4.0]undec-5-ene; TMG = 1,1,3,3-tetramethylguanidine

Thus, a pentacyclic amidine (vinamidine) [6] and biguanide [7] with a vinylogous conjugation system show very strong basicity [8], as expected by the above account (Figure 1.3).

An alternative stabilization effect on the protonation to these two bases leading to their highly potential basicity is through bidentate-type hydrogen bond formation as shown in Figure 1.4. Alder also discussed the effects of molecular strain on the Brønsted basicity of amines [9].

In 1985, Schwesinger [10] introduced phosphazenes (triaminoiminophosphorane skeletons), which contain a phosphorus atom [P(V)] bonded to four nitrogen functions of three amine and one imine substituents, as organobases containing a phosphorus atom. They are classified as P_n bases, based on the number (n) of phosphorus atoms in the molecule [11].

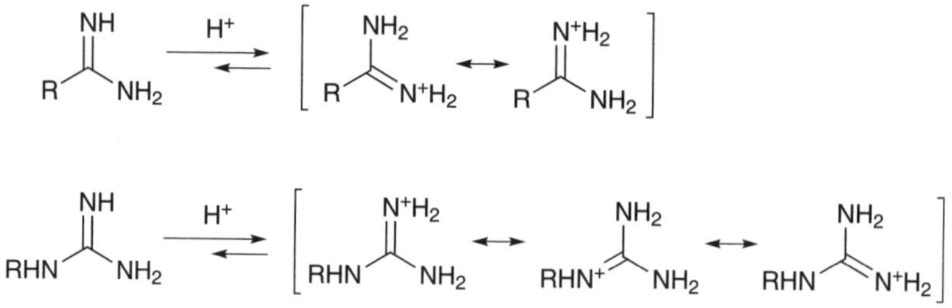

Figure 1.2 Conjugation of amidinium and guanidinium ions

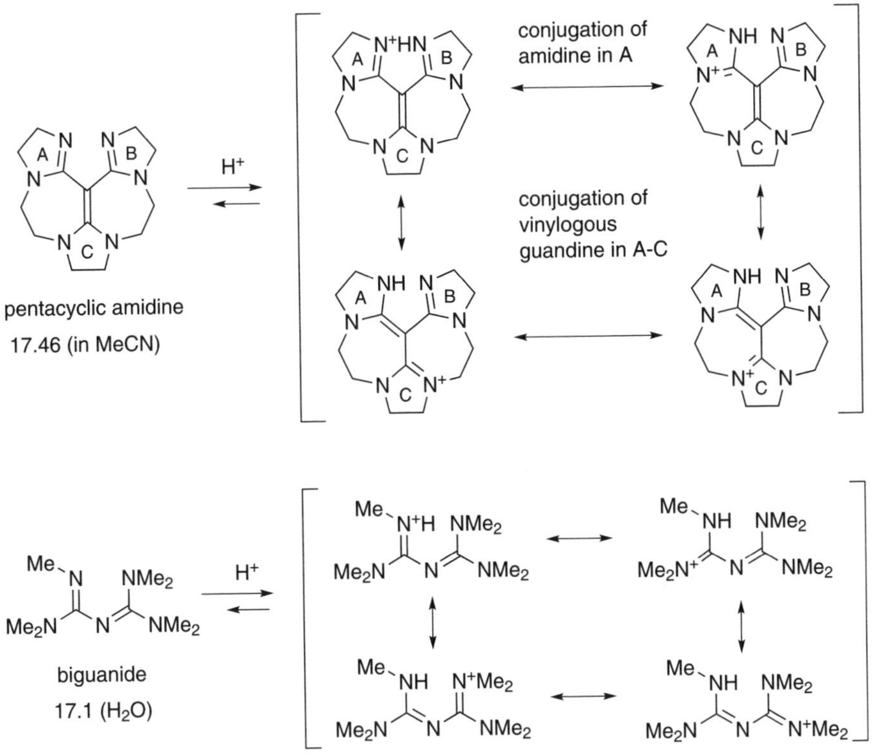

Figure 1.3 Amidine and guanidine derivatives with a vinylogous conjugation system

The examples of simple P1 and P4 bases are shown in Figure 1.5. Their basicity is basically reflected by the number of the triaminoiminophosphorane units and, thus, P4 bases, the strongest phosphazene bases, show basicity comparable to organolithium compounds. Schwesinger *et al.* [12] reported that the strong basicity of phosphazene bases could be caused by the efficient distribution of positive charge through conjugation system in the molecules. However, crystallographic analysis indicates a tetrahedral-like structure around the phosphorus atom in solid state. Phosphazene bases are easily soluble in common organic solvents and stable to not only hydrolysis but also attack by electrophiles owing to their steric bulk [13].

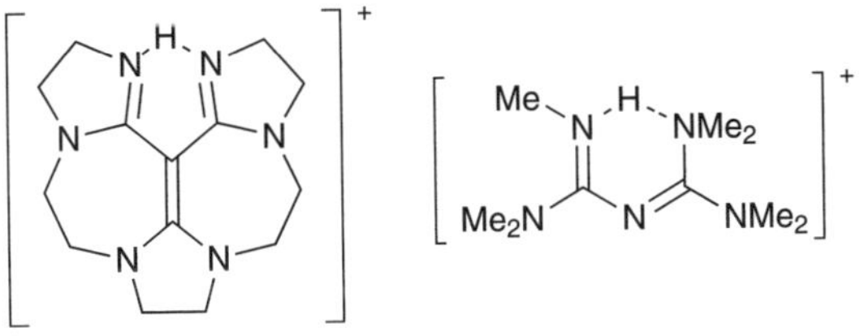

Figure 1.4 Stabilization effect through hydrogen bonding

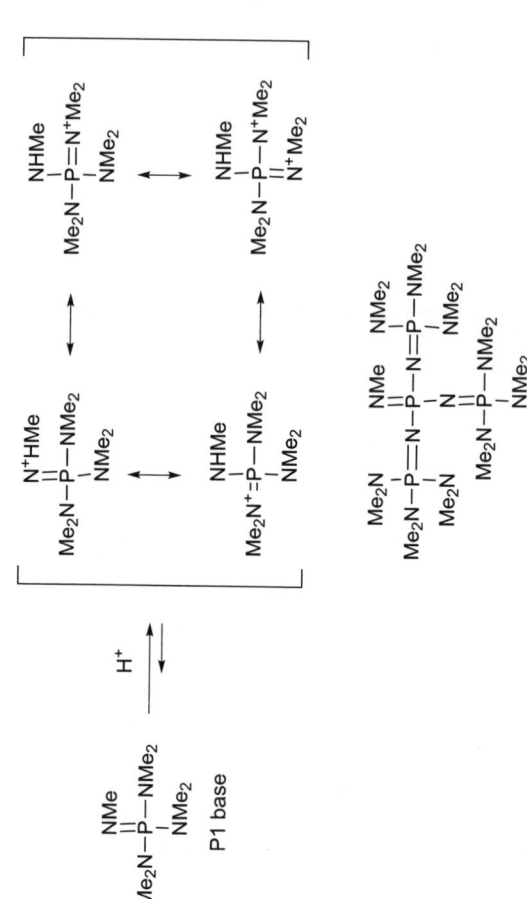

Figure 1.5 Examples of P1 and P4 phosphazene bases

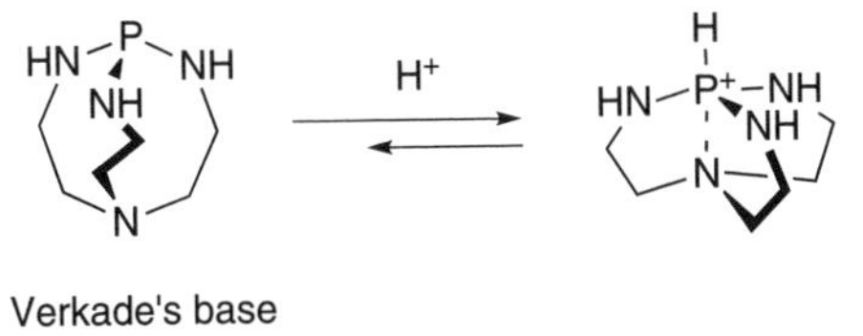

Verkade's base

Figure 1.6 Typical structure of Verkade's base and its basicity due to trans-annular P–N formation

Verkade [14] discovered proazaphosphatranes (cycloazaphosphines) as alternative phosphorus-containing organobases, in which a P(III) atom bonded to three amino groups is located at the bridge head. The basicity of Verkade's bases is comparable to those of P2-type phosphazene bases. The corresponding phosphonium salts formed by protonation on the phosphorus atom are stabilized through effective *trans*-annular N–P bond formation, to which the fourth nitrogen atom located at the alternative bridge head position participates; this result in propellane-type compounds with tricylo[3.3.3]dodecane skeletons, as shown in Figure 1.6.

In 1968, Alder [15] reported the preparation of 1,8-bis(dimethylamino)naphthalene (DMAN) by *N*-methylation of 1,8-diaminonaphthalene. DMAN shows exceptional proton affinity through bidentate-type coordination by the two dimethylamino groups located at *peri* position of the naphthalene skeleton, in spite of being categorized as a weakly basic aromatic amine (Figure 1.7). Thus, DMAN is called a 'proton sponge'.

1,8-Bis(tetramethylguanidino)naphthalene (TMGN) [16] and guanidinophosphazenes [17], such as tris[bis(dimethylamino)methylene]amino-*N-tert*-butylaminophosphorane [(tmg)$_3$N^{t}Bu], are designed as hybrid organobases by the introduction of the guanidine function into the proton sponge and phosphazene skeletons, respectively (Figure 1.8).

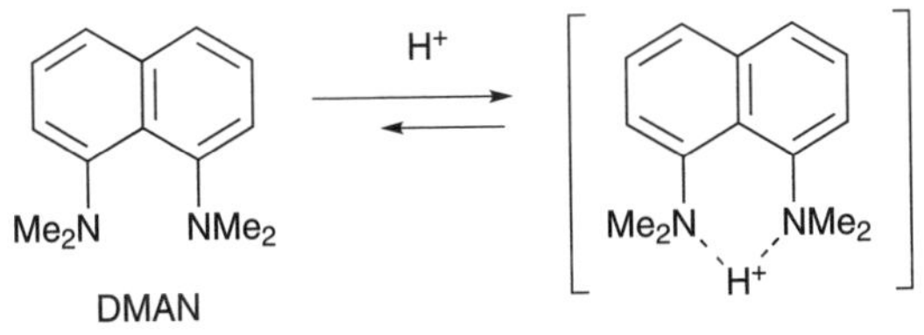

Figure 1.7 Bidentate-type chelation of DMAN

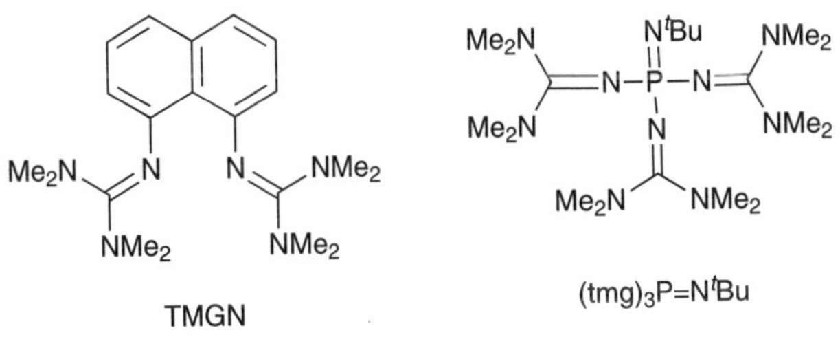

Figure 1.8 Examples of hybrid organobases

Scheme 1.1 *Schematic equation for the definition of superbases proposed by Caubère*

$$\text{base 1} + \text{base 2} + \cdots \rightleftharpoons [\text{super base}]$$

equation 1: unimetal super base
$$A^- M^+ + B^- M^+ \rightleftharpoons [(A^- + B^-)(M^+)_2]$$

equation 2: multimetal super base
$$A^- M_1^+ + B^- M_2^+ \rightleftharpoons [(A^- + B^-)(M_1^+ + M_2^+)]$$

Computational calculation of their proton affinities indicates that these new generations are, as expected, stronger than the original ones [12,18].

Organic chemists often use the words 'strong' or 'super' as the intensive expression of basic property; however, the criteria are ambiguous and dependent upon the chemists who use the expression. Therefore, the expression such as 'strong' or 'super' is ambiguous and causes confusion among organic chemists. Caubère has proposed the definition of superbases as follows in his excellent review [19]: *The term 'superbases' should only be applied to bases resulting from a mixing of two (or more) bases leading to new basic species possessing inherent new properties. The term 'superbase' does not mean a base is thermodynamically and/or kinetically stronger than another, instead it means that a basic reagent is created by combining the characteristics of several different bases*. The general equation for the definition of a 'superbase' is illustrated in Scheme 1.1, in which the examples of 'unimetal superbase' introduced by Caubère and a 'multimetal superbase' by Schlosser [20] are given. Thus, the term superbases in general applies to ionic metal-containing bases acting under irreversible proton abstraction.

One of important and beneficial characteristics of an organic base, especially from the view point of environmental aspects, is the ability of recycling use in repeated reaction, in which reversible proton transfer occurs between the base and a substrate, an acidic counterpart. Thus, powerful organic bases that may be applicable in various organic syntheses as base catalysts have attracted much attention. According to Caubère's definition, organic superbases should be a mixture of two or more different kinds of amine species and show a new property. In this book nonionic powerful amine derivatives of amidines, guanidines, phosphazenes and Verkade's bases with comparable or higher basicity to that of DMAN are *arbitrarily* classified as organic superbases and discussed on their chemistry due to basic characteristics, mainly focusing on their applications to organic synthesis as potentially recyclable base catalysts. Related intelligent molecules are also discussed.

References

1. Haflinger, G. and Kuske, F.K.H. (1991) *The Chemistry of Amidines and Imidates*, **Vol. 2** (eds S. Patai and Z. Rapport), John Wiley & Sons, Chichester, pp. 1–100.
2. Raczynska, E.D., Maria, P.-C., Gal, J.-F. and Decouzon, M. (1994) Superbases in the gas phase. Part II. Further extension of the basicity scale using acyclic and cyclic guanidines. *Journal of Physical Organic Chemistry*, **7**, 725–733.

3. Schlippe, Y.V.G. and Hedstrom, L. (2005) A twisted base? The role of arginine in enzyme-catalysed proton abstractions. *Archives of Biochemistry and Biophysics*, **433**, 266–278.
4. Dawson, R.M.C., Elliott, D.C., Elliott, W.H. and Jones, K.H. (1986) *Data for Biochemical Research*, Oxford Science Publications, pp. 1–31.
5. Raczynska, E.D., Cyranski, M.K., Gutowski, M. *et al.* (2003) Consequences of proton transfer in guanidine. *Journal of Physical Organic Chemistry*, **16**, 91–106.
6. Schwesinger, R., Missssfeldt, M., Peters, K. and von Schnering, H.G. (1987) Novel, very strongly basic, pentacyclic 'proton sponges' with vinamidine structures. *Angewandte Chemie – International Edition*, **26**, 1165–1167.
7. Kurzer, F. and Pitchfork, E.D. (1968) Chemistry of biguanides. *Fortschritte Der Chemischen Forschung*, **10**, 375472.
8. Chamorro, E., Escobar, C.A., Sienra, R. and Perez, P. (2005) Empirical energy–density relationships applied to the analysis of the basicity of strong organic superbases. *Journal of Physical Chemistry A*, **109**, 10068–10076.
9. Alder, R.W. (1989) Strain effects on amine basicities. *Chemical Reviews*, **89**, 1215–1223.
10. Schwesinger, R. (1985) Extremely strong, non-ionic bases: syntheses and applications. *Chimia*, **39**, 269–272.
11. Schwesinger, R. and Schlemper, H. (1987) Penta-alkylated polyaminophosphazenes – extremely strong, neutral nitrogen bases. *Angewandte Chemie – International Edition*, **26**, 1167–1169; Schwesinger, R., Hasenfratz, C., Schlemper, H. *et al.* (1993) How strong and how hindered can uncharged phosphazene bases be? *Angewandte Chemie – International Edition*, **32**, 1361–1363.
12. Schwesinger, R., Schlemper, H., Hasenfratz, C. *et al.* (1996) Extremely strong, uncharged auxiliary bases. Momomeric and polymer-supported polyaminophosphazenes (P2–P5). *Liebigs Annalen*, 1055–1081; Koppel, I.A., Schwesinger, R., Breuer *et al.* (2001) Intrinsic basicities of phophorus imines and ylides: a theoretical study. *Journal of Physical Chemistry A*, **105**, 9575–9586.
13. Allen, C.W. (1994) Linear, cyclic, and polymeric phosphazenes. *Coordination Chemistry Reviews*, **130**, 137–173; Allen, C.W. (1996) Phosphazenes. *Organophosphorus Chemistry*, **27**, 308–351; Kondo, Y., Ueno, M. and Tanaka, Y. (2005) Organic synthesis using organic superbase. *Journal of Synthetic Organic Chemistry*, **63**, 453–463.
14. Kisanga, P.B., Verkade, J.G. and Schwesinger, R. (2000) pK_a Measurements of P($RNCH_2CH_3)_3$N. *The Journal of Organic Chemistry*, **65**, 5431–5432; Verkade, J.G. and Kisanga, P.B. (2003) Proazaphosphatranes: a synthesis methodology trip from their discovery to vitamin A. *Tetrahedron*, **59**, 7819–7858; Verkade, J.G. and Kisanga, P.B. (2004) Recent applications of proazaphosphatranes in organic synthesis. *Aldlichimica*, **37**, 3–14.
15. Alder, R.W., Bowman, P.S., Steele, W.R.S. and Winterman, D.R. (1968) The remarkable basicity of 1,8-bi(dimethylamino)naphthalene. *Chemical Communications*, 723–724;Alder, R. W., Bryce, M.R., Goode, N.C. *et al.* (1981) Preparation of a range of N,N,N′,N′-tetrasubstituted 1,8-diaminonaphthalene, *Journal of the Chemical Society – Perkin Transactions 1*, 2840–2847.
16. Raab, V., Kipke, J., Gschwind, R.M. and Sundermyer, J. (2002) 1,8-Bis(tetramethylguanidino) naphthalene (TMGN): a new, superbasic and kinetically active 'proton sponge'. *Chemistry – A European Journal*, **8**, 1682–1693.
17. Kolomeitsev, A.A., Koppel, I.A., Rodima, T. *et al.* (2005) Guanidinophosphazenes: design, synthesis, and basicity in THF and in the gas phase. *Journal of the American Chemical Society*, **127**, 17656–17666.
18. Kovacevic, B. and Maskic, Z.B. (2002) The proton affinity of the superbase 1,8-bis(tetramethylguanidino)naphthalene (TMGN) and some related compounds: a theoretical study. *Chemistry – A European Journal*, **8**, 1694–1702.
19. Caubère, P. (1993) Unimetal super bases. *Chemical Reviews*, **93**, 2317–2334.
20. Schlosser, M. (1992) Superbases as powerful tools in organic synthesis. *Modern Synthetic Methods*, **6**, 227–271.

2
Physico-Chemical Properties of Organosuperbases

Davor Margetic
Rudjer Bošković Institute, Bijenička c. 54, 10001 Zagreb, Croatia

2.1 Introduction

From the physical-organic point of view, the most interesting physico-chemical property of neutral organosuperbases is their exceptional basicity associated with high kinetic activity in proton exchange reactions. Because of their high basicity and relatively weak nucleophilicity, these nonionic compounds have found wide application as catalysts for organic reactions. By definition [1] the superbases are stronger bases than a 'proton sponge' [1,8-bis (dimethylamino)naphthalene: DMAN], that is, they have an absolute proton affinity (APA) larger than 245.3 kcal mol^{-1} and a gas phase basicity (GB) over 239 kcal mol^{-1}.

The basicity of organic molecules can be measured by various physical methods, in gas phase or condensed media experiments, or calculated by quantum chemical methods. Several solvents, including dimethyl sulfoxide (DMSO), acetonitrile (MeCN), and tetrahydrofuran (THF), have found wide application as media for studies of strong bases. Acetonitrile has been the most popular solvent and a vast number of basicity measurements in acetonitrile have been carried out. Nevertheless, various reference bases have been used for gas phase measurements. Hence, a variety of results obtained by different authors and by different methods has been collected in this review. Material collected in tables has been arranged in descending order of basicity. A number of other organobases approaching a superbasicity threshold have been also measured; however, due to limitations in space here, these molecules have been omitted. This review is divided according to the chemical classes

---

Superbases for Organic Synthesis: Guanidines, Amidines, Phosphazenes and Related Organocatalysts
Edited by Tsutomu Ishikawa
© 2009 John Wiley & Sons, Ltd

2.2 Proton Sponges

2.2.1 'Classical' Proton Sponges

Proton sponges (PS) are organic diamines with unusually high basicity. The exceptional basicity of the very first proton sponge, DMAN (**1**) was reported by Alder in 1968 [2]. This compound has a basicity about 10 million times higher ($pK_a = 12.1$ in water) than other similar organic amines (its experimentally measured proton affinity (PA) in the gas phase is 246.2 [3], while the calculated value [4] is 246.5 and 258.7 kcal mol^{-1}, using 6-31G*//6-31G+ZPE and 6-31G methods). Proton sponges and their complexes have attracted considerable interest from chemists, giving rise to over 70 structural and 100 spectroscopic papers (Table 2.1) [5]. The name proton sponge is given because of the high thermodynamic basicity combined with a kinetic inactivity to deprotonation that resembles the affinity of a sponge for water.

The general feature of all proton sponges is the presence of two basic nitrogen centres in the molecule, which have an orientation that allows the uptake of one proton to yield a stabilized intramolecular hydrogen bond (IMHB). A dramatic increase in basicity of aromatic proton sponge can be achieved on account of: destabilization of the base as a consequence of strong repulsion of unshared electron pairs; formation of an IMHB in the protonated form; and relief from steric strain upon protonation [6]. Two general concepts to raise the thermodynamic basicity or PA are established. One is to replace the naphthalene skeleton by other aromatic spacers, thus influencing the basicity by varying the nonbonding N...N distances of the proton-acceptor pairs. The other concept focuses on the variation of basic nitrogen centres or its adjacent environment ('buttressing effect').

The trend that proton sponges with high thermodynamic basicity typically have a low kinetic basicity (kinetic activity in proton exchange reactions) is a serious limitation of proton sponges: the captured proton does not usually take part in rapid proton exchange reactions, which would allow such neutral superbases to serve as catalysts in base-catalysed reactions. Their further limitations are moderate solubility in aprotic nonpolar solvents and stability towards auto-oxidation.

From a physical organic point of view, there is continuing debate about whether the enhanced basicity in proton sponges is due mainly to strain relief on protonation, or to the special properties of the hydrogen bonds in their monoprotonated ions.

Experimental and theoretical studies have shed light on the structural factors that influence the high basicity of proton sponges. Their abnormally high basicity is accepted to be produced by various contributions: the effective PA for one of the amine groups (assuming asymmetric protonation at one nitrogen); the relief of strain (possibly also accompanied by an increase in aromatic stability) caused by loss of destabilizing lone pair–lone pair repulsion on protonation; the formation of a hydrogen bond which stabilizes the protonated species; the difference in solvation energies of the base and protonated cation in solution.

Table 2.1 Experimentally determined pK_a values of the conjugated acids of naphthalene-type proton sponges [8]

1: $R^1 = R^2 = H$
3: $R^1 = NMe_2, R^2 = H$
4: $R^1 = R^2 = NMe_2$

Compound	R^3	R^4	R^5	R^6	pK_a (solvent)
1					12.00; 12.03; 12.34 (H_2O); 12.10 (30% DMSO/H_2O) 18.28; 18.18 (MeCN); 7.47 (DMSO); 16.8 (THF); 11.5 (20% dioxane)
2	NEt_2	NEt_2	H	H	18.95 (MeCN); 12.7 (20% dioxane)
	NMe_2	NMe_2	OMe	OMe	16.1 (35% DMSO/H_2O); 16.3 (H_2O); 11.5 (DMSO) [10]
	NEt_2	NEt_2	OMe	OMe	16.6 (H_2O)
	Me—N N—Me (piperidine-bridged)		H	H	13.6 (30% DMSO)
	Me—N N—Me (bridged)		H	H	13.0 (30% DMSO)
	Me—N N—Me (O-bridged)		H	H	12.9 (30% DMSO)
	CH_2NMe_2	CH_2NMe_2	H	H	18.26 (MeCN)
	NMe_2	NMeEt	H	H	18.5 (MeCN)
	NEt_2	NMe_2	H	H	18.7 (MeCN)
	NMeEt	NMeEt	H	H	18.7 (MeCN)
	NEt_2	NMeEt	H	H	18.9 (MeCN)
	NMe_2	NMe_2	NH_2	H	10.3 (MeCN)
	NMe_2	NMe_2	NMe_2	H	9.0 (MeCN)
	NMe_2	NMe_2	NMe_2	NMe_2	11.2 [10]; 15.8 [11] (DMSO)
	NMe_2	NMe_2	Me	Me	9.8 (DMSO) [10]
	NMe_2	NMe_2	SMe	SMe	8.1 (DMSO) [10]
	NMe_2	NMe_2	pyrr[a]	pyrr[a]	15.5 (DMSO) [11]
3					8.0 (DMSO)
4					9.8 [12]; 14.4 [11] (DMSO) [13]
5					7.7 (DMSO) [12]; 18.3 (MeCN) [12]

[a] Pyrrolidinyl.

It was shown that the IMHB and the solvation energies in the protonated forms could each increase the basicity by 2–4 pK_a units. The lone pair repulsion was found to be able to increase the pK_a values by up to six pK_a units. Protonation and IMHB reduce steric deformation to the neutral molecule due to repulsion of lone electron pairs of nitrogen atoms of the nearby alkyl groups. Structures particularly favourable for forming an almost linear and, therefore, strong IMHB between nitrogen atoms on protonation (N–H$^+$...N around 175°) are more stabilized. The structural parameters of the IMHB obtained for twenty proton sponges show: the N...N distances are 2.717–2.792 Å [7], while protonated proton sponges generally have N...N, N–H$^+$ and N...H$^+$ distances of 2.541–2.881, 0.86–0.92 and 1.376–1.580 Å, respectively, and N–H$^+$...N angles between 150 and 180°. The strength of the IMHB calculated for proton sponges is in the range of 16–21.5 kcal mol^{-1} [9].

Detailed study by Pozharskii [12] has shown that the so-called 'buttressing effect' represents the complex combination of various interactions of *ortho* substituents with dimethylamino groups in corresponding bases and cations. Basicity is determined by interplay of several factors among which are: the polar effect of *ortho* substituents (electron-releasing substituents increase basicity); electrostatic repulsion between lone electron pairs of the dimethylamino (N(CH$_3$)$_2$) groups (considered as one of the leading in increasing the basicity); reduction of conjugation between *peri*-N(CH$_3$)$_2$ groups and the naphthalene system (may enhance the basicity by 1.2 pK_a); increase of the p-character of the nitrogen lone pairs (enhances the basicity of nitrogen atom); increase of the IMHB strength in the cation; decrease of steric strain in the cation (strain decrease on protonation increases basicity); p,d-interaction of the N(CH$_3$)$_2$ groups with d-elements of such *ortho*-substituents as SCH$_3$, Si(CH$_3$)$_3$ and bromine (slightly reduces basicity); and changes in solvation caused by substituents (hydrophobic groups inhibit solvation of the cation and thus exert some reduction in basicity, while hydrophilic groups favour solvation and bring some increase in pK_a).

Several authors have pointed out the importance of the extent of the p-character of the lone electron pairs of nitrogen and of the distortion of the n,π-conjugation (Korzhenevskaya *et al.*). Flattening of the nearby NR$_2$ groups (planarization caused by sp$^3 \rightarrow$ sp^2 rehybridization) increases their mutual repulsion, which is supplemented by the Coulomb repulsion due to high charge density on the nitrogen atoms, so that the neutral molecule of the proton sponge is structurally destabilized. The consequence of considerable twisting of the N(CH$_3$)$_2$ groups out of the ring plane is loss of conjugation between the NR$_2$ and naphthalene π-system, which is initially reduced by the larger sp^2 character of the nitrogen (N) atom. Both effects provide an essential increase in basicity [14]. The protonation of the proton sponge leads to sp$^2 \rightarrow$ sp^3 rehybridization of the nitrogen atomic orbitals (AOs). Repulsion becomes considerably weaker, the N...N distance shortens, while strong attractive interaction appear, which is favoured by formation of the N–H$^+$...N hydrogen bond due to the high p-character of lone electron pair of the unprotonated nitrogen atom. At the same time, sp$^2 \rightarrow$ sp^3 rehybridization enhances π-conjugation and stabilization of the protonated system.

2.2.2 Proton Sponges with Other Aromatic Backbones

After the discovery of DMAN (**1**), the search started for other, more basic proton sponges possessing aromatic backbones with optimal, even shorter N...N distances to form N–H$^+$...N bridges with the optimal linear geometry. Staab has found that fluorene series

Table 2.2 Experimentally determined pK_a values of aromatic proton sponges

7: X = CH$_2$
10: X = S
11: X = Se
13: X = CH$_2$CH$_2$
14: X = 2 xH
15: X = CH$_2$OCH$_2$
16: X = CH=CH
17: X = CCl=CCl

Compound	pK_a (solvent)	Compound	pK_a (solvent)
7	12.8 (35% DMSO/H$_2$O)	14	7.9 (DMSO)
8	13.6 (60% DMSO/H$_2$O)	15	9.4 (DMSO)
9	14.1 [17] (60% DMSO/H$_2$O)	16	11.5 (DMSO)
10	11.9 (DMSO)	17	10.4 (DMSO)
11	11.8 (DMSO)	18	12.8 (DMSO)
12	12.3 (DMSO)	19	10.3 (DMSO)
13	10.9 (DMSO)		

are stronger bases than DMAN (**1**), with a pK_a value of 12.8 units for 4,5-bis(dimethylamino)fluorene (**7**) (measured in 35% DMSO/H$_2$O) [15] (Table 2.2). It was found that in this molecule the reduction in N...N distance upon protonation is more important for the increase in basicity, that is, the strength of the N−H$^+$...N bond, than the N...N distance in the neutral proton sponge [16]. 4,5-Bis(diethylamino)-9,9-diethylfluorene (**8**) was found to be an even stronger base, as well as a cyclic derivative **9**. The basicity of the corresponding heterocycles dibenzothiophene **10**, dibenzoselenophanes **11** and **12** are between that of fluorene **7** and DMAN (**1**). Although the N...N distances are predicted to be shorter than fluorene **7**, extreme steric strain due to the interactions between N(CH$_3$)$_2$ groups causes a deviation from planarity of the aromatic moiety, as evidenced by X-ray crystallography, with an energy increase of protonated species. Biphenyl structures **14–15** are more conformationally flexible and sterically less strained. Thus, the destabilizing interactions of lone pairs are minimized, but the preconditions of forming a stable, nearly linear hydrogen bridge are reflected in smaller pK_a values. Proton sponges with a phenanthrene structure **16** and **17** have shorter N...N distances than in DMAN (**1**) and fluorene **7**. Because of the larger steric strain, the N−H$^+$...N hydrogen bridge is compressed beyond the energetically optimal N...N distance. There is noticeable bending of peripheral aromatic rings by 36°, making them nonplanar aromatics. Furthermore, the 9,10-dichloro derivative **17** shows a reduction in pK_a due to the inductive effect of the chlorine substituents. As a consequence, compounds **16** and **17** show lower basicity than DMAN (**1**).

Incorporation of a dialkylamino group of proton sponge into an aromatic ring such as in quino[7.8-*h*]quinoline (**18**) results in an increase in basicity [18,19]. In contrast to naphthalene proton sponges, **18** lacks hydrophobic shielding of the basic centres and thus of the hydrogen bond in the monoprotonated cation, which was responsible for the low rate of proton transfer. Therefore, this kinetically active base is stronger than quinoline by 8 pK_a units. A quinoline derivative of **16**, benzo[1,2-*h*:4,3-*h'*]diquinoline [20] (**19**), has a helical deviation of the planar arrangement of aromatic rings, but the N...N distance of 2.705 Å is almost identical to DMAN (**1**) (2.728 Å). Due to lack of conformational flexibility, protonation exchange experiments showed lower basicity than **16**.

Superbasic properties have been predicted by Bucher [21] for a series of bases possessing the *syn*-tris-8-quinolylborane framework. Density functional theory (DFT) calculations employing the B3LYP/6-311 + G(d,p)//B3LYP/6-31G(d) and IPCM/B3LYP/6-31G(d) methods indicate a dramatic increase in the basicity of **20–22** possessing three quinoline nitrogen lone pairs, with additional bridging forcing the nitrogen lone pairs into close proximity. In particular, **21** is predicted to show a basicity approaching that of the most basic known neutral nitrogen bases [PA 268.1 kcal mol^{-1}, pK_a 30.2 (MeCN)], while a smaller PA was calculated for **22** due to the lower flexibility of the benzoquinolines (Figure 2.1).

2.2.3 Polycyclic Proton Sponges

Further extension of the proton sponge concept by anchoring two nitrogen atoms closely in a rigid framework and functionalizing them with substituents has led to the design of a variety of molecular frameworks, which differ from the classical proton sponge framework consisting of condensed phenyl rings. For instance, superbasic properties have been found for bispidines (3,7-diazabicyclo[3.3.1]nonanes) **23–25** by Toom [22] (Figure 2.2). Their respective, spectrophotometrically determined pK_a values in acetonitrile are 21.25, 21.38 and 21.66 units. Previously a considerably lower pK_a of 17.50 was determined for (−)-sparteine in acetonitrile, which might be explained by the presence of moisture that would generate a levelling effect. Measurements in water gave pK_a = 11.96 for **25**, while its gas phase determined GB is 243.4 kcal mol^{-1} [23]. Calculations indicate that the dominant basicity-increasing factor in bispidines **23–25** is the nature and strength of the IMHB.

Structurally related, superbasic proton sponges have been designed by Estrada [24]. A minimal framework was provided by the structures of 3,6,7,8-tetraazatricyclo[3.1.1.1^{2,4}] octane (**26**) and 4,8,9,10-tetraazatricyclo[5.1.1.1^{3,5}]decane (**30**). These molecules, possessing two pairs of nitrogen atoms fixed in a configuration where two nitrogen atoms are in

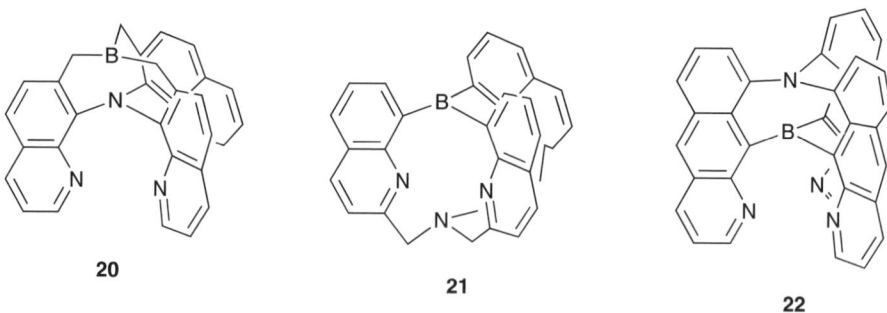

Figure 2.1 Structures of C3-symmetric organic base quinolines **20–22**

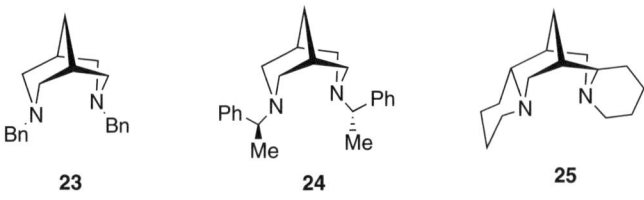

Figure 2.2 Structures of bispidines **23–25**

close proximity, are predicted to have pK_a values in water larger by 5–11 units than DMAN (IPCM/B3PW91/6-311++G** level), with tetraalkylated **29** the largest. Their extreme basicity is ascribed to atypically short N...N distances (average 2.4 Å, while proton sponges have on average 2.5 Å).

The *endo,endo*-8,11-disubstituted pentacyclo[5.4.0.0^{2,6}·0^{3,10}·0^{5,9}]undecanes **32–34** and the tetracyclo derivatives **35** ensure that nitrogen lone pairs of electrons are in close proximity, and the rigid framework of the polycyclic cage guarantees acid–base properties similar to those of a proton sponge [25]. Amine functionalization with alkyl substituents, imines and guanidines, in the line of Sundermeyer and Maksić's work on 1,8-bis(tetramethylguanidino)naphthalene (TMGN) and 1,8-bis(hexamethyltriaminophosphazenyl) naphthalene (HMPN) proton sponges, leads to high basicity. Based on DFT computations, **32–34** are predicted to have gas phase PAs higher than DMAN (**1**) and approaching the TMGN. Compound **34** has the highest PA value among all the aliphatic proton sponges reported (282.7 kcal mol^{-1}, B3LYP/6-311 + G**//B3LYP/6-31G*) (Figure 2.3).

The tetracyclic framework of 11,12-diazasesquinorbornane **36** ($n=1$) is another polycyclic proton sponge system forcing the nitrogen lone pairs of electrons to be present in close proximity. The rigid skeleton leads to basicity properties that are similar to those of the DMAN (**1**), which are partially the consequence of the multiple IMHB stabilization in poly (7-azanorbornane) systems **36** [26]. Furthermore, DFT calculations of the parent tetracyclic compound **37** (which is *endo*-substituted by electron donor groups) have predicted gas phase PAs that exceed that of the DMAN (**1**) by as much as 18.1 kcal mol^{-1} for **39** [27] and equal some of the higher PA values reported for aliphatic proton sponges (at the B3LYP/6-311 + G**//B3LYP/6-31G* level). Furthermore, superbasic proton sponges **38** and **39** are predicted to have a high kinetic activity, with the low-energy proton-transfer barriers (1.2 and 0.86 kcal mol^{-1}, respectively, at the B3LYP/6-311 + G** level). Unsaturation in **40** and

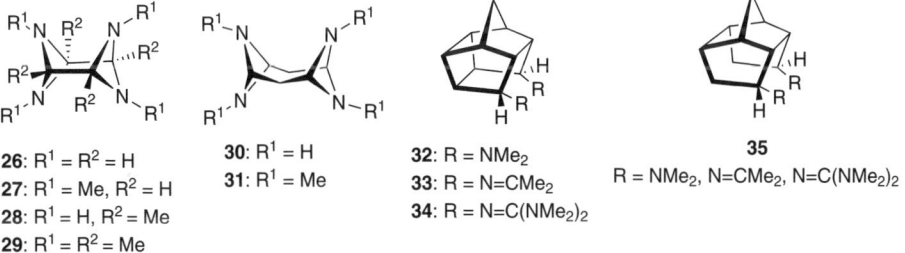

Figure 2.3 Structures of polycyclic amines **26–35**

16 Physico-Chemical Properties of Organosuperbases

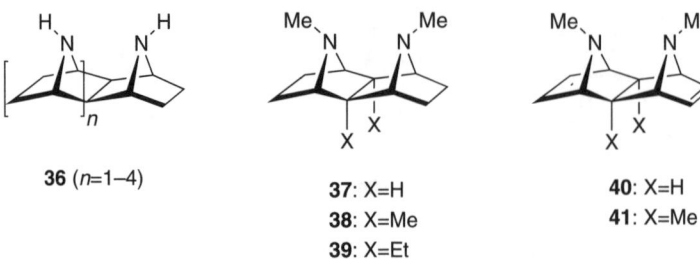

36 (n=1–4)

37: X=H
38: X=Me
39: X=Et

40: X=H
41: X=Me

Figure 2.4 Structures of polycyclic polyamines **36–41**

41 has a minor effect on PAs, while bulkier *endo* substituents enhance basicity, presumably due to the buttressing effect, by forcing the NCH$_3$ groups to be closer to each other, which increases the lone pair repulsions and, thus, destabilizes the base (Figure 2.4).

Potentiometric titration of azatriquineamine trimer **42** [28] in acetonitrile gave a pK_a value of 25.1, which is considerably higher than 1,8-diazabicycloundec-7-ene (DBU) (23.9) and close to phosphazenes. The enhanced basicity of **42** is due to enforced pyramidalization of the apical nitrogen atom in rigid tricycle, together with a tweezer-like structure. Calculations indicate that its protonated form has a strong IMHB with almost ideal geometrical parameters (N...N distance 2.75 Å, and N–H$^+$...N angle 177°), contributing to its pronounced superbasicity (Figure 2.5).

Nitrogen bridgehead cyclic polyamines are another structural class of molecules in which interaction between lone pair electrons is dictated by the carbon framework. Many polyamines in which two or more nitrogen atoms lie in close proximity show increased basicity, some of them larger than DMAN (**1**).

In 1988 Alder reported that the alicyclic diamine **43** was a slightly stronger base than **2** ($R^3 = R^4 = N(CH_3)_2$, $R^5 = R^6 = OCH_3$) in DMSO (pK_a = 11.9) [29]. The 1,6-diazacyclodecane framework provides an ideal geometry for a *trans*-annular hydrogen bond, but there is not much strain relief when **43** is protonated, since the nitrogen lone pairs in the free base are accommodated on opposite sides of a relatively strain-free conformation. Following this work, by use of DFT calculations, Alder designed new, chiral diamines with high pK_a values, in which strain relief on monoprotonation is the main cause of the extreme thermodynamic basicity [30]. A more tightly constrained structure than **43** is that of **44**: the calculated PA of **44** is much higher than that of **43**, due to strain relief on protonation;

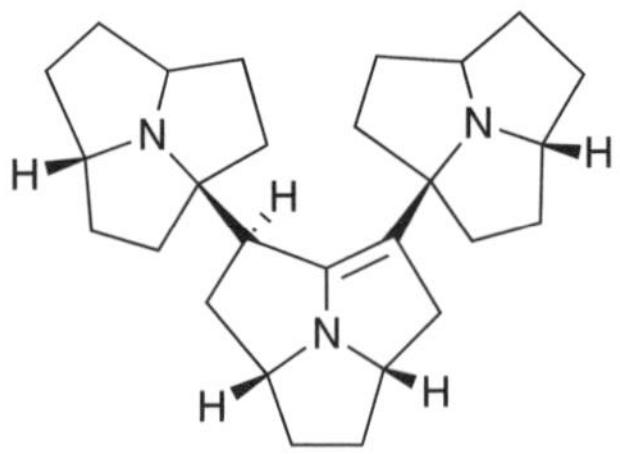

42

Figure 2.5 Structures of azatriquineamine trimer **42**

theoretical PA values are 257.6 (250.2 [31]) and 250 kcal mol^{-1}, respectively, while aqueous pK_a values are estimated to be 21.0 (**44**) and 15.8 (**43**). However, deprotonation experiments have shown that **44** is ineffective as a base, because the inside proton can be neither inserted nor removed by conventional proton transfers [32]. A very low energy barrier for proton transfer between the nitrogen atoms contributes to the high thermodynamic stability of inside protonated **44** [33]. Calculations show that strain in **45** is not effectively relieved by protonation, since the calculated PA is only a little higher than that for **43**. Diamines **46** and **47** are quite flexible, and are able to achieve relatively strain-free conformations. The calculated PAs of **46** and **47** are 253.3 and 243.3 kcal mol^{-1}, presumably due to the electron-withdrawing acetal groups. Diamines **48** and **49**, in which each nitrogen atom is built into a bicyclic framework, retain enough conformational freedom to allow one of the lone pairs to interact with external hydrogen bond donors. Their calculated PAs are 268.6 and 265.8 kcal mol^{-1}, respectively, making them stronger bases than **44**, since the hydrogen bond in protonated **44** (**44H$^+$**) is shorter than ideal. There is greater strain relief when **48** and **49** are protonated than in the case of **44**. The C−N−C angles in **48** average 109.8°, thus signifying perfect sp^3 hybridization. Tricyclic diamines **50** and **51** are only slightly weaker bases than **49**. For the C_2-symmetrical bis tertiary diamine **52** [34] it was found that the proton is buried in the core, leaving a hydrophobic surface. Tertiary amine **52** is more basic than DBU in acetonitrile (pK_a = 24.7) (Figure 2.6).

Bell [35] has developed a series of triamines **53**–**55** which show enhanced basicities. The pK_a values in water for **53**, **55** and **54** are >13.5, 13.1 and 12.8, respectively, due to the exceptional stability of rapidly protonated **53** obtained by strain relief [36]. Cooperation between all three nitrogen atoms of a triamine in stabilization of a single proton increases the pK_a, as evidenced from the crystal structure of **55H$^+$**, where three nitrogen atoms are participating almost equally in a hydrogen-bonded network. The strength of the hydrogen bond of 6.2 kcal mol^{-1} was estimated from dynamic NMR for **55** (Figure 2.7).

Meyer investigated the physico-chemical properties of the tripyrollydinyl-1,4,7-triazacyclononane system **56** [37]. Triamine **56** has a sterically favourable disposition of the three nitrogen lone pairs towards an electrophilic centre, leading to the high Brønsted basicity in aqueous solution (pK_a = 12.8). Due to an effective stabilization of the positive charge, the PA is up to 20 kcal mol^{-1}, higher than the values of noncyclic tertiary aliphatic amines, as estimated by experiment and MP2/6-31 + G*//MP2/6-31 + G* calculations. The stabilizing effect is an energy-lowering interaction between the lone pairs of both the unprotonated nitrogen atoms with the positive charge of the ammonium group via hydrogen bridges (Figure 2.7).

Macrocyclic cryptates **57**–**61** behave as superbasic fast-equilibrating PS due to their flexibility (Figure 2.8). For instance, bicyclo[6.6.2] tetraamine **57** (cross-bridged cyclam) is capable of adopting conformations having all four nitrogen lone pairs pointing towards the cavity centre. The acid dissociation constant for **57H$^+$** is larger than 13.5 in water, and the pK_a = 24.9 in acetonitrile, which is a comparable basicity to DBU (24.32 in acetonitrile) [38].

Cage-adamanzanes such as [3]^{6}adamanzane **58** capture a proton, which in solution rapidly shuttles between all four nitrogen atoms. All the nitrogen lone pairs in the polyaza cage were pointed inward toward the cation, so the complex enjoys both thermodynamic and kinetic stability [39]. As reported by Springborg, these cryptate species are so kinetically resistant to deprotonation that it survives in the presence of strong bases and

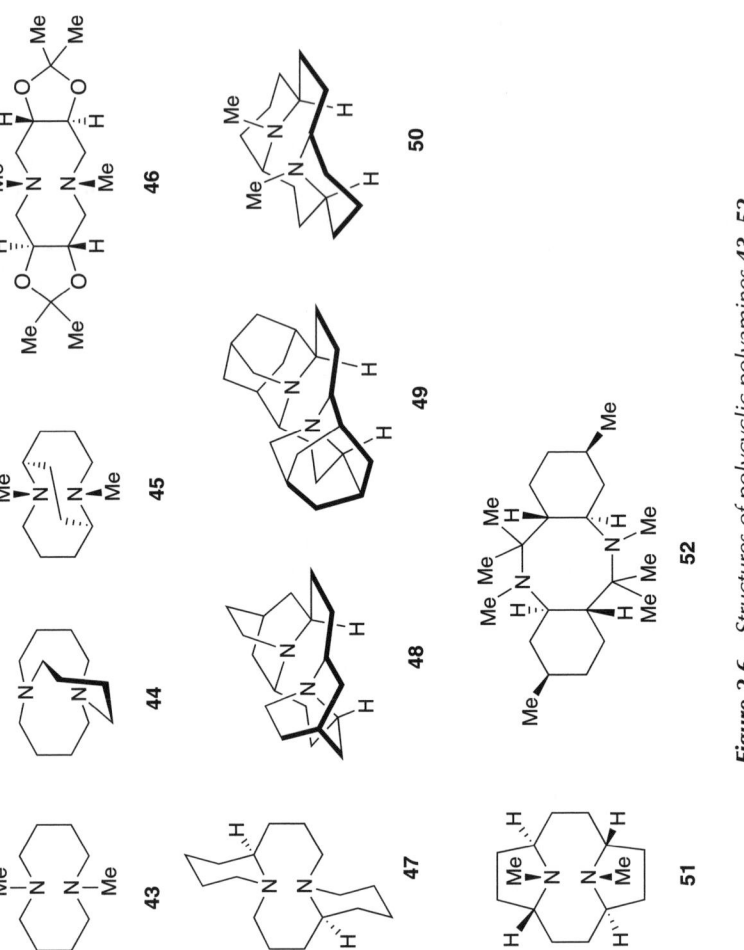

Figure 2.6 Structures of polycyclic polyamines **43–52**

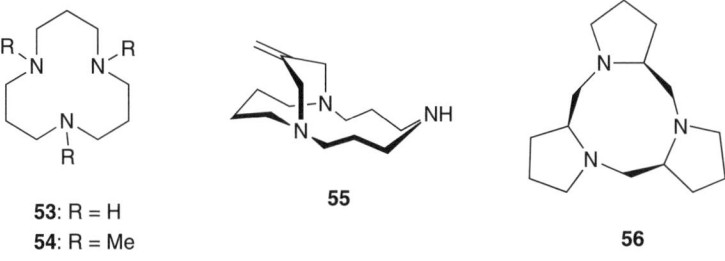

Figure 2.7 Structures of polycyclic polyamines **53–56**

even alkali metal solutions in ammonia and amines, so the pK_a values for protonated cages remains unknown [40]. A common feature for outside protonated adamanzanes is the strong basicity of the free amines. In several cases monoprotonated adamanzanes have a pK_a between 13 and 15 units (in water), while the pK_a of N,N'-dimethyl[$2^3.3^2.2^1$]adamanzane is measured to be 24.9 (acetonitrile), acting as a superbasic proton sponge. The remarkably increased basicity is explained by a hydrogen bonding stabilization of the monoprotonated species between the two bridgehead nitrogen atoms. When the distance between the two bridgehead nitrogen atoms becomes larger, it results in a weakening of the hydrogen bonding stabilization and a concomitant decrease in the base strength, as supported by the B3LYP/6-31 + G**//HF/6-31G** calculations [41].

A number of other medium ring di- and polyamines have been found to have enhanced basicities. For instance, Lehn's [1.1.1]cryptand **59** has an estimated pK_a in water of 17.8 [42] for the internal protonation, while the externally protonation has a much smaller pK_a value. The rates of the proton transfer in and out of the cavity are very slow, the internally protonated species cannot be deprotonated unless the cage is destroyed, making it a thermodynamically very strong and kinetically extremely slow base. Variations on the

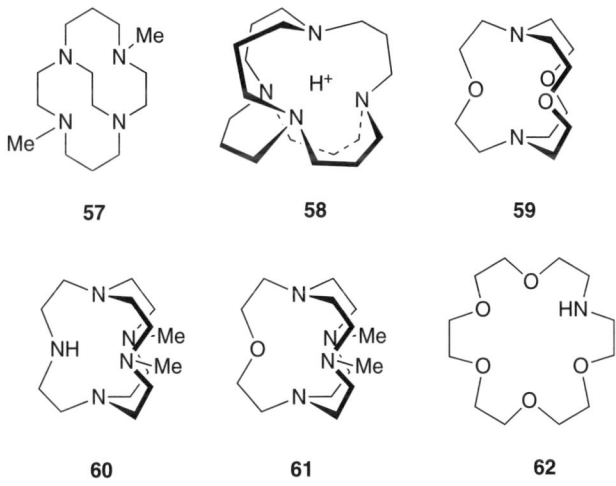

Figure 2.8 Structures of polycyclic amines **57–62**

cryptand structure developed by the Ciampolini and Micheloni groups, such as **60** and **61**, are kinetically much more active [43]. Their respective pK_a values in water are >14, while pK_a(**60**)$_{H_2O/DMSO}$ = 14.8. Protonation of **60** at the secondary nitrogen atom gives IMHB, which is stabilized by interactions with bridgehead atoms. The specially pre-organized structural array of six hydrogen bonds makes the structure particularly stable from a thermodynamic point of view, albeit no single hydrogen bond is particularly strong, accounting for its fast protonation/deprotonation kinetics (Figure 2.8).

The substitution of nitrogen for oxygen greatly increases the proton affinity of aza-18-crown-6 ether (**62**) relative to 18-crown-6; it was detected to be 250 kcal mol^{-1} by the kinetic method. This unusually high basicity is rationalized by a highly symmetrical structure for protonated **62** in which the three most distant oxygen atoms are able to fold back and simultaneously hydrogen bond with the protonated nitrogen [44] (Figure 2.8).

2.3 Amidines

Formamidines are iminoamines built from two nitrogen-containing functionalities, one imine and one amine. Until recently, amidines and guanidines were generally considered the strongest synthetically useful auxiliary bases. Their high basicity is the effect of resonance-stabilized cations. A number of amidine bases are well established reagents in organic synthesis, the strongest of them that are commercially available are 1,5-diazabicyclo[4.3.0]non-5-ene (DBN) and DBU.

Basicity measurements of a great number of amidines have been conducted, both in solution and more recently in the gas phase. It was established that the basicity of amidines depends on the extent and type of substitution at three sites: at the amino and imino nitrogen atoms and at the functional carbon atom. Since the protonation occurs at the imino nitrogen atom, substitution at this site has the largest influence on the pK_a value of amidines, followed by the substitution at the functional carbon [45]. The pK_a values of alkyl-N-substituted amidines measured in ethanol are presented in Table 2.3. Since these substituents show identical electronic effects, pK_as are quite uniform along the series, and the values indicate a modest superbasicity at the lower part of the superbasicity scale.

As shown by Koppel, solution basicity of amidines could be increased to some extent by OCH$_3$ and NCH$_3$ groups positioned at the end of an alkyl chain [47] (Table 2.4). Here, the

Table 2.3 pK_a values of amidines (in 98.5% EtOH) [46]

R	Me$_2$N–C(=NR)–H	Me$_2$N–C(=NR)–Me	Me$_2$N–C(=NR)–Et	Me$_2$N–C(=NR)–i-Pr
^{i}Pr		12.56	12.20	12.26
cyclohexyl		12.55	12.20	12.26
^{n}Pr	12.22	12.46	12.20	12.26
^{n}hexyl		12.37	12.11	11.95
^{n}Bu		12.34	12.13	12.22
^{i}Bu		12.30	11.96	11.93

Table 2.4 pK_a values of amidines with (hetero)ethyl chain

R^1	R^2	R^3	R^4	pK_a (DMSO)	pK_a (H$_2$O)
H	H	H	H	11.6	13.6
H	Me	H	H	11.4	13.6
H	H	Me	H	12.0	13.4
H	H	Me	Me	12.0	13.4
Tos	H	H	H	10.8	13.3
Tos	Me	H	H	11.5	13.3
Tos	H	Me	H	11.9	13.3
Tos	H	Me	Me	11.2	12.8

additional lone pair of the 3-aminoethyl (or 3-methoxyethyl) chain enhances the basicity by stabilization of the conjugate acid (by chelation of the proton and IMHB).

The influence of incorporating an amidine moiety into cyclic structures has also been studied. Simple bicyclic amidines such as DBN and DBU have larger pK_as, compared to alycyclic amidines (Table 2.5). Five-membered amidine DBN is weaker base compared to

Table 2.5 pK_a values for cyclic amidines

Base	pK_a (MeCN)	pK_a (DMSO)	pK_a (H$_2$O)	pK_a (THF)	pK_a (MeCN) (calcd [49])
70	31.94				31.4
69	30.03				
68	29.51				29.8
66	26.95				27.1
65	26.22				
67	25.18				
64			13–14		
DBU	24.33	13.9		16.9 (17.5)	
63	24.0				
DBN	23.79				

six- and seven-membered rings, which is associated with smaller ring strain. The strongest base is DBU, having values close to guanidines. Furthermore, it was found that in water dihydrodiazepine **64** has high basicity, the pK_a value range is 13–14 units, due to the possession of a stable vinamidinium moiety in the conjugate acid [48]. X-ray crystallographic study showed the extensive π-electron delocalization in the vinamidinium portion of the molecule, with stabilization energy of about 19 kcal mol^{-1}.

Schwesinger found a dramatic increase in pK_a by fusion of a five-membered cyclic amidine with a vinamidine moiety, which was additionally attached to a 1,3-imidazolidine ring in **65** [50]. The cation of **65** is planar and has an effective conjugation, which is crucial for effective aromatic stabilization. Joining together two **65** moieties, tricyclic 2,4-diaminovinamidine proton sponges **68** and **69** of extraordinary basicity have been prepared [their pK_a(MeCN) are 30.03 and 29.51, respectively] [51]. In such a molecular system, destabilizing lone pair interactions of the chelate forming nitrogen atoms are diminished by protonation, nitrogen atoms come closer in the protonated form (the N...N distance of 2.54 Å indicates strong IMHB), and pronounced conjugation. Vinamidine proton sponge **67** has slightly less pronounced basicity. The introduction of an additional double bond in **68** gives the most basic vinamidine **70** [52]. Calculations of pK_a (MeCN) values for **66**, **68**, and **70** by IPCM-B3LYP/6-311 + G**//HF/6-31G* method by Kovačević and Maksić [49] have revealed the importance of enlargement of the π-system and aromatic stabilization upon protonation.

Comparisons of solution and gas phase proton transfer thermodynamic data by Raczyńska shed some light on the role played by solvation on amidine basicity. The gas phase basicity and proton affinity values collected in Table 2.6 indicate identical basicity order, as established previously by solution basicity measurements (Tables 2.3–2.5). The following order of gas phase basicities could be established: $GB_{P1phosphazene} > GB_{VIN} > GB_{AAMs} > GB_{Guanidine} > GB_{CAM} > GB_{DBU} > GB_{AMs} > GB_{FAMs} > GB_{BAMs} > GB_{EAMs}$. Vinamidines **66** and **65** are by far the strongest bases containing imidine functionality.

As observed, an increase in the number of amino groups in the molecule has an exceptionally strong influence on the gas phase basicity. The large set of investigated compounds allows some additional comparisons to be made. For simple alkyl amidine derivatives, the basicity in series increases with the polarizability of alkyl groups [56]. Alkylation of the imino nitrogen by increasingly larger groups increases their basicity (1-adamantyl group being the largest, compound **72**). This alkylation effect at the imino nitrogen is estimated to be ~8.4 kcal mol^{-1} [57].

The influence of the aryl substitution for derivatives bearing the aryl group at the functional carbon depends on the electronic nature of the aryl substituent. Smaller basicity is measured than for acetamidines. Bicyclic amidines show larger basicity, compared to the most of simple alkyl derivatives, with the gas phase basicities in the following order: DBU > DBD > PMDBD > DBN > **74**. DBU has the largest gas phase basicity value of 243.4 kcal mol^{-1}. Five-membered ring amidines display smaller basicity than six- and seven-membered cyclic amidines.

Respectable strength has been observed by gas phase basicity and PA measurements for acyclic derivatives containing the dimethylaminopropyl, dimethylaminoethyl and methoxyethyl groups at the imino nitrogen; however, these compounds exhibit rather slow proton transfer rates. The PA value of **71** is 256.7 kcal mol^{-1}, indicating that **71** is a stronger base than TBD and DBU. These groups induce a strong increase in basicity in the following

Table 2.6 Gas phase basicities of amidine derivatives [53,54]

Base	R^1	R^2	R^3	Type[c]	GB	GB[d]	PA
65				VIN	>258	278.0[e]	
66				VIN	254.85		
71	Me	Me	$(CH_2)_3NMe_2$	AAM	246.1	242.9	256.7
72	Me	Me	1-Ad	AAM	244.0	240.4	248.2
DBU				CAM	243.4	239.7	247.5
DBD				CAM	242.5		
amidine	Me	Et	t-Bu	AM	242.15	238.6	246.4
amidine	Me	H	$(CH_2)_3NMe_2$	FAM	241.7	238.0	251.7
PMDBD				CAM	241.15	237.6	251.7
73	Me	H	$(CH_2)_2N(2\text{-py})_2$[f]	FAM	241.1		
DBN				CAM	241.0	237.4	245.1
amidine	Et	Me	n-Pr	EAM	240.75	237.3	245.0
amidine	Me	Me	$(CH_2)_2OMe$	AAM	240.55	236.9	248.3
amidine	Me	Ph	Me	BAM	239.7	236.2	244.0
amidine	Me	H	1-Ad	FAM	239.6	236.2	244.0
amidine	Me	Me	i-Pr	AAM	239.4	235.8	243.5
amidine	Me	Ph	Ph	BAM	235.5		
74				CAM	233.65		
amidine	Me	H	Me	FAM	232.0		

[a] 1,5-Diazabicyclo[4.4.0]dec-5-ene.
[b] PMDBD = 3,3,6,9,9-pentamethyl-2,10-diazabicyclo-[4.4.0]dec-1-ene.
[c] VIN = vinamidine, FAM = formamidine, AAM = acetamidine, AM = amidine, BAM = benzamidine, EAM = Et_2N-acetamidine; CAM = cyclic amidine.
[d] Re-evaluated values with different reference base.
[e] B3LYP/6-31G* gas phase value [55].
[f] py = pyridyl.

order: $(CH_2)_3N(CH_3)_2 > (CH_2)_2N(CH_3)_2 > (CH_2)_2OCH_3$. The basicity increase is associated with the formation of the resonance-assisted strong IMHB. The exceptionally high basicity of **71** suggests that both basic sites participate in the monoprotonation reaction. One site binds the proton and the other one interacts with the protonated function. The chelation of a proton in flexible bidentate nitrogen ligands increases the gas phase basicity value by 5–20 kcal mol^{-1} in comparison to monodentate bases. Strong basicity was also observed for the flexible polyfunctional (2-pyridylethyl)-formamidine **73** (GB = 241.1 kcal mol^{-1}) [58]. The separation of the two basic sites in **71** (and **74**) by the alkane chain increases the chelation effect against the proton. The size of IMHB effect depends on the strength of the basic sites that can chelate the proton and the geometry of monocations, which favours the formation of the IMHB. Direct azinyl substitution on the imino nitrogen in formamidines, leads to proton chelation and IMHB, similar to **74**. However, gas phase basicity values are below the superbasicity threshold, due to less effective cation stabilization by n-π conjugation of the amidine with aza groups [59].

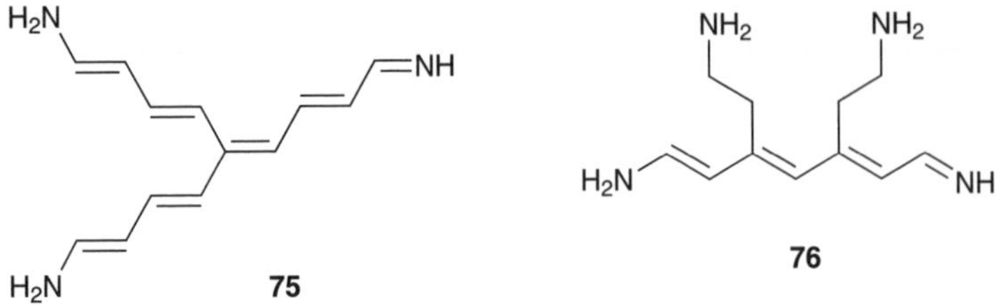

Figure 2.9 Structures of iminoamines **75–76**

Functionally related to amidines are iminoamines, in which imino and amino functionalities are placed at the termini of the conjugated backbone. Howard has published high level calculations of the series of iminoamines employing B3LYP/6-31 + G**//6-31G** method [60]. These studies indicate that the larger the backbone, the larger the conjugation (stabilization of the conjugate acid by the resonance mechanism) and basicity. Theoretically predicted maximum gas phase PAs for linear (unbranched) iminoamines with chain lengths up to six CH=CH units slowly approached saturation point at 266.3 kcal mol^{-1}. Single-branched iminoamines have PAs that are not much larger, slowly approaching the saturation limit at 275.8 kcal mol^{-1} (iminoamine **75**), while double-branched iminoamines are more basic, with the saturation limit at 288.9 kcal mol^{-1} (molecule **76**). Since an unsaturated carbon chain transmits conjugation differently to an alternate =N–C=N–C= structure there is a different relationship between basicity and chain length (Figure 2.9).

2.4 Guanidines

Guanidines have long been recognized as very strong nitrogen organic bases in solution. The measured pK_a(H$_2$O) of guanidine is 13.6, while the most advanced G2 calculations provide a PA of 235–236 kcal mol^{-1}. Furthermore, the biologically important compound arginine has superbasic properties [61]. Guanidine is a stronger base than other nitrogen compounds containing one potentially basic site (*N*-imino or *N*-amino) linked to a carbon atom, for example, pyridines, amines, and amidines, and also those with two basic nitrogen atoms, for example, diamines. Basic properties of guanidines have been used for catalysis of various organic reactions [62]. The unusual thermodynamic stability of acyclic guanidines and their monoprotonated forms is ascribed to the following factors: resonance stabilization; Y-aromaticity [63,64]; favourable distribution of positive charge that leads to a favorable Coulomb interaction; aromaticity; stabilization by intramolecular hydrogen bonding; and the effect of solvation on the stability of the protonated form (strong hydrogen bond between the cation and solvent molecules in solution). All of these factors play an important role in the gas phase and in solution [65].

It was found experimentally that the protonation of guanidine occurs at the imino nitrogen, as the amino nitrogen is less basic than the imino nitrogen atom by ~30 kcal mol^{-1}.

The solution basicity of guanidine is larger than its gas phase basicity. Solvation effects are highly important in determining this difference. The enthalpy of hydration depends on the number of hydrogen atoms linked to the nitrogen atom(s), the size of the neutral and ionic species, and the charge dispersion in the ionic form. The number of positively charged peripheral hydrogen atoms in the guanidinium ion, capable of forming hydrogen bonds with water molecules, and the relatively small size of the cation capable of interacting electrostatically with water dipoles may be the reason for the high enthalpy of hydration of the guanidinium ion. This may account, for the higher basicity of guanidine compared with trimethylamine, which contains the same number of heavy atoms. The pK_a data [45,46,66] experimentally determined are given in Table 2.7.

Substitution of the nitrogen atoms in guanidine by electron-donating groups (such as alkyls) slightly increases its basicity, and substation by electron-accepting groups (e.g., Ph, NH_2, OH, OCH_3, $COCH_3$, CN, NO_2) causes a reverse effect. The linear relationship between the pK_a and the Hammett σ_I constants (or other structural parameters, $\sigma_{m,p}$) describes this behaviour.

An increase in the number of amino groups in the molecule has a strong influence on the gas phase basicity. Generally, biguanides that contain three amino groups are stronger bases than guanidines that have two amino groups (for instance **77** and **79**). Guanidines are stronger bases than amidines which contain only one amino group, while amidines are stronger bases than imines which have no amino group. The same behaviour was found in

Table 2.7 Experimentally determined pK_a values for guanidines

Base	R¹	pK_a (H_2O)	pK_a (EtOH)	pK_a (MeCN)	pK_a (THF)
TBD				25.96	21.0
MTBD		[17.33]^b		25.43	17.9
77	C(NMe$_2$)=NMe		17.1 [67] (benzene)		
PMG^c	Me	15.6 [16.90]		25.00	
78a		[16.45]		24.55	
78b				23.79	
TMG^d	H	[15.2]		23.3	
guanidine	CH$_2$Ph		14.19		
guanidine	Ph	12.18		20.6	14.0
79				19.43 [68]	

^{a}1,5,7-Triazabicyclo[4.4.0]dec-5-ene.
^{b}Values in square brackets are extrapolated from other solvent: ΔpK_a(MeCN-H_2O)~8.1.
^{c}N,N,N',N',N''-Pentamethylguanidine.
^{d}N,N,N',N'-Tetramethylguanidine.

solution. The pK_a values in acetonitrile obtained for biguanides (27–32) were larger than for the corresponding guanidines (23–26) [69].

By combining the proton sponge skeleton, with highly basic guanidine, a superbasic TMGN was obtained by Raab [70]. It represents the one of most basic guanidines experimentally determined, with pK_a(MeCN) = 25.1. This value is comparable with basicities of MTBD (7-methyl-1,5,7-triazabicyclo[4.4.0]dec-5-ene) and PMG in acetonitrile and it is by 5.2 pK_a units higher than DMAN (**1**). In a subsequent paper, Kovačević and Maksić report theoretical PA and pK_a values for TMGN and related proton sponges (Table 2.8), as well as giving an insight into the origin of the high basicity of TMGN [71]. The basicity of these bisguanidines is the combined result of the unfavourable nonbonded repulsions in the initial base, the large PA of guanidine group and strong IMHB present in the protonated species.

The computational results suggest that the origin of high PA and basicity in guanidine proton sponges arises from the inherent basicity of the tetramethyl substituted guanidine fragment and from strong IMHB in the corresponding conjugate acid. The structural and electronic motif given by guanidine fragments undergoes a very strong cationic resonance stabilization that is caused by protonation. Resonance stabilization is found not only in the directly bonded guanidine moiety, but also in the other guanidine fragment, which is more distant from proton (partial protonation). The strength of IMHB is enhanced by this effect and contributes to the IMHB stabilization. Furthermore, angular strain effect and steric repulsion are practically nonexistent in TMGN, contributing to its high basicity. The nonbonded repulsions in the fluorene proton sponge counterpart [4,5-bis(tetramethylguanidino)fluorene (TMGF)] are higher than in TMGN, which in conjunction with a slightly stronger IMHB in the corresponding conjugate acids makes TMGF more basic (PA (MP2) = 263.7 kcal mol^{-1}).

When the guanidine moiety of proton sponge is incorporated in a five-membered ring, such as in bis(dimethylethyleneguanidino)naphthalene (DMEGN), it has been shown by experiment and calculations to be less basic than its TMGN counterpart [72]. The decrease of DMEGN basicity is the consequence of constraints imposed by the geometry of the five-membered imidazoline ring. It leads to considerable pyramidalization of ring nitrogen atoms, thus preventing a perfect π-conjugation of both amino groups with the CN$_3$ unit. In TMGN, N(CH$_3$)$_2$ groups are conformationally less constrained, thus being in better conjugation with the CN3 unit. Basicity could be increased by involving the peripheral nitrogen atoms in an aromatic planar 1,3-dimethylimidazole system (e.g. by introduction of C=C bond in **84**). Induction of some π-conjugation in imidazoline ring by protonation indicates that the IMHB is a complex phenomenon, affecting the whole conjugate acid. IMHB in guanidine and its proton sponge derivatives plays an important role in the basicity increase [73] (Table 2.8). The strength of IMHB could be deduced from the molecular structures of protonated species: **80** and **81** have the shortest H$^+$...N distance, the highest hydrogen bond stabilization and largest APAs. At the same time, the N–H$^+$...N angle assumes maximal value for **80** and **81** (138° and 137°). On the contrary, N...N distance in a protonated base (BH$^+$) is not a good indicator of the strength of IMHB. The 'partial protonation' of other N imino atoms in BH$^+$ is a general feature for all molecules.

Good choice of molecular backbone for attachment of bis(tetramethylguanidine) fragments leads to a strong superbase. The substitution of the aromatic backbone could also affect the basicity, as shown for 1,2-bis(dimethylethyleneguanidino)benzene (**86**) by

Table 2.8 Computationally estimated PAs and pK_a values for guanidine proton sponges

	PA (MP2)[a]	PA (HF$_{SC}$)[b]	pK_a (MeCN)$_{theor}$[c]
80	268.2	264.7	29.0
81	266.8	264.0	28.8
82	263.8	260.0	25.9
TMGF	263.7		27.8
TMGN	257.5		25.4 (25.1)
83	260.5	260.1	24.4
84	256.0	257.8	25.8
85	254.3	253.3	24.0
86 average	240.3–250.3	243.6–252.7	18.0–20.7
87	253.8		24.6
88	252.0	251.8	
DMEGN	250.8	252.3	23 (TMGN > DMEGN)
DMAN	245.5		19.9 (18.2)
TMPhG	245.1		21.1 (20.6)
TMG	244.6		23.7 (23.6)
guanidine	233.7		24.1

[a] MP2(fc)/6-311+G**//HF/6-31G*+ZPVE(HF/6-31G*).
[b] Scaled HF/6-31G*.
[c] IPCM-B3LYP/6-311+G**//HF/6-31G* (experimental values are given in brackets).

Margetić and Ishikawa [74]. The backbone change from benzene to naphthalene and phenanthrene rings has a minor effect on the APA of **86**. Substitution of an aromatic ring in *trans*-**86** with electron-withdrawing halogen atoms leads to decreased APAs in the order 3,4,5,6-tetrafluoro < 3-6-dibromo < 3-bromo < **86**. These estimations are in good accordance with literature data showing that aromatic substitution and electron withdrawing groups decrease the APA of amines. The decrease of basicity going from **86** to **85** is expected, as earlier discussed for the DMEGN/TMEGN basicity difference (PA(MP2)$_{86}$ 250.3, PA(HF$_{SC}$)$_{86}$ = 252.7 kcal mol^{-1}, pK_a(MeCN)$_{86}$ = 20.7). A comparison with the APA of DMAN (**1**) indicates that *o*-bisguanidinobenzenes **86** are of similar basicity or slightly more basic than DMAN (**1**), thus at the lowest part of the superbasicity ladder, however still being strong hydrogen acceptors [75].

The increase in the guanidine basicity scale, going from mono- to polyguanidines has been theoretically studied by Maksić and Kovačević. Replacement of one of the amino groups in guanidine by the N=C(NH$_2$)$_2$ structural fragment forms polyguanidines [76]. It was shown that the increase of APA(HF$_{SC}$) could be as much as 50 kcal mol^{-1}, compared to guanidine itself. The origin of their very highly intrinsic basicity is traced to a large increase in the resonance interaction of the corresponding conjugate bases (in the range of 24–27 kcal mol^{-1}): that is, to the high stability of the corresponding protonated cations. The guanidine group of biguanidine linked by its imino nitrogen to the amidine group has a strong electron-donating character, and the amidine group linked by its functional carbon to the guanidine group has a strong electron-accepting character. Therefore, the n-π conjugation of the guanidine group may be transmitted to the imino nitrogen in the amidine group. It was calculated that the linear chain polyguanides exhibit increased basicity as a function of the number of guanide subunits. The increase of APA is around 7 kcal mol^{-1} per guanidine unit, reaching a limit for $n = 5$ (APA value of 254 kcal mol^{-1}). The APA increase going from mono to biguanide can be explained in terms of π-electron conjugation induced in the conjugate acid (larger number of resonance structures), indicating greater stability of protonated species. At the same time, there is a relaxation effect due to reorganization of the electron density upon protonation. Positive charge is delocalized over the conjugate acid via mobile π-electrons through the π-conjugation mechanism. Branched polyguanides have higher APAs than their linear counterparts, possessing an additional stabilization by IMHB. The largest PA was found in a doubly bifurcated heptaguanide **89**, being as high as 285 kcal mol^{-1}, due to strong resonance effect, partially enhanced by IMHB intramolecular multiple corona effect (on six N sites). *N*-Alkylation further increases APA (for instance, going from **90** to its heptamethyl derivative, or from **89** to **91**), where **91** has the largest calculated APA of the all studied polyguanides (290.0 kcal mol^{-1}) (Figure 2.10).

These predictions are further corroborated by pK_a estimations using the IPCM-B3LYP/6-311 + G**//HF/6-31G* method (heptaguanidine **86** (33.5), triguanidine (31.3), heptamethyl derivative of biguanide **90** (28.1)) [49], and by the B3LYP/6-31G* calculations of polyguanidines recently reported by Chamorro *et al.* [55]. Similarly, Kolomeitsev [77] have calculated very high gas based basicity (PA) values for triguanidine [(H$_2$N)$_2$C=N]$_2$C=NH and its tetramethyl derivative [((CH$_3$)$_2$N)$_2$C=N]$_2$C=NH at the B3LYP/6-311 + G** level – 248.4 (255.1) and 268.4 (276.2) kcal mol^{-1}, respectively.

The importance of the IMHB for achieving larger intrinsic PAs in organosuperbases has been mentioned in several papers and discussed in more details by Kovačević *et al.* in the case of *N*,*N*′,*N*″-tris(3-dimethylaminopropyl)guanidine (**92**) [78]. The aminopropyl chain

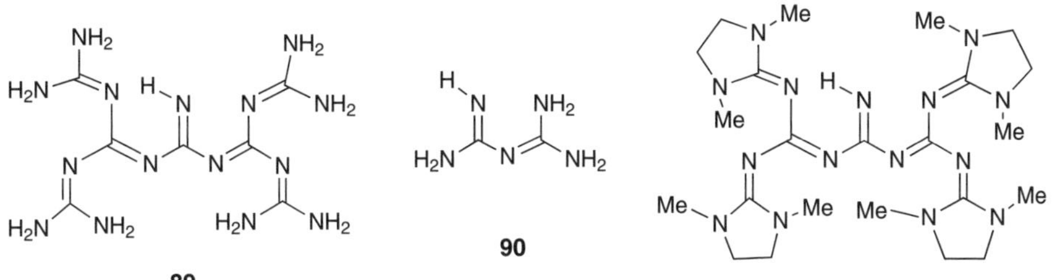

Figure 2.10 Structures of polyguanidines **89–91**

amplifies basicity, if it was linked to the highly basic imine nitrogen, which in turn was a part of the molecular backbone undergoing aromatization upon protonation (Table 2.9). The APA of **94** forming a pseudo eight-membered ring (*two-centre corona effect*, with IMHB on the imine aminopropyl chain with the neighboring amine) is larger than the pseudo six-membered IMHB of **93** (*single-centre corona effect*, the loop starts and ends with the same atom) by about 2 kcal mol^{-1}. This difference is caused by ring strain differences and stronger IMHB of **94** due to better alignment of N–H$^+$...N atoms.

Aminopropyl substitution of guanidines leads to high intrinsic APAs and basicities culminating in the APA(MP2) of **92**. The reason behind a high PA is a strong cationic resonance in the central guanidine moiety (the contribution to the PA is in the range

Table 2.9 Calculated APAs (kcal mol^{-1}) and pK_a values

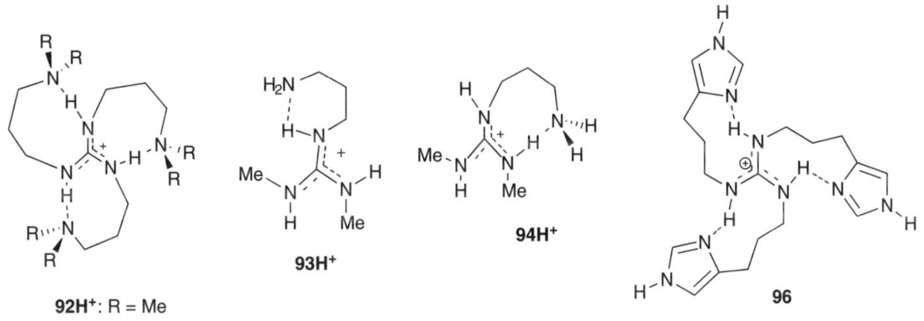

Molecule	PA (MP2)	APA (HF$_{SC}$)	pK_a (MeCN)
92	275.5	266.9	29.4
93	254.2	254.5	25.9
94	256.5	256.7	26.8
95	268.4	267.5	28.8
96		275.1	

MP2(fc)/6-311+G**//HF/6-31G*+ZPVE(HF/6-31G*);IPCM-B3LYP/6-311+G**//HF/6-31G*.

24–27 kcal mol^{-1}) and the strength of the IMHB, which is enhanced upon protonation. In the conjugate acid BH$^+$ the protonated imino nitrogen of guanidine acts as a proton donor, whereas the side chain amino group is a proton acceptor. A cooperative IMHB effect in **95H$^+$** realized by three N(sp^3)–H...N(sp^3) hydrogen bridges (*multiple corona effect*) contributes 18.3 kcal mol^{-1} to the APA of **95**. Guanidine **92** assumes an even higher PA due to an additional relaxation effect caused by the methyl groups. The multiple two-centre corona effect was later found experimentally by Glasovac in the X-ray structure of **92** [79]. The incorporation of stronger hydrogen bonding acceptor functionalities in the propyl chain, such as imidazole in molecule **96**, is predicted to further enhance guanidine basicity (275.1 kcal mol^{-1}) [80]. However, it should be noted that IMHB [81] are almost exclusive to the gas phase, and their relevance to the reactivity in condensed media is much smaller. This is so because the hydrogen bond donor of an IMHB prefers to form IMHBs with the solvent molecules, mainly when they are good hydrogen bond acceptors, rather than being involved in the IMHB itself.

Extended π-electron systems with guanidine and cyclopropenimine structural subunits [82] devised by Maksić offer another structural variation of guanidines capable of effective stabilization of the system upon protonation. It was calculated that the guanidino-cyclopropenimines possess higher PAs than their polyguanide counterparts (PA(HF$_{SC}$): **97**: 251.0; **98**: 262.8; **99**: 267.9, and **100**: 280.1 kcal mol^{-1}). The origin of the increased basicity lies in a large resonance effect triggered by the protonation via aromatization of the three-membered rings in the conjugate acid form. Resonance stabilization of guanidine is within the range of 24–27 kcal mol^{-1}, while those of guanidino-cyclopropenimines contribute to PAs by 41–44 kcal mol^{-1}. Alkyl groups further stabilize protonated forms by an inductive effect through σ- and pseudo-π-hyperconjugation mechanisms (Figure 2.11).

Gas phase basicities of a number of guanidine derivatives with PAs larger than 239 kcal mol^{-1} (1000 kJ mol^{-1}) have been experimentally evaluated by Raczynska *et al.* [53,54]; the results are summarized in Table 2.10. Some trends for various classes of compounds could be identified.

Examination of the gas phase basicity values indicates that for acyclic alkyl derivatives, the basicity increases with the polarizability of alkyl groups in the following order: TEGs > TMGs > ITBD > ETBD > MTBD > TBD, (where TEG is the (Et$_2$N)$_2$C=N– and TMG the ((CH$_3$)$_2$N)$_2$C=N– group). Guanidine derivatives bearing the aryl group at the imino nitrogen have smaller gas phase basicity values than alkyl groups, hence: TMG–CH$_3$ > TMG–Ph. The relative order follows the same substitution trends obtained by

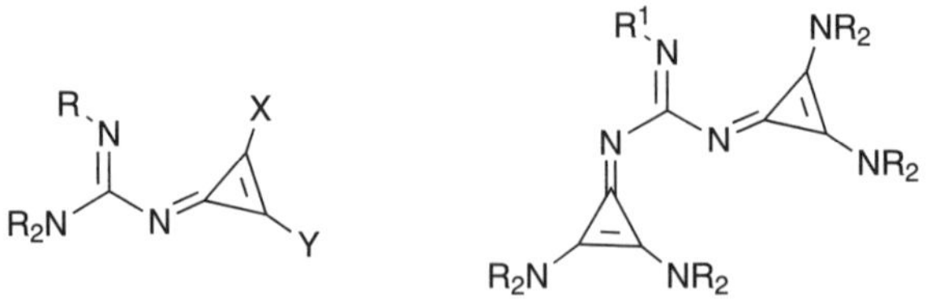

97: X = R = H, Y = NH$_2$
98: X = Y = NMe$_2$, R = Me

99: R = R^1 = Me
100: R = R^1 = Me

Figure 2.11 Structures of guanidino-cyclopropenimines **97–100**

Table 2.10 Gas phase basicities of guanidine derivatives (kcal mol^{-1})

Guanidine base	GB(B)	PA [18]
TMG	234.9 [18]	243.1
(Et$_2$N)$_2$C=N−Me	246.1	
ITBD	248.3	248.3
ETBD	247.5	247.5
MTBD	246.95; 243.3 [18]	250.9
TBD	244.7; 241.2 [18]	249.3
(Me$_2$N)$_2$C=N−Me (PMG)	242.8	
(Me$_2$N)$_2$C=N−Ph (TMPhG)	240.7	
(Me$_2$N)$_2$C=N−(CH$_2$)$_2$OMe	246.6	246.6
(Me$_2$N)$_2$C=N−(CH$_2$)$_2$NMe$_2$	248.4	248.4
(Me$_2$N)$_2$C=N−(CH$_2$)$_3$OMe	248.75	248.75
(Me$_2$N)$_2$C=N−(CH$_2$)$_3$NMe$_2$	250.65	250.65

solution pK_a measurements (Table 2.7). Heteroalkyl groups [(CH$_2$)$_n$OCH$_3$ and (CH$_2$)$_n$N(CH$_3$)$_2$, $n = 2$ or 3] at the imino nitrogen induce the strongest increase in basicity. The order of this increase is: (CH$_2$)$_2$OCH$_3$ > (CH$_2$)$_2$N(CH$_3$)$_2$ > (CH$_2$)$_3$OCH$_3$ > (CH$_2$)$_3$N(CH$_3$)$_2$. Incorporation of the guanidine moiety into cyclic structures such as six-membered rings in TBD increases basicity comparing to acyclic guanidine, the gas phase basicity values are further increased by alkylation in MTBD, ETBD and ITBD.

2.5 Phosphazenes

The exceptional basicity of phosphazenes (iminophosphoranes) has been discovered by Schwesinger [83]. The phosphazene derivatives have been proved to be chemically very stable, kinetically active and highly versatile, and the large number of these bases has been synthesized. The gas phase and solution equilibrium basicity measurements for a large number of phosphazenes are conducted by Kaljurand et al. [84], and their PA and pK_a values are published in various papers. These measurements show that phosphazenes surpass in their basicity the derivatives of acyclic or bicyclic guanidines, amidines and vinamidines (Table 2.11). Extraordinary basicity of phosphazenes was theoretically rationalized by Maksić et al. [85], in terms of effective resonance stabilization of protonated molecules [86].

Condensed phase measurements have been conducted in several solvents. In acetonitrile, for most of the alkylphosphazenes the pK_a values are within the range 26–47 pK_a units. The analysis of the pK_a values in acetonitrile (Table 2.11) shows some general trends. The most significant is that substitution of the hydrogen atoms of the aniline (amine) amino group by the =P(pyrr)$_3$ group increases the basicity approximately by 10–12 pK_a units [69]. Throughout the text Schwesinger's nomenclature will be used (here P1, P2, etc. numbers denote number of P=N- units). There is an increase in basicity with increasing number of phosphorus atoms. Hence, P2 phenyliminophosphoranes are stronger bases than the corresponding P1 phenyliminophosphoranes by about 4–5 pK_a units and so on, where for

Table 2.11 Solution and gas phase basicity data for phosphazenes [90,91]

101: Me$_2$N–P(NMe$_2$)=N–P(NMe$_2$)=N–P(NMe$_2$)=N^{t}Bu

102–107: R$_2$N–P(NR$_2$)=N–P(NR$_2$)(=N–R')–N=P(NR$_2$)–NR$_2$, with R$_2$N–P(NR$_2$)=N–P(NR$_2$)–NR$_2$ branch
- 102: R = Me; R' = ^{t}Bu
- 103: R = Me; R' = CMe$_2$CH$_2$^{t}Bu
- 104: R = Me; R' = CEt$_3$
- 105: R = Me; R'= CMe$_2$^{t}Bu
- 106: NR$_2$ = pyrr ; R' = ^{t}Bu
- 107: NR$_2$ = pyrr ; R' = 2-ClC$_6$H$_4$

108, 109: R$_2$N–P(NR$_2$)=N–P(NR"$_2$)=N–NR$_2$
- R = Me R'P$_3$(dma)R"
- NR$_2$ = pyrr R'P$_3$(pyrr)R"
- 108: R = R" = Me, R' = ^{t}Bu
- 109: R = Me, R" = ^{i}Pr, R' = ^{t}Bu

110, 111: R$_2$N–P(NR$_2$)=N–P(NR$_2$)=N–P(NR$_2$)(=N–R')–NR$_2$ with R$_2$N–P(NR$_2$)–NR$_2$ branch
- 110: R = Me; R' = ^{t}Bu
- 111: NR$_2$ = pyrr ; R' = ^{t}Bu

117: Me$_2$N–P(NMe$_2$)=N–P(NMe$_2$)=N–P(N^{t}Bu)(N^{i}Pr$_2$)–N=P(NMe$_2$)–N=P(NMe$_2$)–NMe$_2$

112–116: R$_2$N–P(NR$_2$)=N–P(NR$_2$)=N–R'
- 112: R = R' = Me   R'P$_2$(dma)
- 113: R = Me, R' = Et
- 114: R = Me, R' = ^{t}Bu
- 115: R = Me, R' = cctyl
- 116: NR$_2$ =pyrr, R' = ^{t}Bu   R'P$_2$(pyrr)

118: Me$_2$N–P(NMe$_2$)=N–P(NMe$_2$)–NMe$_2$ with branched structure (NMe$_2$ groups)

119–121: R$_2$N–P(NR$_2$)(=NR')–NR$_2$
- 119: R = Me; R' = ^{t}Bu   R'P$_1$(dma)
- 120: R = R' = Me
- 121: NR$_2$ = pyrr ; R' = ^{t}Bu   R'P$_1$(pyrr)

122: cyclic phosphazene with Me, Et groups

123, 124: Me$_2$N–P(NMe$_2$)=N–P(cyclic N-R, N-^{t}Bu)–Me
- 123: R = Me; 124: R = Pr

125: BEMP

Base		pK$_a$ (MeCN)	pK$_a$ calc (MeCN)	pK$_a$ (THF)	pK$_a$ DMSO (H$_2$O)	GB$_{exp}$	GB$_{calc}$
111	P$_5$	46.9					
118	P$_7$	(45.3)			>32		
110	P$_5$	45.3			>32		
117	P$_5$	(44.0)					
106	P$_4$	44.0					
105	P$_4$	(42.7)					
104	P$_4$	(42.7)					
103	P$_4$	(42.7)			29.4		
102	P$_4$	(42.7)			30.2		
108	P$_3$	38.6			26.2		
109	P$_3$	38.6			26.1		
4-MeO-C$_6$H$_4$P$_4$(pyrr)				28.9 [76]			
101	P$_4$	36.6					
114	P$_2$	33.49			21.5		
107	EtP$_3$(dma)			25.3			
115	P$_2$	33.27					
124	P$_2$	33.11					
123	P$_2$	33.08					
113	P$_2$	32.94			21.1		
112	P$_2$	32.72					

(continued)

Table 2.11 (Continued)

	Base	pK_a (MeCN)	pK_a calc (MeCN)	pK_a (THF)	pK_a DMSO (H$_2$O)	GB$_{exp}$	GB$_{calc}$
114	P$_2$	33.49			21.5		
107	EtP$_3$(dma)			25.3			
115	P$_2$	33.27					
124	P$_2$	33.11					
	EtP$_1$(pyrr)	28.89	28.8	21.7		259.5	257.8
	t-BuP$_1$(pyrr)	28.35		20.2		258.7	258.2
122	P$_1$	28.27					
	PhP$_2$(pyrr)	27.55		20.9			
120	MeP$_1$(dma)	27.58 [70]		20.7		252.2	252.3
125	BEMP^a P$_1$	27.5 [92]				256.0	255.6
	HP$_1$(pyrr)	27.01		20.8	(13.93)	255.1	255.0
119	t-BuP$_1$(dma)	26.88	27.2	18.9	15.7	253.2	252.1
	HP$_1$(dma)	25.85	25.7	19.7	(13.32)	250.0	249.6
	2,5-Cl$_2$-C$_6$H$_3$P$_1$(pyrr)					248.4	
	2-Cl-C$_6$H$_4$P$_2$(pyrr)	25.42		17.5		260.5	258.7
	4-NMe$_2$-C$_6$H$_4$P$_1$(pyrr)	23.88		17.1	(12.00)	257.5	
	4-OMe-C$_6$H$_4$P$_1$(pyrr)	23.12		16.6	(11.94)	255.2	
	PhP$_1$(pyrr)	22.34	22.8	15.9	(11.52)	252.2	250.9
	4-Br-C$_6$H$_4$P$_1$(pyrr)	21.19		15.3	(11.23)	249.3	
	2-NO$_2$-4-Cl-C$_6$H$_3$P$_1$(pyrr)					245.0	246.4
	2-Cl-C$_6$H$_4$P$_1$(pyrr)	20.17		13.2	(9.98)	251.1	
	4-CF$_3$-C$_6$H$_4$P$_1$(pyrr)	20.16		14.6	(10.65)	254.4	
	4-NO$_2$-C$_6$H$_4$P$_1$(pyrr)	18.51		13.3	(9.22)		
	2,5-Cl$_2$-C$_6$H$_3$P$_1$(pyrr)	18.52		11.9	(9.21)	248.4	244.2
	2,6-Cl$_2$-C$_6$H$_3$P$_1$(pyrr)	18.56		11.8	(9.00)	245.3	247.5

^{a} 2-*tert*-Butylimino-2-diethylamino-1,3-dimethyl-perhydro-1,3,2-diazaphosphorane.

every additional phosphorus unit decreasing basicity increments were found [87]. There is a limit in the homologation, as shown by P7 phosphazene **118** in DMSO, which is not more basic than P5 base **110**. This finding indicates that a leveling effect due to resonance saturation of the cation is operative in large systems. In the gas phase there is currently no answer on the limit, because no measurements have been made with higher phosphazenes. Computational studies on model compounds predict the beginning of the plateau ranges from $n = 4$ to 9, depending on the type of model compound and on the computational method used.

Branching of phosphazene bases leads to higher basicity, as evident in the P3 series where branched base **108** is more basic than linear base **101**. This increase in basicity of branched compared to linear structures has been ascribed mainly to effect of stability differences of the free bases, rather than of the cations, since linear conjugation in neutral phosphazenes is more effective than cross conjugation. Similarly, we could observe that in the P5 series linear base **117** is weaker than the branched **110**, but in this case differences in resonance of the cations are more significant.

The basicity in P1 phosphazenes is affected by phenyl substitution on the imino nitrogen, as expected from inductive and resonance effects of substituents. By substitution of the phenyl ring, the basicity can be varied over a wide pK_a range. The pK_a values are decreasing in the following order: 4-N(CH$_3$)$_2$ > 4-CH$_3$O > H > Br > 4-CF$_3$ > 2-Cl > 2,5-Cl$_2$ > 4-Cl-2-NO$_2$ > 2,4-NO$_2$. Furthermore, it is evident that alkylimino phosphoranes are significantly stronger bases than corresponding arylimino phosphoranes: EtP$_1$(pyrr) > t-BuP$_1$(pyrr) > PhP$_1$(pyrr). As suggested by Rodima, the inductive effect and some delocalization of the lone electron pair of the imino nitrogen into the aromatic ring are most probably the reasons [88]. While in guanidines substitution of imino hydrogen by phenyl group leads to basicity increase, in the phosphazenes small decrease has been found. The low sensitivity of pK_a to steric bulk of the substituents in the phosphazene systems could be established. This behavior is explained by the high flexibility of the P–N–P bridges in the cations, which effectively helps to accommodate steric strain. There is no significant difference in basicity measured for sterically larger groups on the imino nitrogen in P1 bases. In the P2 series a small increase in basicity is noted for a successive increase of bulkiness of alkyl substitution on the nitrogen basic center. However, the replacement of the *tert*-butyl group in **114** by neopentyl group slightly reduces the basicity. Similar basicity reducing effect of neopentyl group was found for phosphazene **103** within the P4 series. Bulky diisopropylamino group does not affect basicity change in P3 series going from **108** to **109**.

The enhancement in basicity was also achieved by the replacement of N(CH$_3$)$_2$ groups by cyclic pyrrolidine groups and their inductive effects, as was previously observed for the guanidine superbases. Small pK_a increase is obtained by incorporation of phosphazene moiety into six-member ring. For instance 2-*tert*-butylimino-2-diethylamino-1,3-dimethy-perhydro-1,3,2-diaza-phosphorane (BEMP) **125** is highly basic ($pK_a = 27.5$ in MeCN) [93]. Similar basicity possesses cyclic **122** [89] (28.27), and these values are amongst the strongest for P1 bases. The representatives of cyclic P2 phosphazene bases **123** and **124** have pK_a values strongest than most of the P2 and even more basic than some of P3 bases. The branched P5 phosphazene **111** is the strongest phosphazene base known to date, with measured pK_a value in acetonitrile of 46.9 units.

Further increase of nitrogen atom basicity was achieved by the combination of Schwesinger's phosphazene base concept and the Alder's idea of the disubstituted 1,8-naphthalene spacer, as reported by Raab [94]. 1,8-Bis(hexamethyltriaminophosphazenyl)naphthalene (HMPN) (**126**), represents the up to date most basic representative of this class of 'proton sponges', as evidenced by the theoretically estimated proton affinity of 274 kcal mol^{-1} and the measured pK_{BH}^+(MeCN) = 29.9 (Figure 2.12). HMPN is by nearly 12 orders

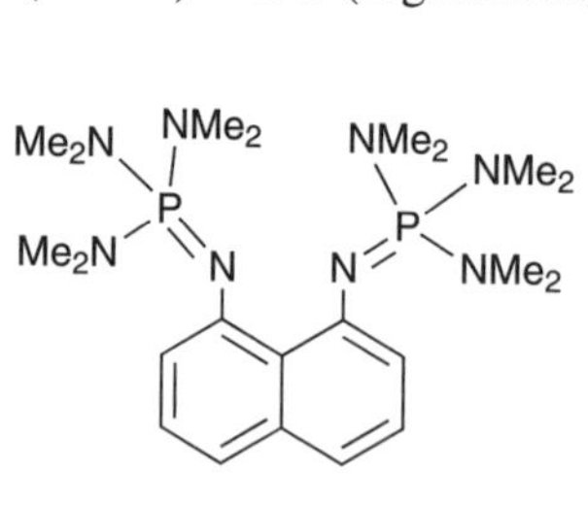

126

Figure 2.12 Structures of HMPN

of magnitude more basic than DMAN (**1**). Its high basicity is ascribed to the high energy content of the base in its initial neutral state and additional stabilization of conjugate acid by strong intramolecular hydrogen bonding. This cooperative proton chelating effect renders the bisphosphazene more basic than Schwesinger's P1 phosphazene bases.

2.6 Guanidinophosphazenes

As described in the previous section, replacement of phosphazene alkyl substituents with aminosubstituents, and especially with the phosphazo substitutents, leads to a significant enhancement in the gas phase basicity. However, Schwesinger has shown that the progressive increase of the basicity of phosphazene bases with increasing number of phosphorus atoms reaches a limit at $n = 5$. The further increase in basicity of phosphazenes (and also guanidines) has been accomplished by Koppel et al. [77] by the introduction of guanidino or substituted guanidino (cyclic or acyclic) fragments into the phosphazene structure. In such a way, a new family of super strong, uncharged bases of the guanidinophosphazene type has been synthesized, and their basic strengths in THF have been measured (Table 2.12).

The pattern of basicity change has been observed in THF solution by incorporation of tetramethylguanidino (tmg) groups in PhP$_1$(dma) phosphazene. In THF, changing all three dma groups in ^{t}Bu-P$_1$(dma) and P$_1$(dma) phosphazenes to tmg groups increases the basicity enormously: by 10.2, 9.1 and 8.9 powers of 10, respectively. The consecutive introduction of tmg fragments, instead of dma substituents, contributes equally to the total basicity increase of 9.1 pK_a units in **131**, by an increment of 3.0 pK_a units per tmg group. The observed additive basicity increase in THF is in contrast to the findings in the gas phase,

Table 2.12 Guanidinophosphazenes: experimental pK_a values in THF

Guanidinophosphazenes	GB$_{calc}$	pK_a (THF)
127: EtP$_1$(tmg)		29.7
128: ^{t}Bu-P$_1$tmg)		29.1
129: P$_1$(tmg)	276.1	28.6
130: ^{t}Bu-P$_1$(tmg)$_2$(NEt$_2$)		26.8
131: PhP$_1$(tmg)		24.3
132: PhP$_1$(tmg)$_2$(dma)		21.5
133: PhP$_1$(tmg)(dma)$_2$		18.4

where the consecutive introduction of guanidino and tmg fragments seems to be nonadditive. Further increase going from **131** was achieved by replacement of the phenyl imino group with hydrogen and alkyl substituents (^{t}Bu and Et) in **129, 128** and **127** by 4.3, 4.8 and 5.4 pK_a units, respectively. The most basic guanidinophosphazene experimentally measured so far is **127** with a pK_a value of 29.7 in THF. There is an increase of about 2.3 pK_a units for sterically more demanding *tert*-butyl groups than phenyl on the imino nitrogen.

As seen from the data collected in Table 2.11, the differences in measured pK_a values of phosphazenes in acetonitrile and THF, Δ[pK_a(MeCN–THF)] are in the range 7–10 pK_a units, with an average being eight. This allows an estimate to be made of the acetonitrile basicities of guanidinophosphazenes (Table 2.12) and compared with those of phosphazenes (Table 2.11). The estimates for **127** and **128** in acetonitrile indicate their extreme basicity comparable to these of P3 and P4 phosphazenes. Furthermore, Raab *et al*. [70] found that in acetonitrile solution the basicity of DMAN (**1**) increased by about 6.9 pK_a units when the dma substituents were replaced with tmg groups to form TMGN. In THF solution a similar change of the substituents in the naphthalene ring results in a somewhat more modest increase in the basicity of TMGN (by 5.8 pK_a units). These results allow a prediction of a further increase of 1 pK_a unit going from THF to MeCN.

Furthermore, DFT calculations have been used for quantum-chemical studies of gas phase basicities of some guanidinophosphazene bases (B3LYP/6-311 + G** method). Their extremely high basicity has been rationalized by electronic reasons. The strongly electron-donating tmg groups as the building units lead to a significant delocalization of the positive charge over the tmg groups, due to the conjugation of guanidino moieties with either a phosphorus atom of guanidinophosphazene or a phosphonium centre. These electronic effects reduce the electrophilic character of phosphorus and provide the respective bases with increased thermodynamic stability.

The calculated gas phas basicities indicate that the protonation at the =NH group of guanidinophosphazene is the most favourable site in the molecule (Figure 2.13). The replacement of amino groups by the cyclic guanidino fragment (1,3-dimethylimidazolidin-

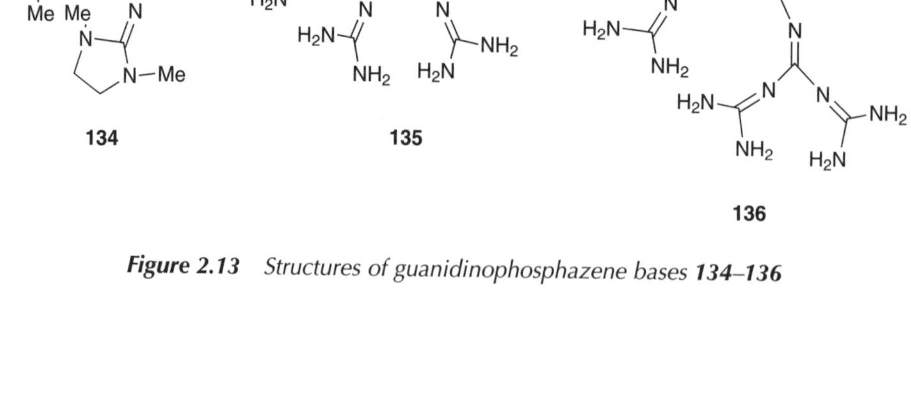

Figure 2.13 Structures of guanidinophosphazene bases **134–136**

2-ylidene)amino is predicted to have a somewhat stronger basicity-increasing effect than replacement by the less polarizable guanidino, 1,3-dihydro-2H-imidazol-2-ylideneamino, and imidazolidin-2-ylideneamino fragments. All these trisubstituted cyclic or open chain guanidino phosphazenes are expected to be extremely strong superbases, such as **134** (GB = 280 kcal mol^{-1}). However, replacement of guanidinophosphazene NH$_2$ groups by a tmg group gives the largest increase. The introduction of alkyl groups into the guanidine moiety strongly increases the basicity due to the change in charge delocalization effects. Significant (9.6 kcal mol^{-1}) basicity-increasing alkylation effects are expected to accompany the transfer from the tris-guanidino to the respective tris-tmg guanidinophosphazene derivatives. Smaller basicity increases due to imino N-alkylation were predicted: CH$_3$, 2–3 and ^{t}Bu, 5.2–6.5 kcal mol^{-1}.

The replacement of only one NH$_2$ group in (H$_2$N)$_3$P=NH phosphazene with the triguanidine fragment, yielding P1 guanidinophosphazene, increases the gas phase basicity by 30.9 kcal mol^{-1}. Replacement of all three NH$_2$ groups leads to non-alkylated P1 triguanidophosphazene **136**, with a gas phase basicity of 296.2 kcal mol^{-1}. Concomitant increase by alkylation of the imino nitrogen and the amino groups is predicted to result in gas phase basicity values well beyond 300 kcal mol^{-1}.

The estimated increase of the intrinsic basicity for the transfer from P1 to P2 guanidinephosphazenes is 13–18 kcal mol^{-1}. Gas phase basicity of the novel guanidino substituted P2 phosphazenes is expected to be above 290 kcal mol^{-1}. Calculations show that the replacement of three NH$_2$ groups with three guanidino groups in the simple P2 phosphazene produces the superbase **135** with an GB = 290.8 kcal mol^{-1}. Further basicity increase of **135** is predicted due to alkylation of guanidino NH$_2$ and of the imino basicity centre, with the basicity higher than 300 kcal mol^{-1}.

Another extraordinarily strong base has been constructed by the combination of Alder's DMAN proton sponge concept and guanidinophosphazenes [95]. Considerable increase in basicity was obtained upon replacement of N(CH$_3$)$_2$ groups in the phosphazene base HMPN by dimethylguanido groups yielding **137** (Figure 2.14). One of the highest calculated gas phase proton affinities calculated so far (305.4 kcal mol^{-1}, B3LYP/6-311 + G**//B3LYP/6-31G* method) and pK_a value in acetonitrile (44.8, IPCM) is 26 orders of magnitude more powerful than DMAN.

2.7 Other Phosphorus Containing Superbases: Verkade's Proazaphosphatranes

Other phosphorus containing superbases are phosphines and phosphorus ylides. The most detailed studied class is Verkade's proazaphosphatranes (cyclic azaphosphines). Because of the extraordinary basicity and relatively weak nucleophilicity of nonionic proazaphosphatranes, they have been found to be efficient catalysts and promoters for many reactions [96]. Proazaphosphatranes **138** have a cage framework in which one end is flattened owing to van der Waals interactions among the methylene protons adjacent to the axial nitrogen [97]. In contrast to proazaphosphatranes, azaphosphatranes **139** have oblate frameworks that feature a trans-annular N–P bond in the [3.3.3] tricyclic cage, as in the case of cations in which the PN$_3$ end of the proazaphosphatrane is flattened and its bottom is puckered upward. The approximately 3 Å bridgehead–bridgehead transannular N$_{ax}$–P distance in **138**

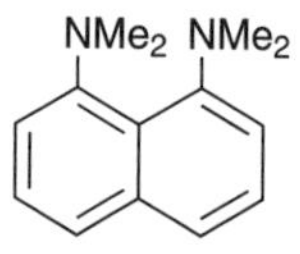

DMAN
APA 245.1 calc
pK_a 18.2 expt

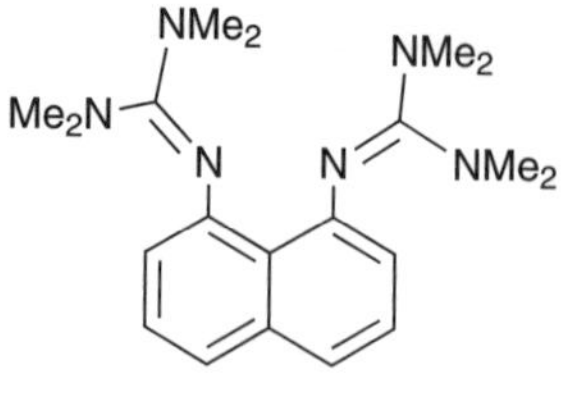

DMEGN
APA 263.8 calc
pK_a 25.1 expt

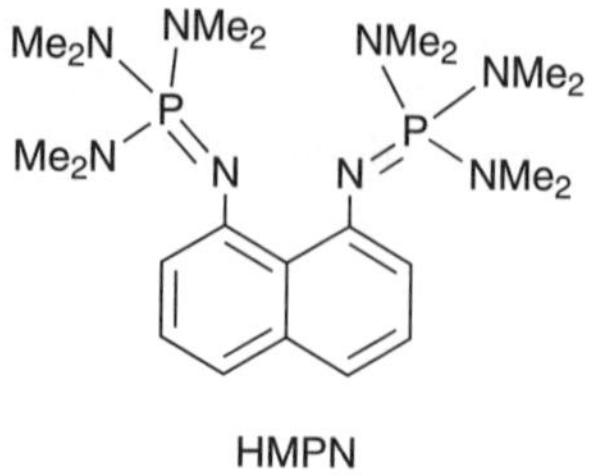

HMPN
APA 273.9 calc
pK_a 29.9 expt

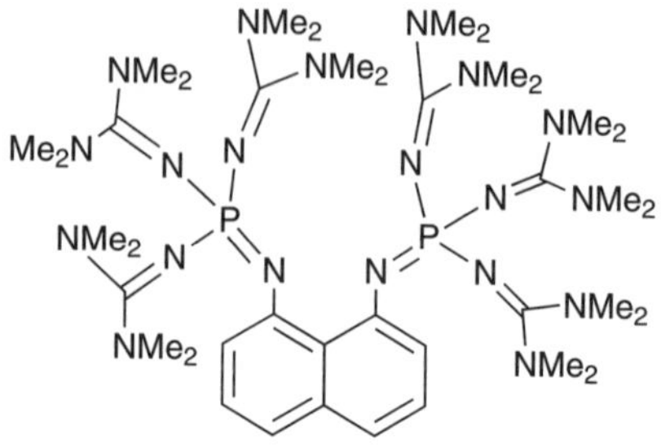

137
APA 305.4 calc
pK_a 44.8 calc

Figure 2.14 Structures of DMAN-related guanidinophosphazene bases

thus shortens to about 2 Å in azaphosphatranes [98]. Basicity measurements have shown that proazaphosphatranes are exceedingly strong nonionic Brønsted bases. They are unique in that, unlike all of the commonly used nonionic bases (including the phosphazene series), they are protonated on their phosphorus atom rather than on one of their nitrogen atoms. High basicity is the combination of the degree of delocalization of the bridgehead nitrogen lone pair into a three-centre, four-electron bond system along the molecular axis and partial donation of electron density from the axial nitrogen enhances electron density on phosphorus atom [99].

The originally reported pK_a value of 41.2 for **139**, which was in the vicinity of that for P3 and P4 phosphazene bases, was recently re-evaluated at 32.82 in acetonitrile by UV–Vis spectrophotometric titration, as shown in Table 2.13 [100]. This value is similar to the phosphazene base P$_2$-Et (pK_a = 32.74). The pK_a values for **139–141** determined in acetonitrile by ^{31}P NMR spectroscopy fall in the same range (32.84–33.63). X-ray-determined P–N$_{ax}$ bond distances of **141** and **142** are 2.037 and 1.958 Å. An even larger pK_a value for **143** was determined by ^{31}P NMR spectroscopy in acetonitrile (34.49) [101], while the parent compound **138** has a pK_a value of 29.6 in DMSO [102]. The greater stability and therefore weaker acidity of cation **138H$^+$** compared with **139** (R = CH$_3$) and **144**

Table 2.13 Basicity values for proazaphosphatranes

	R¹	R²	R³	pK_a (MeCN)	pK_a (DMSO)	pK_a (THF)	GB [85]
138	H	H	H		29.6		
143	^{i}Pr	H	H	34.49			
140	^{i}Pr	^{i}Pr	^{i}Pr	33.63			
142	^{i}Bu	^{i}Bu	^{i}Bu	33.53			260.8 (261.7)
139	Me	Me	Me	32.90; 32.82 (41.2)	26.8	26.6 [102]	259.1
141	Piv	Piv	Piv	32.84			
144	CH$_2$Ph	CH$_2$Ph	CH$_2$Ph		26.8		

(R = CH$_2$Ph) is rationalized as the consequence of the dominant electronic stabilization effect associated with greater delocalization and hence greater charge balance in the phosphorus orbitals involved in the three-centre, four-electron H–P–N$_{ax}$ bond. This idea is supported by X-ray: the short P–N$_{ax}$ bond distance in **138H$^+$** is 2.078 Å.

Theoretical predictions have been made that ylides of phosphorus, nitrogen and sulfur are potentially superstrong neutral organic bases. Limited experimental results: showed that the Ph$_3$P=CH$_2$ ylide in DMSO has pK_a = 22.5 [103], while the pK_a of (Ph(CH$_3$)$_2$N)$_3$P = C(CH$_3$)$_2$ ylide in THF was estimated to be in the range between 26 and 28 units [104].

General predictions are the following: phosphazenes and phosphorane can reach high superbasicity levels of about 300 kcal mol^{-1} (for P7 or higher number of phosphorus atoms in the system, $n \geq 7$), whereas the strongest organic phosphazene ylide superbases are estimated to have (at $n \geq 5$) gas phase basicities around or beyond 310–320 kcal mol^{-1}. The phosphine superbases are predicted to have basicity comparable to P2 phosphazenes or P1 phosphorus ylides, whereas the respective proazaphosphatrane imines and ylides are expected to be the strongest organic superbases which contain only a single phosphorus atom.

Calculations have revealed that the proazaphosphatrane bases are approximately equal in thermodynamic basicity to the Schwesinger P2 phosphazene bases (Table 2.14). The RHF/ 6-31G* calculations of proton affinities indicate that phosphazene (Z = NH) and phosphorus ylide (phosphorane, Z = CH$_2$) counterparts (**146** and **145**) are stronger than the parent Verkade superbase **139**. Higher basicity is associated with the higher degree of delocalization of the positive charge in the protonated iminophosphoranes and phosphorus ylides as a result of the more electropositive character of phosphorus atom [67]. On the other hand, phosphorus oxides **147** and **150** are weaker than **139**, but still above the superbasicity borderline. The strained polycycles **138**, **149** and **148**, as representatives of superbases

Table 2.14 Theoretical basicity values for phosphorus imines, ylides and phosphines[a]

Base	R	Z	PA[b] [106]	PA[c]	GB	pK_a(MeCN)[d] [86]
139	Me	—	274.6		255.0[e]	
145	Me	=CH$_2$	296.6			
144	Me	=NH	281.8			
147	Me	O	249.4			
138	H	—	267.3	244.6		
148	H	=CH$_2$	304.7	273.3		
149	H	=NH	281.8	258.9		
150	H	O	252.2			
151	Me	=NMe			265.2[f]	29.0
152	Me	tmg			272.4[f]	31.0
	phosphines [78]					
153	[(NH$_2$)$_3$P=N]$_3$P			283.3	275.0[c]	
154	[(dma)$_2$C=N]$_3$P			276.7	267.1[c]	
155	[(NH$_2$)$_2$C=N]$_3$P			263.7	258.9[c]	

[a] PA and GB values are given in kcal mol^{-1}.
[b] 6-31G*.
[c] B3LYP/6-311+G**.
[d] IPCM/B3LYP/6-311+G**//B3LYP/6-31G*.
[e] B3LYP/6-311+G**.
[f] B3LYP/6-311+G(2df,p)//B3LYP/6-31G*.

which contain only one phosphorus atom, are significantly stronger bases than the respective open chain derivatives of phosphines, phosphorus imines and phosphorus ylides, which also contain only one phosphorus atom. However, P2 and P3 iminophosphoranes and phosphoranes, which contain several phosphorus atoms, are stronger bases than **138**, **149** or **148**. Although theory indicated that **146** would be more basic than **139**, experiments by Verkade in acetonitrile showed the opposite [105]. This was rationalized by the more distant polarizing proton from the lone pair on axial nitrogen in **143H$^+$** than in **139H$^+$** which is fully transannulated. Solvation effects probably play a significant role. The same authors confirmed experimentally that **145** is more basic than **139**.

Triguanidinophosphines **154** and **155** are expected by calculations to be rather strong superbases. The modification from **155** to **154** results in a much stronger base; the gas phase basicity of **154** is expected to exceed that of the landmark cyclic phosphatrane superbase **139** by 12 kcal mol^{-1}. Even stronger phosphine superbases are expected to be designed by replacement of the tmg groups in **154** by (dma)$_3$P=N— groups to give [(dma)$_3$P=N]$_3$P. The basicity of its simple NH$_2$ analogue **153** is calculated to be about equal to the predicted basicity of the respective imino base [(H$_2$N)$_3$P=N]$_3$P=NH (GB = 273.2 kcal mol^{-1}). Calculations estimate the gas phase basicity of [(dma)$_3$P=N]$_3$P at around 280 kcal mol^{-1}, which is close to the gas phase basicity value for **153**.

2.8 Theoretical Methods

There have been extensive efforts put into the accurate calculation of PAs, gas phase basicities and solution pK_a values in recent years. Results vary depending on the level of sophistication of the applied calculations, and the fit to experimental values varies with the theoretical model employed. Although the calculated PA, gas phase basicity and pK_a values differ from the experimentally determined, these calculations appear reliable enough that the major effects found in superbases can be clearly demonstrated. To generalize, more high level the theory method the better the results, but it is not always, dependent on the system under study. Large molecular systems need some trade-off between accuracy and computational effort (CPU time).

The proton affinity of a base is defined as the negative change of enthalpy (PA = $-\Delta H_{prot}$) associated with the protonation reaction B + (H$^+$ → BH$^+$), while the absolute gas phase basicity corresponds to the negative of the Gibbs free energy change, that is, GB(B) = $-\Delta G°_{prot}$. For estimation of PAs and GBs in the gas phase, high level Hartree-Fock (HF), post-HF and DFT quantum-chemical calculations were employed. It has been found that the quality of obtained geometry is not crucial if PAs are estimated by single-point calculations at higher levels of theory, which is the approach generally used. For instance, ZPE(HF/6-31G*) + MP2/6-31 + G*//MP2/6-31 + G*; [37] ZPE + MP2/6-311++G**//HF/6-311++G**; MP2(fc)/6-311 + G**//RHF/6-31G*; [71] BP86/TZVP//BP86/TZVP + ZPE; [107] B3LYP/6-311 + G(3df,3pd); [108] B3LYP/6-31 + G**//HF/6-31G**; [8] BP/DZVP [6] and B3PW91/6-311++g(d,p)//B3PW91/6-31G* methods have been the most successful. For larger molecular systems, a somewhat less accurate but computationally more efficient model, the scaled Hartree-Fock (HF$_{SC}$), scheme has been developed by Maksić et al., based on the linear correlation of experimental results and calculations [109]: APA(B) = 0.8924 $\Delta E_{el(HF/6-31G*)}$ + 10.4 kcal mol^{-1}.

High level computations such as CBS-QB3, CPCM/MP2/6-311 + G**//CPCM/HF/6-31 + G* and CPCM/B3LYP/6-311 + G**//B3LYP/6-31 + G** gave gas phase basicity and absolute aqueous pK_a values with chemical accuracy [110]. Unfortunately, these methods are computationally too intensive to be applied to larger molecules. Maksić and Kovačević [49] have used a computationally more economical model for pK_a estimation, in which solvation energies were calculated by IPCM/B3LYP/6-311 + G**//HF/6-31G* method. Then, the pK_a values were estimated by a linear regression model of the experimental pK_a data and calculated APA(solv). The correlation between APA(MeCN) calculated in acetonitrile and the experimental pK_a values for series of bases containing imino groups is: pK_a(MeCN)$_{imine}$ = 0.4953, APA(MeCN) = 119.7. It should be noted that such linear relations of pK_a are valid for each computational level and each family of compounds separately, hence they should be derived in each case [85,111–113].

2.9 Concluding Remarks

In summary, guanidinophosphazenes belong to the most basic, experimentally determined class of superbases, followed by phosphazenes, proazaphosphatranes and guanidines. Amidines and classical proton sponges generally show less pronounced basicity.

Computational predictions indicate that extended π-systems could be even more potent superbases. On the basis of experimental evidence and theoretical results, the set of general rules for the molecular design of exceedingly strong superbases has been established by Kovačević and Maksić [114]: the choice of the appropriate skeleton subunit (molecular backbone + highly basic functional group); modulation (insertion of substituents at strategic sites); and inclusion of additional effects (destabilizing in initial base, stabilizing in the conjugate acid). In addition, application of various physical phenomena, such as host-mediated basicity shifts [115] may even further increase amine basicity. Following these premises, novel classes of superbases have been designed: polyimines, cyclopropeneimines, quinodiimines, polycyclic quinines, carbonyl polyenes, [3]carbonylradialenes, iminopolyenes, poly-2,5-dihydropyrrolimines, acenes, zethrenes, pyrones and pyrone-like structures, N-substituted azacalix[n](2,6) pyridines, subporphyrins [116], polypyridine macrocycles and bases possessing S=N functional group (Chapter 11). The success of the molecular design will be completed by synthesis of novel bases, preferably by simple synthetic procedures, possessing chemical stability, solubility in common organic solvents and, above all, high kinetic activity.

References

1. Raczyńska, E.D., Decouzon, M., Gal, J.-F. *et al.* (1998) Superbases and superacids in the gas phase. *Trends in Organic Chemistry*, **7**, 95–103.
2. Alder, R.W., Bowmann, P.S., Steels, W.R. and Winterman, D.R. (1968) The remarkable basicity of 1,8-bis(dimethylamino)naphthalene. *Chemical Communications*, 723–724.
3. Lau, Y.K., Saluya, P.P.S., Kebarle, P. and Alder, R.W. (1978) Gas phase basicities of N-methyl substituted 1,8-diaminonaphthalenes and related compounds. *Journal of the American Chemical Society*, **100**, 7328–7333.
4. Platts, J.A. and Howard, S.T. (1994) *Ab initio* studies of proton sponges: 1,8-bis(dimethylamino) naphthalene. *The Journal of Organic Chemistry*, **59**, 4647–4651.
5. Mallinson, P.R., Woźniak, K., Smith, G.T. and McCormack, K.L. (1997) A charge density analysis of cationic and anionic hydrogen bonds in a 'proton sponge' complex. *Journal of the American Chemical Society*, **119**, 11502–11509.
6. Guo, H. and Salahub, D.R. (2001) Origin of the high basicity of 2,7-dimethoxy-1,8-bis (dimethylamino)naphthalene: implications for enzyme catalysis. *Journal of Molecular Structure (Theochem)*, **547**, 113–118.
7. Llamas-Saiz, A.L., Foces-Foces, C. and Elguero, J. (1994) Proton sponges. *Journal of Molecular Structure*, **328**, 297–323.
8. Howard, S.T. (2000) Relationship between basicity, strain, and intramolecular hydrogen bond energy in proton sponges. *Journal of the American Chemical Society*, **122**, 8238–8244.
9. Korzhenevskaya, N.G., Schroeder, G., Brzezinski, B. and Rybachenko, V.I. (2001) Concept of superbasicity of 1,8-bis(dialkylaminio)naphthalenes ('proton sponges'). *Russian Journal of Organic Chemistry*, **37**, 1603–1610.
10. Pozharskii, A.F., Ryabtsova, O., Ozeryanskii, V.A. *et al.* (2003) Organometallic synthesis, molecular structure, and coloration of 2,7-disubstituted 1,8-bis(dimethylamino)naphthalenes. How significant is the influence of 'buttressing effect' on their basicity? *The Journal of Organic Chemistry*, **68**, 10109–10122.
11. Staab, H.A., Kirsch, A., Barth, T. *et al.* (2000) New 'proton sponges', 14. Isomeric tetrakis (dimethylamino)naphthalenes: syntheses, structure-dependence of basicities, crystal structures and physical properties. *European Journal of Organic Chemistry*, 1617–1622.

12. Pozharskii, A.F., Ozeryanskii, V.A. and Starykova, Z.A. (2002) Molecular structure of 5,6-bis(dimethylamino)acenaphthene, 5,6-bis(dimethylamino)acenaphthylene, and their monohydrobromides: a comparison with some naphthalene proton sponges. *Journal of the Chemical Society – Perkin Transactions 2*, 318–322.
13. Barth, T., Krieger, C., Neugebauer, F.A. and Staab, H.A. (1991) 1,4,5,8-Tetrakis(dimethylamino)naphthalene: synthesis, structure, 'proton sponge' and electron donor properties. *Angewandte Chemie – International Edition*, **30**, 1028–1030.
14. Korzhenevska, N.G., Rybachenko, V.I. and Schroeder, G. (2002) The basicity of 1,8-bis(dimethylamino)naphthalene and the hybrid state of the nitrogen atoms of its dimethylamino groups. *Tetrahedron Letters*, **43**, 6043–6045.
15. Staab, H.A. and Saupe, T. (1988) 'Proton sponges' and the geometry of hydrogen bonds: aromatic nitrogen bases with exceptional basicities. *Angewandte Chemie – International Edition*, **27**, 865–1008.
16. Platts, J.A. and Howard, S.T. (1996) Ab initio studies of proton sponges 3. 4,5-Bis(dimethylamino)fluorene and 4,5-bis(dimethylamino)phenanthrene. *The Journal of Organic Chemistry*, **61**, 4480–4482.
17. Rentzea, M., Brox, W. and Staab, H.A. (1992) Proximity effects in the mass spectra of crowded bis(dialkylamino)arenes. Part IV. A tandem mass spectrometry study of fluorenodiazonine and fluorenodiazecine derivatives. *Organic Mass Spectrometry*, **27**, 521–522.
18. Krieger, C., Newsom, I., Zirnstein, M.A. and Staab, H.A. (1989) Structures of quino[7,8-*h*]quinoline and quino[8,7-*h*]quinoline. *Angewandte Chemie – International Edition*, **28**, 84–86.
19. Zirnstein, M.A. and Staab, H.A. (1987) Quino[7,8-h]quinoline, a new type of proton "sponge". *Angewandte Chemie (International Edition in English)*, **26**, 460–462.
20. Staab, H.A., Zirnstein, M.A. and Krieger, C. (1989) Benzo[1,2-*h*:4,3-*h*′]diquinoline ('1,14-diaza[5]helicene'): synthesis, structure, and properties. *Angewandte Chemie – International Edition*, **28**, 86–88.
21. Bucher, G. (2003) DFT calculations on a new class of C_3-symmetric organic bases: highly basic proton sponges and ligands for very small metal cations. *Angewandte Chemie – International Edition*, **42**, 4039–4042.
22. Toom, L., Kütt, A., Kaljurand, I. et al. (2006) Substituent effects on the basicity of 3,7-diazabicyclo[3.3.1]nonanes. *The Journal of Organic Chemistry*, **71**, 7155–7164.
23. Rõõm, E.-I., Kütt, A., Kaljurand, I. et al. (2007) Brønsted basicities of diamines in the gas phase, acetonitrile and tetrahydrofuran. *Chemistry – A European Journal*, **13**, 7631–7643.
24. Estrada, E. and Simn-Manso, Y. (2006) Rational design and first principles studies toward the discovery of a small and versatile proton sponge. *Angewandte Chemie – International Edition*, **45**, 1719–1721.
25. Singh, A. and Ganguly, B. (2007) DFT studies toward the design and discovery of a versatile cage-functionalized proton sponge. *European Journal of Organic Chemistry*, 420–422.
26. Margetić, D., Warrener, R.N. and Butler, D.N. (2002) Proximity effects: A computational study of proton affinities of sesqui- and sester-7-azabicyclo[2.2.1]heptanes, Article 23,. The eight electronic computational chemistry conference (ECCC-8), Robert Q. Topper (Ed.), http://eccc8.cooper.edu.
27. Singh, A. and Ganguly, B. (2007) Novel tetracyclic proton sponge. *Journal of Physical Chemistry A*, **111**, 6468–6471.
28. Mascal, M., Lera, M., Blake, A.J. et al. (2001) The azatriquineamine trimer – a novel proton chelate. *Angewandte Chemie – International Edition*, **40**, 3696–3698.
29. Alder, R.W., Eastment, P., Hext, N.M. et al. (1988) Strongly basic medium-ring diamines which mimic gas phase behaviour in solution: 1,6-dimethyl-1,6-diazacyclodecane. *Journal of the Chemical Society. Chemical Communications*, 1528–1530.
30. Alder, R.W. (2005) Design of C_2-chiral diamines that are computationally predicted to be a million-fold more basic than the original proton sponges. *Journal of the American Chemical Society*, **127**, 7924–7931.
31. DuPré, D.B. (2003) The compressed hydrogen bond in a molecular proton cage. *Journal of Physical Chemistry A*, **197**, 10142–10148.

32. Alder, R.W., Casson, A. and Sessions, R.B. (1979) Inside- and outside-protonated ions from 1,6-diazabicyclo[4.4.4]tetradecane. *Journal of the American Chemical Society*, **101**, 3652–3653.
33. Peräkylä, M. (1996) *Ab initio* quantum mechanical study on the origin of the pK_a differences of the proton sponges 1,8-bis(dimethylamino)naphthalene, 1,8-bis(dimethylamino)-2,7-dimethoxynaphthalene, 1,6-dimethyl-1,6-diazacyclodecane, and 1,6-diazabicyclo[4.4.4]tetradecane. *The Journal of Organic Chemistry*, **61**, 7420–7425.
34. Snider, B.B., Grabowski, J.F., Alder, R.W., Foxman, B.M. and Yang, L. (2006) Synthesis of a hindered C_2-symmetric hydrazine and diamine by a crisscross cycloaddition of citronellal azine. *Canadian Journal of Chemistry*, **84**, 1242–1249.
35. Bell, T.W., Choi, H.-J. and Harte, W. (1986) Unusually basic, rapidly protonated tricyclic triamine: 11-methylene-1,5,9-triazabicyclo[7.3.3]pentadecane. *Journal of the American Chemical Society*, **108**, 7427–7438.
36. Bell, T.W., Choi, H.-J., Harte, W. and Drew, M.G.B. (2003) Syntheses, conformations, and basicities of bicyclic triamines. *Journal of the American Chemical Society*, **125**, 12196–12210.
37. Meyer, N.C., Bolm, C., Raabe, G. and Kölle, U. (2005) Proton affinities and relative basicities of two 1,4,7-triazacyclononanes, Me$_3$TACN and TP-TACN. Quantum-chemical *ab initio* calculations, solution measurements, and the structure of [TP-TACN.2H]$^{2+}$ in the solid state. *Tetrahedron*, **61**, 12371–12376.
38. Weisman, G.R., Rogers, M.E., Wong, E.H. *et al.* (1990) Cross-bridged cyclam. Protonation and Li$^+$ complexation in a diamond-lattice cleft. *Journal of the American Chemical Society*, **112**, 8604–8605.
39. Redko, M.Y., Vlassa, M., Jackson, J.E. *et al.* (2002) 'Inverse sodium hydride': a crystalline salt that contains H$^+$ and Na$^-$. *Journal of the American Chemical Society*, **124**, 5928–5929.
40. Springborg, J. (2003) Adamanzanes – bi- and tricyclic tetraamines and their coordination compounds. *Journal of the Chemical Society – Dalton Transactions*, 1653–1665.
41. Howard, S.T. (2000) Relationship between basicity, strain and intramolecular hydrogen-bond energy in proton sponges. *Journal of the American Chemical Society*, **122**, 8238–8244.
42. Smith, P.B., Dye, J.L., Cheney, J. and Lehn, J.-M. (1981) Proton cryptates. Kinetics and thermodynamics of protonation of the [1.1.1]macrobicyclic cryptand. *Journal of the American Chemical Society*, **103**, 6044–6048.
43. Ciampolini, M., Nardi, N., Valtancoli, B. and Micheloni, M. (1992) Small aza cages as 'fast proton sponges' and strong lithium binders. *Coordination Chemistry Reviews*, **120**, 223–236.
44. Julian, R.R. and Beauchamp, J.L. (2002) The unusually high proton affinity of aza-18-crown-6 ether: implications for the molecular recognition of lysine in peptides by lariat crown ethers. *Journal of the American Society for Mass Spectrometry*, **13**, 493–498.
45. Häfelinger, G. and Kuske, F.K.H. (1991) General and theoretical aspects of amidines and related compounds, in *The chemistry of amidines and imidates*, Vol. 2 (ed. S. Patai and Z. Rappoport), John Wiley & Sons Ltd, Chichester, pp. 3–100.
46. Oszczapowicz, J. (1991) Basicity, H-bonding, tautomerism and complex formation of imidic acid derivatives, in *The chemistry of amidines and imidates*, Vol. 2 (eds S. Patai and Z. Rappoport), John Wiley & Sons Ltd, Chichester, pp. 623–688.
47. Koppel, I., Koppel, J. and Leito, I. (1996) Basicity of 3-aminopropionamidine derivatives in water and dimethyl sulphoxide. Implication for a pivotal step in the synthesis of distamycin A analogues. *Journal of Physical Organic Chemistry*, **9**, 265–268.
48. Brisander, M., Harris, S.G., Lloyd, D. *et al.* (1998) Diazepines. Part 30. A comparison between the extent of delocalization of electrons in a vinamidine and its protonated form. Crystal and molecular structure of two 2,3-dihydro-1,4-diazepines. *Journal of Chemical Research, Synopses*, **2**, 72–73.
49. Kovačević, B. and Maksić, Z.B. (2001) Basicity of some organic superbases in acetonitrile. *Organic Letters*, **3**, 1523–1526.
50. Schwesinger, R. (1987) Tricyclic 2,4-diaminovinamidines – a readily accessible, very strong CHN base. *Angewandte Chemie – International Edition*, **26**, 1164–1165.
51. Schwesinger, R., Mißfeldt, M., Peters, K. and von Schnering, H.G. (1987) Novel, very strongly basic, pentacyclic 'proton sponges' with vinamidine structure. *Angewandte Chemie – International Edition*, **26**, 1165–1167.

52. Schwesinger, R. (1990) Strong uncharged nitrogen bases. *Nachrichten aus Chemie Technik und Laboratorium*, **38**, 1214–1226.
53. Decouzon, M., Gal, J.-F., Maria, P.-C. and Raczyńska, E.D. (1993) Superbases in the gas phase: amidine and guanidine derivatives with proton affinities larger than 1000 kJ mol^{-1}. *Rapid Communications in Mass Spectrometry*, **7**, 599–602.
54. Raczyńska, E.D., Decouzon, M., Gal, J.-F. *et al.* (2001) Gas-phase structural (internal) effects in strong organic nitrogen bases. *Journal of Physical Organic Chemistry*, **14**, 25–34.
55. Chamorro, E., Escobar, C.A., Sienra, R. and Pérez, P. (2005) Empirical energy-density relationships applied to the analysis of the basicity of strong organic superbases. *Journal of Physical Chemistry A*, **109**, 10068–10076.
56. Decouzon, M., Gal, J.-F., Maria, P.C. and Raczyńska, E.D. (1991) Gas-phase basicity of N^1,N^1-dimethyl-N^2-aklylformamidine: substituent polarizability effects. *The Journal of Organic Chemistry*, **56**, 3669–3673.
57. Raczyńska, E.D., Maria, P.C., Gal, J.-F. and Decouzon, M. (1992) Gas-phase basicity of N^1,N^1-dimethylformamidine: substituent polarizability and field effects and comparison with Brønsted basicity in solution. *The Journal of Organic Chemistry*, **57**, 5730–5735.
58. Raczyńska, E.D., Dąrowska, M., Dabkowska, I. *et al.* (2004) Experimental and theoretical evidence of basic site preference in polyfunctional superbasic amidinazine: N^1,N^1-dimethyl-N^2-β-(2-pyridylethyl) formamidine. *The Journal of Organic Chemistry*, **69**, 4023–4030.
59. Raczyńska, E.D., Decouzon, M., Gal, J.-F. *et al.* (2000) Gas phase basicity of polyfunctionalized amidinazines: experimental evidence of preferred site(s) of protonation. *The Journal of Organic Chemistry*, **65**, 4635–4640.
60. Howard, S.T., Platts, J.A. and Coogan, M.P. (2002) Relationships between basicity, structure, chemical shift and the charge distribution in resonance-stabilized iminoamines. *Journal of the Chemical Society – Perkin Transactions 2*, 899–905.
61. Maksić, Z.B. and Kovačević, B. (1999) Neutral *vs.* zwitterionic form of arginine – an *ab initio* study. *Journal of the Chemical Society – Perkin Transactions 2*, 2623–2629.
62. Ishikawa, T. and Isobe, T. (2003) Modified guanidines as chiral auxiliaries. *Chemistry – A European Journal*, **8**, 552–557.
63. Gund, P.J. (1972) Guanidine, trimethylenemethane and 'Y-delocalization'. Can acyclic compounds have 'aromatic' stability? *Journal of Chemical Education*, **49**, 100–106.
64. Gobbi, A. and Frenking, G. (1993) Y-conjugated compounds: The equilibrium geometries and electronic structures of guanidine, guanidinium cation, urea and 1,1-diaminoethylene. *Journal of the American Chemical Society*, **115**, 2363–2372.
65. Raczyńska, E.D., Cyrański, M.K., Gutowski, M. *et al.* (2003) Consequences of proton transfer in guanidine. *Journal of Physical Organic Chemistry*, **16**, 91–106.
66. Yamamoto, Y. and Kojima, S. (1991) Synthesis and chemistry of guanidine derivatives, in *The chemistry of amidines and imidates*, **Vol. 2** (eds S. Patai and Z. Rappoport), John Wiley &Sons Ltd, Chichester, pp. 485–526.
67. Flynn, K.G. and Nenortas, D.R. (1963) Kinetics and mechanism of the reaction between phenyl isocyanate and alcohols. Strong base catalysis and deuterium isotope effects. *The Journal of Organic Chemistry*, **28**, 3527–3530.
68. Kaljurand, I., Rodima, T., Leito, I. *et al.* (2000) Self-consistent spectrophotometric basicity scale in acetonitrile covering the range between pyridine and DBU. *The Journal of Organic Chemistry*, **65**, 6202–6208.
69. Kaljurand, I., Kütt, A., Sooväli, L. *et al.* (2005) Extension of the self-consistent spectrophotometric basicity scale in acetonitrile to a full span of 28 pK_a units: unification of different basicity scales. *The Journal of Organic Chemistry*, **70**, 1019–1028.
70. Raab, V., Kipke, J., Gschwind, R.M. and Sundermeyer, J. (2002) 1,8-Bis(tetramethylguanidino) naphthalene (TMGN): a new, superbasic and kinetically active 'proton sponge'. *Chemistry – A European Journal*, **8**, 1682–1693.
71. Kovačević, B. and Maksić, Z.B. (2002) The proton affinity of the superbase 1,8-bis(tetramethylguanidino)naphthalene and some related compounds – a theoretical study. *Chemistry – A European Journal*, **8**, 1694–1702.

72. Raab, V., Harms, K., Sundermeyer, J. et al. (2003) 1,8-Bis(dimethylethyleneguanidino) naphthalene, DMEGN: tailoring the basicity of bisguanidine 'proton sponges' by experiment and theory. *The Journal of Organic Chemistry*, **68**, 8790–8797.
73. Kovačević, B., Maksić, Z.B., Vianello, R. and Primorac, M. (2002) Computer aided design of organic superbases: the role of intramolecular hydrogen bonding. *New Journal of Chemistry*, **26**, 1329–1334.
74. Margetić, D., Nakanishi, W., Kumamoto, T. and Ishikawa, T. (2007) Quantum-chemical study of 1,2-bis(dimethylethyleneguanidino)benzenes. *Heterocycles*, **71**, 2639–2658.
75. Kawahata, M., Yamaguchi, K. and Ishikawa, T. (2005) *o*-Bisguanidinobenzene, a powerful hydrogen acceptor: crystal structures of organic complexes with benzoic acid, phenol and benzyl alcohol. *Crystal Growth & Design*, **5**, 373–377.
76. Maksić, Z.B. and Kovačević, B. (2000) Absolute proton affinity of some polyguanides. *The Journal of Organic Chemistry*, **65**, 3303–3309.
77. Kolomeitsev, A.A., Koppel, I.A., Rodima, T. et al. (2005) Guanidinophosphazenes: design, synthesis, and basicity in THF and in the gas phase. *Journal of the American Chemical Society*, **127**, 17656–17666.
78. Kovačević, B., Glasovac, Z. and Maksić, Z.B. (2002) The intramolecular hydrogen bond and intrinsic proton affinity of neutral organic molecules: N,N',N''-tris(3-aminopropyl) guanidine and some related systems. *Journal of Physical Organic Chemistry*, **15**, 765–774.
79. Glasovac, Z., Kovačević, B., Meštrović, E. and Eckert-Maksić, M. (2005) Synthesis and properties of novel guanidine bases. N,N',N''-Tris(3-dimethylaminopropyl)guanidine. *Tetrahedron Letters*, **46**, 8733–8736.
80. Margetić, D. Manuscript in preparation.
81. Alcamí, M., Mó, O. and Yáñez, M. (2001) Modeling intrinsic basicities and acidities. *Journal of Physical Organic Chemistry*, **15**, 174–186.
82. Kovačević, B., Maksić, Z.B. and Vianello, R. (2001) The proton affinity of some extended π-systems involving guanidine and cyclopropenimine subunits. *Journal of the Chemical Society – Perkin Transactions 2*, 886–891.
83. Schwesiger, R., Schlemper, H., Hasenfratz, C. et al. (1996) Extremely strong, uncharged auxiliary bases; monomeric and polymer-supported polyaminophosphazenes (P2-P5). *Liebigs Annalen*, 1055–1081.
84. Kaljurand, I., Koppel, I.A., Kütt, A. et al. (2007) Experimental gas phase basicity scale of superbasic phosphazenes. *Journal of Physical Chemistry A*, **111**, 1245–1250.
85. Kovačević, B., Barić, D. and Maksić, Z.B. (2004) Basicity of exceedingly strong non-ionic bases in acetonitrile – Verkade's superbase and some related phosphazenes. *New Journal of Chemistry*, **28**, 284–288.
86. Kovačević, B. and Maksić, Z.B. (2006) High basicity of phosphorus-proton affinity of tris-(tetramethylguanidylphosphine) and tris(hexamethyltriaminophosphazenyl)phosphine by DFT calculations. *Chemical Communications*, 1524–1526.
87. Schwesinger, R., Hasenfratz, C., Schlemper, H. et al. (1993) How strong and how hindered can uncharged phosphazene bases be? *Angewandte Chemie – International Edition*, **32**, 1361–1364.
88. Rodima, T., Kaljurand, I., Pihl, A. et al. (2002) Acid–base equilibria in nonpolar media. 2. Self-consistent basicity scale in THF solution ranging from 2-methoxypyridine to EtP1(pyrr) phosphazene. *The Journal of Organic Chemistry*, **67**, 1873–1881.
89. Schwesinger, R. and Schlemper, H. (1987) Peralkylated polyaminophosphazenes - extremely strong, neutral nitrogen bases. *Angewandte Chemie – International Edition*, **26**, 1167–1170.
90. Kaljurand, I., Rodima, T., Pihl, A. et al. (2003) Acid–base equilibria in nonpolar media. 4. Extension of the self-consistent basicity scale in THF medium. Gas phase basicities of phosphazenes. *The Journal of Organic Chemistry*, **68**, 9988–9993.
91. Sooväli, L., Rodima, T., Kaljurand, I. et al. (2006) Basicity of some P1 phosphazenes in water and in aqueous surfactant solution. *Organic and Biomolecular Chemistry*, **4**, 2100–2105.
92. Xu, W., Mohan, R. and Morrissey, M.M. (1998) Polymer supported bases in solution phase synthesis. 2. A convenient method for *N*-alkylation reactions of weakly acidic heterocycles. *Bioorganic & Medicinal Chemistry Letters*, **8**, 1089–1092.

93. Rodima, T., Mäemets, V. and Koppel, I. (2000) Synthesis of *N*-aryl substituted imino phosphoranes and NMR spectroscopic investigation of their acid-base properties in acetonitrile. *Journal of the Chemical Society – Perkin Transactions 1*, 2637–2644.
94. Raab, V., Gauchenova, E., Merkoulov, A. *et al.* (2005) 1,8-Bis(hexamethyltriaminophosphazenyl)naphthalene, HMPN: A superbasic bisphosphazene 'proton sponge'. *Journal of the American Chemical Society*, **127**, 15738–15743.
95. Kovačević, B. and Maksić, Z.B. (2006) High basicity of tris(tetramethylguanidinyl)phosphine imide in the gas phase and acetonitrile – a DFT study. *Tetrahedron Letters*, **47**, 2553–2555.
96. Verkade, J.G. and Kisanga, P.B. (2004) Recent applications of proazaphosphatranes in organic synthesis. *Aldrichimica Acta*, **37**, 3–14.
97. Verkade, J.G. and Kisanga, P.B. (2003) Proazaphosphatranes: a synthesis methodology trip from their discovery to vitamin A. *Tetrahedron*, **59**, 7819–7858.
98. Kingston, J.V. and Verkade, J.G. (2005) P[N(*i*Bu)CH$_2$CH$_2$]$_3$N: A versatile non-ionic base for the synthesis of higher coordinate silicates. *Inorganic Chemistry Communications*, **8**, 643–646.
99. Verkade, J.G. (1993) Atranes: New examples with unexpected properties. *Accounts of Chemical Research*, **26**, 483–489.
100. Koppel, I.A., Schwesinger, R., Breuer, T., Burk, P. *et al.* (2001) Intrinsic basicities of phosphorus imines and ylides: a theoretical study. *Journal of Physical Chemistry A*, **105**, 9575–9586.
101. Kisanga, P.B. and Verkade, J.G. (2000) pK_a measurements of P(RNCH$_2$CH$_3$)$_3$N. *The Journal of Organic Chemistry*, **65**, 5431–5432.
102. Kisanga, P.B. and Verkade, J.G. (2001) Synthesis of new proazaphosphatranes and their application in organic synthesis. *Tetrahedron*, **57**, 467–475; Laramay, M.A.H. and Verkade, J.G. (1990) The "anomalous" basicity of P(NHCH$_2$CH$_2$)$_3$N relative to P(NMeCH$_2$CH$_2$)$_3$N and P(NBzCH$_2$CH$_2$)$_3$N: a chemical consequence of orbital charge balance? *Journal of the American Chemical Society*, **112**, 9421–9422.
103. Bordwell, F.G. (1988) Equilibrium acidities in dimethyl sulfoxide solution. *Accounts of Chemical Research*, **21**, 456–463.
104. Goumri-Magnet, S., Guerret, O., Gornitzka, H. *et al.* (1999) Free and supported phosorus ylides as strong neutral Brønsted bases. *The Journal of Organic Chemistry*, **64**, 3741–3744.
105. Liu, X., Thirupathi, N., Guzei, I.A. and Verkade, J.G. (2004) An investigation of Staudinger reactions involving cis-1,3,5-triazidocyclohexane and tri(alkylamino)phosphines. *Inorganic Chemistry*, **43**, 7431–7440.
106. Windus, T.L., Schmidt, M.W. and Gordon, M.S. (1994) Theoretical investigation of azaphosphatranes. *Journal of the American Chemical Society*, **116**, 11449–11455.
107. Raabe, G., Wang, Y. and Fleischbauer, J. (2000) Calculation of the proton affinities of primary, secondary, and tertiary amines using semiempirical and *ab initio* methods. *Zeitschrift fur Naturforschung*, **55a**, 687–694.
108. Burk, P., Koppel, I.A., Koppel, I. *et al.* (2000) Critical test of performance of B3LYP functional for prediction of gas-phase acidities and basicities. *Chemical Physics Letters*, **323**, 482–489.
109. Hillebrand, C., Klessinger, M., Eckert-Maksić, M. and Maksić, Z.B. (1996) Theoretical model calculations of the proton affinities of aminoalkanes, aniline and pyridine. *The Journal of Physical Chemistry*, **100**, 9698–9702.
110. Liptak, M.D. and Shields, G.C. (2001) Accurate pK_a calculations for carboxylic acids using complete basis set and gaussian-*n* models combined with CPCM continuum solvation methods. *Journal of the American Chemical Society*, **123**, 7314–7319.
111. Despotović, I., Kovačević, B. and Maksić, Z.B. (2007) Pyridine and *s*-triazine as building blocks of nonionic organic superbases – density functional theory B3LYP study. *New Journal of Chemistry*, **31**, 447–457.
112. Bucher, G. (2003) DFT calculations on a new class of C_3-symmetric organic bases: highly basic proton sponges and ligands for very small metal cations. *Angewandte Chemie-International Edition*, **42**, 4039–4042.
113. Estrada, E. and Simn-Manso, Y. (2006) Rational design and first principles studies toward the discovery of a small and versatile proton sponge. *Angewandte Chemie-International Edition*, **45**, 1719–1721.

114. Kovačević, B. and Maksić, Z.B. (2003) Computational design of highly potent superbases, NIC Symposium Proceedings. in John von Neumann Institute for Computing, Jülich, NIC series 20, (eds D. Wolf, G. Münster and M. Kremer), 71–80.
115. Pluth, M.D., Bergman, R.G. and Raymond, K.N. (2007) Making amines strong bases: thermodynamic stabilization of protonated guests in a highly-charged supramolecular host. *Journal of the American Chemical Society*, **129**, 11459–11467.
116. Glasovac, Z., Vazdar, M., Eckert-Maksić, M. and Margetić, D. (2007) Proton affinities of dehydroporphyrin and subporphyrin in ground and excited states obtained by high level computations. The eleventh electronic computational chemistry conference (ECCC-11), Robert Q. Topper Ed., April 1–30, http://eccc.monmouth.edu.

3
Amidines in Organic Synthesis

Tsutomu Ishikawa and Takuya Kumamoto
Graduate School of Pharmaceutical Sciences, Chiba University,
1-33 Yayoi, Inage, Chiba 263-8522, Japan

3.1 Introduction

Amidines are the nitrogen analogues of carboxylic acids and contain two nitrogen atoms in amino and imino groups. Amidines are widely used for the constructions of nitrogen-containing heterocycles because of the functional units of biologically and medicinally important compounds [1]. The $n - \pi$ conjugated hetero allylic systems, isoconjugatable to the allyl ions resulting in cross-conjugated (or Y-conjugated) hetero π-systems [2], control the total functionality of amidine as bases and/or nucleophiles in organic reactions (Figure 3.1).

Benzamidine reacts with *p*-nitrophenyl acetate in chlorobenzene at least 15 000 times faster than *n*-butylamine, which has a basicity similar to that of benzamidine. This reactivity is attributable to bifunctional nature of the nucleophile, which can concertedly attack the carbonyl carbon of the ester and deliver a proton to the carbonyl oxygen (Scheme 3.1) [3].

The bifunctional character allows for amidines to catalyse the transfer of two hydrogen atoms in allylic rearrangement and enolization (epimerization) through formal intermolecular 1,3-sigmatropic shifts (Scheme 3.2a). However, monofunctional 1,3-rearrangement is possible, too, and thus amidines can also operate as monofunctional catalysts (Scheme 3.2b). These modes of reaction are dependent upon the conditions and/or the amidine substrates used [1a].

On the other hand, in the 1960s sterically hindered bicyclic amidines, 1,8-diazabicylo [5.4.0]undec-7-ene (DBU) (**1**) and 1,5-diazabicylo[4.3.0]non-5-ene (DBN) (**2**), were introduced as useful dehydrohalogenation reagents in the synthesis of vitamin A. Treatment

---

Superbases for Organic Synthesis: Guanidines, Amidines, Phosphazenes and Related Organocatalysts
Edited by Tsutomu Ishikawa
© 2009 John Wiley & Sons, Ltd

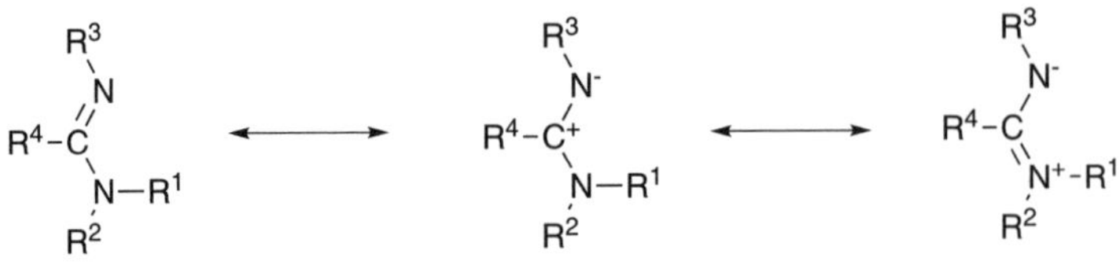

Figure 3.1 Possible conjugation forms of amidine

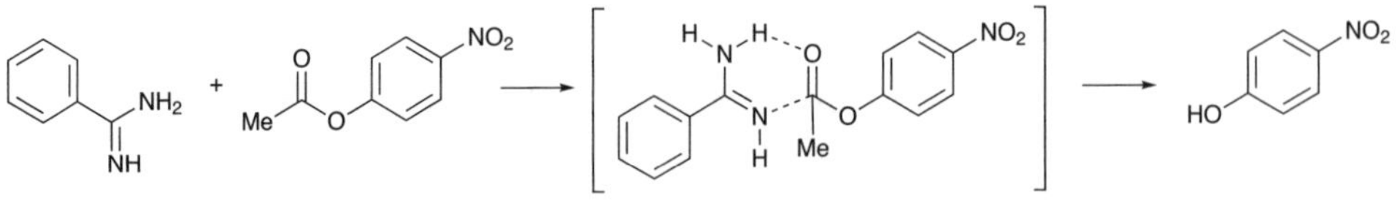

Scheme 3.1 Benzamidine-catalysed hydrolysis of p-nitrophenyl acetate

of halotetraene with DBN (**2**) in benzene gave a conjugated pentaene, in which a newly introduced double bond has Z-configuration [4] (Scheme 3.3). DBN (**2**) was found to be the most suitable dehydrohalogenation reagent among organobases examined.

DBU (**1**) and DBN (**2**) are originally synthesized from cyclic lactams by three steps of Michael addition of acrylonitrile, reduction with Raney nickel, and treatment with p-toluenesulfonic acid (TsOH). Thus, DBN (**2**) was prepared in 69% overall yield [4] (Scheme 3.4).

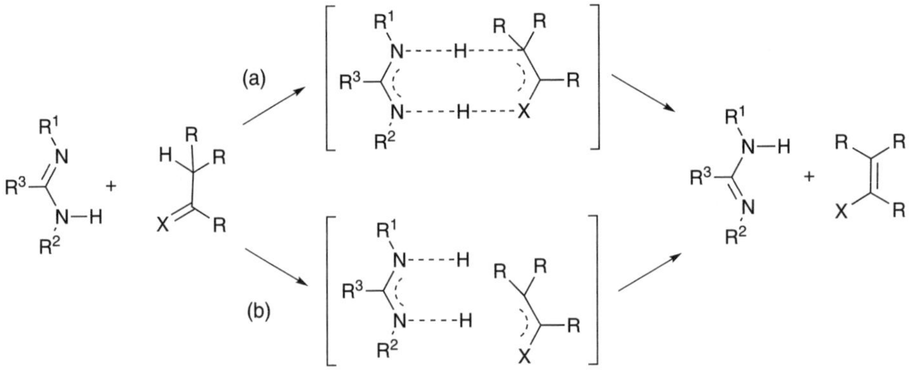

Scheme 3.2 Possible mechanisms for amidine-catalysed allylic rearrangement

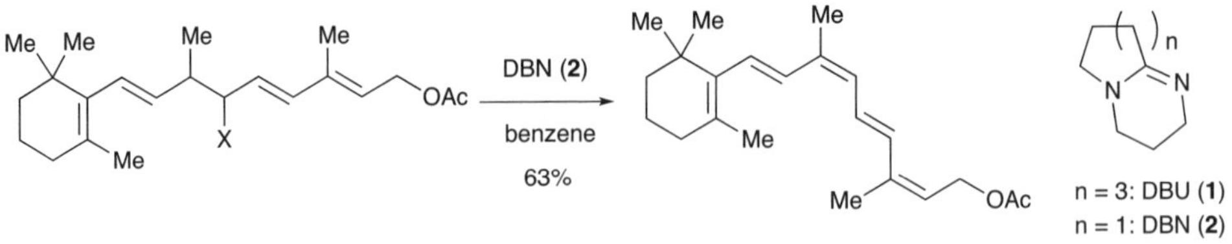

n = 3: DBU (**1**)
n = 1: DBN (**2**)

Scheme 3.3 Dehydrohalogenation wrih DBN (**2**) in the synthesis of vitamin A

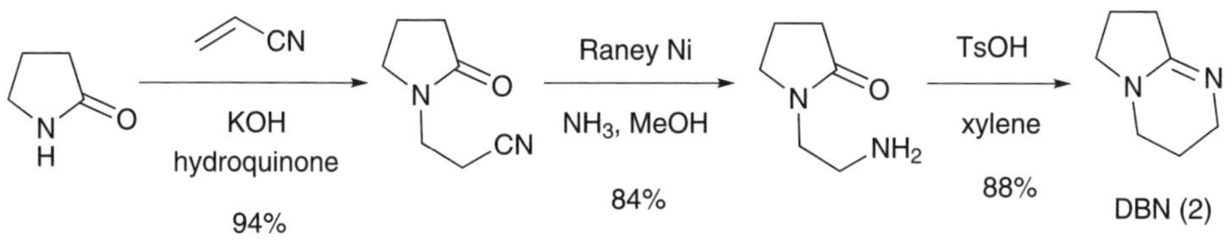

Scheme 3.4 Preparation of DBN (2)

These amidines have been extensively applied to dehydrohalogenation in organic synthesis and in some cases DBU (**1**) is more effective than DBN (**2**) [5]. A double bond can be also introduced into organic molecules by elimination of sulfonate ester instead of the halogen atom (i.e. dehydrosulfonation in addition to dehydrohalogenation). Furthermore, these amidines can be applied to the Wittig reaction [6], aldol condensation [6], 1,3-allyl rearrangement [7] and epimerization of the β-lactam skeleton (at C_6 of the penicillic acid derivatives). Sterically hindered phenols (e.g. 2,6-di(*tert*-butyl)-4-fluorophenol) are *O*-acetylated with DBU (**1**), which is superior to sodium hydroxide in the synthesis [8].

The related 6-6 bicyclic system, 2,10-diazabicyclo[4.4.0]dec-1-ene (**3**), was prepared from *trans*-decahydro-1,8-naphthyridine [9] and an alternative method via direct cyclization of bis(3-aminopropyl)malonic acid was developed for large scale operation [10]. Heinzer *et al.* [11] reported the preparation of a sterically strongly hindered bicyclic amidine, 3,3,6,9,9-pentamethyl-2,10-diazabicyclo[4.4.0]dec-1-ene (Eschenmoser amidine) (**4**) (Figure 3.2) and its *N*-alkylated analogues and their potential uses in the formations of salts of carboxylic acids and related proton complexes of bidentate ligands (Section 3.3.8) [11b].

The efficiencies of DBU (**1**) and DBN (**2**) as sterically hindered (non-nucleophilic) and strong organobase catalysts have been widely demonstrated [5]. However, Reed *et al.* [12] claimed that they could behave as strong nucleophiles in the reaction of chlorobis(diisopropylamino)phosphane and DBU (**1**) or DBN (**2**).

Modification of the amidine function to chiral versions has been examined. For example, C_2-symmetrical chiral bicylic amidine **5** was prepared for studies on molecular recognition and were proven to differentiate analytically between the enantiomers of chiral carboxylic acids [13]. Near the same time, a mannose-based amidine **6** was synthesized as a potential mannosidase inhibitor, but not a chiral auxiliary [14]. Three enantiopure hydroxyl substituted amidines **7** of the DBN-type were synthesized from 5-(phenylsulfonyl)pyrrolidine-2-one by an oxazaborolidine-catalysed reductive desymmetrization of meso-imide followed by functionalization through *N*-acyliminium ion [15] (Figure 3.3).

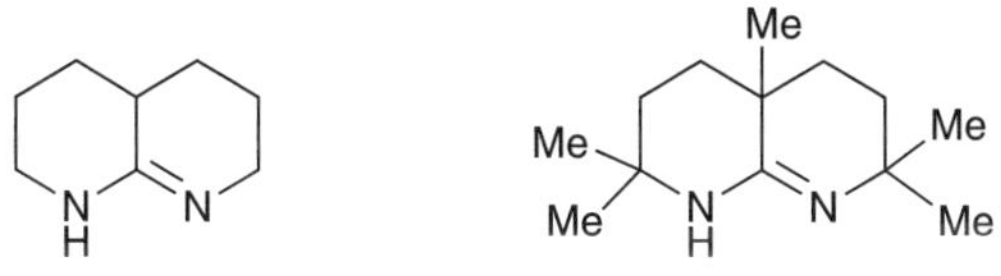

Figure 3.2 A 6-6 bicyclic amidine system (**3**) and the Eschenmoser amidine (**4**)

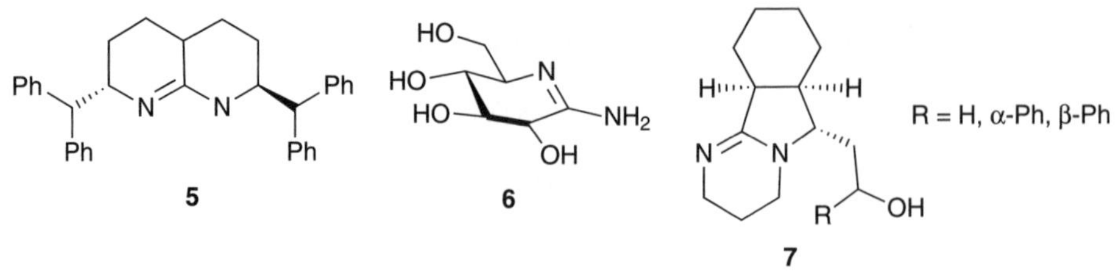

52 Amidines in Organic Synthesis

Figure 3.3 Structures of some chiral amidines

Numerous applications have been found for the uses of imidazole derivatives as ionic liquids and *N*-heterocyclic carbenes and their use in organic chemistry has been well discussed in books or reviews. Thus, in this chapter, the use of non-heteroaromatic amidine compounds as functional tools in asymmetric synthesis and the related chemistry after presentation of the preparation method of amidines will mainly be discussed.

3.2 Preparation of Amidines

Amidines can be basically synthesized by manipulation of carboxaminde or its analogues. In this section, preparation methods based on the mode of the reactions are given in alphabetical order.

3.2.1 Alkylation of Amidines

Cyclic amidines are prepared by alkylation of acylic amidines [16]. Bromoamination of olefins by *N*-bromosuccinimide (NBS) and cyanamide (NH_2CN) affords β-bromoalkyl-cyanamides. Chemoselective hydrogenation of the nitrile function leads to acyclic amidines followed by spontaneous cyclization to yield cyclic amidines, which are hydrolyzed under basic condition to give 1,2-diamines (Scheme 3.5).

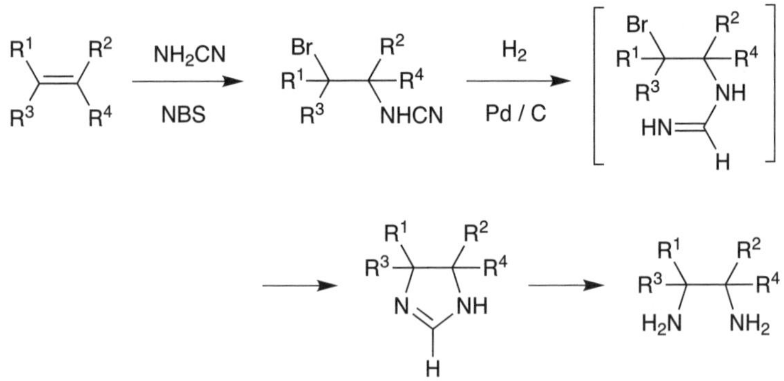

Scheme 3.5 Bromoamination of olefins and reductive cyclzation of bromocyanamide for the synthesis of cyclic amidine

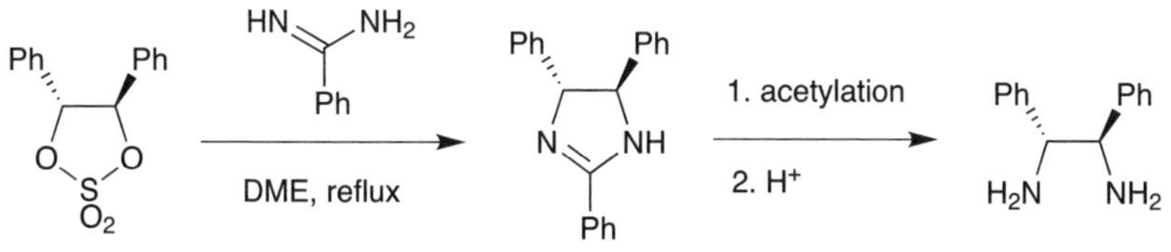

Scheme 3.6 *Alkylation of phenylamidine by cyclic sulfate for the synthesis of amidine*

Cyclic sulfate is transformed to cyclic amidine [isoamarine (**8**)] by reaction with phenylamidine [17]. The cyclic amidine is hydrolyzed after acetylation to give chiral 1,2-diphenylethylenediamine, a useful chiral building block (Scheme 3.6).

3.2.2 Condensation of 1,2-Diamine

Cyclic amidines are prepared by the condensation of 1,2-diamines and β-ketoester derivatives [18]. Reaction of *N*-monomethyl-1,2-diamine and formylacetate acetal in the presence of an acid catalyst such as hydrochloric acid (HCl) or *p*-TsOH initially forms amidines by cyclocondensation. Elimination of ethanol to ethoxyvinyl amidine, followed by incorporation of a different 1,2-diamine, furnishes 1,3-dimethyl- and 1-butylimidazolidines (Scheme 3.7). This reaction is applied to the preparation of several kinds of diamines as a key step [19].

3.2.3 Coupling of Imines (Isoamarine Synthesis)

A sequence of hydrobenzamide–amarine–isoamarine (**8**) is one of the useful synthetic routes for C_2-symmetric 1,2-diamines through cyclic amidines (Scheme 3.8). The reaction course is proposed as follows: (i) formation of hydrobenzamide from benzaldehyde and ammonia, (ii) trimeric condensation to amarine, and (iii) isomerization to isoamarine (**8**) under basic condition [20] (path A). Corey and Kuhnle [21] proposed an alternative path for the reaction course (path B) based on characterization of each intermediate. This amidine is found to be rather stable to acid-catalysed hydrolysis. Thus, reduction of isoamarine (**8**) to imidazolidine with aluminum amalgam in wet tetrahydrofuran (THF) followed by acid hydrolysis yields the corresponding 1,2-diamine.

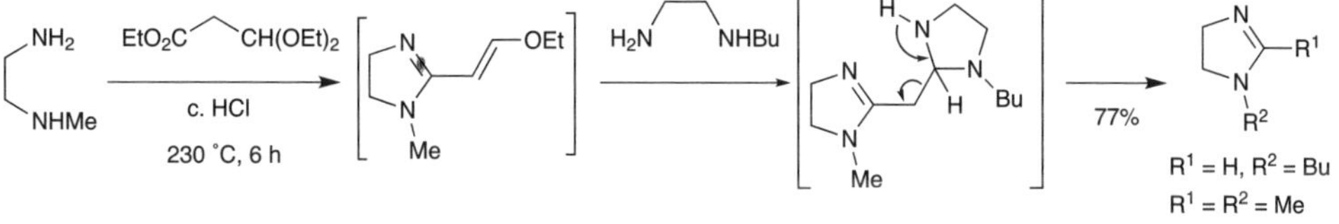

Scheme 3.7 *Amidine formation from β,β-diethoxypropionate and N-methyl- and N-butylethylenediamine*

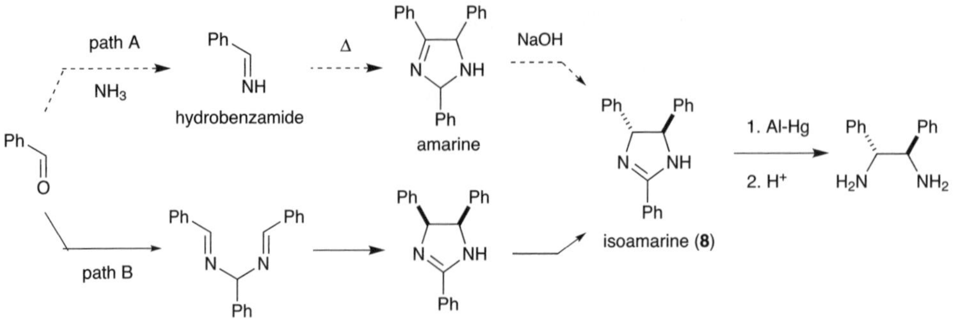

Scheme 3.8 Originally proposed (path A) and revised pathways (path B) for synthesis of isoamarine (**8**)

3.2.4 Modification of Amide Derivatives

3.2.4.1 Amide

As described in the preparation of DBN (**2**) in Scheme 3.4, the most fundamental method for the synthesis of amidines is the dehydration reaction of amide and amines. However, severe conditions such as high temperature and/or pressure are normally required.

3.2.4.2 Imidate

Imidates derived from nitriles and alcohols are also effective precursors for the preparation of amidines. For example, furamidine, known as an active compound towards parasitic microorganisms and DNA minor groove binder [22], and its derivative DB-181, were synthesized from nitrile via the corresponding imidate [23] (Scheme 3.9).

Chiral *N*-sulfinylamidines are prepared from chiral sulfonamides through *N*-sulfinylimidates [24]. The resultant amidines react with excess amounts of imidates to be able to furnish iminoamidines (Scheme 3.10). Cyclic amidines are also synthesized from the same

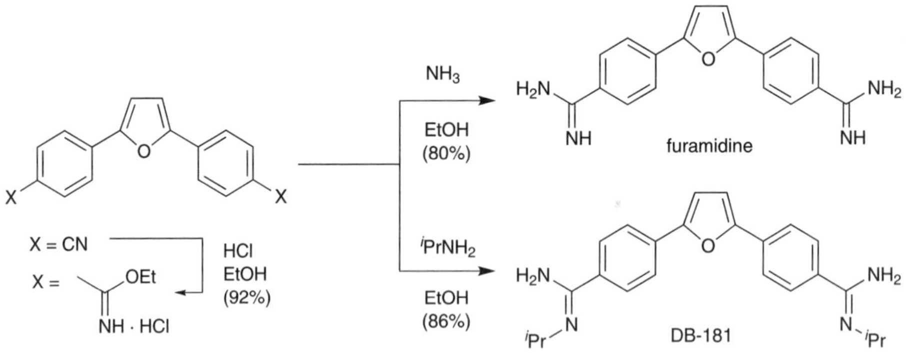

Scheme 3.9 Synthesis of furamamide and DB-181 via imidate

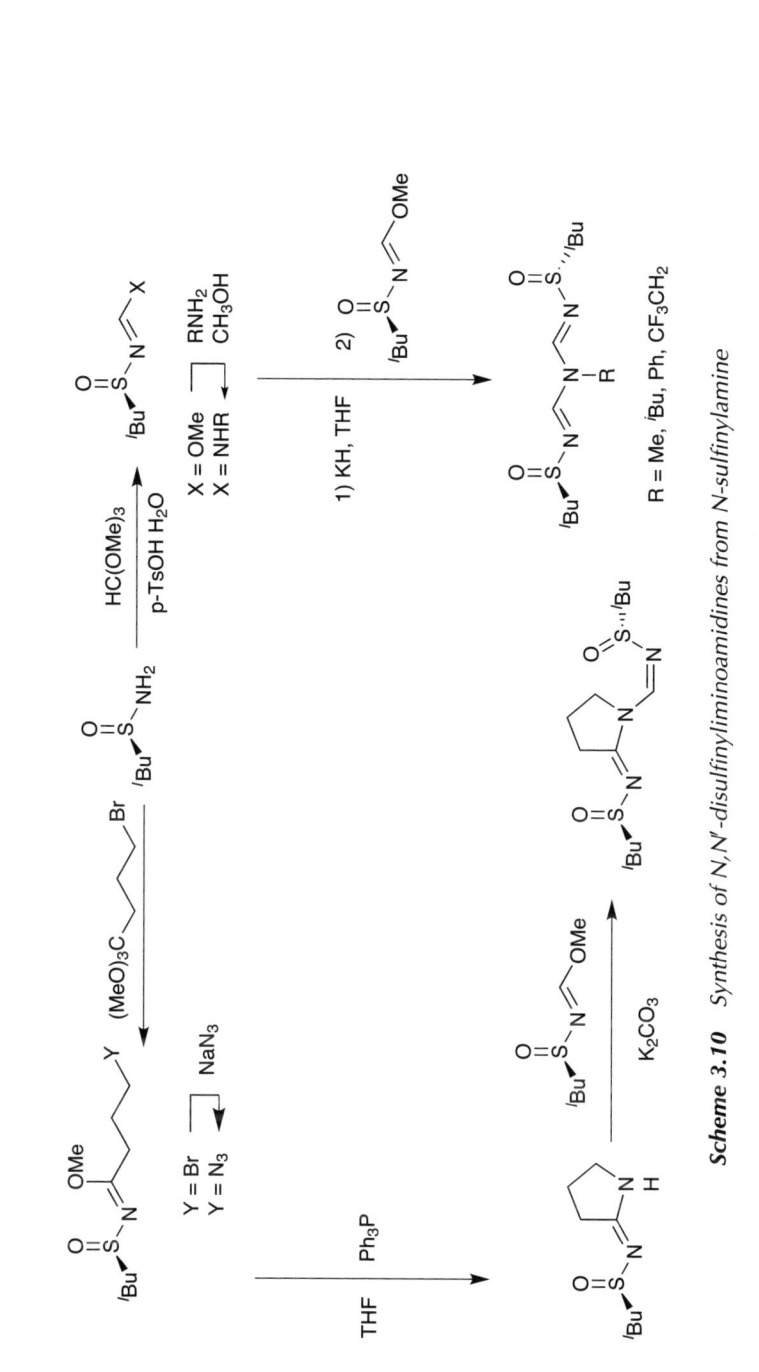

Scheme 3.10 Synthesis of N,N'-disulfinyliminoamidines from N-sulfinylamine

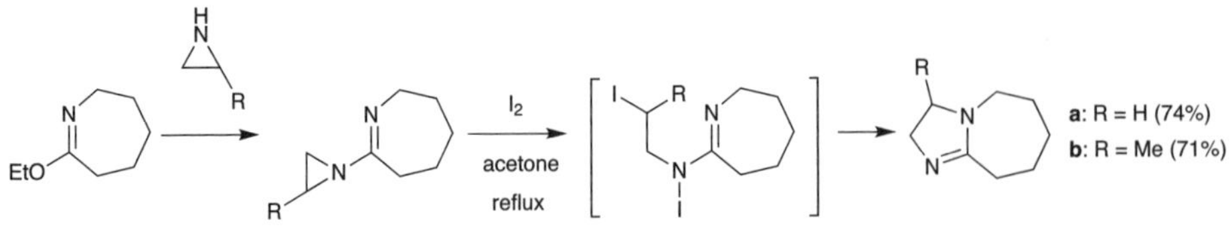

Scheme 3.11 Synthesis of cyclic amidines from imidate and aziridines

sulfonamide by treatment with γ-bromoorthobutanoate followed by azidation and reduction. These bisamidines are used as chiral ligands for the copper-catalysed enantioselective Diels–Alder reaction [24a].

Bicyclic amidines are synthesized from aziridines and cyclic imidates [25]. Thus, NH-aziridines react with cyclic imidates in the presence of a small amount of ammonium bromide to give aziridinylamidine, which is treated with iodine (I_2) to give bicyclic amidines. The use of 2-methylaziridine results in the introduction of methyl group at position 2 of the bicyclic amidine product. The mechanism proposed is shown in Scheme 3.11, which involves iodine-induced ring opening of aziridine and recyclization of the resultant iodoethylamidine.

3.2.4.3 Haloiminium Salt

One-step conversion of N-(ω-azidoalkyl)lactams to bicyclic amidines, avoiding the protection–deprotection sequence on the amine part, is explored by applying the intramolecular Staudinger-type reaction [26]. Oxalyl chloride [$(COCl)_2$] and bromide [$(COBr)_2$] are found to be effective trigger reagents and the corresponding bicyclic amidines are produced in high yield (Table 3.1).

DBN (**2**) is prepared in 92% yield by treatment of N-(3-azidopropyl)-γ-lactam with oxalyl bromide after quenching with anisole (run 3). A trace of the reaction by IR spectrum suggests the formation of a bromoiminium intermediate, which spontaneously cyclizes to the bicyclic system through either 1,2-addition (path A) or [3 + 2]cycloaddition (path B) (Scheme 3.12).

Table 3.1 Synthesis of DBN (**2**) from N-(3-azidopropyl)-γ-lactam

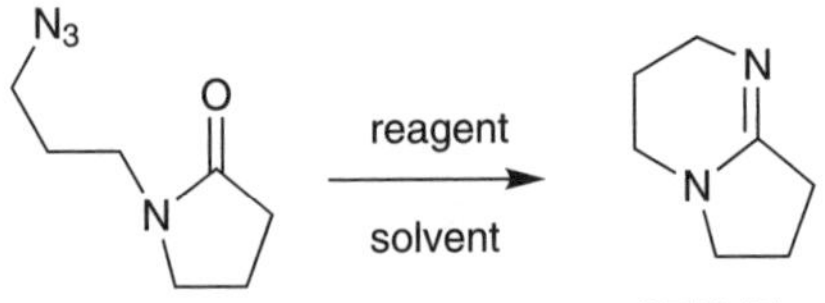

Run	Reagent	Solvent	Yield (%)
1	Ph_3P or Bu_3P	xylene	<10
2	$(COCl)_2$	DCM	81
3	$(COBr)_2$	$(CH_2Cl)_2$	92

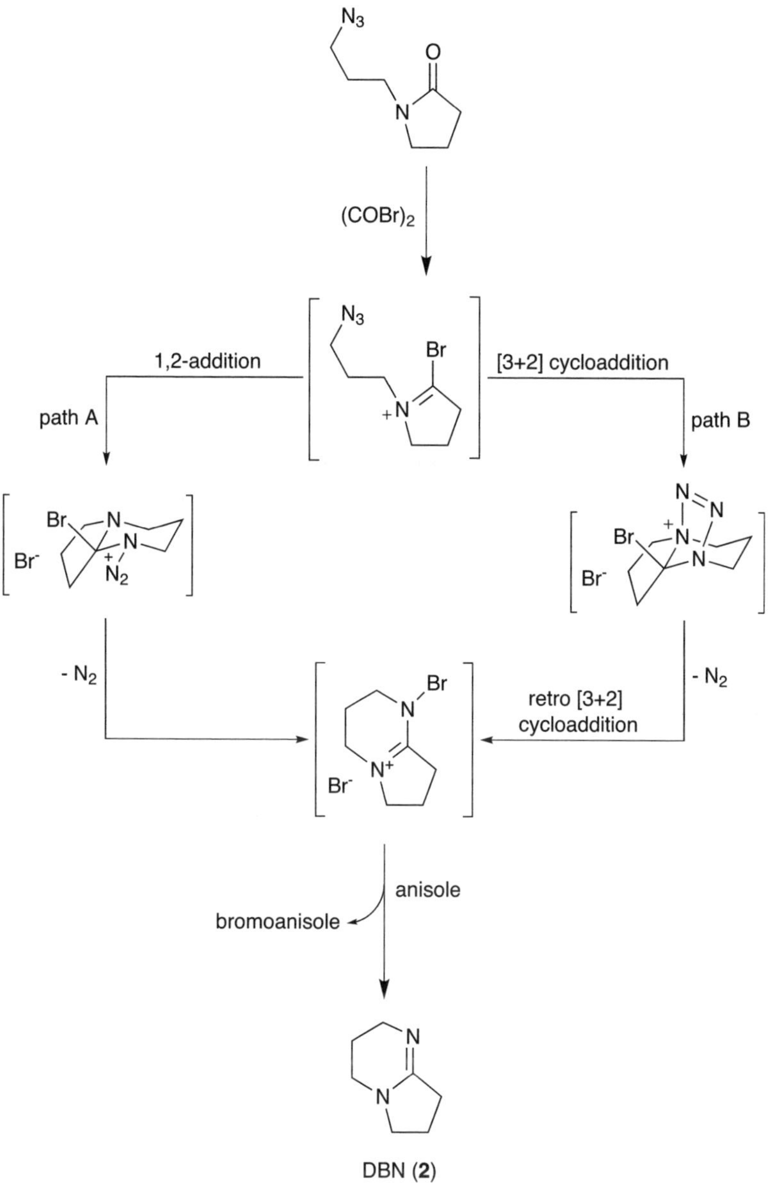

Scheme 3.12 *Proposed reaction mechanisms for the synthesis of bicyclic amidine from haloiminium salt*

3.2.4.4 Thioamide

Thioamides are used as more reactive precursors than amides for amidine synthesis. Lawesson reagent is normally used for conversion of amide to thioamide. Preparation of an amidine-type mannosidase inhibitor **6** is shown as a representative example [14] (Scheme 3.13).

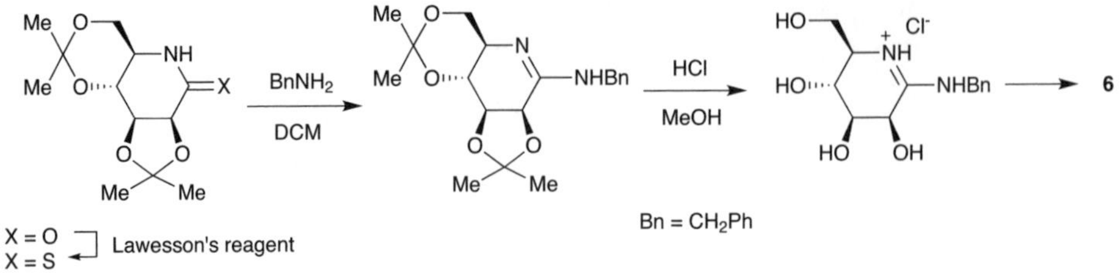

Scheme 3.13 An example of conversion of thioamides to amidines

Diazabicyclo[4.3.0]nonene-based peptidomimetics with a quaternary chiral centre are prepared via intramolecular condensation of N-aminopropyl-γ-lactam [27]. Reductive amination of oxazolidinone aldehydes with N-monoprotected propylenediamine give N-phthalimidopropyl lactams; however, trials of cyclization to bicyclic amidines after deprotection under dehydration conditions are unsuccessful. To solve this problem, the phthalimides are converted to thiolactams with Lawesson reagent. Deprotection followed by treatment with mercury (II) chloride ($HgCl_2$) yields desired cyclic amidines (Scheme 3.14).

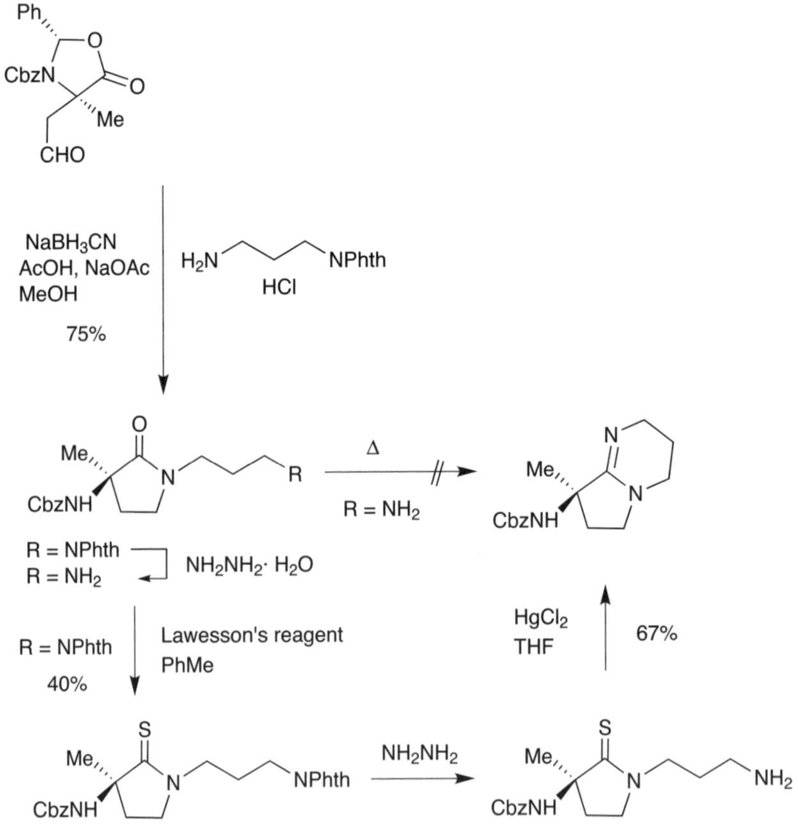

Scheme 3.14 Synthesis of bicyclic amidine on cyclization of N-aminopropyl-γ -thiolactam

3.2.4.5 Thioimidate

Activation of thioamides to thioimidates with alkylation is also applied to the synthesis of amidines. Dijkink et al. [28] applied this method to the synthesis of chiral bicyclic amidines from (S)-malic acid as a key step. However, the amidine was found to be unstable due to isomerization of the imine double bond followed by elimination of the silyloxy group (Scheme 3.15).

γ-Lactam, which is derived from pyroglutamic acid by reduction, protection of the hydroxy group and conjugate addition to acrylonitrile, is subjected to a similar synthetic route to afford a desired amidine [15] (Scheme 3.16). Intermolecular hydrogen bonding between hydrogen on the hydroxy group and imino nitrogen was observed by X-ray crystallographic analysis.

3.2.5 Multi-Component Reaction

Multi-component reaction for the preparation of a variety of compounds in a single step is a powerful tool in combinatorial chemistry and drug discovery. Reaction of cyclohexylisonitrile and isobutyraldehyde in methanol in the presence of dimethylamine hydrochloride as a weak acid catalyst produces N-cyclohexyl-α-dimethylaminoisovaleramide and the corresponding α-hydroxy compound [29] (Scheme 3.17). α-Aminoamidine is obtained as a sole product when dimethylamine is added as a nucleophile. Keung et al. [30] optimized the reaction conditions using various metal catalysts. Scandium (III) triflate was found to be the best catalyst and tolerant to a wide variety of amine and aldehyde units.

N-Sulfonylamidines can be prepared by three-component coupling [31] of alkynes (R^1 = alkyl, aryl or silyl), sulfonyl azide and amine, which is known as 'click chemistry.' [32] The use of alkyl azides in place of sulfonyl azide without a copper catalyst results in the formation of 1,2,3-triazoles (Scheme 3.18). This reaction shows substrate tolerance to each component. Reaction with an optically active amino ester is performed without racemization. N-Boc-ynamide (R^1 = NPhBoc) can act as the alkyne component in the synthesis of N-Boc-aminoamidines [33].

A step-wise procedure for the preparation of N-sulfonylamidines via N-sulfonylimidates has also been reported [34]. Similar multi-component reaction using alcohols gives N-sulfonylimidates, which are converted to N-sulfonylamidines by treatment with primary and secondary amines in the presence of catalytic amount of sodium cyanide (NaCN) [24a] (Scheme 3.19). This indirect process sometimes shows a better yield than the above direct method: for example, the one-pot reaction of phenylacetylene (R^1 = Ph), p-toluenesulfonyl azide (R^2 = Ts) and morpholine ($2R^3$ = $CH_2CH_2OCH_2CH_2$) provides amidines in 19% yield, whereas product is obtained in 72% yield in the step-wise procedure via imidate (R^4 = Me).

ω-Alkynylamines can be used as a combined component of alkyne and amine, in which two kinds of cyclic amidines are formed dependent upon the conditions used [35] (Table 3.2). When 5-amino-1-pentyne is reacted in the presence of copper catalyst, a five-membered amidine is obtained as a major product with the loss of one carbon (run 1). On the other hand, reaction using 6-amino-1-hexyne in the presence of rhodium [Ru(III)] catalyst affords a six-membered amidine as a major product (run 2).

This selectivity could be rationalized as follows: a stable *exo*-methylene pyrrolidine from alkynylamines (n = 1) reacts with TsN_3 to give spirotriazoline **9**, which liberates diazomethane (CH_2N_2) to furnish a five-membered amidine, whereas a bicyclic triazoline

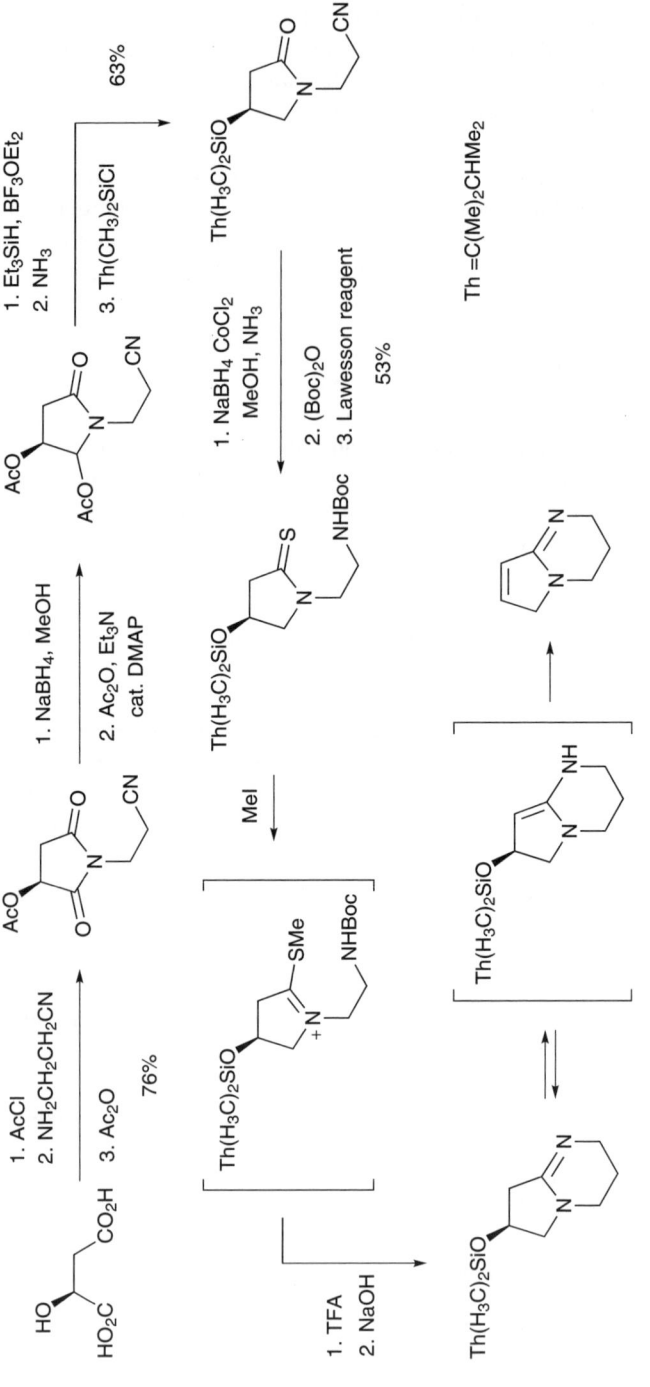

Scheme 3.15 Synthesis of bicyclic amidine from (S)-malic acid via thioimidate

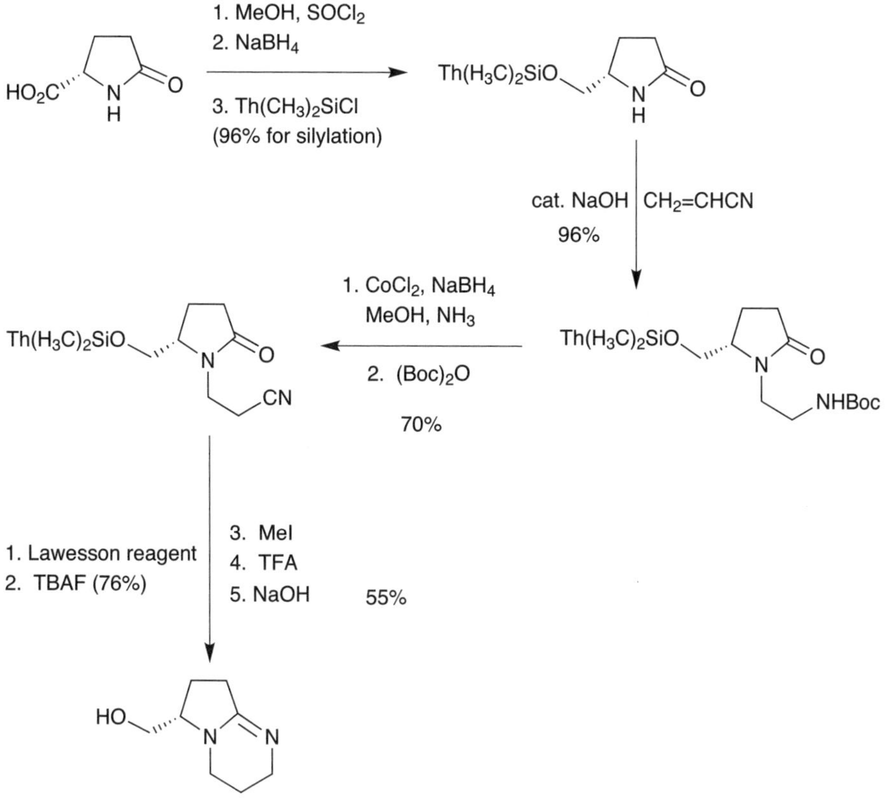

Scheme 3.16 Synthesis of bicyclic amidine from (S)-pyroglutamic acid

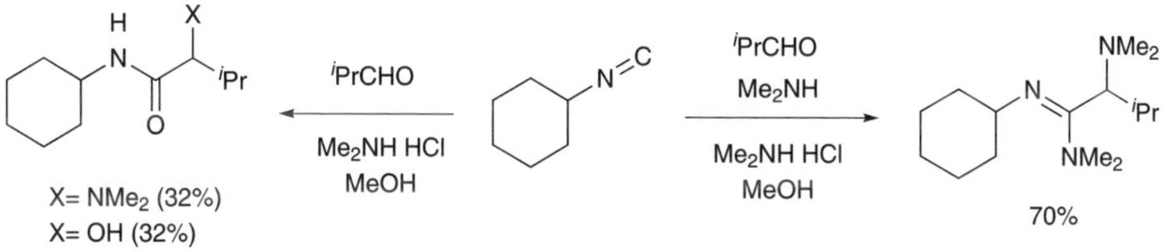

Scheme 3.17 Multicomponent reaction of isocyanide, dimethylamine and aldehyde

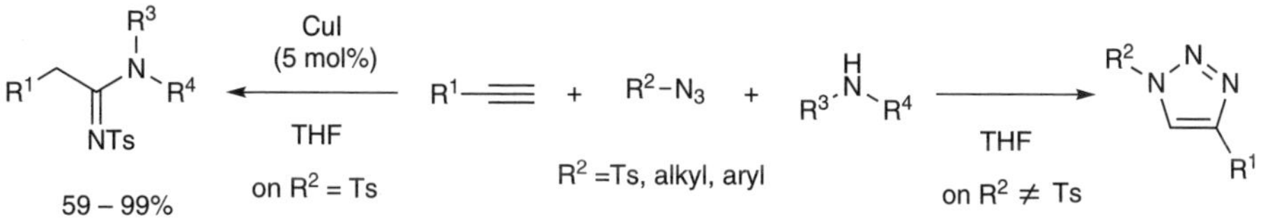

Scheme 3.18 Multicomponent reaction of alkynes, N-sulfonylazides, and amines

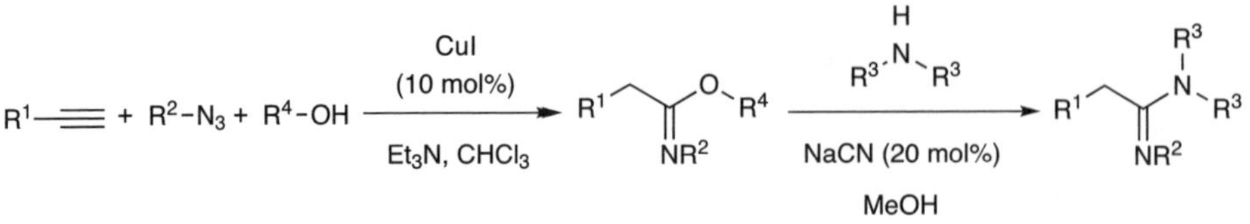

Scheme 3.19 Step-wise procedure for the preparation of N-sulfonylamidines via N-sulfonylimidates

Table 3.2 Preparation of cylic amidines by ω-alkynylamines and tosyl azide

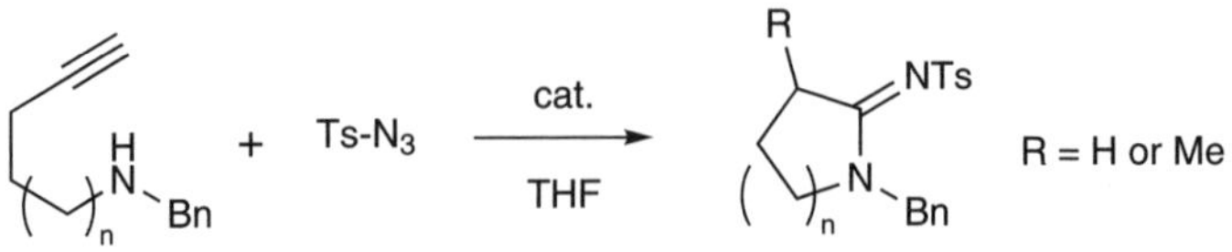

Run	n	Catalyst	Yield (%)	R=H/R=Me
1	1	CuI	63	3.1/1
2	2	Ru$_3$(CO)$_{12}$	75	1/8

10, derived from alkynylamines (n = 2) through dehydropiperidine, is transformed to a six-membered amidine by sequential reactions of N—N bond cleavage and methyl-migrated rearrangement together with the release of nitrogen N$_2$ (Scheme 3.20).

3.2.6 Oxidative Amidination

On the development of copper-catalysed olefin aziridination, Evans et al. [36] reported oxidative amidine formation. Treatment of cyclohexene, a less reactive substrate towards

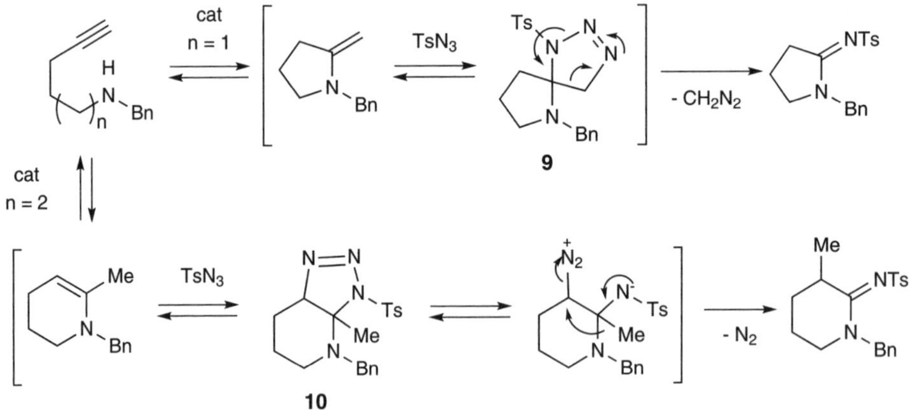

Scheme 3.20 Proposed mechanisms for the selectivity on formation of amidines

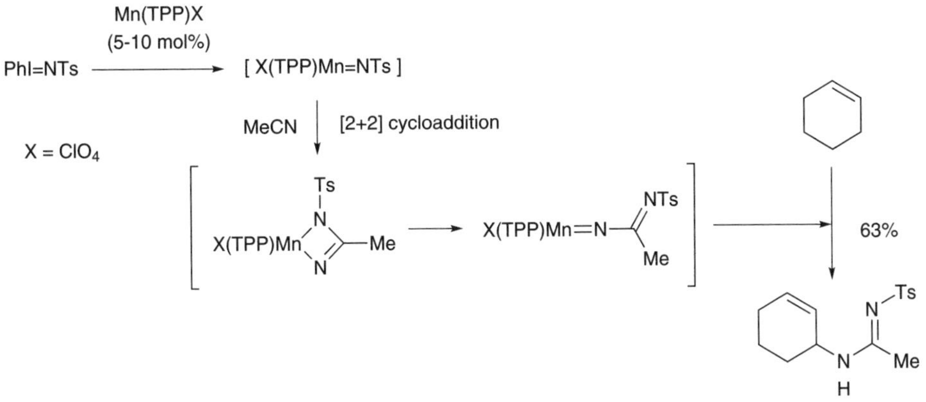

Scheme 3.21 Oxidative amidination of cyclohexene with PhI=NTs and acetonitrile catalysed by Mn(TPP)ClO$_4$

aziridination, and a stoichiometric amount of (N-tosylimino)phenyliodinane (PhI = NTs) in acetonitrile in the presence of a catalytic amount of manganese (III) tetraphenylporphyrin perchlorate [Mn(TPP)ClO$_4$] gives N-tosylamidine. In this reaction a metal–imido intermediate is initially formed from Mn(TPP)ClO$_4$ and PhI = NTs, and then three sequential reactions – [2 + 2] cycloaddition with acetonitrile (MeCN) [37], rearrangement to iminonitrenide and insertion at the allylic position of cyclohexene – occur (Scheme 3.21). Use of Mn(TPP)ClO$_4$ gives the desired N-tosylamidine in 63% yield, whereas the corresponding chloride [Mn(TPP)Cl] gives a lower yield (27%).

3.2.7 Oxidative Cyclization to Bisamidine

Bicyclic bisamidines can be prepared from *trans*-N,N'-dimethyl-1,2-diaminocyclohexane and a palladium bis(arylisocyanide) complex [38] (Scheme 3.22). Oxidation of the initially-formed palladium bis(acyclic diaminocarbene) complex with air or idosobenzene (PhI = O) followed by treatment with excess amounts of methylisonitrile (MeNC) yields bicyclic bisamidines. The structure of the product obtained in each step is unequivocally determined by X-ray crystallographic analysis.

3.2.8 Ring Opening of Aziridine

3.2.8.1 Aziridine

Lewis acid catalysed [3 + 2] cycloaddition of N-alkoxycarbonylaziridines and cyanoalkane such as MeCN furnishes 2-imidazolidines [39] (Scheme 3.23). Although ring-fused aziridines are useable as substrates, *cis*-stereochemistry in the ring juncture is isomerized to *trans*-stereochemistry in the bicyclic amidine products. Benzonitrile also serves as a nitrile source.

Application to chiral N-acylaziridine-2-carboxylate gives the corresponding chiral 2-imidazolidine with the retention of configuration [40] (Scheme 3.24). This implies that the

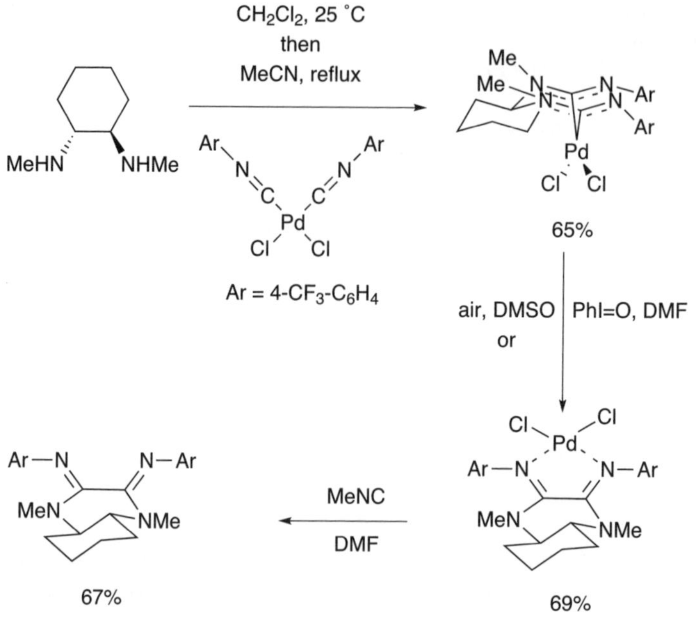

Scheme 3.22 Synthesis of bicyclic bisamidine from diamine and palladium isonitrile complex

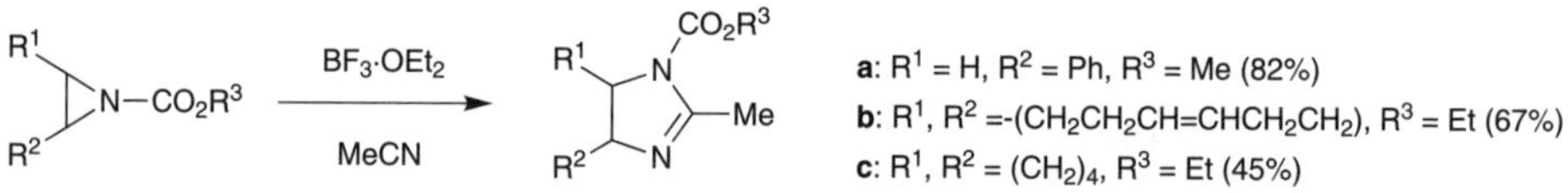

Scheme 3.23 Synthesis of cyclic amidines from aziridines and acetonitrile

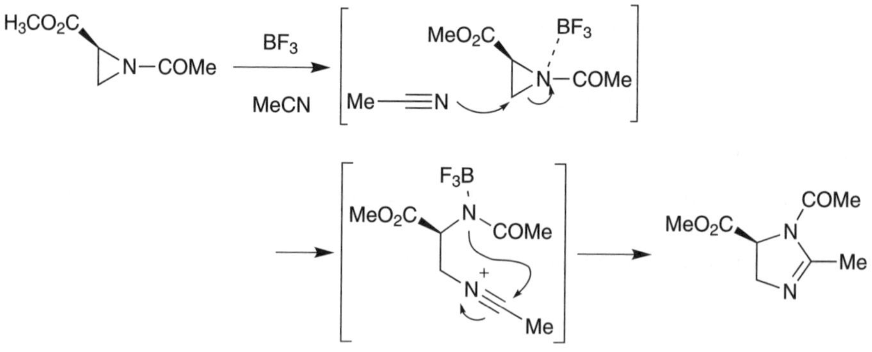

a: R^1 = H, R^2 = Ph, R^3 = Me (82%)
b: R^1, R^2 = -(CH$_2$CH$_2$CH=CHCH$_2$CH$_2$), R^3 = Et (67%)
c: R^1, R^2 = (CH$_2$)$_4$, R^3 = Et (45%)

Scheme 3.24 Chiral cyclic amidine from N-acylaziridine-2-carboxylate and acetonitrile

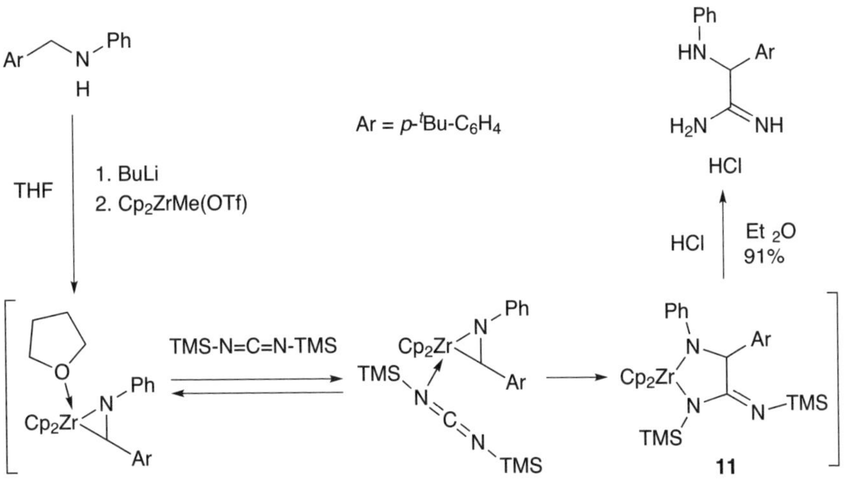

Scheme 3.25 Insertion of carbodiimides to zirconaaziridines for the preparation of aminoamidines

aziridine ring is opened by nucleophilic attack of nitrile at the less-hindered position 3 and the formed ynimium is recycled. As a Lewis acid catalyst, trimethyloxonium tetrafluoroborate (Meerwein reagent) ($Me_3O^+BF_4^-$) [41], scandium triflate [$Sc(OTf)_3$] [42] and cupric triflate [$Cu(OTf)_2$] [43] are also effective, among which $Cu(OTf)_2$ could be recommended because of easy handling and wide applicability to aziridine substrates.

3.2.8.2 Zirconaaziridine

Carbodiimides are potential nitrogen sources for amidines [44]. Zirconaaziridines, generated *in situ* from amines, butyllithium (BuLi) and bis(η^5-cylopentadienyl)methyl (trifluoromethanesulfonyl)zirconium [$Cp_2ZrMe(OTf)$], are efficiently trapped by carbodiimides. Zirconacycles **11**, produced by insertion of carbodiimides into the Zr–C bond of zirconaaziridines, are supposed to be key intermediates, which are hydrolyzed to give α-aminoamidines (Scheme 3.25).

3.3 Application of Amidines to Organic Synthesis

3.3.1 Acetoxybromination

Isoamarine (**8**), a cyclic amidine, is used for the transfer of electrophilic bromine from NBS to vinylarenes [45]. Thus, styrene is acetoxybrominated with NBS in the presence of catalytic isoamarine (0.01 equiv.) in acetic acid (AcOH) to afford a bromoacetate in 95% yield. A single anti-diastereoisomer is obtained when the 2-substituted derivative is used (Scheme 3.26).

The catalytic cycle shown in Scheme 3.27 is proposed, in which isoamarine acts as an electrophilic bromine carrier from NBS. A related brominated 2-phenylamidine, which is

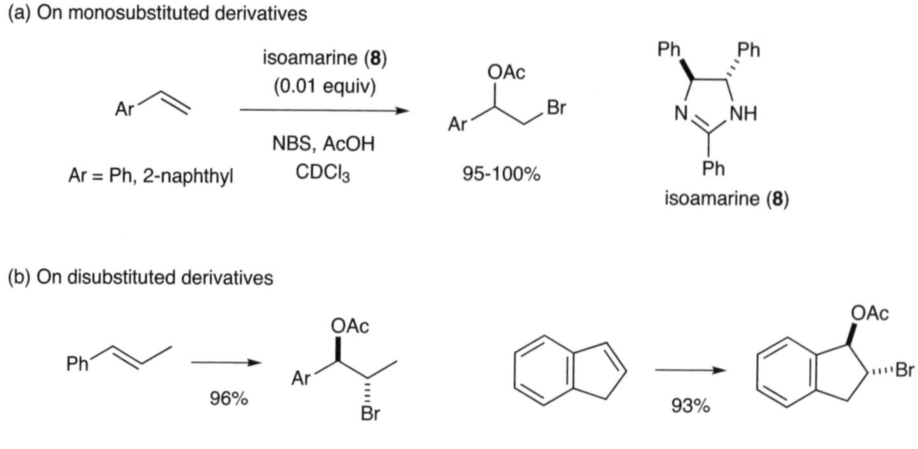

66 Amidines in Organic Synthesis

(a) On monosubstituted derivatives

Scheme 3.26 *Isoamarine (8) catalysed acetoxybromination of olefins by NBS*

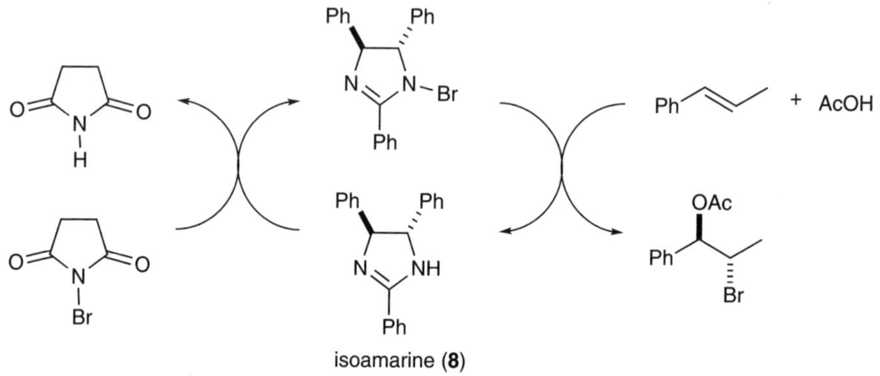

Scheme 3.27 *Proposed catalytic cycle of isoamarine (8) catalysed acetoxybromination*

also proven to be a bromine donor, is analyzed by X-ray crystallography. Unfortunately, chiral induction has never been observed in the reaction using enantiopure isoamarine.

3.3.2 Aldol-Like Reaction

DBU (**1**) and DBN (**2**) promote extremely the reaction between chloroform and benzaldehyde to give trichloromethylcarbinol [46] (Scheme 3.28). An equimolar amount of base is required practically. The reaction is applied to a range of aromatic and aliphatic aldehydes and ketones and is also promoted by a guanidine base.

The DBU (**1**)-promoted intramolecular aldol condensation of two partially protected L-*lyxo*-hexos-5-ulose derivatives, in turn obtained from methyl β-D-galactopyranoside, takes place with fairly good yield and complete diastereoselectivity to give β-hydroxyinososes [47] (Scheme 3.29).

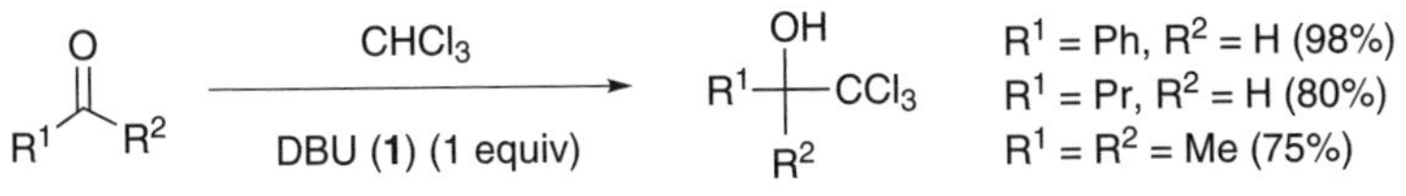
CHCl3
DBU (1) (1 equiv)
R1 = Ph, R2 = H (98%)
R1 = Pr, R2 = H (80%)
R1 = R2 = Me (75%)

Scheme 3.28 *Amidine mediated trichloromethylcarbinol synthesis*

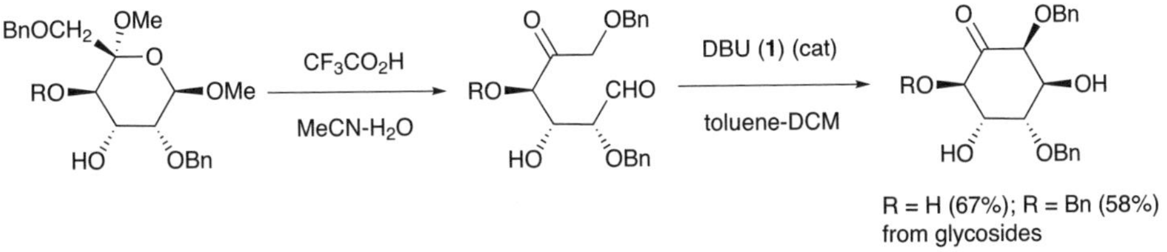
CF3CO2H
MeCN-H2O
DBU (1) (cat)
toluene-DCM
R = H (67%); R = Bn (58%)
from glycosides

Scheme 3.29 *DBU (**1**) promoted intramolecular aldol reaction*

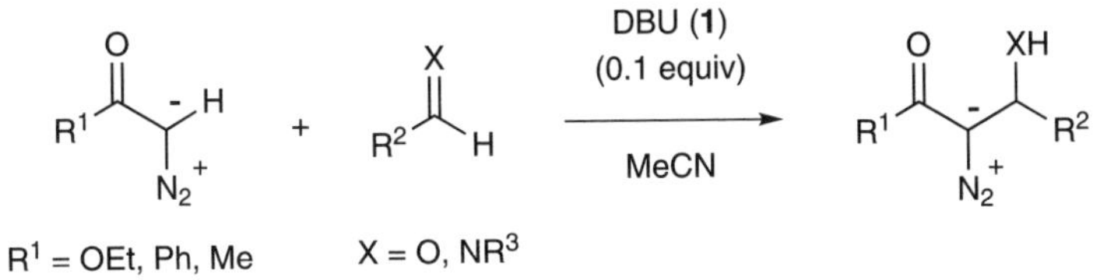
DBU (1)
(0.1 equiv)
MeCN
R1 = OEt, Ph, Me
X = O, NR3

Scheme 3.30 *DBU (**1**) catalysed addition of acyldiazomethane to aldehyde or imine*

Nucleophilic addition of acyldiazomethane to aldehydes or imines is one of the methods for preparing α-hydroxy- or α-aminoacyldiazomethanes. Stoichiometric amounts of rather strong bases such as BuLi, lithium diisopropylamide, sodium hydride (NaH), potassium hydroxide and so on are frequently required. DBU (**1**) was found to be an effective catalyst in this reaction under milder conditions without using anhydrous conditions [48] (Scheme 3.30). Moderate to high yields (58–97%) are achieved on the reaction of α-diazoacetate ($R^1 = OEt$) with aromatic and aliphatic aldehydes. The reaction with electron-rich aromatic aldehydes such as *p*-anisaldehyde was sluggish. α-Diazoacetate ($R^1 = Ph$, Me) reacts only with electron-poor aromatic aldehydes ($R^2 = m\text{-}CF_3\text{-}C_6H_4$, $m\text{-}NC\text{-}C_6H_4$). *N*-Tosylimides ($R^3 = Ts$) are also promising electrophiles and the corresponding β-(*N*-tosylamino)-α-diazoesters or ketones ($R^3 = Ts$) are obtained. This reaction system can be carried out under aqueous conditions [49].

3.3.3 Azidation

Evans *et al.* [50] examined the DBU (**1**)-mediated azidation of α-hydroxy ester with phosphoryl azides and found that the amount of DBU (**1**) is critical for asymmetric induction (Table 3.3). In the reaction with bis(*p*-nitrophenyl)phosphoryl azide in DMF, the use of 1.2 equiv. of **1** resulted in the production of azide with 80% ee (run 5), while product was obtained with less than 2% racemization and in good yield when 0.95 equiv. of **1** was used.

Table 3.3 DBU (1) catalysed azidation

Run	Azide[a]	Solvent	1 (equiv.)	Yield (%)	ee (%)
1	DPPA	toluene	1.2	63	0
2	DPPA	toluene	0.95	61	20
3	$(NO_2)_2$DPPA	toluene	1.2	86	83
4	$(NO_2)_2$DPPA	THF	0.95	60	93
5	$(NO_2)_2$DPPA	DMF	1.2	81	80
6	$(NO_2)_2$DPPA	DMF	0.95	72	>95

[a] DPPA = diphenyl phosphoryl azide; $(NO_2)_2$DPPA = bis(p-nitorphenyl) phosphoryl azide.

3.3.4 Aziridination

Chiral cyclic and acyclic allylsulfoxonium ylides are generated from sulfoxonium-substituted γ,δ-unsaturated α-amino acids (method A) and 1-alkenylsulfoxonium salts (method B) upon treatment with DBU (1) [51] (Scheme 3.31). Their application to the asymmetric aziridination of N-tert-butylsulfonyl imine ester, generated either in situ (method A) or externally added (method B), affords the corresponding alkenylaziridinecarboxylate with medium to high diastereoselectivity and enantioselectivity.

3.3.5 Baylis–Hillman Reaction

Amidines catalyse the Baylis–Hillman reaction [52]. A novel one-pot synthesis-kinetic resolution process involving a DBU (1)-catalysed Baylis–Hillman reaction and a subsequent pyridine catalyst/DBU (1)-mediated enantioselective acylation has been developed [52a] (Scheme 3.32).

3.3.6 Cycloaddition

Enantioselective [3 + 2] cycloaddition of nitrile imines, which are generated in situ by dehydrobromination of hydrazonyl bromides with N-crotonyloxazolidinone, has been developed. On N-arylhydrazonyl bromides, tertiary amines such as triethylamine (Et_3N), diisopropylethylamine ($^i Pr_2NEt$) and N-methylmorpholine (NMM) give excellent yields (90%) and selectivity (94–99% ee). 1,4-Diazabicyclo[2.2.2]octane (DABCO) gives good selectivity (98% ee) but reduced yield (51%), while both yield and enantioselectivity are inferior with DBU (1) (60%, 80% ee) and pyridine (37%, 79% ee). However, dehydrobromination of N-benzylhydrazonyl bromide did not proceed in the presence of $^i Pr_2NEt$. Use of DBU (1) enables dipole formation, giving the cycloadduct in 57% yield and 94% ee [53] (Scheme 3.33).

(a) Synthesis by method A

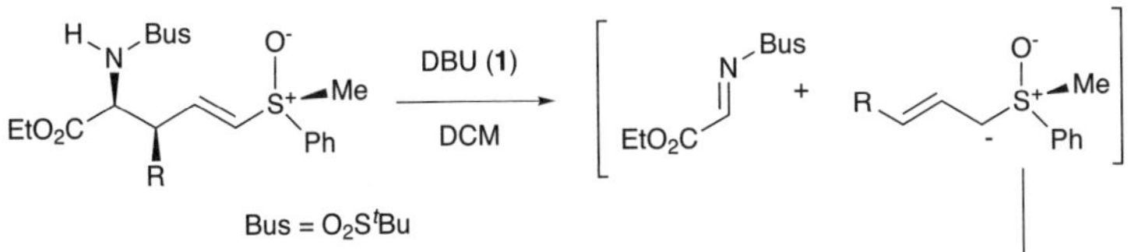

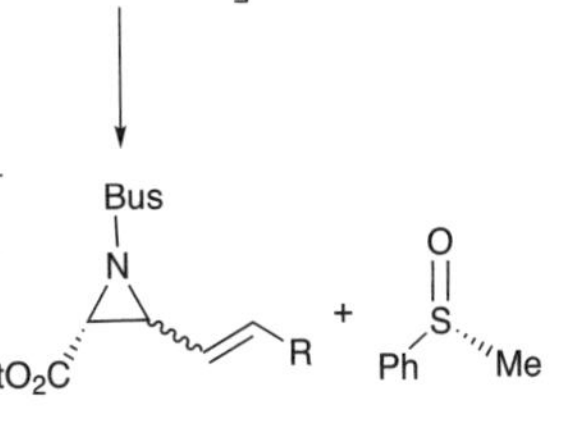

Bus = O₂S^{t}Bu

R	cis-aziridine (%)	trans-aziridine (%)	sulfoxide (%)
Ph	82 (29% ee)	5 (30% ee)	79
^{i}Pr	76 (92% ee)	6 (26% ee)	71
cylohexyl	75 (71% ee)	4 (48% ee)	70

(a) Synthesis by method B

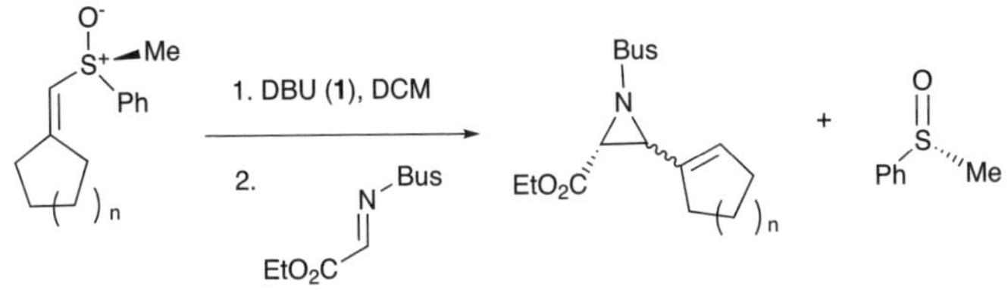

n	cis-aziridine (%)	trans-aziridine (%)	sulfoxide (%)
2	42 (76% ee)	28 (56% ee)	81
3	41 (78% ee)	26 (57% ee)	76
4	75 (71% ee)	30 (25% ee)	70

Scheme 3.31 *DBU (1) mediated aziridination through allylsulfoxonium ylides*

A combination of lithium bromide (LiBr) and DBU (**1**) catalyses regio- and stereospecific cycloaddition [54]. Imines of aminopyradazino[1,2-a][1,2]diazepine react with a range of achiral and chiral dipolarophiles in the presence of LiBr and DBU (**1**) in MeCN to afford enantiopure spiro-cycloadducts in excellent yield via lithio azomethine ylides [54b] (Scheme 3.34).

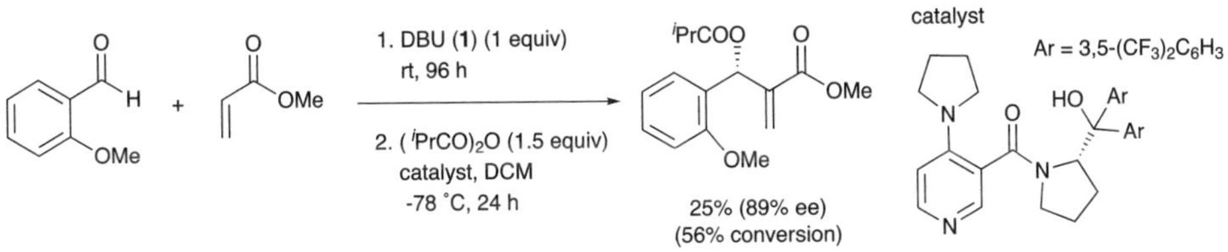

Scheme 3.32 *DBU (1) catalysed Baylis–Hillman reaction and kinetic resolution*

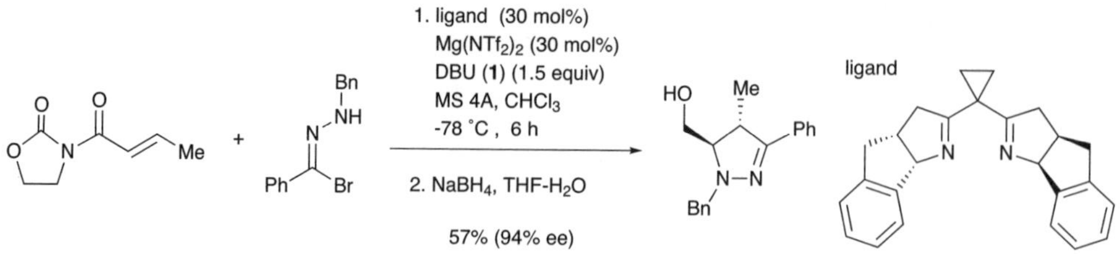

Scheme 3.33 DBU (1) mediated [3 + 2] cycloaddition

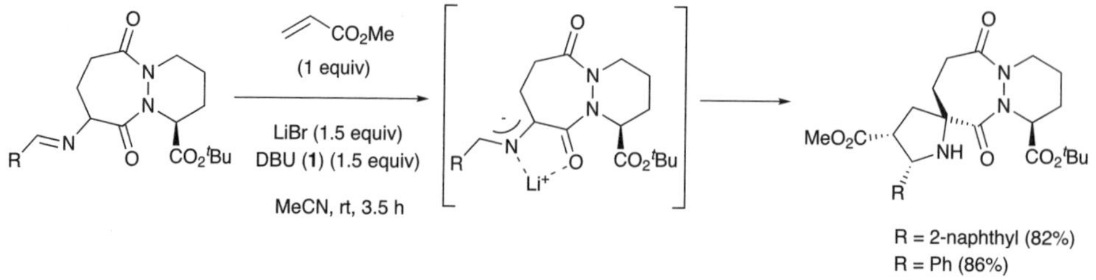

Scheme 3.34 DBU (1) mediated 1,3-dipolar cycloaddition

3.3.7 Dehydrohalogenation

Regioselective introduction of a bromine atom to a double bond in the substituted vinyl sugar is achieved by bromination with pyridinium tribromide and debromination with DBU (1) [55] (Scheme 3.35). *E*-configuration of the product is expected from a specific *anti*-addition in the bromination of the *E*-alkene followed by on E2 (*anti*-elimination) process.

3.3.8 Deprotection

In the chemistry of β-lactam antibiotics, isolations of carboxylic acid derivatives are successfully achieved by formation of amidinium salts [56]. Lewis acid catalysed reaction of 4-substituted 1-trimethylsilyloxyfurans with 4-acetoxyazetidinone chiron leads to highly enantioselective construction of tricyclic carbapenam and penems, in which DBU (1) and Eshenmoser amidine (4) were used for the introduction of the *exo* double bond on the β-lactam skeleton by demesylation (A route) and the isolation of carboxylic acids as

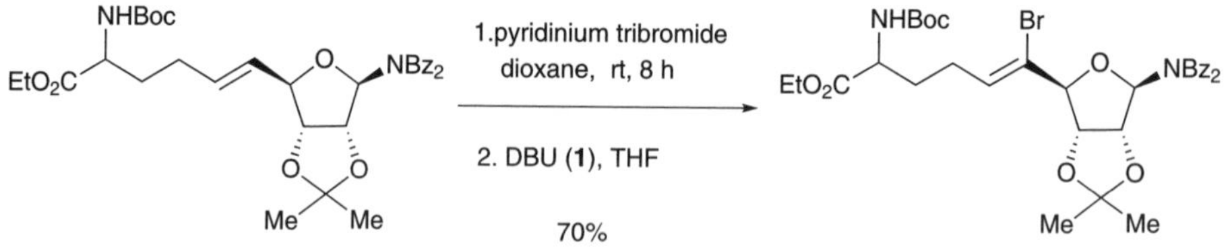

Scheme 3.35 DBU (1) mediated regioselective introduction of bromine atom to double bond

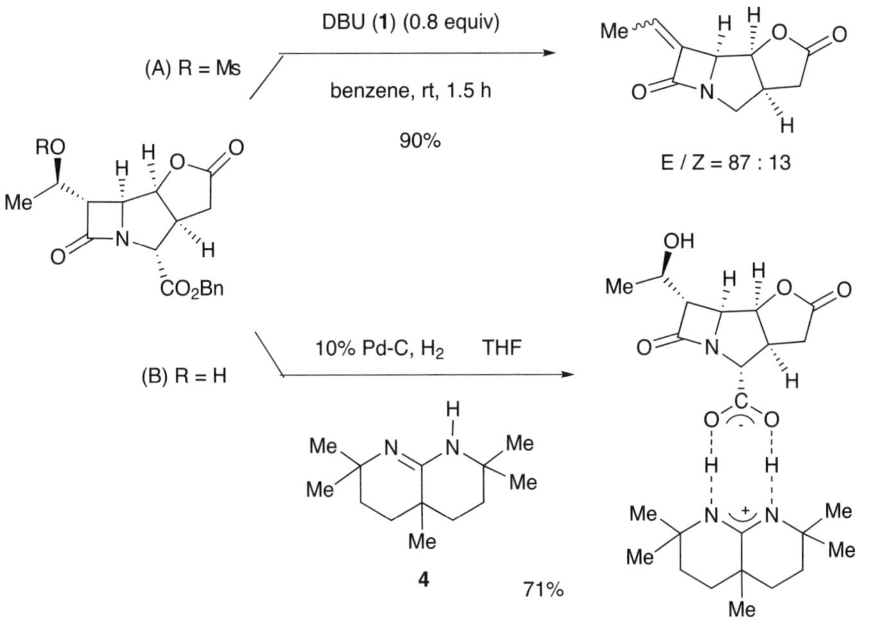

Scheme 3.36 *The use of amidines in the β-lactam chemistry*

crystalline amidinium salts with **4** in the deprotection of the benzyloxycarbonyl function (B route), respectively [56b] (Scheme 3.36).

3.3.9 Deprotonation

Chiral allyl alcohols are obtained from *meso*-epoxides by treatment with bases [57–59]. Addition of amidines like DBU (**1**) alters the reactivity and the enantioselectivity in the epoxide rearrangements, in which **1** is lithiated and works as a bulk base (a catalyst-regenerating base) as well as being a strong solvating agent [59] (Figure 3.4). NMR studies using isotopically labeled chiral lithium amide and lithiated DBU show the formation of a mixed dimer.

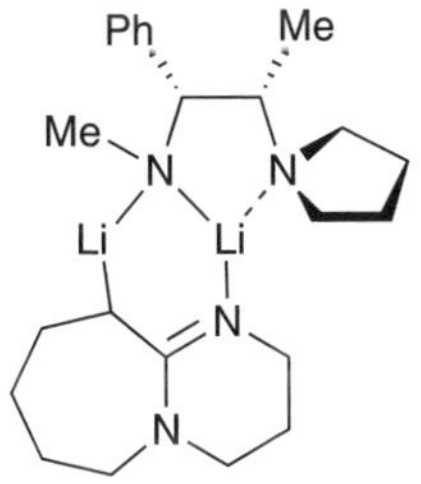

Figure 3.4 *NMR-supported mixed dimer containing DBU (**1**)*

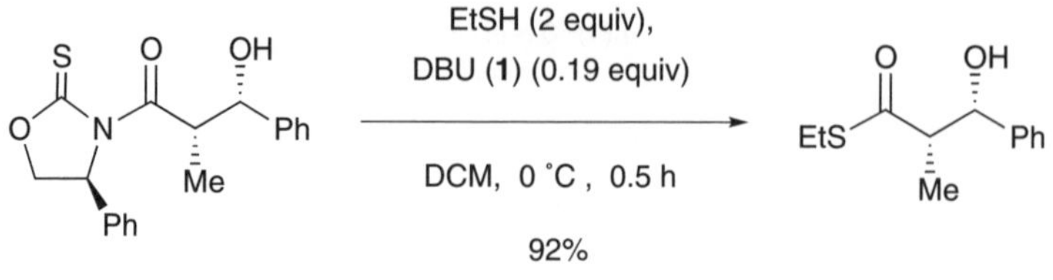

Scheme 3.37 *DBU (1) catalysed removal of chiral auxiliary*

3.3.10 Displacement Reaction

DBU (**1**) was used as a base in the dehydrochlorination/ring closure of chiral chlorohydrins with high retention of optical purity [60]. *N*-Acyl-β-hydroxy-4-phenyloxazolidinethiones are rapidly converted into the corresponding ethyl thioesters in high yields by treatment with ethanethiol (EtSH) in the presence of a catalytic DBU (**1**) [61] (Scheme 3.37). Thus, the chiral auxiliary could be removed cleanly and non-destructively.

3.3.11 Horner–Wadsworth–Emmons Reaction

The Wittig reaction of 4-oxopiperidine with ethoxycarbonyl triphenylphosphonium methylide either did not occur or conversion was extremely low depending on the reaction conditions; for example, potassium *tert*-butoxide (*t*BuOK) (excess). The use of *t*BuOK (2.8 equiv.) gives the best isolation of 15% of the *E*-derivative after 43 h reflux in toluene. On the other hand, the Horner–Wadsworth–Emmons (HWE) reaction using excess amounts of triethyl phosphonoacetate and DBU (**1**) [or DBN (**2**)] occurred in the presence of lithium chloride to give a diastereoisomeric *E*/*Z* mixture, in which the *E*-alkene is predominant and unexpected epimerization at position 2 in the product caused by possible deprotonation at position 3 with the organobase followed by ring opening and recyclization is observed [62] (Scheme 3.38).

3.3.12 Intramolecular Cyclization

Alkylidene phthalides are produced from 5-*exo-dig* cyclization of *o*-alkynylbenzoic acids. However, concomitant generation of isocoumarins via 6-*endo*-dig cyclization is normally problematic [63].

Cyclization of *o*-alkenylbenzoic acid catalysed with organobases affords the phthalide through 5-*exo* mode regioselectivity in good to excellent yields [64]. Among the bases examined, DBU (**1**) exhibits the highest catalytic activity, 5 mol% of **1** is sufficient to promote completion of the reaction and **1** displays an excellent performance in highly polar solvents such as MeCN and DMSO (Table 3.4).

3.3.13 Isomerization

DBU (**1**) effectively works in the isomerization of the condensation products from L-menthone with salicylamide, which could be a potential chiral 1,3-benzoxazinone auxiliary [65] (Scheme 3.39).

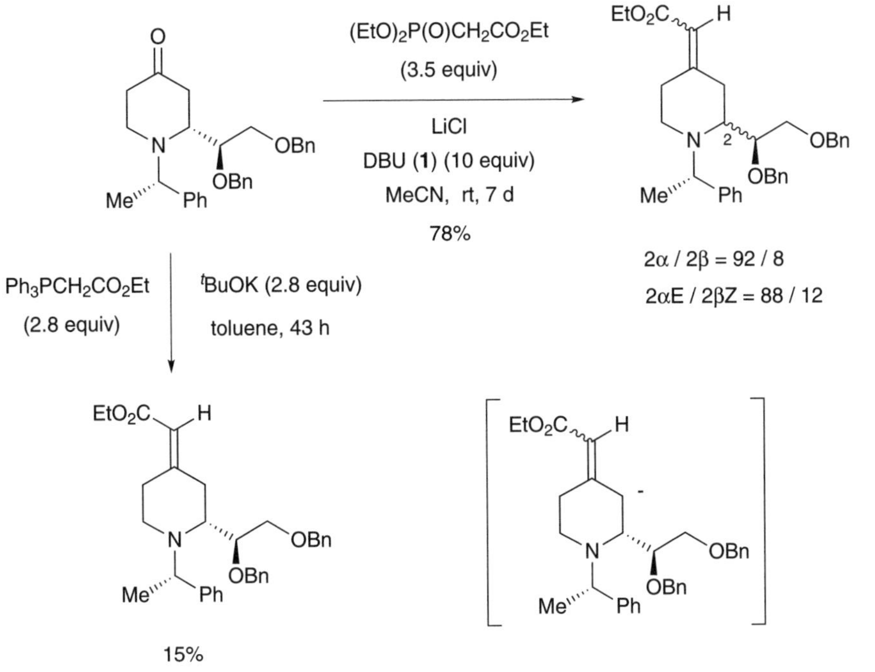

Application of Amidines to Organic Synthesis 73

Scheme 3.38 Comparison of olefination of piperidone under the basic conditions

Table 3.4 DBU (1) catalysed intramolecular cyclization

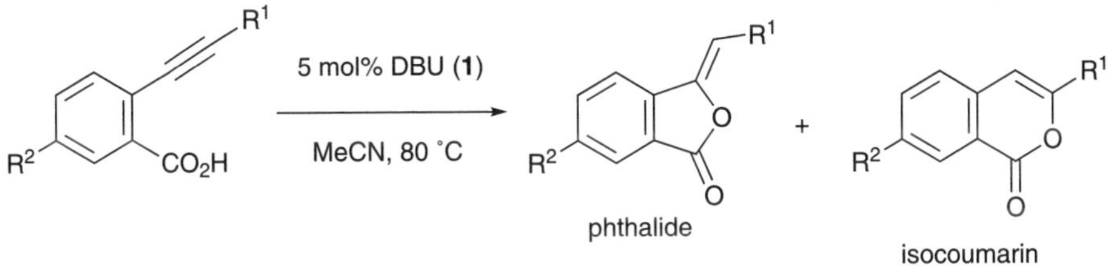

Run	R^1	R^2	Time (h)	Phthalide (%)	Isocoumarin (%)
1	Ph	H	2	94	nd[a]
2	p-MeOC$_6$H$_4$	H	4	96	nd
3	p-CF$_3$C$_6$H$_4$	H	24	80	nd
4	1-naphthyl	H	5	97	nd
5[b]	Ph	Ac	12	79	8
6	Ph	OMe	4	99	nd
7[c]	Pr	H	12	58	36
8	2-propenyl	H	3	65	nd
9	H	H	6	83	nd

[a] Not determined. [b] DMSO as solvent. [c] 10 mol% **1**.

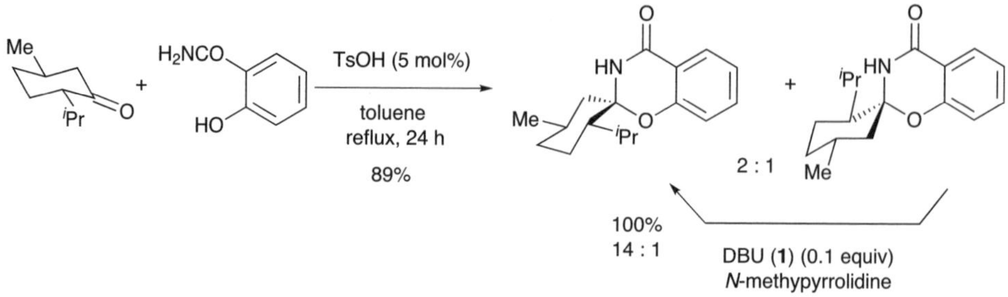

Scheme 3.39 DBU (**1**) catalysed isomerization of spiro system

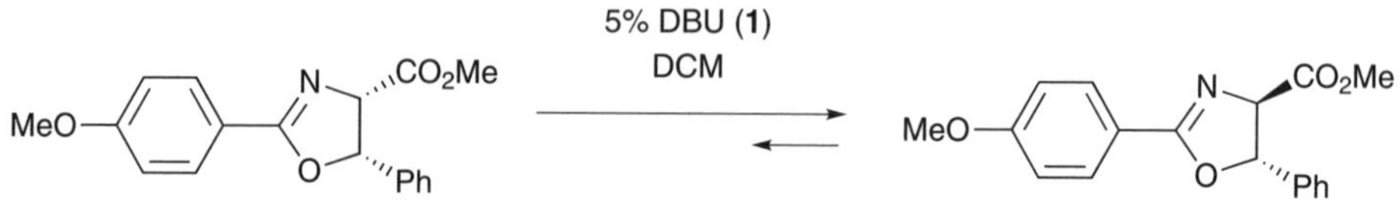

Scheme 3.40 DBU (**1**) catalysed epimerization

cis-Oxazole (>99% ee) is also epimerized to trans-oxazole by treatment with a catalytic amount of DBU (**1**) (cis:trans = 5 : 95, >99% ee) [66] (Scheme 3.40).

3.3.14 Metal-Mediated Reaction

3.3.14.1 Cobalt

Ketoiminatocobalt complexes catalysed enantioselective Henry reaction in the presence of organobase to give β-nitro alcohols have been reported. Although ^{i}Pr$_2$NEt was the most suitable amine for this reaction, the use of DBU (**1**) as base accelerated the reaction but no enantioselection was observed [67]. Bis(triphenylphophoranylidene)ammonium fluoride (PPNF) was found to be an effective base co-catalyst in the cobalt (III)–salen complex catalytic asymmetric addition of carbon dioxide (CO_2) to propylene oxide giving propylene carbonate [68] (Scheme 3.41). DBU (**1**) or the N-methyl analogue also afford good enantioselectivities (75 and 72% ee, respectively). Thus, these strong and sterically hindered organobases can act as co-catalysts in this metal-mediated reaction system.

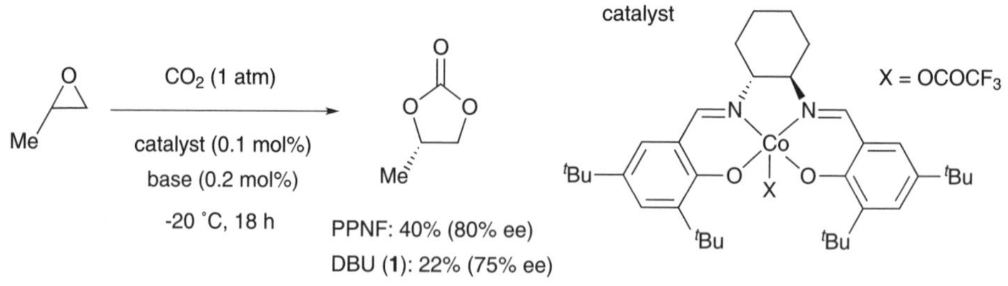

Scheme 3.41 Asymmetric addition of CO_2 to propylene oxide

3.3.14.2 Copper

Copper(II) catalysed enantioselective decarboxylative aldol-type addition of malonic acid hemithioesters to aldehydes in the presence of tartaric acid-derived bisbenzimidazole and an achiral base was examined. The use of DBU (**1**) as an achiral base resulted in low enantioselectivity [69].

3.3.14.3 Iridium

A combination of bis(iridiumcyclooctadienyl chloride) [Ir(COD)Cl]$_2$, a chiral phosphoramidite ligand, and DBU (**1**) as a base in THF effects the iridium (I) catalysed intermolecular allylic amidation of ethyl allyl carbonates with soft nitrogen nucleophiles under completely salt-free conditions [70]. The reaction is quite general, accommodating a wide variety of substrates and nucleophiles, and proceeds with excellent regio- and enantioselectivities to afford the branched *N*-protected allyl amines.

3.3.14.4 Molybdenum

The reaction of molybdenumcyclopentadienyltricarbonyl chloride [CpMo(CO)$_3$Cl] with optically active amidines affords separable diastereoisomers of the Cp(CO)$_2$Mo-amidinato complexes, which could act as chiral catalysts, by fractional recrystallization [71]. The molybdenum configuration is equilibrated at 70 °C in acetone.

3.3.14.5 Nickel

DBU (**1**) is screened as base co-catalyst for enantioselective Michael additions of malononitrile catalysed with the aqueous complex of 4,6-dibenzofurandinyl-2,2'-bis(4-phenyloxazoline) and nickel perchlorate hexahydrate [72].

3.3.14.6 Palladium

Phosphorous-containing amidine was prepared through several steps from L-valine and evaluated as a new ligand for asymmetric palladium (Pd) catalysed allylic alkylation of 1,3-diphenylprop-2-enyl acetate and pivalate [73]. The results with the nucleophile derived from dimethyl malonate are summarized in Table 3.5 [73a]. Excellent asymmetric inductions up to 95% ee were achieved along with an efficient conversion.

A new class of chiral amidine-phosphine and -sulfide hydrid ligands with a variety of modifications is used for the palladium mediated allylic substitutions of both acyclic and cyclic compounds [74] (Figure 3.5). High levels of asymmetric induction were achieved for both substrates.

Some ferrocenylphosphine-amidine ligands (Figure 3.6) with central and planer chirality were prepared and their efficiency and diastereomeric impact in the palladium catalysed asymmetric allylic substitution were examined [75]. Up to 96% ee with 98% yield was achieved by the use of a ligand with a methyl-substituted ligand.

DBU (**1**) is often screened in the utility as co-catalyst (or base) in the palladium mediated coupling reactions [76].

3.3.14.7 Rhodium

DBN (**2**) is used as a strong, sterically hindered base in the asymmetric hydrogenation of acetophenone and styrene by a combination of rhodium (I) complex and chiral ligands

Table 3.5 Asymmetric allyic alkylation catalysed by the palladium complex of amidine ligand

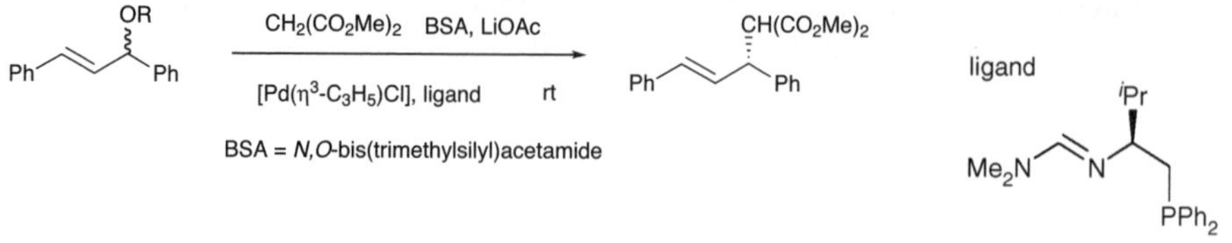

run	R	Pd (mol equiv)	solvent	time (h)	yield (%)	ee (%)
1	COMe	0.05	DCM	48	85	92 (R)
2	CO^{t}Bu	0.05	DCM	24	99	93 (R)
3	CO^{t}Bu	0.05	THF	24	91	91 (R)
4	CO^{t}Bu	0.025	DCM	24	87	94 (R)

^{e} Molar ratio:
Pd/ligand/substrate/CH$_2$(CO$_2$Me)$_2$/BSA/LiOAc = 1–5 : 4–20 : 100 : 300 : 300 : 5.

derived from *N*-substituted diphenylphosphinoacetamides, in which low to moderate hydrogenating activity and enantioselectivity were obtained [77].

3.3.14.8 Tin

Chiral allylating reagents, readily generated *in situ* from tin (II) catecholate [Sn(II)(O$_2$C$_6$H$_4$)], allyl halides, chiral dialkyl tartarates and DBU (**1**), react smoothly with

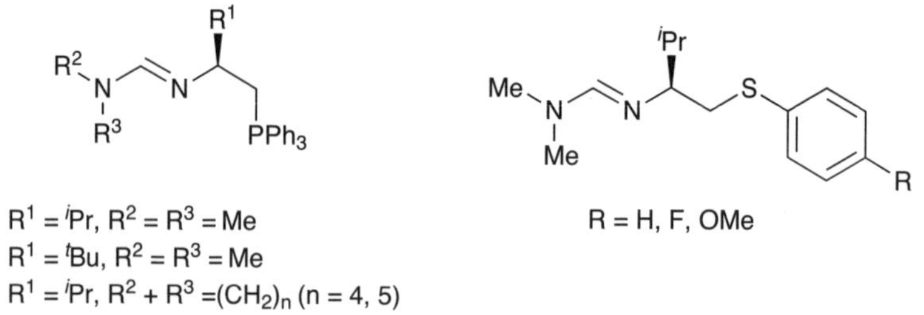

R^1 = ^{i}Pr, R^2 = R^3 = Me
R^1 = ^{t}Bu, R^2 = R^3 = Me
R^1 = ^{i}Pr, R^2 + R^3 =(CH$_2$)$_n$ (n = 4, 5)

R = H, F, OMe

Figure 3.5 Structures of amidine-phosphine and -sulfide hybrid ligand for palladium mediated couling reaction

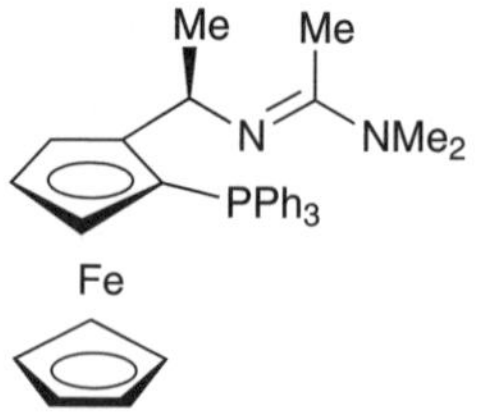

Figure 3.6 Structure of ferrocenylphosphine-amidine ligand for palladium coupling reaction

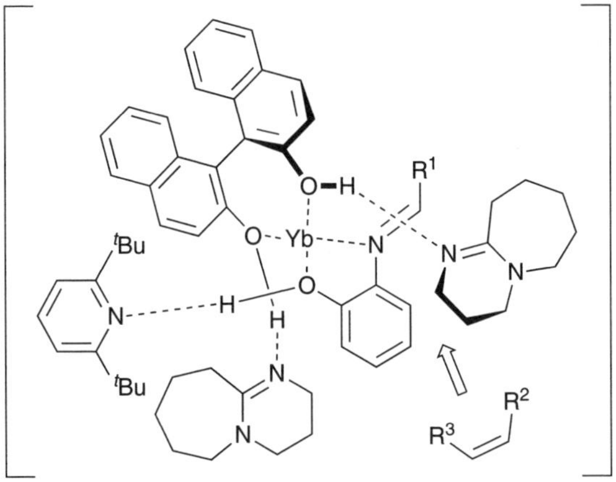

Figure 3.7 *Supposed transition state without triflate anion*

aldehydes or reactive ketones at −78 °C in the presence of a catalytic amount of copper salts to afford the corresponding optically active homoallyl alcohols in high yield (81–99%) and high enantioselectivities (89–94%ee) [78].

3.3.14.9 Ytterbium

Achiral ytterbium Lewis acid was prepared from ytterbium triflate [Yb(OTf)$_3$], (R)-(+)-1,1′-bis(2-naphthol) (BINOL) and DBU (**1**), and subjected to aza Diels–Alder reactions of achiral imines (N-benzylidene-2-hydroxyanilines) and achiral dienophiles [79]. In this reaction the use of a chiral Lewis acid containing 1,3,5-trimethylpiperidine instead of **1** resulted in a lowering of the enantiomeric excess of adduct. Thus, the phenolic hydrogen of the imine interacts with DBU (**1**) in transition state, as shown in Figure 3.7, to increase the selectivity.

3.3.15 Michael Reaction

Four enantiopure hydroxyamidines were prepared from (S)-pyroglutamic acid by coupling of an (S)-malic acid derived N-allyliminium ion with β-naphthol, and from an (S)-serine-derived imide [80] (Figure 3.8). Unfortunately, their application to the catalytic Michael

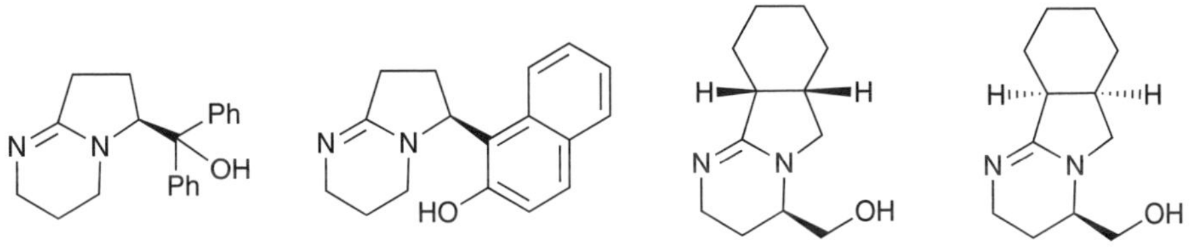

Figure 3.8 *Structures of chiral hydroxyamidines*

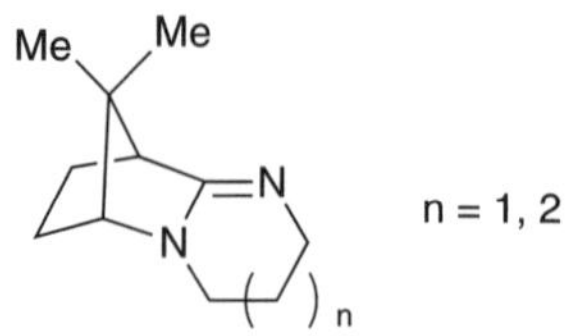

Figure 3.9 Camphor-derived amidines

addition of cyclohexenone with thiophenol and methyl vinyl ketone with 1-carbomethoxy-2-indanone resulted in low asymmetric induction even though the chemical yield was high.

Sterically hindered chiral DBU/DBN-related molecules designed based on (+)-camphor lactam were applied to the Michael addition of β-keto ester with methyl vinyl ketone. However, disappointedly low asymmetric induction was observed [81] (Figure 3.9).

DBU (**1**) catalyses the formation of 2H-1-benzopyran from salicylic aldehydes and allenic carbonyl compounds [82]. This reaction could be categorized as a tandem reaction composed of Michael and aldol-type reactions. Wide substrate tolerance on the aldehyde unit is observed. The introduction of a large phenyl group on R^2 and use of allenyl ester diminishes the yield of benzopyrans (Table 3.6).

3.3.16 Nef Reaction

The conversion of primary or secondary nitroalkanes to aldehydes and ketones is known as the Nef reaction. Strong acidic conditions are normally necessary for this reaction. Ballini *et al.* [83] reported that treatment of secondary nitroalkanes with DBU (**1**) yielded the corresponding ketone in moderate yield (54–80%) (Scheme 3.42). No reaction occurred in the use of primary nitroalkanes. Lower yields were observed when DBN (**2**) (50%) and tetramethylguanidine (25%) were used.

Rearrangement in the *aci*-nitro form to the hydroxynitroso derivative via *N*-hydroxyoxaziridine followed by elimination of hyponitrous acid is proposed as the reaction mechanism (Scheme 3.43).

Table 3.6 Synthesis of benzopyrans from salicylic aldehydes and allenylic carbonyl compounds

Run	R^1	R^2	R^3	Yield (%)	anti:syn
1	H	H	Me	99	74:25
2	5-Me	H	Me	>99	70:30
3	3-MeO	H	Me	>99	81:19
4	3,5-Cl$_2$	H	Me	>99	76:23
5	H	Ph	Me	84	anti
6	H	H	OEt	59	anti

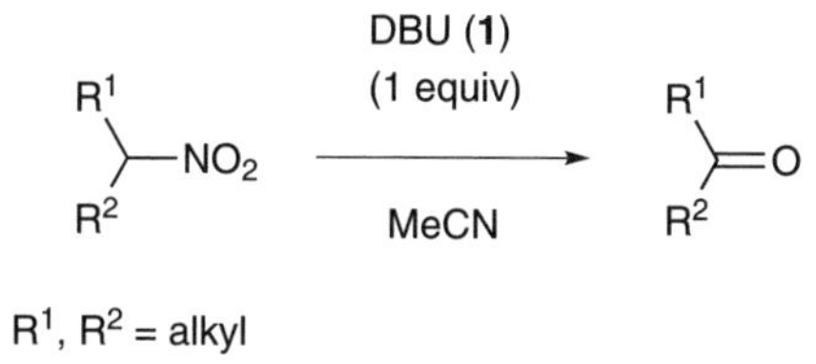

Scheme 3.42 DBU (**1**) mediated Nef reaction of nitroalkanes

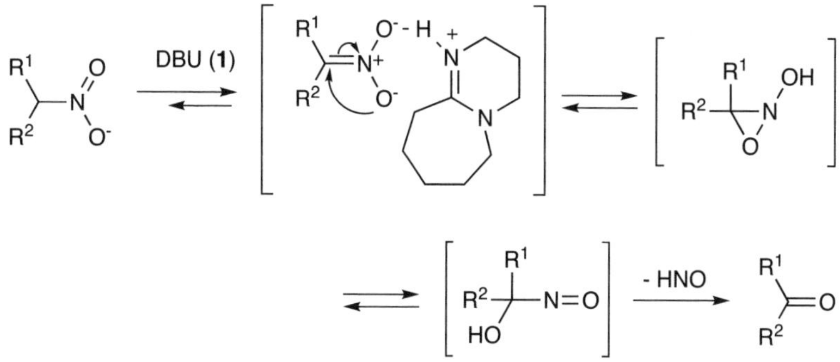

Scheme 3.43 Proposed reaction mechanism for DBU (**1**) mediated Nef reaction

3.3.17 Nucleophilic Epoxidation

DBU (**1**) effectively catalyses the epoxidation of a range of enones derived from tetralone or related cyclic ketones using poly-L-leucine and urea-hydrogen peroxide (H_2O_2) in isopropyl acetate [84]. Epoxides were obtained in 63–85% yield and 59–96% ee [84b] (Table 3.7).

Table 3.7 DBU (**1**) catalysed asymmetric epoxidation

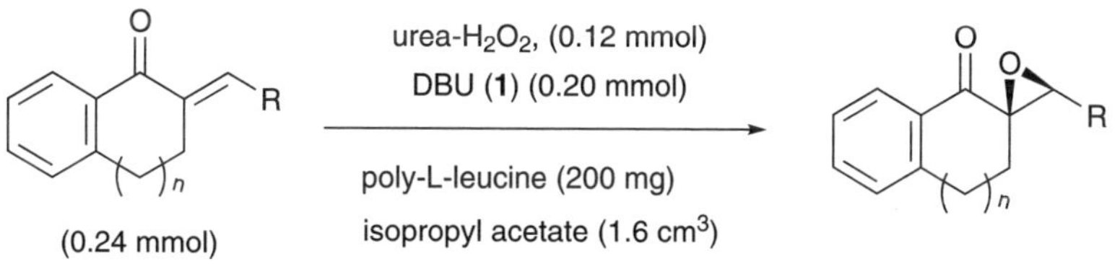

Run	R	n	Time (min)	Yield (%)	ee (%)
1	Ph	1	90	76	84
2	4-BrC$_6$H$_4$	1	72	81	82
3	Me	1	60	66	92
4	^{t}Bu	1	192	63	83
5	H	1	7	64	94
6	Ph	0	48	72	88
7	Ph	2	168	74	59

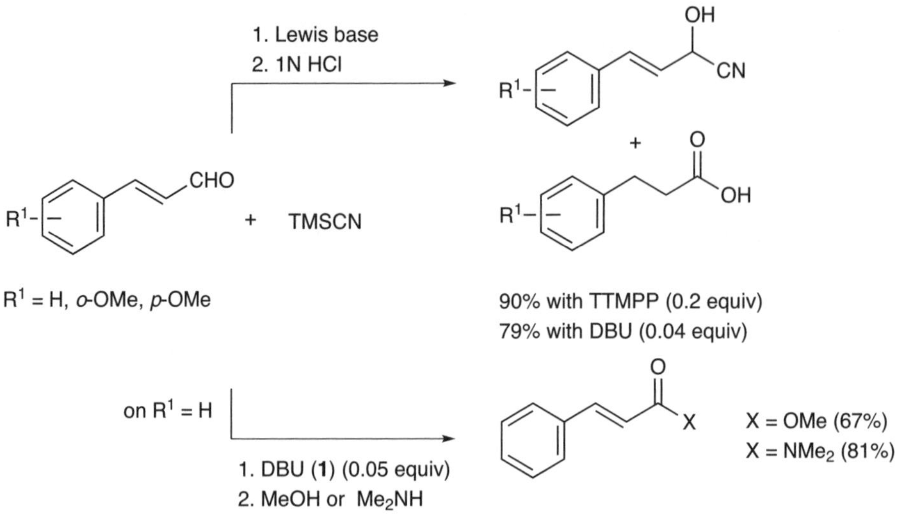

Scheme 3.44 Oxidation of α,β-unsaturated arylaldehydes under basic conditions

3.3.18 Oxidation

Strong Lewis bases such as tris(2,4,6-trimethoxyphenyl)phosphine (TTMPP) and DBU (**1**) can catalyse the transformation of cinnamaldehyde to saturated carboxylic acids, whereas mild bases such as triphenylphosphine or tris(4-methoxyphenyl)phosphine give simple cyanohydrin products [85]. When quenching the reaction with alcohol and amine, the corresponding ester and amide are produced, respectively (Scheme 3.44).

3.3.19 Pudovik-phospha-Brook Rearrangement

The addition reaction of hydrogen phosphites to aldehydes and ketones is a well known method for the synthesis of α-hydroxyphosphonate (Pudovik reaction) [86]. When carbonyl compounds possessing electron-withdrawing groups at the α position (α-dicarbonyl compounds, perfluoroalkyl aldehydes and ketones, benzophenones, etc.) are used, rearrangement of hydroxyphosphonates to phosphates occurs via base catalysed phospha-Brook rearrangement. El Kaïm et al. [87] found that DBU (**1**) catalysed the Pudovik-phospha-Brook reaction of 2-nitrophenyl, 2-pyridiyl or even 1-naphthyl aldehydes (R^3 = H) to give directly the corresponding phosphates. Although no reaction occurred from acetophenone, methyl 2-pyridiyl ketone (R^2 = 2-pyridyl, R^3 = Me) gives the corresponding phosphonate in 58% yield (Scheme 3.45).

3.3.20 [1,4]-Silyl Transfer

In the course of total synthesis of (−)-rasfonin, skillful DBU (**1**) mediated rearrangement of furanol with concomitant [1,4]-silyl transfer was used for the preparation of pyranol as a key synthetic fragment [88] (Scheme 3.46).

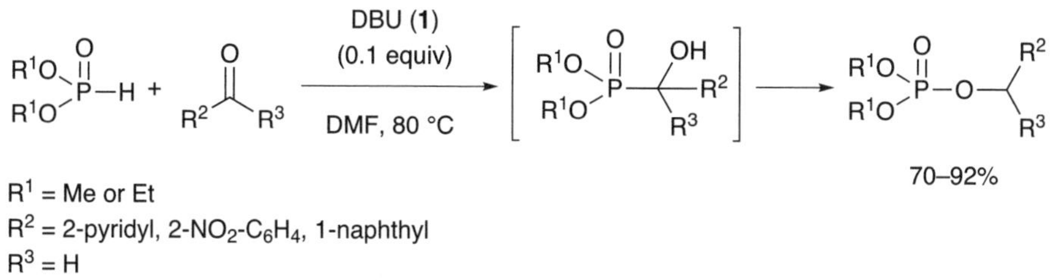

Application of Amidines to Organic Synthesis 81

Scheme 3.45 DBU (1) catalysed Pudovik-phospha-Brook reaction

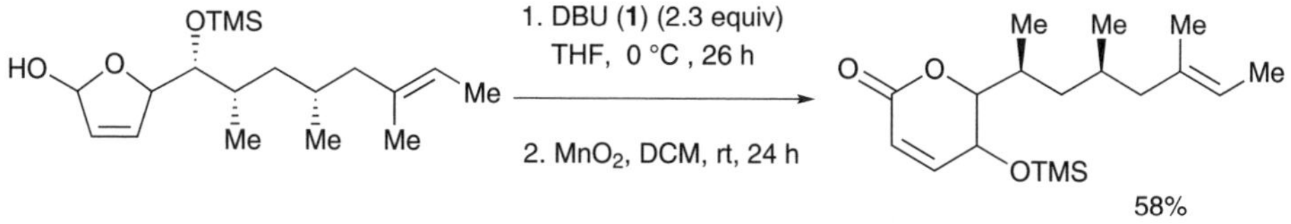

R¹ = Me or Et
R² = 2-pyridyl, 2-NO₂-C₆H₄, 1-naphthyl
R³ = H

Scheme 3.46 DBU (1)-induced [1,4]-silyl transfer with ring expansion

3.3.21 Tandem Reaction

Chiral decalin systems could be stereoselectively prepared by tandem oxy Cope-ene reaction of 1,2-divinylcyclohexenol, derived from (+)-limonene, under microwave irradiation, which is accelerated by an organobase, including DBU (1), tetramethylethylenediamine (TMEDA) and *tert*-butyltetramethylguanidine (BTMG) [89] (Table 3.8). The ee is

Table 3.8 Tandem oxy Cope-ene reaction with various bases.

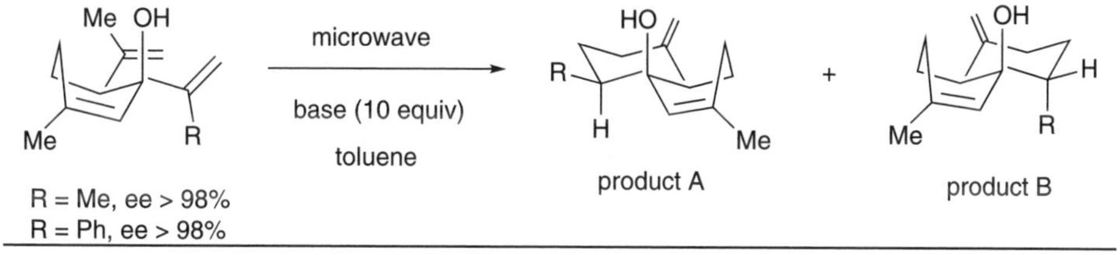

R = Me, ee > 98%
R = Ph, ee > 98%

Run	Base	Product A: yield (ee)[a]	
		R = Me	R = Ph
1	DBU (1)	60% (93%)	93% (35%)
2	TMEDA	48% (>98%)	86% (>98%)
3	Et₃N	36% (96%)	76% (>98%)
4	pyridine	39% (>98%)	36% (>98%)
5	2,6-di(*tert*-butyl)pyridine	decomp.	65% (>98%)
6	DMAP	28% (97%)	75% (>98%)
7	sparteine	57% (>98%)	91% (>98%)
8	BTMG	46% (96%)	98% (>98%)

[a] R = Me: product A/product B = 15:1, R = Ph: product A/product B = ≥25:1

Table 3.9 DBU (**1**) catalysed cyclopropanation of nitroolefins

$$R\text{-CH=CH-}NO_2 \xrightarrow[\substack{2.\ HMPA\ (0.1\ M) \\ DBU\ (\mathbf{1})\ (1\ equiv),\ rt,\ 24\ h}]{\substack{1.\ CHCl(CO_2Me)_2\ (1\ equiv) \\ DBU\ (\mathbf{1})\ (5\ mol\%),\ THF,\ rt,\ 24\ h}} \text{cyclopropane with }MeO_2C,\ CO_2Me,\ R,\ NO_2$$

Run	R	Yield (%)	de (%)
1	Ph	75	>98
2	4-Br-C$_6$H$_4$	73	>98
3	2-NO$_2$-C$_6$H$_4$	72	>98
4	1-naphthyl	72	>98
5	2-thiopheny	71	>98
6	hexyl	70	>98

dependent on the base used in the cascade process and the electronic nature of the vinylic substituent on the starting cyclohexenol. On the other hand, the de of the process is controlled by the conformational preference of macrocycles at the transition state for the ene reaction.

A convenient and novel one-pot organocatalytic methodology for the stereoselective synthesis of highly functionalized nitrocyclopropenes has been developed [90] (Table 3.9). DBU (**1**) catalyses the addition of dimethyl chloromalonate to a variety of nitroolefins to afford a Michael adduct, which cyclizes to form the cyclopropane in the presence of **1** under carefully controlled reaction conditions with outstanding diastereoselectivity.

3.4 Amidinium Salts: Design and Synthesis

3.4.1 Catalyst

A novel C$_2$-chiral bis(amidinium) salt can be synthesized from 5-(*tert*-butyl)isophthalic acid. The salt [tetrakis(3,5-bistrifluoromethylphenyl)borate (TFPB)] contributes to not only rate acceleration but also asymmetric induction in the Diels–Alder reaction of 1-vinyl-3,4-dihydronaphthalene and cylopentendione, owing to hydrogen bond mediated association of chiral auxiliary with dienophile [91] (Scheme 3.NaN).

3.4.2 Molecular Recognition

Yashima *et al.* [92] have designed and synthesized novel artificial double helixes, consisting of two complementary *m*-terphenyl-based strands intertwined through chiral amidinium–carboxylate salt bridges. Due to the chiral substituents on the amidine groups, the double

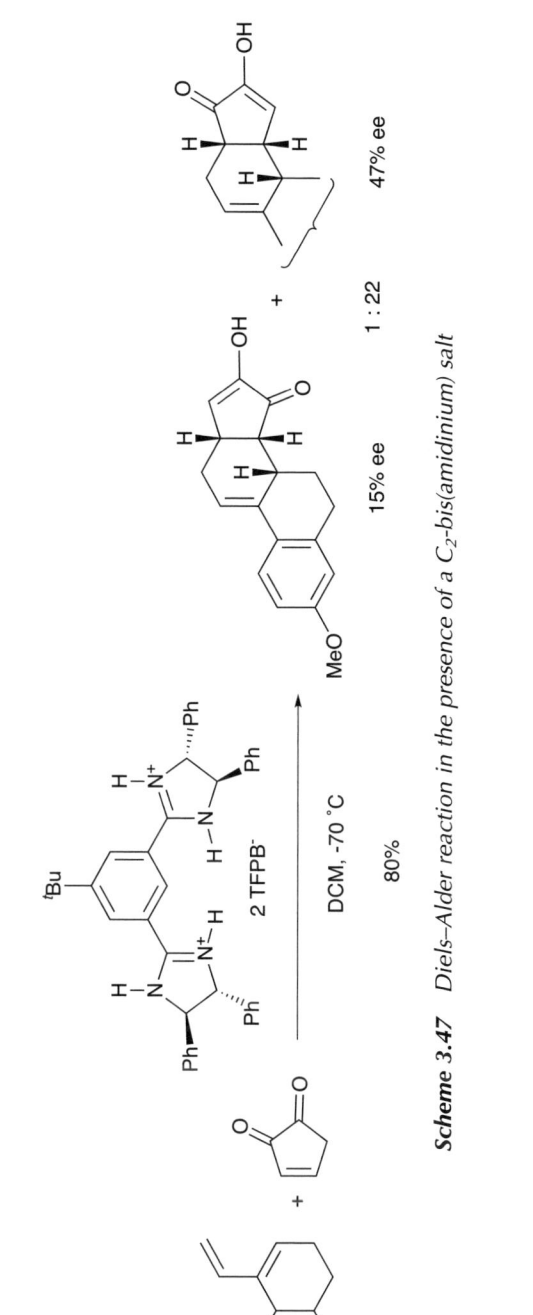

Scheme 3.47 Diels–Alder reaction in the presence of a C_2-bis(amidinium) salt

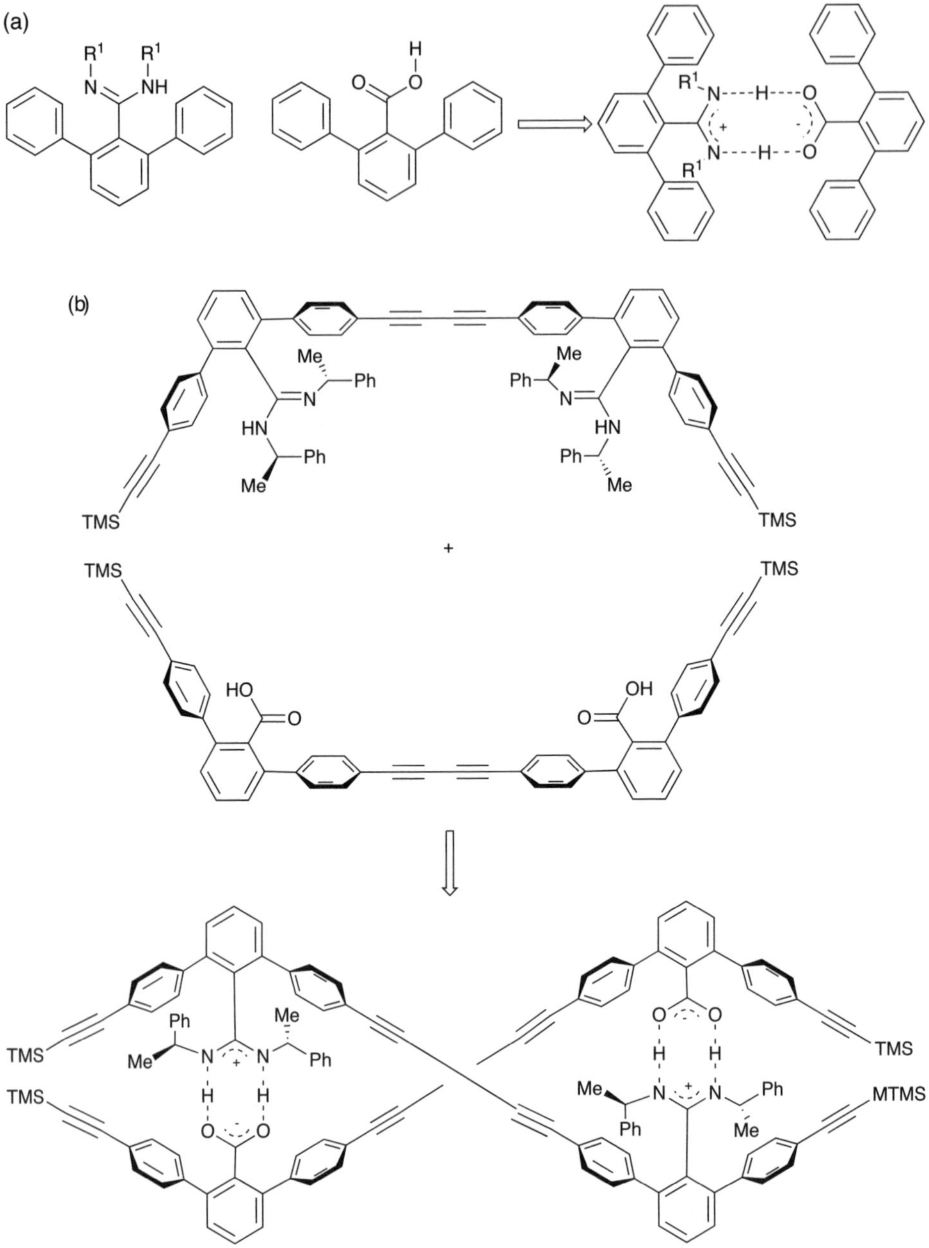

Scheme 3.48 Schematic illustration: (a) supramolecular complementary duplex; (b) artificial double helix

helices adopted an excess one-handed helical conformation in solution as well as in the solid state. By extending this molecular strategy, double helices bearing platinum(II) linkers, which undergo the double helix-to-double helix transformations through the chemical reactions of the platinum(II) complex moieties, were synthesized (Scheme 3.48).

Furthermore, optically active double helixes are synthesized through a twist-sense bias induced by a chiral phosphine ligand on one of the complementary metallostrands followed by a ligand exchange reaction with an achiral amidine ligand, which replaces the chiral ligand, to bridge the two strands [93]. The catalyst can efficiently induce asymmetry in the cyclopropanation reaction of styrene and ethyl diazoacetate.

3.4.3 Reagent Source

A seven-membered amidinium compound can be synthesized from 2,2′-diaminobiphenyl as a seven-membered heterocyclic carbine precursor [94] (Figure 3.10).

Cyclic imidazolinium salts are prepared from acylic amidines through intramolecular hydroamidiniumation of alkene. Thus, heating alkenyl amidinium salts, obtained by treating alkenylamidines with a stoichiometric amount of hydrogen chloride, induces ring closure regioselectively via exo addition of the nitrogen–hydrogen bond to the pendent carbon–carbon double bond, to give cyclic imidazolinium salts [95] (Scheme 3.49). This synthetic method is easily accessible to 4,4-disubstituted imidazolidinium salts, potential precursors for novel N-heterocyclic carbenes and applicable to the synthesis of cyclic iminium salts.

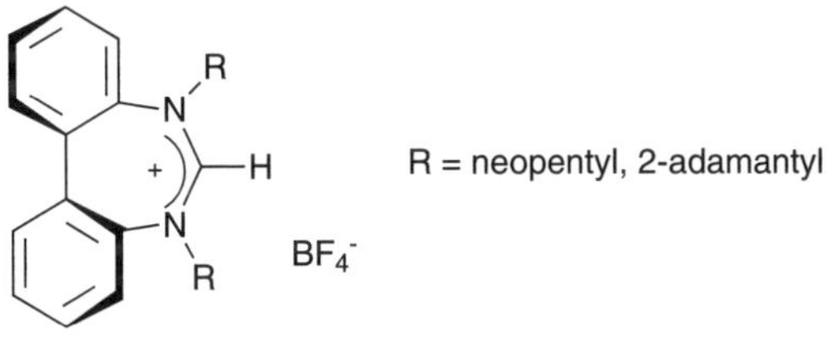

Figure 3.10 Structures of seven-membered amidinum compounds

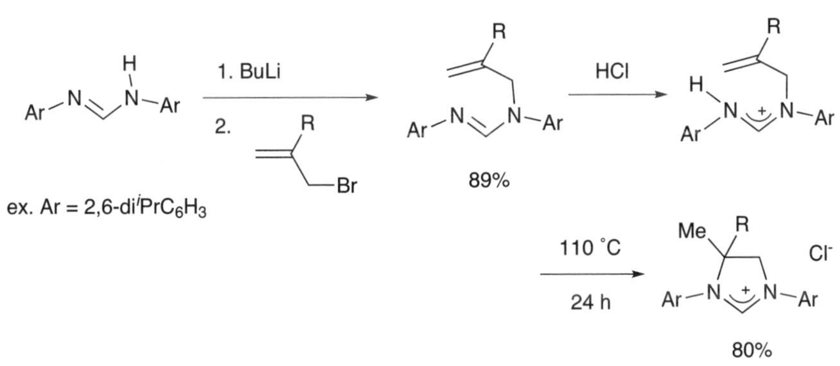

Scheme 3.49 Preparation of cyclic imidazolidinium salts from acyclic amidines

3.5 Concluding Remarks

The functionality of amidine compounds is attributable to the formation of stable amidinium species due to the conjugation of a two-nitrogen framework in the molecules after protonation. It is possible to structurally modify the amidine skeleton by introduction of substituents on the nitrogen atoms or incorporation of the amidine unit into cyclic systems. Thus, more sophisticated molecules with unique function could be designed in the future.

References

1. Höfelinger, G. and Kushe, F.K.H. (1991) *Chemistry of Amidines and Imidates*, **Vol 2** (eds S. Pathai and Z. Rappoport), John Wiley & Sons, Chichester, pp. 1–100;Fersht, A. (1985) *Enzyme Structure and Mechanism*, 2nd edn, Freeman, New York;Greehills, K.J.V. and Lue, P. (1993) Amidines and Guanidines in Medicinal Chemistry, **Vol. 30** (eds G.P. Ellis and D.K. Luscombe), Elsevier Science Publishers, New York, chapter 5, pp. 203–326.
2. Gund, P. (1972) Guanidine, trimethylenemethane and y-delocalization, Can arylic compounds have aromatic stability. *Journal of Chemical Education*, **49**, 100–103; Klein, J. (1983) Directive effects in allylic and benzylic polymetalations: the question of u-stabilization, y-aromaticity and cross-conjugation. *Tetrahedron*, **39**, 2733–2759.
3. Menger, F.M. (1966) The aminolysis and amidinolysis of *p*-nitrophenyl acetate in chlorobenzene. A facile bifunctional reactivity. *Journal of the American Chemical Society*, **88**, 3081–3084.
4. Oediger, H., Kabbe, H.J., Möller, F. and Eiter, K. (1966) 1,5-Diazabicylco[4. 3. 0]-5-nonene. A new reagent for the introduction of double bonds. *Chemische Berichte*, **99**, 2012–2016.
5. Oediger, H., Möller, F. and Eiter, K. (1972) Bicyclic amidines as reagents in organic synthesis. *Synthesis*, 591–598.
6. Corey, E.J., Anderson, N.H., Carlson, R.M. *et al.* (1968) Total synthesis of prostaglandins. Synthesis of the pure dl-E1, -F1-α, F1-β, -A1, and –B1 hormones. *Journal of the American Chemical Society*, **90**, 3245–3247.
7. Corey, E.J. and Achiwa, K. (1969) Oxidation of primary amines to ketones. *Journal of the American Chemical Society*, **91**, 1429–1432.
8. Rundel, W. and Kohler, H. (1972) Synthesis and rearrangement of *N,N*-dimethyl-*O*-(4-substituted-2,6-di-*tert*-butylphenyl)thiocarbamates. *Chemische Berichte*, **105**, 1087–1091.
9. Janne, K. and Ahlberg, P. (1972) Synthetic routes to a new bicyclic amidine, 1,2,3,4,4a,5,6,7,-octahydro-1,8-naphthyridine (2,10-diazabicyclo[4.4.0]dec-1-ene). *Synthesis*, 452–453.
10. Löfas, S. and Ahlberg, P. (1984) Spiro- and bicyclic azalactams by hydrolysis of α-chlorinated bicyclic amidines. *Journal of Heterocyclic Chemistry*, **21**, 583–586.
11. Heinzer, F., Soukup, M., and Eschenmoser, A. (1978) Uber 3,3,6,9,9-pentamethyl-2,10-diazabicyclo[4.4.0]-1-decen und einige seiner derivate. *Helvetica Chimica Acta*, **61**, 2851–2874; Sternbach, D., Shibuya, M., Jaishi, F. *et al.* (1979) A fragmentational approach to macrolides: (5-*E*,8-*Z*)-6-methyl-5,8-undecadien-11-olide. *Angewandte Chemie – International Edition*, **18**, 634–636.
12. Reed, R., Réau, R., Dahan, F. and Bertrand, G. (1990) DBU and DBN are strong nucleophiles: X-ray crystal structures of onio- and dionio-substituted phosphanes. *Angewandte Chemie – International Edition*, **32**, 399–401.
13. Convery, M.A., Davis, A.P., Dunne, C.J. and MacKinnon, J.W. (1995) Synthesis and properties of enantiopure bicyclic amidines. *Tetrahedron Letters*, **36**, 4279–4282.
14. Blériot, Y., Gerne-Grandpierre, A. and Tellier, C. (1994) Synthesis of a benzylamidine derived from D-mannose. A potential mannosidase inhibitor. *Tetrahedron Letters*, **35**, 1867–1870.

15. Ostendorf, M., van der Neut, S., Rutjes, F.P.J.T. and Hiemstra, H. (2000) Enantioselective synthesis of hydroxy-substituted DBN-type amidines as potential chiral catalysts. *European Journal of Organic Chemistry*, 105–113.
16. Jung, S. and Kohn, H. (1984) A new reductive procedure for the preparation of vicinal diamines and monoamines. *Tetrahedron Letters*, **25**, 399–402; Jung, S.H. and Kohn, H. (1985) Stereoselective synthesis of vicinal diamines from alkenes and cyanamide. *Journal of the American Chemical Society*, **107**, 2931–2943.
17. Oi, R. and Sharpless, K.B. (1991) Stereospecific conversion of chiral 1,2-cyclic sulfates to chiral imidazolines. *Tetrahedron Letters*, **32**, 999–1002.
18. Baganz, H. and Rabe, S. (1965) Reaktionen von β- unt γ- keto-carbonsaureestern mit Aethylendiaminen. *Chemische Berichte*, **98**, 3652–3658.
19. Yamamoto, H. and Maruoka, K. (1981) Regioselective carbonyl amination using diisobutylaluminum hydride. *Journal of the American Chemical Society*, **103**, 4186–4194.
20. Saigo, K., Kubota, N., Takebayashi, S. and Hasegawa, M. (1986) Improved optical resolution of (±)-1,2-diphenylethylenediamine. *Bulletin of the Chemical Society of Japan*, **59**, 931–932.
21. Corey, E.J. and Kühnle, F.N.M. (1997) A simplified synthesis of (±)-1,2-diphenyl-1,2-diaminoethane (1) from benzaldehyde and ammonia. Revision of the structures of the long-known intermediates, 'hydrobenzamide' and 'amarine'. *Tetrahedron Letters*, **38**, 8631–8634.
22. Laughton, C.A., Tanious, F., Nunn, C.M. et al. (1996) A crystallographic and spectroscopic study of the complex between d(CGCGAATTCGCG)$_2$ and 2,5-bis(4-guanylphenyl)furan, an analogue of berenil. Structural origins of enhanced DNA-binding affinity. *Biochemistry*, **35**, 5655–5661; Neidle, S. and Nunn, C.M. (1998) Crystal structures of nucleic acids and their drug complexes. *Natural Product Reports*, **15**, 1–15.
23. Boykin, D.W., Kumar, A., Xiao, G. et al. (1998) 2,5-Bis[4-(N-alkylamidino)phenyl]furans as anti-*Pneumocystis carinii* Agents. *Journal of Medicinal Chemistry*, **41**, 124–129; Chaires, J.B., Ren, J., Hamelberg, D., Kumar, A. et al. (2004) Structural selectivity of aromatic diamidines. *Journal of Medicinal Chemistry*, **47**, 5729–5742.
24. Owens, T.D., Souers, A.J. and Ellman, J.A. (2003) The preparation and utility of bis(sulfinyl) imidoamidine ligands for the copper-catalysed Diels–Alder reaction. *The Journal of Organic Chemistry*, **68**, 3–10; Kochi, T. and Ellman, J.A. (2004) Asymmetric α-alkylation of N'-tert-butanesulfinyl amidines. Application to the total synthesis of (6R,7S)-7-amino-7,8-dihydro-α-bisabolene. *Journal of the American Chemical Society*, **126**, 15652–15653.
25. Bormann, D. (1973) Reaction of imidic esters with aziridines. *Angewandte Chemie – International Edition*, **12**, 768–769.
26. Kumagai, N., Matsunaga, S. and Shibasaki, M. (2004) An efficient synthesis of bicyclic amidines by intramolecular cyclization. *Angewandte Chemie – International Edition*, **43**, 478–482.
27. Hutton, C.A. and Bartlett, P.A. (2007) Preparation of diazabicyclo[4.3.0]nonene-based peptidomimetics. *The Journal of Organic Chemistry*, **72**, 6865–6872.
28. Dijkink, J., Eriksen, K., Goubitz, K. et al. (1996) Synthesis and X-ray crystal structure of (S)-9-hydroxymethyl-1,5-diazabicyclo[4.3.0]non-5-ene, an enantiopure DBN-analogue. *Tetrahedron: Asymmetry*, **7**, 515–524.
29. McFarland, J.W. (1963) Reactions of cyclohexylisonitrile and isobutyraldehyde with various nucleophiles and catalysts. *The Journal of Organic Chemistry*, **28**, 2179–2181.
30. Keung, W., Bakir, F., Patron, A.P. et al. (2004) Novel α-amino amidine synthesis via scandium(III) triflate mediated 3CC Ugi condensation reaction. *Tetrahedron Letters*, **45**, 733–737.
31. Bae, I., Han, H. and Chang, S. (2005) Highly efficient one-pot synthesis of N-sulfonylamidines by Cu-catalysed three-component coupling of sulfonyl azide, alkyne, and amine. *Journal of the American Chemical Society*, **127**, 2038–2039.
32. Rostovtsev, V.V., Green, L.G., Fokin, V.V. and Sharpless, K.B. (2002) A stepwise Huisgen cycloaddition process: copper(I)-catalysed regioselective "ligation" of azides and terminal alkynes. *Angewandte Chemie – International Edition*, **41**, 2596–2599.

33. Kim, J.Y., Kim, S.H. and Chang, S. (2008) Highly efficient synthesis of (α-amino amidines from ynamides by the Cu-catalysed three-component coupling reactions. *Tetrahedron Letters*, **49**, 1745–1749.
34. Yoo, E.J., Bae, I., Cho, S.H. *et al.* (2006) A facile access to *N*-sulfonylimidates and their synthetic utility for the transformation to amidines and amides. *Organic Letters*, **8**, 1347–1350.
35. Chang, S., Lee, M., Jung, D.Y. *et al.* (2006) Catalytic one-pot synthesis of cyclic amidines by virtue of tandem reactions involving intramolecular hydroamination under mild conditions. *Journal of the American Chemical Society*, **128**, 12366–12367.
36. Evans, D.A., Faul, M.M. and Bilodeau, M.T. (1994) Development of the copper-catalysed olefin aziridination reaction. *Journal of the American Chemical Society*, **116**, 2742–2753.
37. Walsh, P.J., Hollander, F.J. and Bergman, R.G. (1988) Generation, alkyne cycloaddition, arene carbon-hydrogen activation, nitrogen-hydrogen activation and dative ligand trapping reactions of the first monomeric imidozirconocene (Cp$_2$Zr:NR) complexes. *Journal of the American Chemical Society*, **110**, 8729–8731.
38. Wanniarachchi, Y.A. and Slaughter, L.M. (2007) One-step assembly of a chiral palladium bis (acyclic diaminocarbene) complex and its unexpected oxidation to a bis(amidine) complex. *Chemical Communications*, 3294–3296.
39. Hiyama, T., Koide, H., Fujita, S. and Nozaki, H. (1973) Reaction of *N*-alkoxycarbonylaziridines with nitriles. *Tetrahedron*, **29**, 3137–3139.
40. Bucciarelli, M., Forni, A., Moretti, I. *et al.* (1995) Regioselectivity of ring-opening reactions of optically active N-acetyl-2-methoxycarbonylaziridine. *Tetrahedron: Asymmetry*, **6**, 2073–2080.
41. Prasad, B.A.B., Pandey, G. and Singh, V.K. (2004) Synthesis of substituted imidazolines via [3 + 2]-cycloaddition of aziridines with nitriles. *Tetrahedron Letters*, **45**, 1137–1141.
42. Wu, J., Sun, X. and Xia, H. (2006) Sc(OTf)$_3$-Catalysed [3 + 2]-cycloaddition of aziridines with nitriles under solvent-free conditions. *Tetrahedron Letters*, **47**, 1509–1512.
43. Ghorai, M.K., Ghosh, K. and Das, K. (2006) Copper(II) triflate promoted cycloaddition of α-alkyl or aryl substituted *N*-tosylaziridines with nitriles: a highly efficient synthesis of substituted imidazolines. *Tetrahedron Letters*, **47**, 5399–5403.
44. Tunge, J.A., Czerwinski, C.J., Gately, D.A. and Norton, J.R. (2001) Mechanism of insertion of carbodiimides into the Zr–C bonds of zirconaaziridines. Formation of α-amino amidines. *Organometallics*, **20**, 254–260.
45. Ahmad, S.M., Braddock, D.C., Cansell, G. *et al.* (2007) Amidines as potent nucleophilic organocatalysts for the transfer of electrophilic bromine from *N*-bromosuccinimide to alkenes. *Tetrahedron Letters*, **48**, 5948–5952.
46. Aggarwal, V.K. and Mereu, A. (2000) Amidine-promoted addition of chloroform to carbonyl compounds. *The Journal of Organic Chemistry*, **65**, 7211–7212.
47. Catelani, G., Corsaro, A., Andrea, F.D. *et al.* (2002) Intramolecular aldol cyclization of L-*lyxo*-hexos-5-ulose derivatives: a new diastereoselective synthesis of D-*chiro*-inositol. *Bioorganic & Medicinal Chemistry Letters*, **12**, 3313–3315.
48. Jiang, N. and Wang, J. (2002) DBU-promoted condensation of acyldiazomethanes with aldehydes and imines under catalytic conditions. *Tetrahedron Letters*, **43**, 1285–1287.
49. Xiao, F., Liu, Y. and Wang, J. (2007) DBU-catalysed condensation of acyldiazomethanes to aldehydes in water and a new approach to ethyl β-hydroxy α-arylacrylates. *Tetrahedron Letters*, **48**, 1147–1149.
50. Evans, D.A., Tregey, S.W., Burgey, C.S. *et al.* (2000) C$_2$-Symmetric copper(II) complexes as chiral Lewis acids. Catalytic enantioselective carbonyl-ene reactions with glyoxylate and pyruvate esters. *Journal of the American Chemical Society*, **122**, 7936–7943.
51. Iska, V.B.R., Gais, H.-J., Tiwari, S.K. *et al.* (2007) Asymmetric aziridination with chiral allyl aminosulfonium ylides: synthesis of alkenyl aziridine carboxyaltes and palladium-catalysed *E*, trans/*E*, cis-isomerization of an alkenyl aziridine. *Tetrahedron Letters*, **48**, 7102–7107.
52. Dalaigh, C.O. and Connon, S.J. (2007) Nonenzymatic acylative kinetic resolution of Baylis–Hillman adducts. *The Journal of Organic Chemistry*, **72**, 7066–7069; Min, S., Wen, G.Y. and Hong-Bin, L. (2007) Baylis-Hillman reaction of sulfonyl aldimines or aryl aldehydes with

3-methylpenta-3,4-dien-2-one or 3-benzylpenta-3,4-dien-2-one. *Chinese Journal of Chemistry*, **25**, 828–835.
53. Sibi, M.P., Stanley, L.M. and Jasperse, C.P. (2005) An entry to a chiral dihydropyrazole scafford: enantioselective [3 + 2] cyloaddition of nitrile imines. *Journal of the American Chemical Society*, **127**, 8276–8277.
54. Dondas, H.A., Grigg, R. and Thornton-Pett, M. (1996) Spiro(pyrrolidinyl-2,3-benzodiazepines) related to MK-329. *Tetrahedron*, **52**, 13455–13466; Dondas, A., Grigg, R. and Kilner, C. (2003) X = Y = ZH Systems as potential 1,3-dipoles. Part 58: Cycloaddition route to chiral conformationally constrained (*R*)-pro- (*S*)-pro peptidomimetics. *Tetrahedron*, **59**, 8481–8487.
55. Andrei, D. and Wnuk, S.F. (2006) S-Adenosylhomocycteine analogues with the carbon-5′ and sulfur atoms replaced by a vinyl unit. *Organic Letters*, **8**, 5093–5096.
56. Shibuya, M., Kuretani, M. and Kubota, S. (1982) Synthesis of 6-(1-hydroxyethyl)-1,1-dimethyl-1-carba-penem derivatives via Dieckmann-type cyclization. *Tetrahedron Letters*, **38**, 2659–2655; Hanessian, S. and Reddy, B. (1999) Total synthesis of tricyclic β-lactams. *Tetrahedron*, **55**, 3427–3443.
57. Asami, M., Ishizuka, T. and Inoue, S. (1994) Catalytic enantioselective deprotonation of *meso*-epoxides by the use of chiral lithium amide. *Tetrahedron:Asymmetry*, **5**, 793–796; Seki, A. and Asami, M. (2002) Catalytic enantioselective rearrangement of *meso*-epoxides mediated by chiral lithium amides in the presence of excess cross-linked polymer-bound lithium amides. *Tetrahedron*, **58**, 4655–4663.
58. Södergren, M.J. and Anderson, P.G. (1998) New and high enantioselective catalysts for the rearrangement of *meso*-epoxides into chiral allylic alcohols. *Journal of the American Chemical Society*, **120**, 10760–10761;Södergren, M.J., Bertilsson, S.K. and Anderson, P.G. (2002) Allylic alcohols via catalytic asymmetric epoxide rearrangement. *Journal of the American Chemical Society*, **122**, 6610–6618; Bertilsson, S.K. and Anderson, P.G. (2002) Asymmetric base-promoted epoxide rearrangement: achiral lithium amides revisited. *Tetrahedron*, **58**, 4665–4668.
59. Patterson, D., Amedjkouh, M. and Ahlberg, P. (2002) Improved enantioselectivity by using novel bulk bases in chiral lithium amide catalysed deprotonations: mixed dimers as reagents and catalysts. *Tetrahedron*, **58**, 4669–4673.
60. Otera, J., Niibo, Y. and Nozaki, H. (1991) Oxirane ring-opening with alcohol catalysed by organotin phosphate condensates. Complete inversion at tertiary and benzylic centers. *Tetrahedron*, **47**, 7625–7634; Corey, E.J. and Helal, C.J. (1993) A catalytic enantioselective synthesis of chiral monosubstituted oxiranes. *Tetrahedron Letters*, **34**, 5227–5230.
61. Wu, Y., Hu, Q., Sum, Y.-P. and Yang, Y.-Q. (2004) Facile removal of 4-phenyl-oxazolidinethione auxiliary with EtSH mediated by DBU. *Tetrahedron Letters*, **45**, 7715–7717.
62. Etayo, P., Badorrey, R., Diéz-de-Villegas, M.D. and Gálvez, J.A. (2006) Unexpected epimerization at C2 in the Horner-Wadsworth-Emmons reaction of chiral 2-substituted-4-oxopiperidines. *Chemical Communications*, 3420–3422.
63. Bellina, F., Ciucci, D., Vergamini, P. and Rossi, R. (2000) Regioselective synthesis of natural and unnatural (*Z*)-3-(1-alkylidene)phthalides and 3-substituted isocoumarins starting from methyl 2-hydroxybenzoates. *Tetrahedron*, **56**, 2533–2545;Yao, T. and Larock, R.C. (2003) Synthesis of isocoumarins and α-pyrones via electrophilic cyclization. *The Journal of Organic Chemistry*, **68**, 5936–5942; Subramanian, V. Batchu, V.R., Barange, D. and Pal, M. (2005) Synthesis of isocoumarins via Pd/C-mediated reactions of *o*-iodobenzoic acid with terminal alkynes. *The Journal of Organic Chemistry*, **70**, 4778–4783.
64. Kanzawa, C. and Terada, M. (2007) Organic-base-catalysed synthesis of phthalides via highly regioselective intramolecular cyclization reaction. *Tetrahedron Letters*, **48**, 933–935.
65. Miyake, T., Seki, M., Nakamura, Y. and Ohmizu, H. (1996) Synthesis of a novel chiral 1,3-benzoxazinone auxiliary and its application to highly diastereoselective aldol reaction. *Tetrahedron Letters*, **37**, 3129–3132.
66. Evans, D.A., Janey, J.M., Magomedov, N. and Tedrow, J.S. (2001) Chiral salen–aluminum complexes as catalysts for enantioselective aldol reactions of aldehydes and 5-alkoxyoxazoles: an efficient approach to the asymmetric synthesis of *syn* and *anti* β-hydroxy-α-amino acid derivatives. *Angewandte Chemie – International Edition*, **40**, 1884–1888.

67. Kogami, Y., Nakajima, T., Ashizawa, T. et al. (2004) Enantioselective Henry reaction catalysed by optically active ketoiminatocobalt complexes. *Chemistry Letters*, **33**, 614–615.
68. Berkessel, A. and Brandendurg, M. (2006) Catalytic asymmetric addition of carbon dioxide to propylene oxide with unprecedented enantioselectivity. *Organic Letters*, **8**, 4401–4404.
69. Orlandi, S., Benaglia, M. and Cozzi, F. (2004) Cu(II)-catalysed enantioselective aldol condensation between malonic acid hemithioesters and aldehydes. *Tetrahedron Letters*, **45**, 1747–1749.
70. Singh, O.V. and Han, H. (2007) Iridium(I)-catalysed regio- and enantioselective allylic amidation. *Tetrahedron Letters*, **48**, 7094–7098.
71. Brunner, H., Agrifoglic, G., Bernal, I. and Creswick, M.W. (1980) Conformational analysis of 1-phenylethyl substituents in metal complexes and their importance for asymmetric catalysis. *Angewandte Chemie – International Edition*, **19**, 641–642.
72. Itoh, K. and Kanemasa, S. (2002) Enantioselective Michael additions of nitromethane by a catalytic double activation method using chiral Lewis acid and chiral amine catalysts. *Journal of the American Chemical Society*, **124**, 13394–13395; Itoh, K., Oderaotoshi, Y. and Kanemasa, S. (2003) Enantioselective Michael addition reactions of malononitrile catalysed by chiral Lewis acid and achiral amine catalysts. *Tetrahedron: Asymmetry*, **14**, 635–639.
73. Saitoh, A., Morimoto, T. and Achiwa, K. (1997) A phosphorus-containing chiral amidine ligand for asymmetric reactions: enantioselective Pd-catalysed allylic alkylation. *Tetrahedron: Asymmetry*, **8**, 3567–3570; Saitoh, A., Achiwa, K. and Morimoto, T. (1998) Enantioselective allylic substitutions using ketene silyl acetals catalysed by palladium-chiral amidine complexes. *Tetrahedron: Asymmetry*, **9**, 741–744.
74. Saitoh, A., Achiwa, K., Tanaka, K. and Morimoto, T. (2000) Versatile chiral bidentate ligands derived from α-amino acids: synthetic applications and mechanistic considerations in the palladium-mediated asymmetric allylic substitutions. *The Journal of Organic Chemistry*, **65**, 4227–4240.
75. Hu, X., Chen, H., Hu, X. et al. (2002) Synthesis of novel ferrocenylphosphine-amidine ligands and their application in Pd-catalysed asymmetric allylic alkylation. *Tetrahedron Letters*, **43**, 9179–9182; Hu, X., Chen, H., Dai, H. et al. (2003) Synthesis of novel ferrocenylphosphine-amidine ligands with central and planar chirality and their diastereomeric effect in Pd-catalysed asymmetric allylic alkylation. *Tetrahedron: Asymmetry*, **14**, 2073–2080.
76. Oshiki, T. and Imamoto, T. (1992) Unprecedented stereochemistry of the electrophilic arylation at chiral phophorus. *Journal of the American Chemical Society*, **114**, 3975–3977; Trost, B.M., Heinemann, C., Ariza, X. and Weigand, S. (1999) Chiral recognition for control of alkene geometry in a transition metal catalysed allylic alkylation. *Journal of the American Chemical Society*, **121**, 8667–8668.
77. Joo, F. and Trocsányi, E. (1982) Asymmetric hydrogenation of acetophenone with rhodium(I) complexes of new chiral phosphines derived from amiono acids. An usual modification of the catalyst system. *Journal of Organometallic Chemistry*, **231**, 63–70.
78. Nishida, M., Tozawa, T., Yamada, K. and Mukaiyama, T. (1996) Asymmetric allylation of aromatic aldehydes using chiral allylic tin reagents. *Chemistry Letters*, **25**, 1125–1126; Yamada, K., Tozawa, K., Nishida, M. and Mukaiyama, T. (1997) Asymmetric allylation of carbonyl compounds with tartrate-modified chiral allylic tin reagents. *Bulletin of the Chemical Society of Japan*, **70**, 2301–2308.
79. Ishitani, H. and Kobayashi, S. (1996) Catalytic asymmetric aza Diels–Alder reactions using a chiral lanthanide Lewis acid. Enantioselective synthesis of tetrahydroquinoline derivatives using a catalytic amount of a chiral source. *Tetrahedron Letters*, **37**, 7357–7360.
80. Ostendorf, M., Dijkink, J., Rutjes, F.P.J.T. and Hiemstra, H. (2000) (*S*)-Pyroglutamic acid, (*S*)-malic acid, and (*S*)-serine as useful starting materials in the synthesis of enantiopure hydroxyamidines. *European Journal of Organic Chemistry*, 115–124.
81. Kotsuki, H., Sugino, A., Sakai, H. and Yasuoka, H. (2000) A novel synthesis of chiral DBU/DBN-related molecules for use in asymmetric base catalysis. *Heterocycles*, **53**, 2561–2567.
82. Zhao, G., Shi, Y. and Shi, M. (2005) Synthesis of functionalized 2*H*-1-benzopyrans by DBU-catalysed reactions of salicylic aldehydes with allenic ketones and esters. *Organic Letters*, **7**, 4527–4530.

83. Ballini, R., Bosica, G., Fiorini, D. and Petrini, M. (2002) Unprecedented, selective Nef reaction of secondary nitroalkanes promoted by DBU under basic homogeneous conditions. *Tetrahedron Letters*, **43**, 5233–5235.
84. Ray, P.C., and Roberts, S.M. (1999) Overcoming intrinsic diastereoselection using polyleucine as a chiral epoxidation catalyst. *Tetrahedron Letters*, **40**, 1779–1784; Bentley, P.A., Bickley, J.F., Roberts, S.M. and Steiner, A. (2001) Asymmetric epoxidation of a geminally-disubstituted and some trisubstituted enones catalysed by poly-L-leucine. *Tetrahedron Letters*, **42**, 3741–3743.
85. Kawabata, H. and Hayashi, M. (2002) Lewis base-catalysed transformation of α,β-unsaturated aldehydes to saturated carboxylic acids, esters and amides. *Tetrahedron Letters*, **43**, 5645–5647.
86. Pudovik, A.N. and Konovalova, I.V. (1979) Addition reactions of esters of phosphorus(III) acids with unsaturated systems. *Synthesis*, 81–96.
87. El Kaim, L., Gaultier, L., Grimaud, L. and Dos Santos, A. (2005) Formation of new phosphates from aldehydes by a DBU-catalysed phospha-Brook rearrangement in a polar solvent. *Synlett*, 2335–2336.
88. Boeckman, R.K., Jr, Pero, J.E. and Boehmler, D.J. (2006) Toward development of a general chiral auxiliary. Enantioselective alkylation of a new catalytic asymmetric addition of silyloxyfurnas: application to a total synthesis of (−)-rasfonin. *Journal of the American Chemical Society*, **128**, 11032–11033.
89. Gauvreau, D. and Barriault, L. (2005) Conservation of the planar chiral information in the tandem oxy-Cope/ene reaction. *The Journal of Organic Chemistry*, **70**, 1382–1388.
90. McCooey, S.H., McCabe, T. and Connon, S.J. (2006) Stereoselective synthesis of highly functionalized nitrocyclopropanes via organocatalytic conjugate addition to nitroalkenes. *The Journal of Organic Chemistry*, **71**, 7494–7497.
91. Tsogoeva, S.B., Dürner, G., Bolte, M. and Göbel, M.W. (2003) A C_2-chiral bis(amidinium) catalyst for a Diels–Alder reaction constituting the key step of the Quinkert-Dane estrone synthesis. *European Journal of Organic Chemistry*, 1661–1664.
92. Ikeda, M., Tanabe, Y., Hasegawa, T. *et al.* (2006) Construction of double-stranded metallosupramolecular polymers with a controlled helicity by combination of salt bridges and metal coordination. *Journal of the American Chemical Society*, **128**, 6806–6807; Furusho, Y., Tanabe, Y. and Yashima, E. (2006) Double helix-to-double helix transformation, using platinum(II) acetylide complexes as surrogate linkers. *Organic Letters*, **8**, 2583–2586; Furusho, Y. and Yashima, E. (2007) Molecular design and synthesis of artificial double helices. *Chemical Record*, **7**, 1–11.
93. Hasegawa, T., Furusho, Y., Katagiri, H. and Yashima, E. (2007) Enantioselective synthesis of complementary double helical molecules that catalyse asymmetric reaction. *Angewandte Chemie – International Edition*, **46**, 5885–5888.
94. Scarborough, C.C., Popp, B.V., Guzei, I.A. and Stahl, S.S. (2005) Development of 7-membered *N*-heterocyclic carbine ligands for transition metals. *Journal of Organometallic Chemistry*, **690**, 6143–6155.
95. Jazzar, R., Bourg, J.-B., Dewhurst, R.D. *et al.* (2007) Intramolecular 'hydroiminiumation and -amidiniumation' of alkenes: a convenient, flexible, and scalable route to cyclic iminium and imidazolinium salts. *The Journal of Organic Chemistry*, **72**, 3492–2499.

4
Guanidines in Organic Synthesis

Tsutomu Ishikawa

Graduate School of Pharmaceutical Sciences, Chiba University, 1-33 Yayoi, Inage, Chiba 263-8522, Japan

4.1 Introduction

Guanidines can be categorized as organic superbases [1] due to the resonance stability of their conjugated acids [2] and, thus, are expected to catalyse various types of base-participated organic reactions. Among guanidine compounds, 1,1,3,3-tetramethylguanidine (or N,N,N',N'-tetramethylguanidine; TMG) (**1**) is regarded as a typical and fundamental guanidine compound and, in fact, has been used in many kinds of base-catalysed reaction. Barton *et al.* [3] reported the preparation of pentaalkylguanidines **2** and their application to organic synthesis as sterically-hindered organic bases, which are called 'Barton's bases'. Bicyclic guanidines **3**, 1,5,7-triazabicycle[4.4.0]dec-5-ene (TBD) and the N-methyl analogue (MTBD) were introduced by Schwesinger [4] (Figure 4.1).

Guanidine participating organic reactions could be schematically classified into two types of reactions: catalytic and stoichiometric, in which a guanidinium salt composed of guanidine like **2** and either an acid or nucleophile plays an important role as a common active complex. In the former type of reaction, **2** is repeatedly used as a free base catalyst, whereas a guanidinium salt is formed in the latter (Figure 4.2).

As expected from a Barton's base (**2**), the guanidine skeleton can be widely and easily modified to a chiral base by introducing chirality into the molecule, in which five chiral centres can be theoretically incorporated in the three nitrogen atoms, indicating that the TMG (**1**) participating organic reactions could be theoretically expanded to asymmetric

---

Superbases for Organic Synthesis: Guanidines, Amidines, Phosphazenes and Related Organocatalysts
Edited by Tsutomu Ishikawa
© 2009 John Wiley & Sons, Ltd

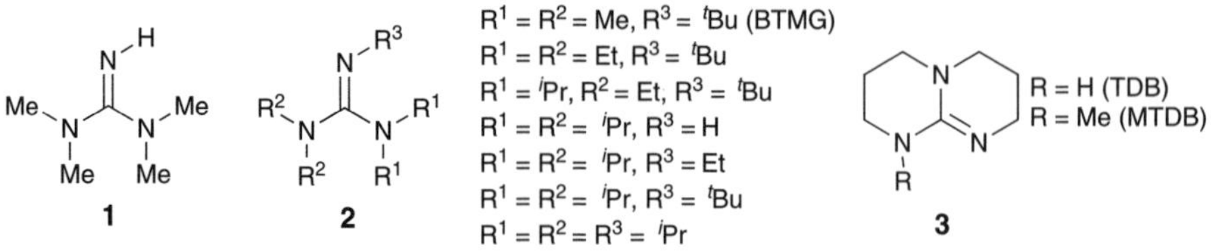

Figure 4.1 Structures of TMG (**1**), Barton's bases (**2**) and bicyclic guanidines **3**

synthesis after modification of the guanidine skeleton to the chiral version according to concept for the role of modified guanidines as chiral auxiliaries [5].

Preparation and use of supported TMG (**1**) as a novel base catalyst is discussed in a review elsewhere [6]. Heterogeneous guanidines are provided as environmentally friendly base catalysts and, thus, precise discussion on supported superbases is given in Chapter 6. Guanidine chemistry has been excellently surveyed in books [7]. This chapter focuses on the synthetic utility of TMG (**1**) and its analogues in organic synthesis and application of modified guanidine to asymmetric reactions.

4.2 Preparation of Chiral Guanidines

Guanidines are classified structurally into three types of compounds dependent upon whether the guanidinyl function is incorporated into ring systems or not. Thus, in addition to acyclic guanidines such as **1** and **2**, monocyclic **4** and bicyclic guanidines **5** including **3** are nominated as modified guanidines (Figure 4.3).

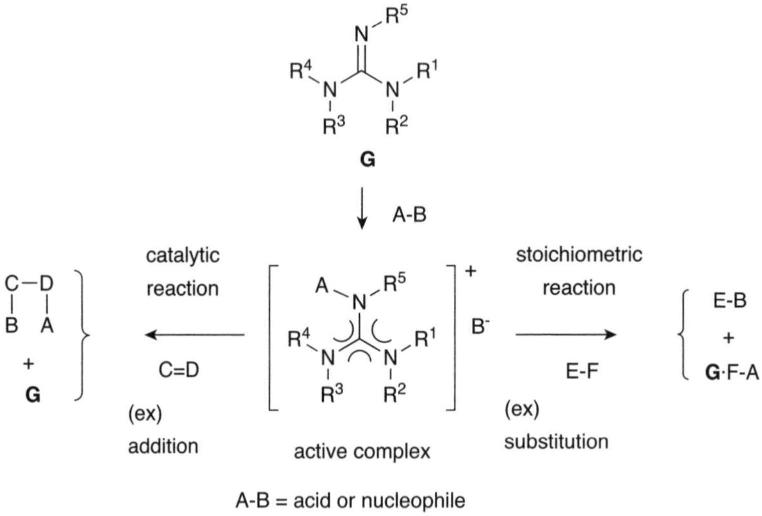

Figure 4.2 Classification of guanidine-participating reactions

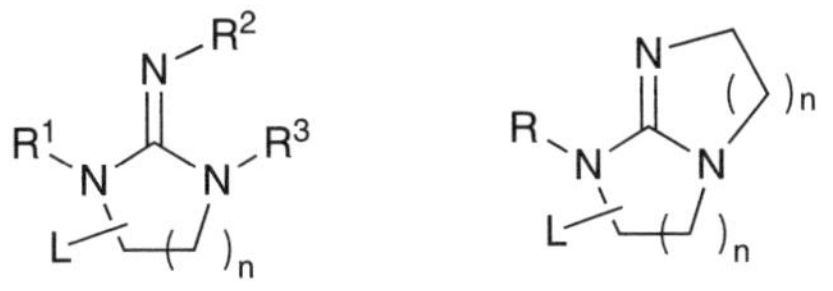

Figure 4.3 General structures of monocyclic **4** and bicyclic guanidines **5**

4.2.1 Polysubstituted Acyclic and Monocyclic Guanidines

Schneck, in 1912, reported the synthesis of TMG (**1**) by the reaction of 1,1,3,3-tetramethyl-2-methylthioamidinium salt **7**, derived from the corresponding thiourea **6**, and ammonia [8] (Scheme 4.1). Later, the original method was improved using alternative reactions of dimethylamine with cyanogen iodide and of 1,1-dimethyl-2-thiomethylamidine and dimethylamine [9].

Chiral di-or trisubstituted acyclic guanidines [10] are, in the literature, prepared from the corresponding amines through thiourea or carbodiimide intermediates (Scheme 4.2a); the corresponding monocyclic guanidines could be supplied by application of these methods. The cyano moiety of cyanogen bromide serves as a source of the core of guanidinyl functions, when reacted with primary or secondary amines. Thus, N-cyanation reaction of amines with cyanogen bromide was applied to the construction of not only acyclic [11] but also monocyclic symmetrical systems [10b] (Scheme 4.2b).

Recently, polysubstituted guanidines were prepared in good yield by catalytic bismuth (only 5 mol%)-promoted synthesis through the guanidinylation reaction of N-benzoyl or N-phenylthioureas with primary and secondary amines [12]. Furthermore, it was found that half-sandwich rare earth metal complexes, which were prepared from [Ln(CH$_2$Si-Me$_3$)$_3$(thf)$_2$] (Ln = Y, Yb, and Lu) and Me$_2$Si(C$_3$Me$_4$H)NHR′ (R′ = Ph, 2,4,6-Me$_3$C$_6$H$_2$, ^{t}Bu), serve as excellent catalyst precursors for the catalytic addition of various primary and secondary amines to carbodiimides, efficiently yielding a series of guanidines [13]. A possible mechanism for catalytic addition of secondary amine to carbodiimides is proposed as shown in Scheme 4.3.

4.2.2 Monosubstituted Guanidines (Guanidinylation)

Guanidine **8** with remote chiral centres, a monosubstituted acyclic guanidine, was prepared by guanidinylation of an amine function with 3,5-dimethylpyrazole-1-carboxamidine [14]

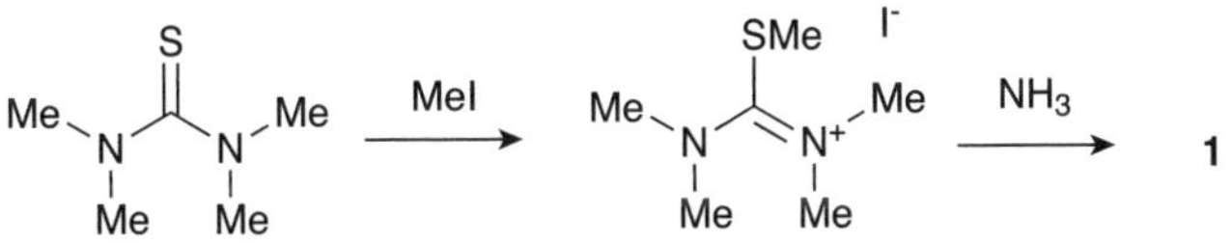

Scheme 4.1 Preparation of TMG (**1**) by Schenck

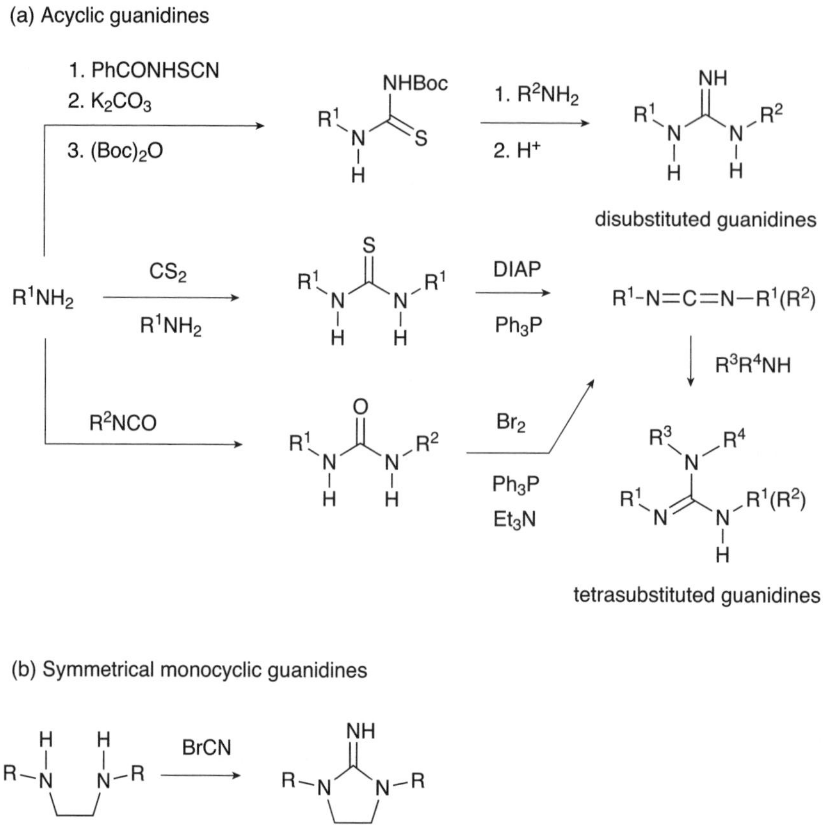

96 Guanidines in Organic Synthesis

(a) Acyclic guanidines

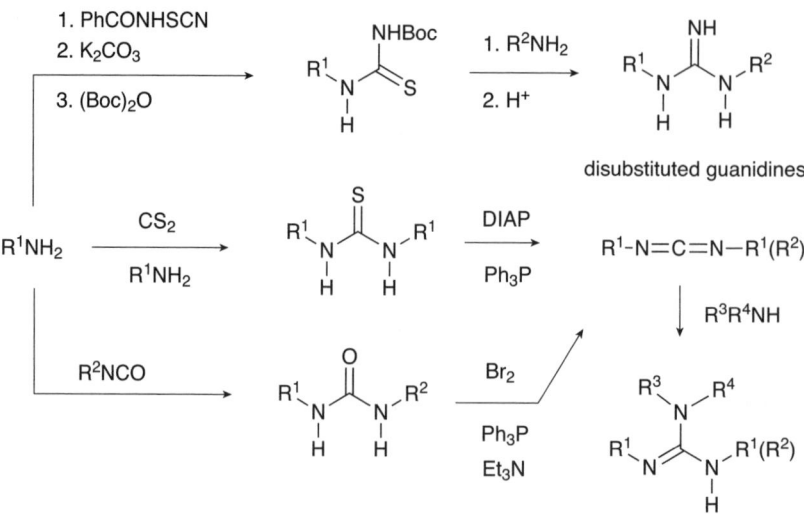

(b) Symmetrical monocyclic guanidines

Scheme 4.2 *Methods of preparing substituted guanidines*

(Scheme 4.4). Other reagents useful for terminal guanidinylation of primary and secondary amines have been reported, in which *S*-methylisothiourea [15], protected guanidine [16], pyrazole-1-carboxamidine [17] and benzotriazole-1-carboxamidine derivatives [18] are explored. A solid support-linked guanidinylating reagent, consisted of a urethane-protected

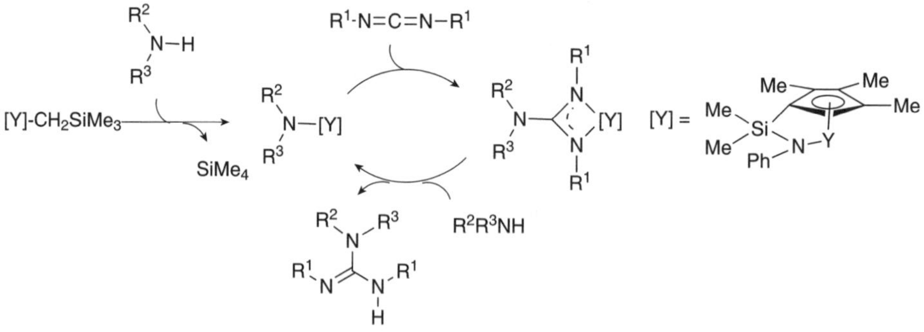

Scheme 4.3 *A possible mechanism for half-sandwich catalyst-participating guanidine synthesis*

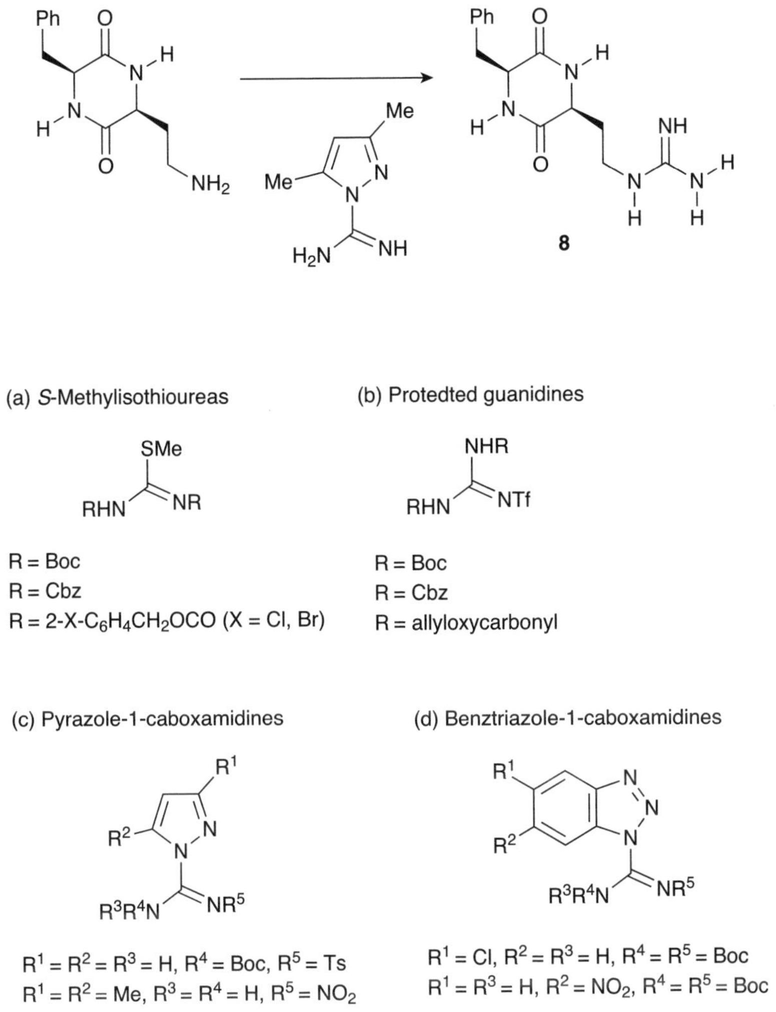

(a) *S*-Methylisothioureas

R = Boc
R = Cbz
R = 2-X-C₆H₄CH₂OCO (X = Cl, Br)

(b) Protedted guanidines

R = Boc
R = Cbz
R = allyloxycarbonyl

(c) Pyrazole-1-caboxamidines

R¹ = R² = R³ = H, R⁴ = Boc, R⁵ = Ts
R¹ = R² = Me, R³ = R⁴ = H, R⁵ = NO₂

(d) Benztriazole-1-caboxamidines

R¹ = Cl, R² = R³ = H, R⁴ = R⁵ = Boc
R¹ = R³ = H, R² = NO₂, R⁴ = R⁵ = Boc

Scheme 4.4 *Preparation of monosubstituted acyclic guanidine 8 and selected guanidinylating reagents of amines*

triflyl guanidine attached to the resin via a carbamate linker, allows for rapid synthesis of guanidines from a variety of amines [19].

N,N',N''-Tri(Boc)-guanidine and N,N',N''-tri(Cbz)-guanidine allow for the facile conversion of alcohols to substituted guanidines under Mitsunobu condition [20].

4.2.3 Bicyclic Guanidines

Bicyclic guanidines [21] are basically prepared according to the method through the thiourea intermediate shown in Scheme 4.2 (Scheme 4.5).

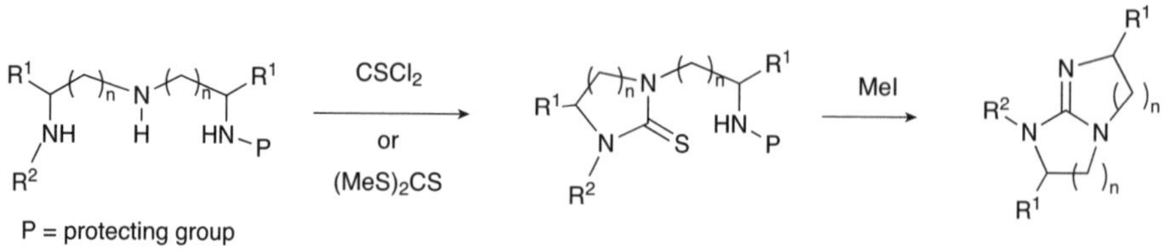

Scheme 4.5 Preparation of bicyclic guanidines

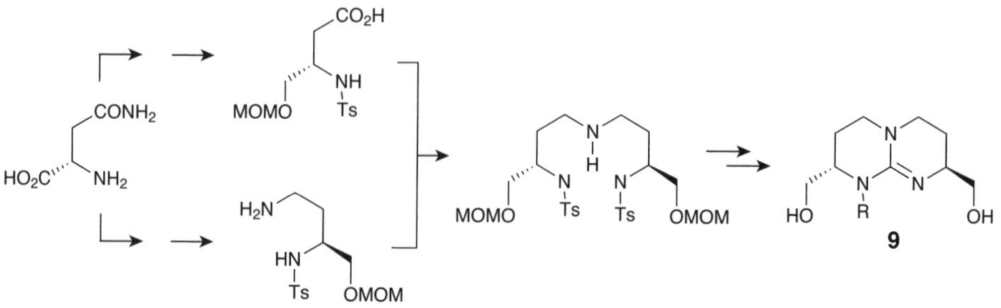

Scheme 4.6 Outline of the preparation of C_2-symmetrical (S,S)-guanidine **9** from asparagine

C_2-Symmetrical (or pseudo-symmetrical) chiral bicyclic guanidines are synthesized from amino acid derivatives [21b,22]. The typical synthetic route for a 2,7-disubstituted bicyclic system, (S,S)-2,7-di(hyroxymethyl)-1,5,7-triazabicyclo[4.4.0]dec-1-ene **9**, from asparagine is shown in Scheme 4.6 [22a].

The polycyclic systems with spiro rings such as **10** have been nominated as modified guanidines [23,24]. Murphy et al. [24a,24c] prepared tetracyclic guanidines (Murphy's guanidines) **10** by double sequential intramolecular cyclizations, as shown in Scheme 4.7.

A series of chiral guanidines, either symmetrical or nonsymmetrical, was newly synthesized from commercial amino alcohols using a concise and efficient aziridine-based synthetic methodology [25].

4.2.4 Preparation Based on DMC Chemistry

2-Chloro-1,3-dimethylimidazolium chloride (DMC) [26] ($R^1 = R^2 = Me$, L = H in **11**, Scheme 4.8a) not only acts as a powerful dehydration agent but also has unique and versatile abilities to chlorinate primary alcohols, to oxidize primary and secondary alcohols and to reduce sulfoxides and so on. In addition, DMC easily reacts with amines to yield the corresponding guanidines. Thus, methods of preparing monocyclic and bicyclic systems by application of DMC chemistry in the key steps have been developed [27]: the reaction of DMC-type chloroamidine compounds with amines for trisubstituted monocyclic guanidines [27a] (Scheme 4.8a), the intramolecular cyclization of thiourea derivatives after activation with DMC for monosubstituted or disubstituted monocyclic and bicyclic guanidines [27b] (Scheme 4.8b), and the DMC mediated cyclization of

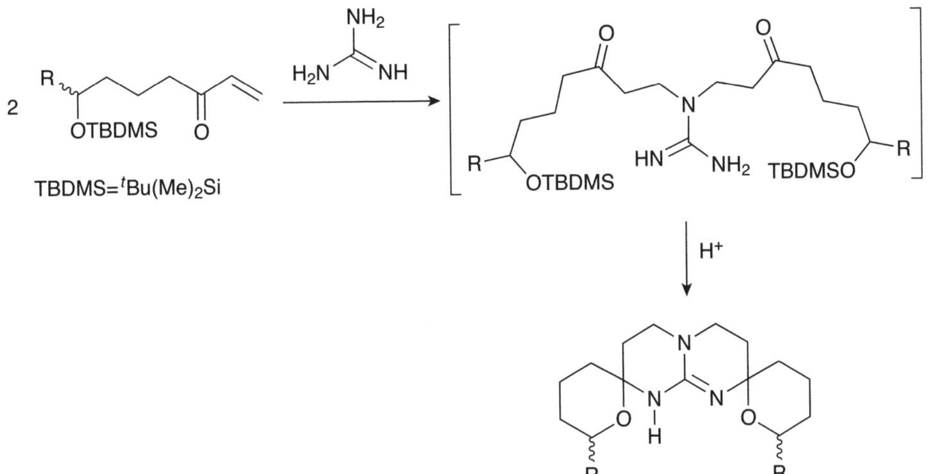

Scheme 4.7 Example of the preparation of bicyclic guanidines with spiro rings

2-hydroxyethyl-substituted guanidines to 2-amino-1,3-imidazolidine systems, in which chlorination of the primary alcohol function followed by intramolecular substitution reaction occurs, for other types of disubstituted monocyclic and bicyclic guanidines [27c] (Scheme 4.8c).

4.3 Guanidines as Synthetic Tools

There are many reports on the synthetic uses of TMG (**1**) and its analogues such as Barton's base (**2**). In this section, their synthetic roles in organic synthesis will be discussed according to their tentative classification into three categories [addition (catalytic reaction), substitution (stoichiometric reaction) and others] from the view points of a landmark for guanidine mediated asymmetric synthesis.

4.3.1 Addition

4.3.1.1 Aldol-Type Reaction

Carbonyl Substrate
Alkyl phosphonates are prepared smoothly by TMG (**1**) catalysed aldol-type addition of dialkyl phosphites to ketones and imines under mild conditions [28] (Scheme 4.9). Dialkyl phosphites can also serve as good nucleophiles for Michael addition (phospha-Michael Reaction).

TBD (**3a**) and its 7-methyl derivative (**3b**) were proven to be powerful catalysts, in many cases superior to TMG (**1**), in the addition of dialkyl phosphites to a variety of carbonyl compounds [29]. The polymer-supported (PS) TBD was also proven to be an efficient catalyst.

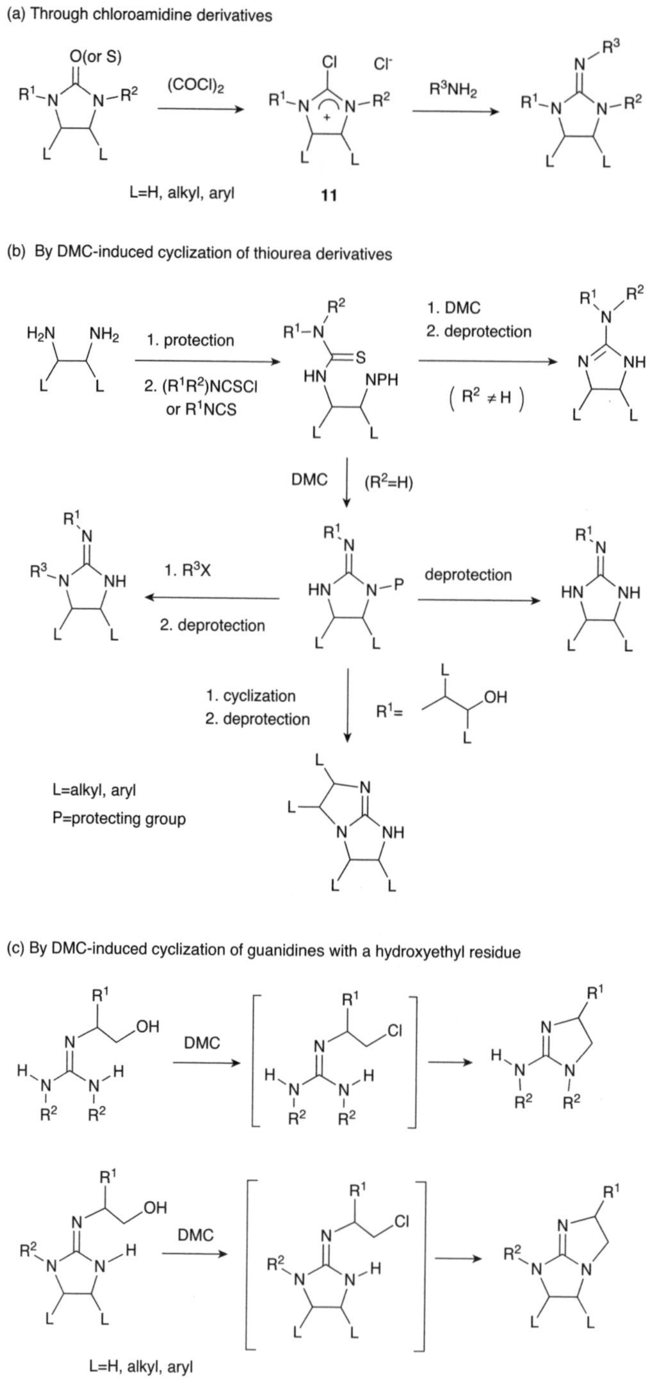
(a) Through chloroamidine derivatives
O(or S)
(COCl)2
Cl
Cl-
R3NH2
L=H, alkyl, aryl
11
(b) By DMC-induced cyclization of thiourea derivatives
1. protection
2. (R1R2)NCSCl
or R1NCS
1. DMC
2. deprotection
(R2 ≠ H)
DMC
(R2=H)
1. R3X
2. deprotection
deprotection
1. cyclization
2. deprotection
L=alkyl, aryl
P=protecting group
(c) By DMC-induced cyclization of guanidines with a hydroxyethyl residue
DMC
DMC
L=H, alkyl, aryl

Scheme 4.8 *Preparation of guanidines based on DMC chemistry*

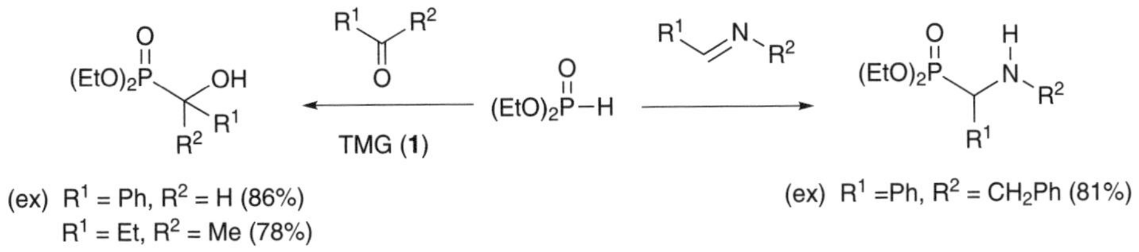

Scheme 4.9 Preparation of alkyl phosphonate by TMG (**1**) catalysed aldol-type addition

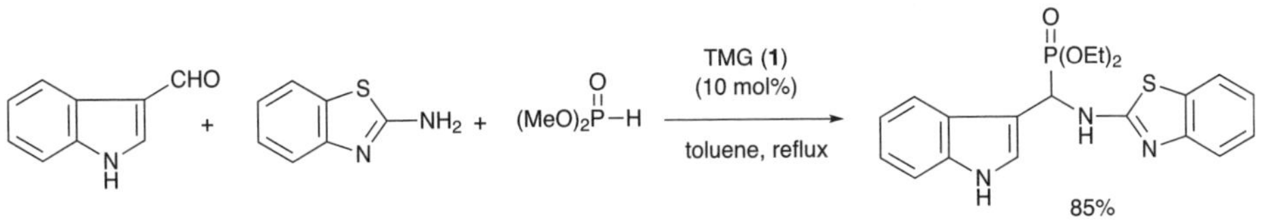

Scheme 4.10 TMG (**1**) catalysed one-pot synthesis of aminophosphonate

Imine Substrate

TMG (**1**) catalyses one-pot simultaneous reaction of indole-3-carboxaldehyde, a dialkyl- or diphenyl phosphite, and a primary amine to give the corresponding α-aminophosphonates in good yield (60–85%) (Scheme 4.10) [30]. It is known as Kabachnik–Fields reaction.

4.3.1.2 Aza-Henry (Nitro-Mannich) Reaction

The aza-Henry (nitro-Mannich) reaction of N-diphenylphosphinoylimines with nitroalkanes under solvent-free conditions was efficiently catalysed with TMG (**1**) to give a series of β-nitroamines in excellent yields and high diastereoselectivity [31] (Table 4.1).

Table 4.1 Solvent-free aza-Henry reaction catalysed with TMG (**1**)

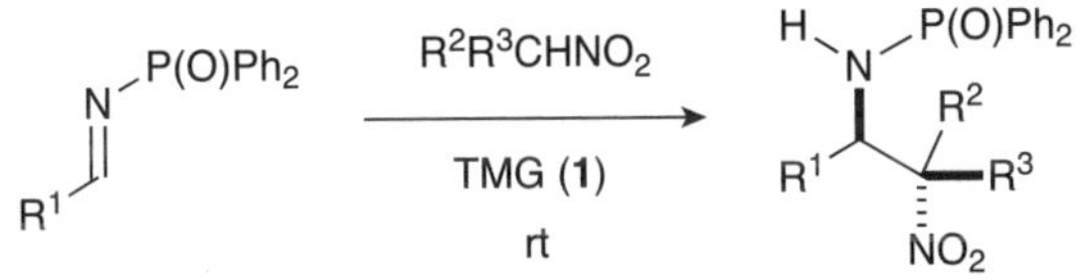

Run	R^1	R^2	R^3	Time (h)	Yield (%)	anti : syn
1	Ph	Et	H	2	96	94 : 6
2	4-ClPh	Et	H	2	95	84 : 16
3	4-(MeO)Ph	Et	H	24	92	95 : 5
4	2-furanyl	Et	H	2	90	95 : 5
5	^{t}Bu	Et	H	170	95	>98 : 2
6	Ph	Me (CH$_2$)$_4$	H	24	90	>98 : 2
7	Ph	Me	Me	3	96	—

Table 4.2 Selected Baylis–Hillman reaction of methyl acrylate with various aldehydes

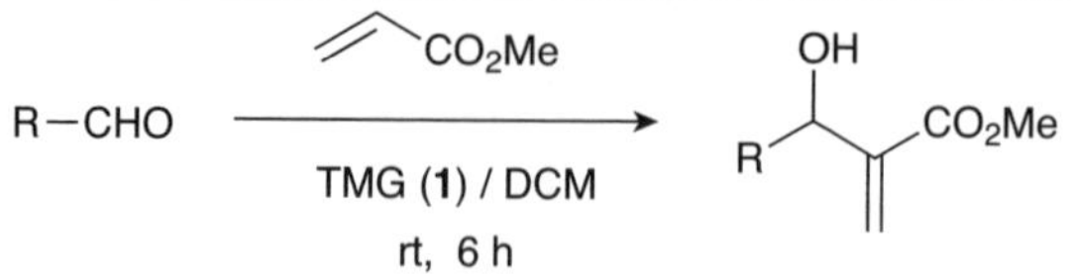

Run	R	Mol% 1	Yield (%)
1	Ph	12.5	49
2	Me	12.5	46
3	Pr	12.5	85
4	Ph(CH$_2$)$_3$	12.5	46
5	(E,E)-Me(CH=CH)$_2$	5	85

TMG (**1**), TBD (**3a**) and MTBD (**3b**) act as effective catalysts for the aza-Henry (nitro-Mannich) reaction of N-diphenylphosphinoyl ketimines and nitromethane. The addition product is given in good to high yields [32]. A phophazene (^{t}Bu-P1) can be also workable as a superior base.

4.3.1.3 Baylis–Hillman Reaction

TMG (**1**) catalyses the Baylis–Hillman reaction [33]. Selected results for the reaction of aldehydes and methyl acrylate are given in Table 4.2 [33a]. The reactions using aromatic aldehydes accelerate when either phenol as co-catalyst was added or reaction was carried out in alcoholic solvent [33b]. The asymmetric version of this reaction remains unexplored.

4.3.1.4 Concerted-Like Reaction

1,3-Dipolar Addition

A wide range of imines has been reacted with 5-menthyloxyfuranone in acetonitrile (MeCN) at ambient temperature in a combination of silver acetate with 1,8-diazabicyclo[5.4.0]undec-7-ene (DBU) or BTMG (**2**) to afford 1,3-cycloaddition product in good yields (71–91%) and high selectivity (de = ≥95%) [34] (Scheme 4.11). BTMG (**2**) is superior to DBU.

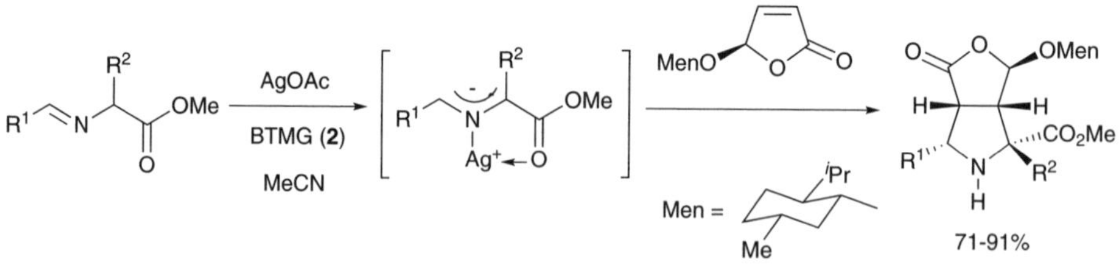

Scheme 4.11 BTMG (**2**) catalysed 1,3-cycloaddition of imine with 2(5H)-furanone

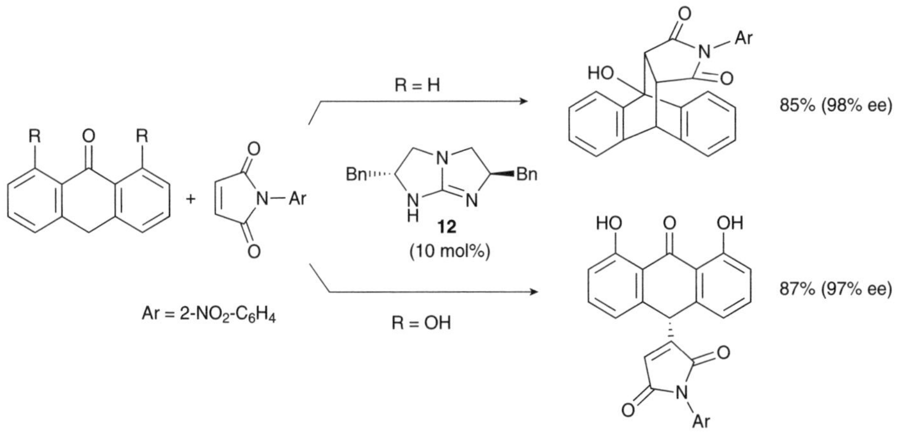

Scheme 4.12 *TMG (**1**) catalysed reaction between anthrone and succinimide*

Diels–Alder Reaction

A chiral bicyclic guanidine **12** was found to be an excellent catalyst for Diels–Alder reactions between anthrones and various dienophiles [35]. The catalyst can torelate a range of substituents and substitution patterns, making several anthrone derivatives suitable for this reaction. The use of 1.8-dihydroxy-9-anthrone as an anthrone substrate led to the production of 5-substituted anthrone as a Michael adduct in excellent yields, high regioselectivities and high enantioselectivities (Scheme 4.12).

4.3.1.5 Cyanosilylation

TMG (**1**) acts as a highly effective catalyst for the cyanosilylation of various ketones and aldehydes to the corresponding adduct in high yields. The reaction proceeds smoothly with 0.1 mol% catalyst loading at 25 °C without solvent [36].

Treatment of aldehydes and trimethylsilyl cyanide (TMSCN) in toluene at −78 °C in the presence of the catalytic amount of symmetrical bicyclic guanidine **13** smoothly afforded an (*S*)-excess adduct with good to moderate ee [37]. (Table 4.3) A ketonic 3-phenyl-2-butanone can work as an acceptor under the above conditions, but reactivity is low (23% yield, 39% ee).

Table 4.3 *Asymmetric TMS cyanation of aldehydes catalysed with guanidine **13***

R-CHO (1 equiv) + TMSCN (1.5 equiv) → [**13** (0.1 equiv), toluene −78 °C]

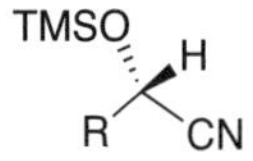

Run	R	Time (h)	Yield (%)	ee (%)
1	$Ph(CH_2)_2$-	6	85	50
2	cyclohexyl	1.5	93	70
3	^{t}Bu-	7	92	43

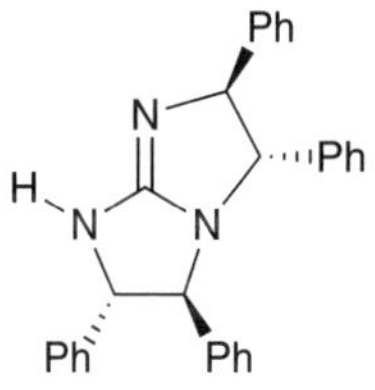

13

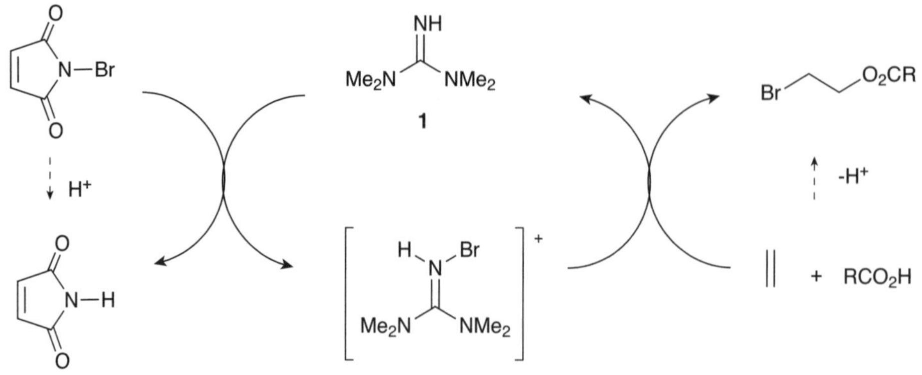

Scheme 4.13 Proposed catalytic cycle for TMG (**1**) catalysed bromolactonization

4.3.1.6 Halolactonization

Three organocatalysts, DMF, DMA and TMG (**1**), were examined for their catalytic roles for the bromolactonization of γ,δ- and δ,ε-unsaturated carboxylic acids with *N*-bromosuccinimide (NBS). TMG (**1**) was found to be a superior catalyst for this bromination (1–10 mol% loading, 100% conversion after 15 min) and to catalyse an intermolecular bromoacetoxylation of alkenes with acetic acid and NBS. The catalytic cycle is proposed (Scheme 4.13) [38].

4.3.1.7 Henry (Nitroaldol) Reaction

TMG (**1**) and TBD derivatives (**3**), in many cases superior to **1**, are proven to be powerful catalysts for Henry (nitro-aldol) reaction [29,39]. The X-ray structure of the TBD-phenylnitromethane complex has been reported [39]. Smooth reactions of aldehydes or ketone with nitroalkanes were observed to afford 2-nitroalkanols under mild conditions when TMG (**1**) was used as not only a base but also a solvent [40] (Table 4.4). However,

Table 4.4 TMG (**1**) catalysed Henry reaction for 2-nitroalkanols

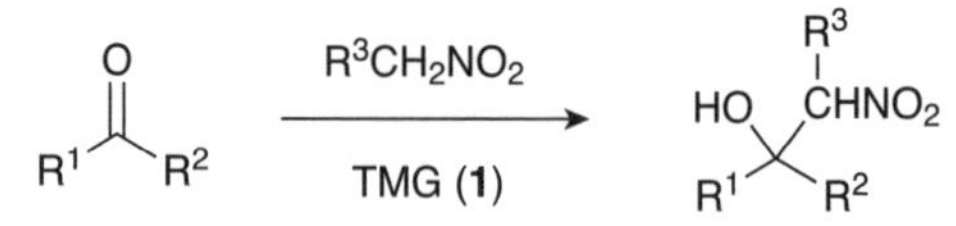

Run	R^1	R^2	R^3	Time/temp (h/°C)	Yield (%)
1	Ph	H	H	0.5/0	94
2	4-(NO_2)Ph	H	H	0.25/rt	97
3	4-(MeO)Ph	H	H	1/0	73
4	Pr	H	H	1/rt	67
5	4-(NO_2)Ph	H	Me	0.25/0	88
6	4-(MeO)Ph	H	Me	1/0	74
7	–(CH_2)$_8$–		Me	48/rt	71

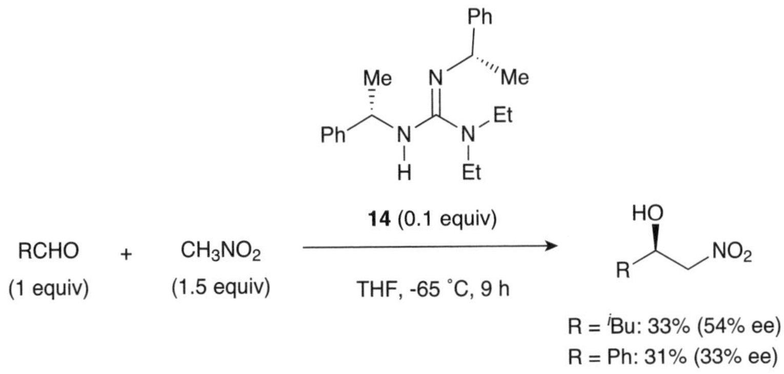

Scheme 4.14 Examples of guanidine catalysed asymmetric nitroaldol reactions

acetophenone or acyclic ketones do not work as electrophiles because of no reaction or predominant self condensation, respectively. This method has been applied to Henry reactions using sugar derivatives [41]. PS-TBD was also proven to be an efficient catalyst.

In 1994, the first guanidine catalysed asymmetric nitroaldol reaction was reported [10a]. Treatment of pivalaldehyde with nitromethane in tetrahydrofuran (THF) afforded an adduct in 33% yield with 54% ee when N,N-diethyl-N',N''-bis[(1S)-1-phenylethyl]guanidine (**14**) was used. (Scheme 4.14).

The Murphy's guanidine **10** (R = Me in Scheme 4.7) also catalyses the nitroaldol reaction of isobutylaldehyde and nitromethane to give an adduct in 52% yield but with low ee (20%) [24c].

Several enantiopure guanidines were studied as the catalysts for the Henry reaction of dibenzylamino aldehydes with nitromethane. (R)-1-(1-Naphthyl)ethylamine-derived guanidine catalysed the reactions of L-isoleucine-derived aldehydes with good diastereoselectivity [42].

Catalytic enantio- and diastereoselective nitroaldol reactions were explored by using designed guanidine–thiourea bifunctional organocatalysts like **15** (Figure 4.4) under mild and operationally simple biphasic conditions. These catalytic asymmetric reactions have a broad substrate generality with respect to the variety of aldehydes and nitroalkanes [43]. On the basis of studies of structure and catalytic activity relationships, a plausible guanidine–thiourea cooperative mechanism and a transition state of the catalytic reactions are proposed.

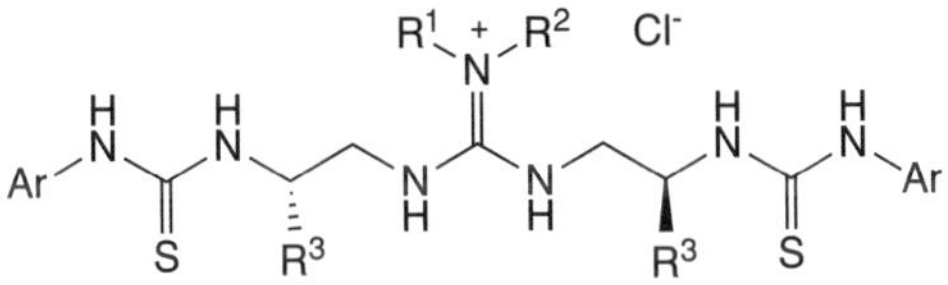

Figure 4.4 Structure of guanidine–thiourea bifunctional organocatalyst **15**

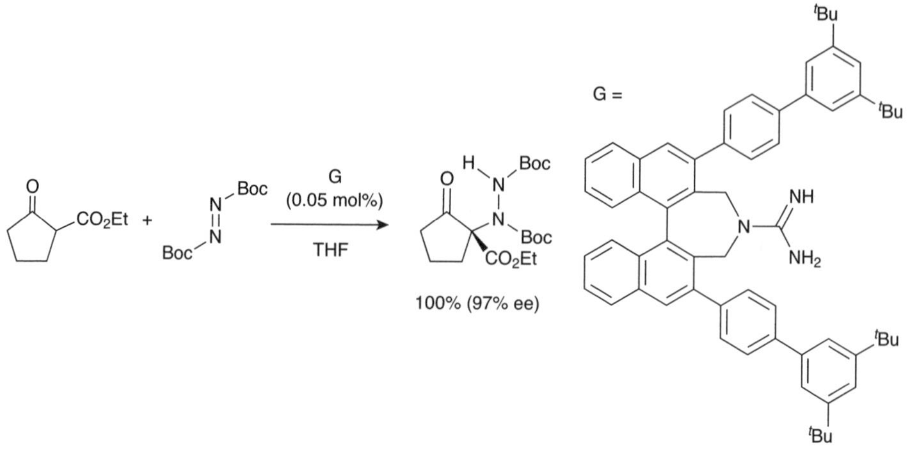

Scheme 4.15 *Axial guanidine catalysed aza-Michael reaction*

4.3.1.8 Michael Reaction

Aza-Michael Reaction

Axially chiral guanidines with an external guanidine unit such as dinaphthoazepineamidine are effective catalysts for the enantioselective addition of β-oxoesters and a 1,3-diketone to di-*tert*-butyl azodicarboxylates to yield α-hydrazino-β-oxoesters and α-hydrazino-β-diketones in 54–99% yields and in 15–98% ee [44]. For example, stirring 2-oxocyclopentanecarboxylate and di(tert-butyl) azodicarboxylate in THF in the presence of 0.05 mol% catalyst for four hours at −60° C provides an adduct in quantitative yield and in 97% ee (Scheme 4.15). The (*R*)-catalyst was prepared from (*R*)-2,2′-dimethyl-3,3′-binaphthalenediol ditriflate, 4-methoxyphenylboronic acid and 3.5-di(*tert*-butyl)phenylboronic acid in six steps.

Carba-Michael Reaction

In 1962, Nysted and Burtner [45] reported that the TMG (**1**) catalysed Michael reaction of methyl acrylate and 17-nitroandrostane derivative (Scheme 4.16) produced adduct in 84% yield, whereas the use of Triton B or sodium alkoxide in place of **1** led to low conversion. This may be the first application of TMG (**1**) to organic synthesis as an organic base. After

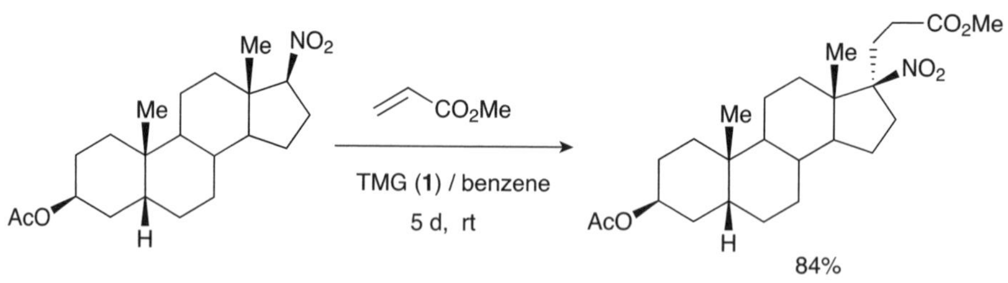

Scheme 4.16 *TMG (**1**) catalysed Michael reaction of 17-nitroandrostane and acrylate*

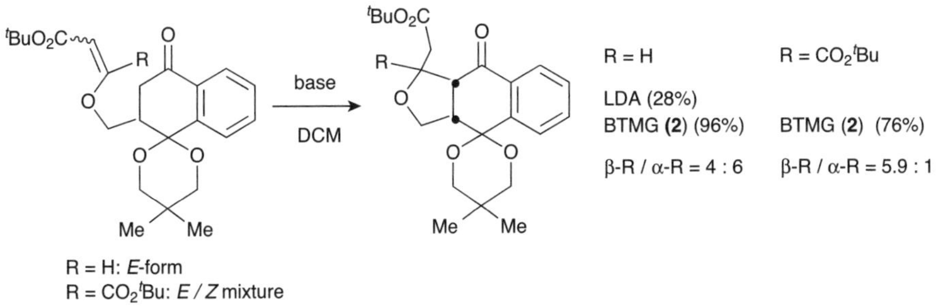

Scheme 4.17 BTMG (**2**) catalysed intramolecular Michael addition

this, nitroalkanes have been often used as nucleophiles in TMG (**1**) catalysed Michael reactions [46].

Of course, BTMG (**2**) [47] and TBD (**3**) [48] similarly work in Michael addition as powerful catalysts. In the synthesis of lactonamycin with a hexacyclic ring system, the BCD tricyclic ring system was effectively constructed by BTMG catalysed intramolecular Michael addition [49] (Scheme 4.17) (see Scheme 7.6).

A C_2-symmetrical pentacyclic guanidinium salt like **16** (Figure 4.5) was used for the conjugate addition of pyrrolidine to γ-crotonolactone, in which structural requirement such as the size of the cavities and substituents on tetrahydropyran rings of the guanidine catalysts is critical for asymmetric induction [23].

Asymmetric nitro-Michael reactions of methyl vinyl ketone (MVK) in the presence of bicyclic guanidine with a benzhydryl group led, disappointedly, to low asymmetric induction (9–12%) [21a] Trials for the reaction of α,β-unsaturated γ- or δ-lactones with pyrrolidine in the presence of the conjugate acids of a bicyclic guanidine [50] or the Murphy's guanidine [24a] (R = Me in Scheme 4.7) resulted in the production of racemic compounds. The latter phase transfer catalyst (PTC) catalyses the nitro-Michael addition of chalcone but with limited range (70% yield, 23% ee) [24c].

Ma *et al.* examined guanidine catalysed Michael reaction of *tert*-butyl glycinate Schiff base with ethyl acrylate in THF and observed 30% ee as the asymmetric induction when an acyclic guanidine (**2**) was used as a catalyst [10b]. Ishikawa *et al.* succeeded in greatly improving the asymmetric induction by the use of guanidine **17a**, originally prepared based

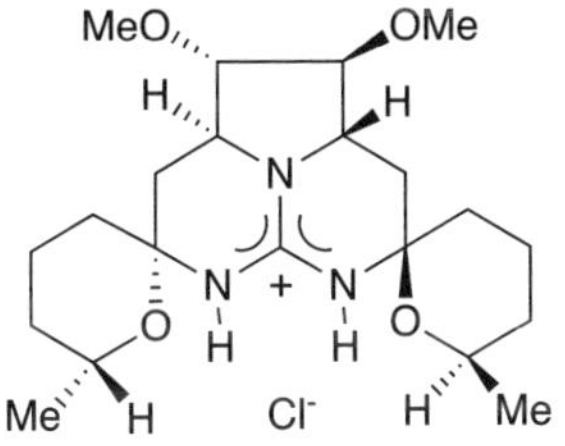

Figure 4.5 The structure of a pentacyclic guanidinium salt **16**

Table 4.5 Asymmetric Michael reaction between t-butyl glycinate Schiff base and active vinyl compounds in the presence of the guanidine **17**

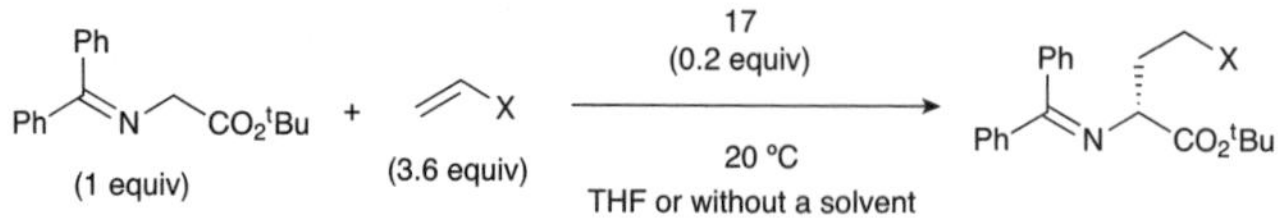

Run	17	X	Solvent	Time	Yield (%)	ee (%)
1	17a	COMe	THF	6 d	90	96
2	17a	COMe	—	15 h	90	80
3	17a	CO_2Et	THF	7 d	15	79
4	17a	CO_2Et	—	3 d	85	97
5	17b	CO_2Et	THF	7 d	62	90
6	17b	CO_2Et	—	5 d	79	97
7	17a	CN	THF	5 d	NR[a]	—
8	17a	CN	—	5 d	79	55

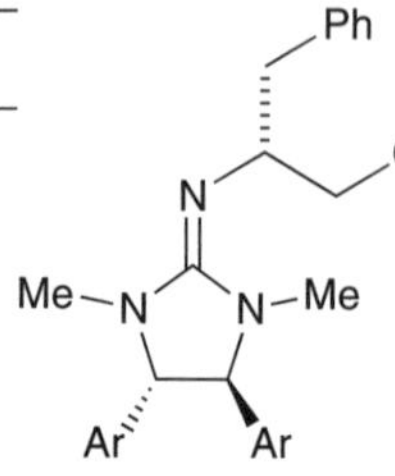

[a] No reaction.

17a: Ar = Ph
17b: Ar = 2-methylphenyl

on DMC chemistry, especially under solvent-free condition (Table 4.5). Thus, an (R)-adduct was given in 85% yield with 97% ee [51]. It is noted that MVK is reactive enough even in solution [51] and that modification of the phenyl pendant in the guanidine skeleton **17b** to 2-methylphenyl ones accelerates the addition reaction [52]. A bicyclic network system through two hydrogen bonds and one CH-π interaction in the transition state is proposed as playing an important role for high asymmetric induction [52].

The same guanidine **17a** also works as a catalyst in the Michael reactions of cyclopentenone and benzyl malonate (or α-methylmalonate). However, moderate selectivity was observed even under solvent-free conditions [53].

An axially chiral and highly hindered binaphthyl-derived guanidine catalyst **18a** with an internal guanidine unit (Figure 4.6) facilitates the highly enantioselective 1,4-addition

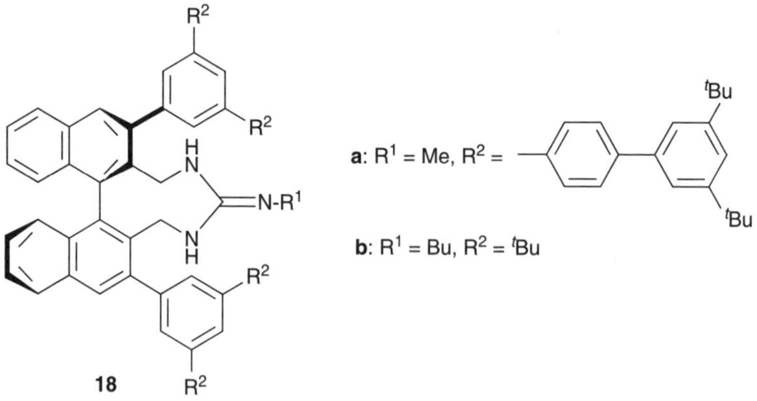

a: R^1 = Me, R^2 = (4-tBu,3-tBu-biphenyl)

b: R^1 = Bu, R^2 = ^tBu

Figure 4.6 The structures of representative axial chiral guanidines **18** with an internal guanidine system

reaction of 1,3-dicarbonyl compounds with a broad range of conjugated nitroalkanes and shows extremely high catalytic activity [54]. The catalyst showed high catalytic efficiency in evaluation by a gram-scaled experiment with low catalyst loading and was recovered in a nearly quantitative yield as an hydrochloride salt by acidic work up following column purification.

A chiral bicyclic guanidine, which corresponded to the *tert*-butyl analogue of guanidine **12** used in Scheme 4.12, effectively catalysed Michael reactions of dithiomalonates and β-keto thioesters using a range a of acceptors including maleimides, cyclic enones, furanone and acyclic 1,4-dicarbonyl butenes [55]. TMG (**1**) immobilized silica gel was used as a basic agent for the Michael addition of cyclopentenone and nitromethane. Addition product was obtained in good yield under mild conditions and the catalysis activity was maintained after fifteen cycles (98% conversion) [56].

Oxa-Michael Reaction
A 2,2-disubstituted chromane system was asymmetrically constructed by application of intramolecular oxa-Michael addition reaction through 6-*exo-trig* mode cyclization [57]. Good asymmetric induction at the quaternary carbon was observed when Z-alkene was treated with the same guanidine **17** used in asymmetric carba-Michael reaction in Table 4.5 (Scheme 4.18).

Phospha-Michael Reaction
Similar to aldol-type reaction, dialkyl phosphites [28] can also serve as good nucleophiles in Michael reactions in the presence of TMG (**1**) (Scheme 4.19). The reaction proceeds smoothly under mild conditions and shows tolerance to variety of functional groups.

A highly enantioselective 1,4-addition reaction of nitroalkene with diphenyl phosphite was successfully accomplished using an alternative axial guanidine catalyst **18b** with an internal guanidine unit [58] (Figure 4.6). A broad range of nitroalkenes, bearing not only aromatic but also aliphatic substituents, is applicable to the present enantioselective reaction. A chiral bicyclic guanidine, which corresponds to the *tert*-butyl analogue shown in the reaction (Scheme 4.12), has been used to catalyse the phospha-Michael reaction of diarylphosphine oxide to nitroalkenes with high enantioselectivities, offering a direct methodology to prepare chiral β-aminophosphine oxides and β-aminophosphines [59].

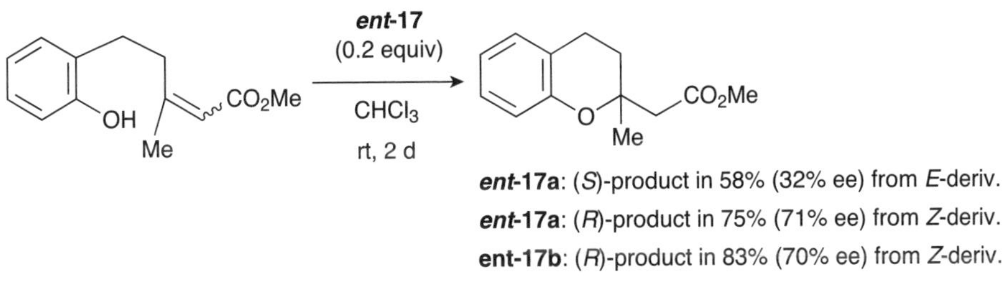

ent-17a: (*S*)-product in 58% (32% ee) from *E*-deriv.
ent-17a: (*R*)-product in 75% (71% ee) from *Z*-deriv.
ent-17b: (*R*)-product in 83% (70% ee) from *Z*-deriv.

Scheme 4.18 *Guaniodine catalysed intramolecular oxa-Michael addition*

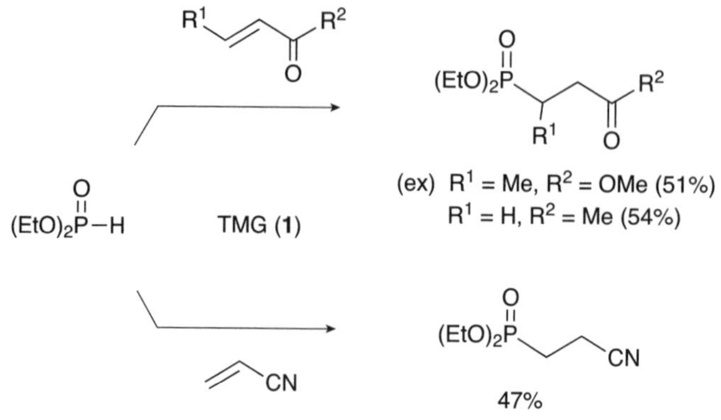

Scheme 4.19 TMG (1) catalysed Michael addition with phosphites

Thio-Michael Reaction

γ-Phenylthio-β-nitro alcohols were smoothly prepared by TMG (**1**) catalysed one-pot reaction of nitroolefines, thiophenol and aldehydes [60]. During the course of the synthesis of ecteinascidins [61] (Scheme 7.20), the ten-membered lactone bridge through the sulfide bond formation, based on Corey's original method, was achieved by BTMG (**2**) promoted intramolecular Michael type addition of thiolate ion to quinone methides, which were produced by treatment with Tf$_2$O in DMSO followed by Hünig base (Scheme 4.20).

4.3.1.9 Nucleophilic Epoxidation

Novel guanidine bases supported on silicas and micelle-templated silicas have been prepared and investigated in the base-catalysed epoxidation of election-deficient alkenes.

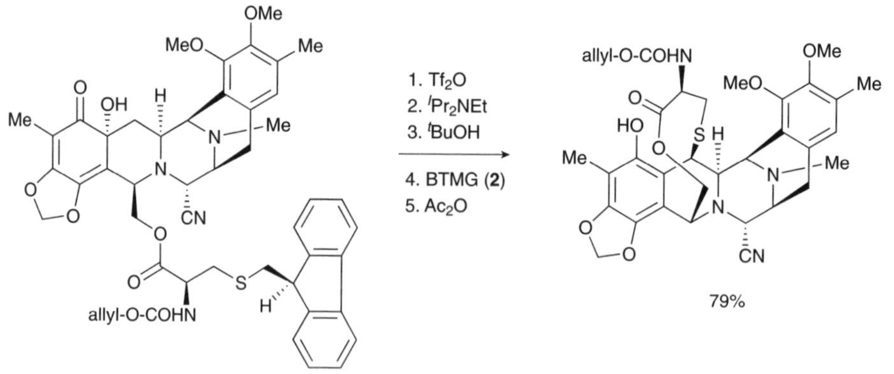

Scheme 4.20 The use of the BTMG (**2**) catalysed intramolecular Michael addition as a key step for ecteinascidin synthesis

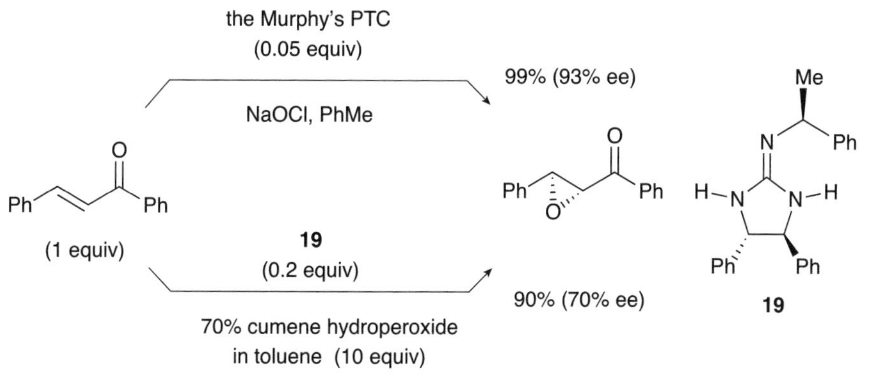

Scheme 4.21 Examples of nucleophilic epoxidation of chalcone

Excellent conversions and selectivities were observed both with respect to the alkene and the primary oxidant [62]. In nitro-Michael reaction it was noted that the Murphy's PTC does not work as a good chiral catalyst for the Michael reaction of chalcone, but the same PTC effectively catalyses the epoxidation of chalcones with sodium hypochlorite (NaOCl) [24c]. Trials for the epoxidation of chalcone in the combination of hydroperoxides and modified guanidines **19** [27b] resulted in less effective asymmetric induction compared to the Murphy's PTC [53] (Scheme 4.21).

4.3.1.10 Strecker Reaction

A diketopiperazine **8** with an external guanidine function is proven to be an effective catalyst for the Strecker reaction of benzhydrylimine [14]. Corey's bicyclic guandine **20** [12b], which is the original of **12** (Scheme 4.12), also works well in the same reaction and the mode of action has been elucidated by density functional theory [63] (Scheme 4.22).

A review elsewhere discusses catalytic Strecker reactions including guanidine catalysts [64].

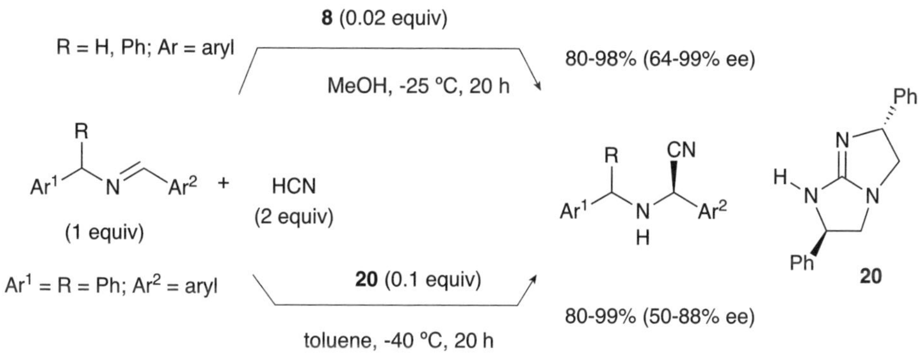

Scheme 4.22 Guanidine catalysed Strecker reaction

Scheme 4.23 Alkylation of β-keto ester in the presence of BTMG (**2**)

4.3.2 Substitution

4.3.2.1 Alkylation

C-Alkylation
β-Keto ester has been alkylated in high yield by the use of BTMG (**2**), whereas the use of collidine or 1,5-diazabicyclo[4.3.0]non-5-ene (DBN) led to ineffective conversions (collidine: 63%; DBN: 31%) [3] (Scheme 4.23).

Asymmetric alkylation of *tert*-butyl glycinate Schiff base with a range of alkyl halides was successfully carried out by use of pentacyclic chiral guanidinium salt **16** and the conjugate acids of the Murphy's guanidines (Scheme 4.7) as PTC [24]. Excellent to good asymmetric inductions were obtained [24b,24c] and selected results using **16** are given in Table 4.6 [24b].

O-Alkylation (Esterification and Alcoholysis)
Alkylative esterification of carboxylic acids with alkyl halides are effected by action with TMG (**1**) [65]. An ester is given by the TMG (**1**) mediated reaction of γ-hydroxy-α,β-unsaturated carboxylic acid with methyl iodide without lactone formation after isomerization [65a]. Barton's base effectively works in the alkylation of sterically hindered carboxylic acid [3]. Ethanolysis of the acetate of tertiary alcohol occurred easily in 86% yield in the presence of BTMG (**2**) [66] (Scheme 4.24).

The kinetic resolution of racemic 1-phenylethyl bromide was examined in the alkylation of benzoic acid using trisubstituted monocyclic guanidines as chiral sources [67]. (R)-Excess ester was obtained in 96% yield even with 15% ee, when the reaction was carried out in benzene with 1,3-dimethyl-(4S,5S)-diphenyl-2-[(1S)-phenylethylimino]imidazolidines [27a] (Scheme 4.25).

Table 4.6 Selected a symmetric alkylation of tert-butyl glycinate Schiff base with alkyl halides in the presence of **16**

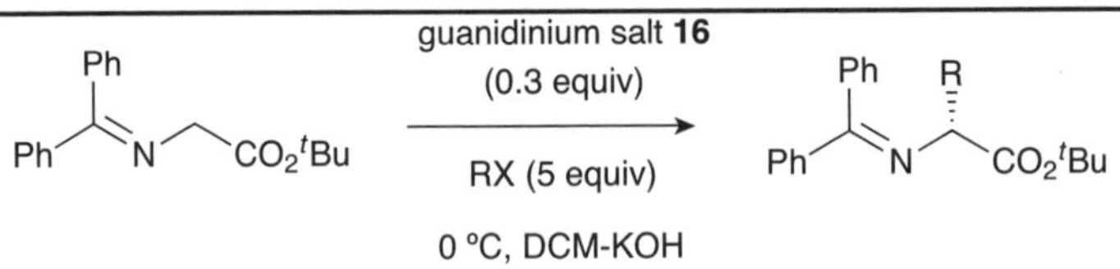

Run	RX	Time (h)	Yield (%)	ee (%)
1	MeI	145	80	76
2	CH$_2$=C(Me)CH$_2$Br	145	85	81
3	2-(Naphthyl)CH$_2$Br	95	81	90

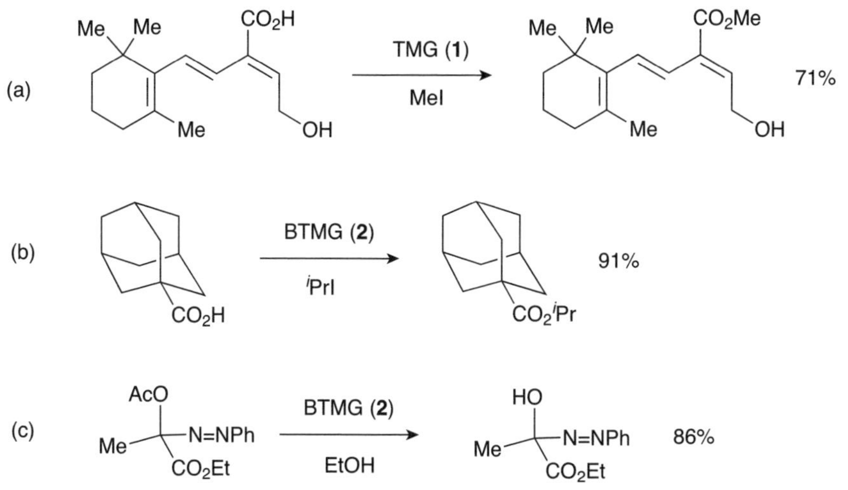

Scheme 4.24 Guanidine mediated esterification and alcoholysis

4.3.2.2 Azidation

In the kinetic resolution of 1-indanol with diphenylphosphoryl azide in dichloromethane (DCM) using modified guanidine [68], (R)-excess azide compound was produced in 58% yield with 30% ee after six hours when a C_2-symmetrical bicyclic guanidine **13** [37] was used as a chiral auxiliary (Scheme 4.26).

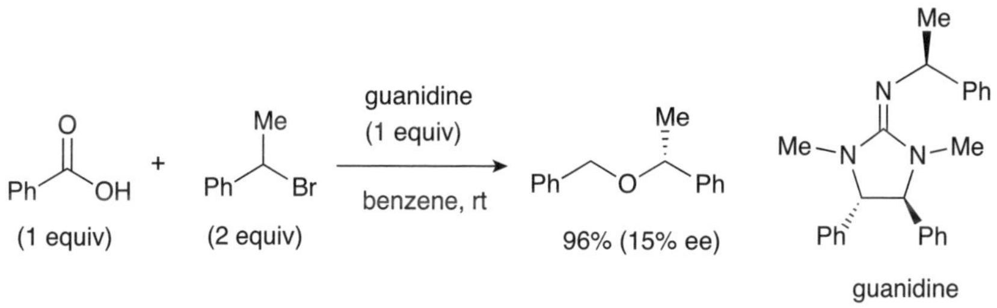

Scheme 4.25 Example of guanidine mediated asymmetric alkylation of carboxylic acid

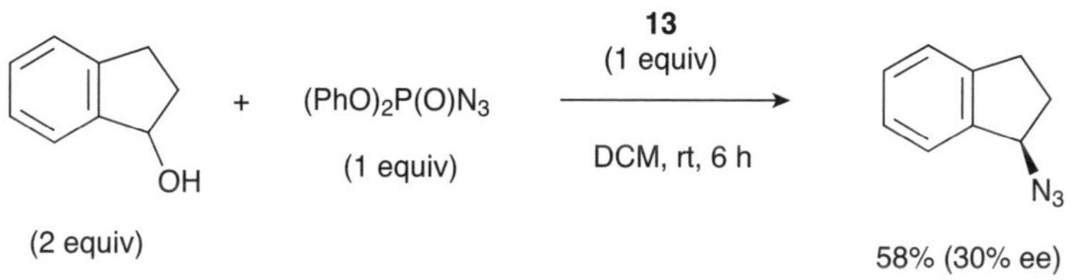

Scheme 4.26 Guanidine **13** mediated asymmetric azidation of 1-indanol

Table 4.7 Glycosidation of phenols with 1,2-anhydroglucose derivative

Run	X	Conditions	Yield (%)	α:β
1	MeO	$ZnCl_2$, DCE	62	55:45
2	MeO	$ZnCl_2$, TMG (**1**), DCE	73	20:80
3	MeO	K_2CO_3, THF, 18-crown-6	73	5:95
4	NO_2	$ZnCl_2$, DCE	60	78:22
5	NO_2	$ZnCl_2$, TMG (**1**), DCE	83	60:40
6	NO_2	K_2CO_3, THF, 18-crown-6	60	5:95

4.3.2.3 Glycosidation

The zinc chloride ($ZnCl_2$) catalysed glycosidation of *para*-substituted phenols with 1,2-anhydro-3,4,6-tri-*O*-methyl-α-D-glucopyranose gives predominantly the corresponding α-anomer [69]. Addition of TMG (**1**) enhances the β-selectivity, even to practical completion under the conditions of potassium carbonate (K_2CO_3) and 18-Crown-6, in THF (Table 4.7).

4.3.2.4 Intramolecular Substitution (Cyclopropanation)

TMG (**1**), as well as benzyltrimethylammonium hydroxide (Triton B) in pyridine and sodium ethoxide in ethanol, was found to work as base catalyst in the cyclopropanation of steroid skeletons controlled by intramolecular S_N2 reaction [70]. Thus, 6-oxo-3α,5-cyclo-5α-steroids were given in high yields for the reaction of 3β-tosyloxy (or -chloro)-6-oxo derivatives (Table 4.8).

4.3.2.5 Silylation of alcohols

A catalytic amount of TMG (**1**) effectively works for the silylation of primary and secondary alcohols with the help of reagents such as *tert*-butyldimethylchlorosilane (TBDMCS) in acetonitrile in the co-presence of a stoichiometric amount of tertiary amine as an acid scavenger [71] (Table 4.9). In the reaction of secondary alcohols, DMF is superior to acetonitrile as solvent.

Guanidine participating kinetic resolution of 1-indanol with chlorosilane reagents, TBDMCS or triisopropylchlorosilane (TIPCS) was investigated [72] (Table 4.10). An (*R*)-excess silyl ether was afforded as a major enantiomer with moderate ee. The bulkiness of silylating reagent, as expected, affects the asymmetric induction and 70% ee was observed in the case of 1-tetrahydrodecanol but yield is low.

4.3.2.6 SNAr Reaction

Barton's bases are used for the formation of diaryl ether by S_NAr reactions [73]. In the comparison of several bases BTMG (**2**) was found to be an excellent and mild alternative for promoting S_NAr reactions [73a] (Table 4.11).

Table 4.8 Base mediated cyclopropanations of steroid compounds

			Yield (%) by method[a]		
Run	R	X	A	B	C
1	Me, Me, Me (isoprenyl-type)	TsO	98	90	85
2		Cl	95	93	95
3	Me, OAc, Me, Et, OAc	TsO	88	85	—
4		Cl	90	80	—

[a] A: TMG (1), 60 °C, 5 min; B: PhCH$_2$Me$_3$N$^+$OH$^-$, 60 °C, 10 min; C: NaOEt/EtOH, reflux, 10 min.

4.3.2.7 Sulfide Formation

Thiol is converted to a nucleophilic thiolate anion, which reacts with an epoxide to give a ring-opened sulfide, in the presence of TMG (1) [74]. A carbapenem antibiotic carrying a sulfide function was synthesized practically by the addition–elimination reactions of a thiol to the enol phosphate of β-keto ester in the presence of TMG (1) as a key step [75] (Scheme 4.27).

Table 4.9 TMG (1) catalysed silylation of alcohols

R-OH $\xrightarrow[\text{TMG (1) (0.2 equiv)}]{\text{TBDMCS (1.2 equiv), Et}_3\text{N (1.2 equiv)}}$ R-OTBDMS
solvent

Run	R	Solvent	Time (h)	Yield (%)
1	Ph(CH$_2$)$_2$	MeCN	0.3	96
2		DMF	0.3	95
3	4-^{t}Bu-cyclohexyl	MeCN	5	96
4		DMF	0.3	95
5	PhCH(Me)	MeCN	12	96
6		DMF	1	95

Table 4.10 Guanidine mediated asymmetric silylation of cyclic secondary alcohols

Run	n	Silyl	Guanidine	Time (d)	Yield (%)	ee (%)
1	1	TBDMS	R = H	11	34	37
2	1	TBDMS	R = Me	9	50	39
3	1	TIPS	R = H	6	36	59
4	1	TBDMS		10	78	31
5	1	TIPS		6	79	58
6	2	TIPS		6	15	70

Table 4.11 Effect of base in the S_NAr reaction of naphthol and fluoronaphthalene

Run	Conditions	Yield (%)
1	NaH, DMSO, 25 °C	49
2	K_2CO_3, AcNMe$_2$, 150 °C	54
3	DBU, MeCN, 70 °C	46
4	TMG(1), MeCN, 70 °C	59
5	BTMG (2), MeCN, 70 °C	85

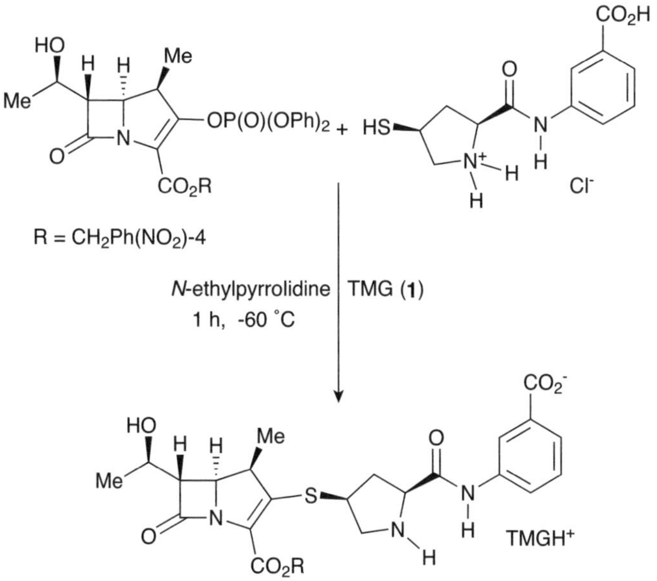

Scheme 4.27 TMG (**1**) assisted substitution reaction for the practical preparation of ertapenem sodium

4.3.3 Others

4.3.3.1 Construction of Heterocycles

Benzimidazole
Copper and palladium catalysed intramolecular C–N bond formation between an aryl halide and a guanidine moiety affords 2-aminobenzimidazoles. Inexpensive copper salts such as copper iodide (CuI) are generally superior to the use of palladium catalysts [76] (Table 4.12).

Furan
Effective cyclization of 2-trimethylsilylethynylphenol to a 2-trimethylsilylbenzfuran was carried out by refluxing in toluene in the presence of TMG (**1**) (>90%) (Scheme 4.28). The co-presence of silicon dioxide (SiO_2) led to a desilylated benzofuran in one pot [77].

Oxazolidinone and Oxazole
N-Alkylprop-2-ynylamines readily react with carbon dioxide (CO_2) in the presence of catalytic strong bases and undergo intramolecular cyclization to 5-methylene-1,3-oxazolidin-2-ones in good yields [78]. The type and strength of the base is of paramount importance to the success of the reaction. DBU, TBD, tetra-alkyl- and penta-alkylguanidines and phosphazene bases are effective, whereas 1.8-bis(dimethylamine)naphthalene (proton sponge), carbodiimide and pyridine did not work. The presence of a triple bond and an amino group, which can react intramolecularly, allow the catalytic incorporation of the intermediate carbamate into an oxazolidine ring by reaction with the triple bond even without any metals (Scheme 4.29a).

Table 4.12 Intramolecular aryl guanidinylation of aryl bromides with Pd(PPh$_3$)$_4$ or CuI

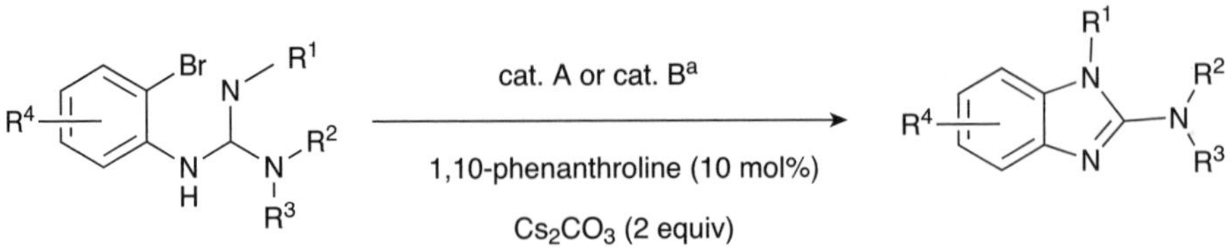

Run	R^1	NR^{2}R^3	R^4	Yield (%) [conversion (%)] cat. A	cat. B
1	Bn	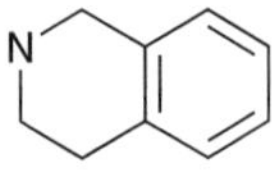 tetrahydroisoquinolinyl	H	88 (>98)	83 (>98)
2	Ph	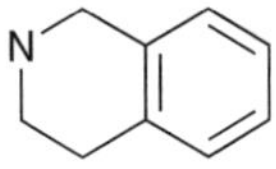 tetrahydroisoquinolinyl	H	84 (>95)	58 (>95)
3	Bn	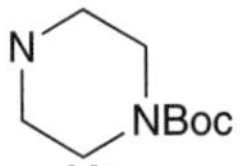 N-Boc piperazinyl	H	93 (>98)	96 (>98)
4	Bn	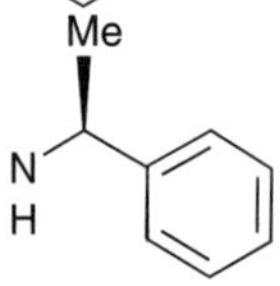 NH-CH(Me)Ph	H	– (85)	97 (>98)
5	Bn	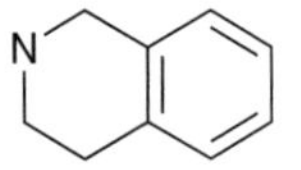 tetrahydroisoquinolinyl	4-Me	66 (70)	90 (95)
6	Bn	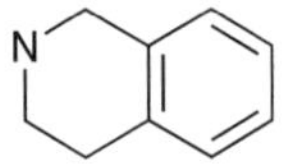 tetrahydroisoquinolinyl	6-Br 4-Me	76 (76)	98 (>98)

^{a}cat. A: Pd(PPh$_3$)$_4$ (10 mol%); cat. B: CuI (5 mol%).

PS-p-toluenesulfonylmethyl isocyanide (TosMIC) reagent, developed by Barrett et al. was found to be effective for the conversion of a range of aryl aldehydes into highly pure 4-aryl oxazoles in the presence of BTMG (**2**) [79] (Scheme 4.29b). A typical procedure involved the reaction of aldehyde with the gel (4 equiv.) in acetonitrile (0.2 M sol) and BTMG (**2**) (4 equiv.) for 12 h at 65 °C.

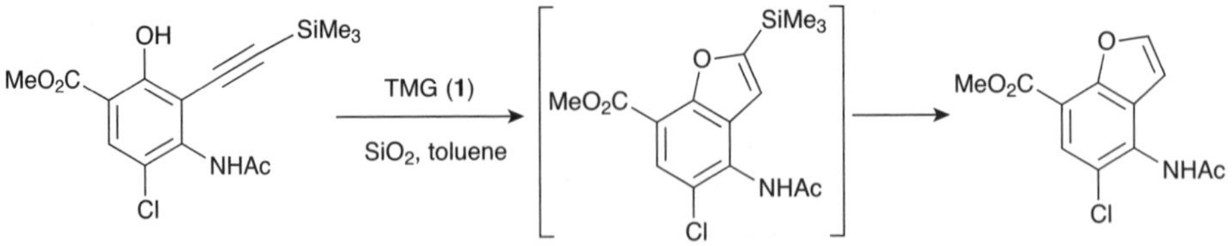

Scheme 4.28 TMG (**1**) mediated intramolecular cyclization to furan ring system

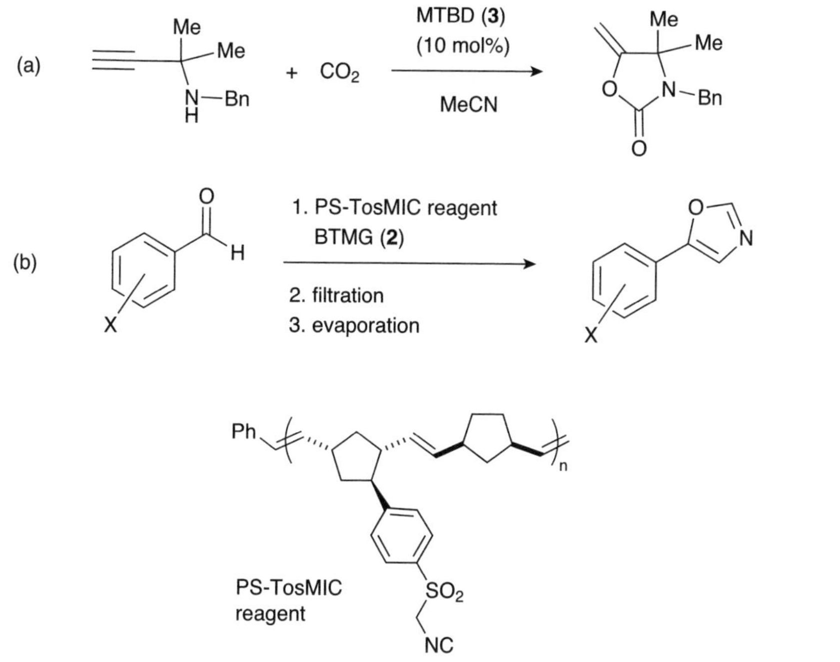

Scheme 4.29 Guanidine catalysed formation of oxazolidinone and oxazole ring systems

Pyrrole

Pyrrole derivatives were given in 80% yield by treatments of β-nitrostyrene with isocyanoacetate and DBU [80] (Scheme 4.30). The reaction was slightly faster and the yield better (90%) when DBU was replaced by BTMG (2), in which Michael addition, cyclization through internal attack of the nitronate on the isocyano group, elimination of a nitronate ion through a vinylogous $E1_{CB}$ mechanism after proton exchange and aromatization by a [1.5]-sigmatropic shift of hydrogen could be successively occurred.

4.3.3.2 Horner–Wadsworth–Emmons Reaction

TMG (1) is used as a base in Horner–Wadsworth–Emmons (HWE) reactions [81]. For example, the sequential reaction of 1,3-diformylbenzene with different phosphorylglycines

Scheme 4.30 BTMG (2) catalysed formation of pyrrole ring system

affords unsymmetrical product as a single diastereisomer [81a] (Scheme 4.31a). Formylmethyl-substituted pyranoses react with phosphorylglycinate at lower temperature ($-78\,^\circ$C) in the presence of TMG (**1**) to give products in satisfactory yields [81b] (Scheme 4.31b).

Barrett *et al.* [82] prepared high-loading polymer phosphonate resins, a related resin used in oxazole formation (Oxazolidinone and Oxazole), by polymerization, which in combi-

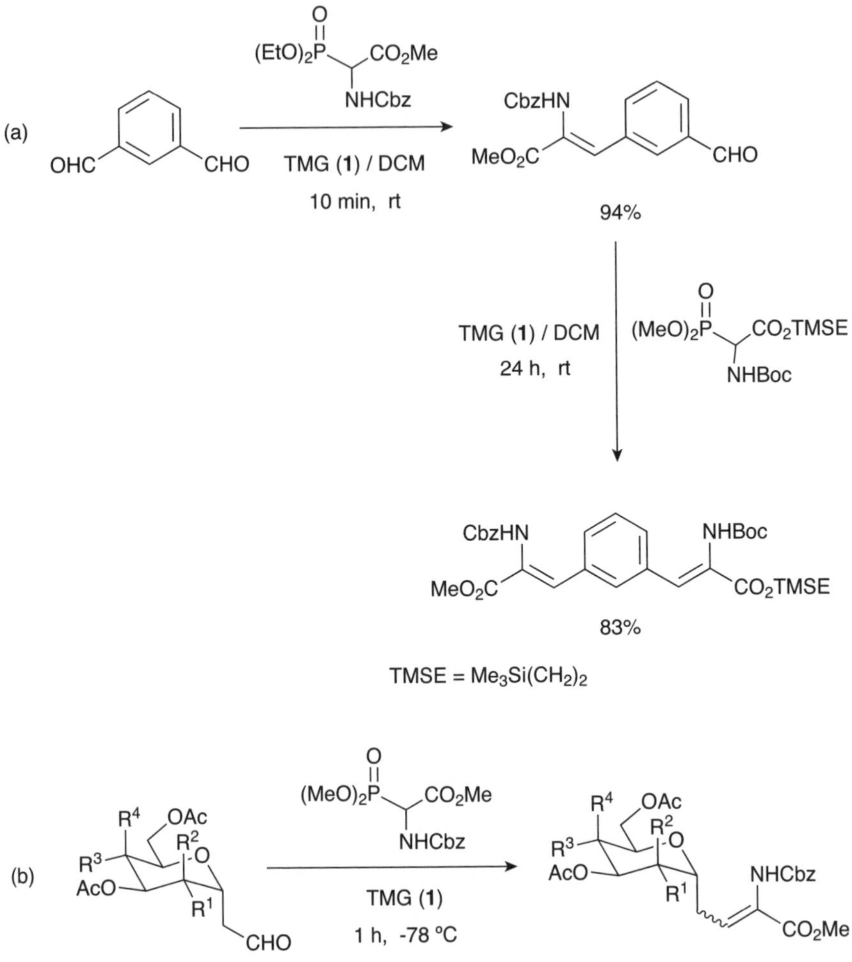

R^1	R^2	R^3	R^4	yield (%)
OAc	H	OAc	H	79
OAc	H	H	OAc	85
H	OAc	OAc	H	88

Scheme 4.31 *Examples of TMG (**1**) mediated HWE reactions*

Table 4.13 TMG (**1**) assisted α-alkenylation of β-keto esters with bismuth reagents

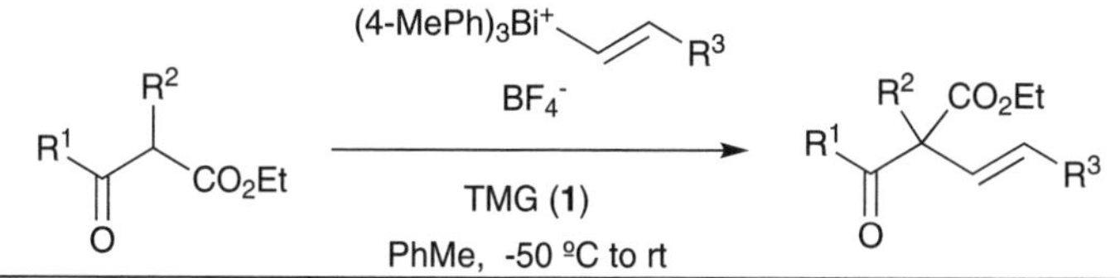

Run	R^1	R^2	R^3	Yield (%)
1		-(CH$_2$)$_4$-	n-C$_6$H$_{13}$	86
2		-(CH$_2$)$_3$-		90
3[a]	Me	H	Ph	96
4	Me	Me		90

[a] 2 Equiv. of the bismuth salt/TMG (**1**) were used.

nation with a Barton base allows a general purification-free HWE synthesis of α,β-unsaturated esters and nitriles from both aromatic and aliphatic aldehydes.

4.3.3.3 Metal Mediated Reaction

Bismuth
TMG (**1**) assisted alkene transfers from alkenyl bismuth reagents to reactive electrophiles have been reported [83]. Treatment of β-keto esters, β-diketones and phenols with alkenyltriarylbismuthonium salts in the presence of TMG (**1**) smoothly affords α-alkenylated products [83b] (Table 4.13).

Palladium
An organobase including guanidine is often used as co-catalyst (or base) in palladium coupling reactions [84]. The 2-methylenepropane-1,3-diol diacetate reacts with 7,8-dihydroquinoline derivative in the presence of palladium acetate [Pd(OAc)$_2$], TMG (**1**) and triphenylphosphine (PPh$_3$) to give the methylene bridged compound in 92% yield, which can be converted to a diamino analogue of huperzine A, an inhibitor of acetylcholine esterase [84a](Scheme 4.32).

O-Allylic urethanes and carbonates are afforded from amines/alcohols, carbon dioxide and allylic chlorides by palladium catalysed reaction in the presence of an organobase. The choice of added base in the generation of carbamates/carbonates was critical for high yields

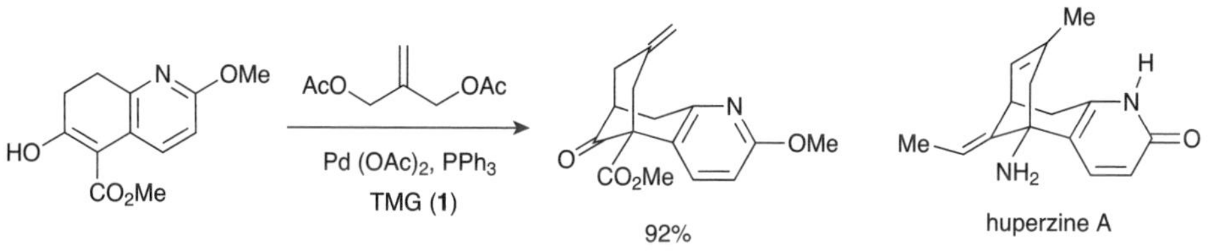

Scheme 4.32 TMG (**1**) catalysed palladium coupling reaction

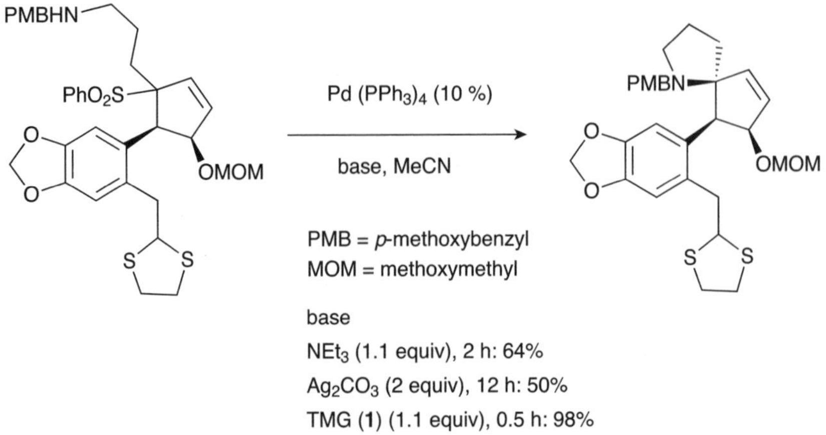

Scheme 4.33 *TMG (1) catalysed cyclization to spiro compound*

of *O*-allylic products. *N*-Cyclohexyl-1,1,3,3-tetramethylguanidine (Barton's base) and DBU were found to be optimal for this system. Tertiary allylic sulfones bearing a secondary aminopropyl moiety afforded a spiro system by palladium(0) catalysed cyclization. The use of TMG (**1**) as a companion base is required for high yielding reactions [84c] (Scheme 4.33).

Thus, *trans*-3-alkyl-6-(phthalimido)cyclopentenes were prepared in excellent to modest yields from the corresponding *trans*-chloroalkene by the palladium coupling reaction [84d]. Inexpensive and efficient Pd–TMG systems, Pd(OAc)$_2$–TMG or PdCl$_2$–TMG, have been developed for the Heck reaction of an olefin with an aryl halide, in which TMG (**1**) acts as a ligand [84e]. In the reaction of iodobenzene with butyl acrylate the turnover numbers were up to 1 000 000. TMG (**1**) was used as a base for the palladium catalysed asymmetric Wagner–Meerwein shift of nonchiral vinylcyclopropane and cyclobutane derivatives leading to asymmetric synthesis of cyclobutanones, cyclopentenones, γ-butyrolactones and δ-valerolactones [85] (Scheme 4.34). Replacement of TMG (**1**) with an inorganic bases such as lithium or cesium carbonate resulted in little effect.

A highly efficient Pd(OA)$_2$/guanidine aqueous system for the room temperature Suzuki cross coupling reaction was developed. The new water-soluble and air-stable catalyst from Pd(OA)$_2$ and 2-butyl-1,1,3,3-tetramethylguanidine was synthesized and characterized by X-ray crystallography [86]. The catalyst catalyses reaction of arylboronic acids

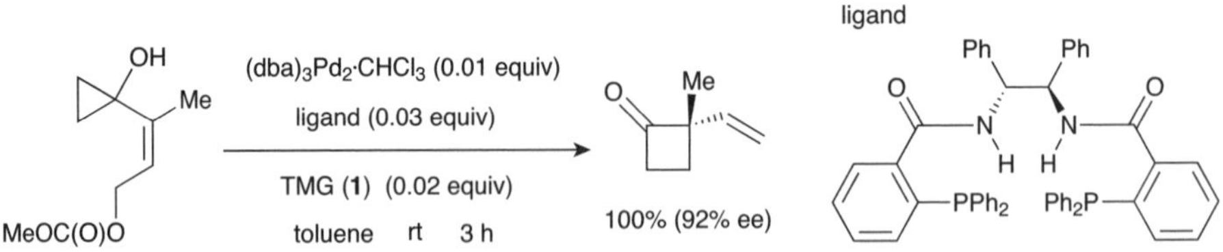

Scheme 4.34 *Asymmetric Wagner–Meerwein shift in the presence of TMG (1)*

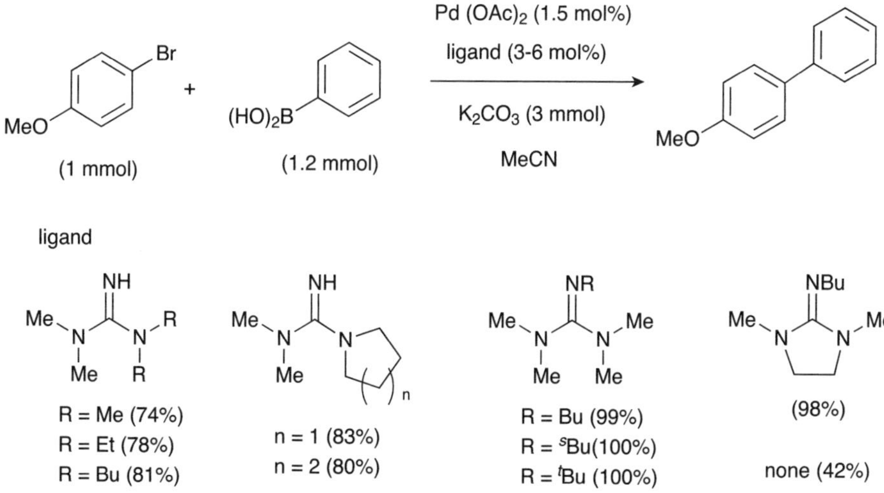

Scheme 4.35 Screening of the guanidine ligands on the Suzuki coupling reaction

with a wide range of aryl halides in aqueous solvent to give the coupling products in good to excellent yields and high turnover numbers. 1-Iodo-4-nitrobenzene was reacted with phenylboronic acid to afford biphenyl in 85% and its turnover number was high (up to 850 000). Results for the preliminary screening of the guanidine ligands in the reaction of 1-bromo-4-methoxybenzene and phenylboronic acid are summarized in Scheme 4.35. Barton's bases were found to be the best ligands, which gave the coupling product in nearly quantitative yield.

The Heck reaction of olefins with aryl halides proceeds successfully in the presence of palladium catalyst supported on TMG (**1**) modified molecular sieves without solvent. The TMG–Pd was found to be much more active and stable than the palladium catalyst without modification with TMG (**1**) [87]. An ionic liquid, tetramethylguanidinium lactate, was used as the TMG source.

Zinc

Chiral zinc-guanidine complexes were designed as possible chiral auxiliaries in organic synthesis [88]. TMG (**1**) was successfully reacted with diethylzinc (Et_2Zn) in a 4 : 3 and a 1 : 1 ratio to yield the corresponding linear $[Zn_3(\mu\text{-}TMG)_4(Et)_2]$ and cyclic $[Zn(\mu\text{-}TMG)(Et)]_3$ complexes. The cyclic complex was further reacted with alcohol and/or phenol to give alternative complexes such as **21**, which can catalyse the ring-opening polymerization of lactide into poly-lactide [89] (Scheme 4.36).

4.3.3.4 Oxidation

Tetraphenylbismuth trifluoroacetate under neutral or slightly acidic conditions phenylates primary alcohol in reasonable yields (65–75%), but gives only moderate yields with secondary alcohols. In contrast, the reaction of bismuth [Bi(V)] reagents with alcohols under basic conditions [BTMG (**2**) or TMG (**1**)] gives, exclusively, oxidation [90].

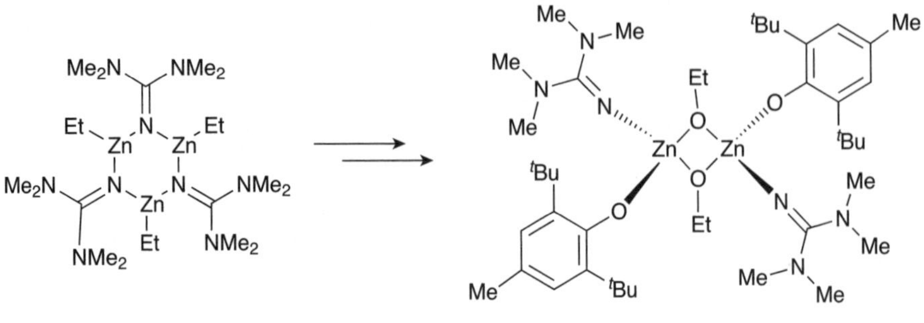

Scheme 4.36 A complex of Et$_2$Zn and TMG (**1**) for polymerization

4.3.3.5 Proton Transfer (Isomerization)

Tricyclic sulfoxides carrying an iminoester function can isomerize to alternative imines in the presence of catalytic amidine or guanidine under mild conditions through [1,3]-proton transfer [91]. Modest asymmetric induction (up to 45% ee) was observed in the use of the Corey's guanidine (**20**) (Table 4.14).

4.3.3.6 Reduction

(5S)-1,3-Diaza-2-imino-3-phenylbicyclo[3.3.0]octane has been successfully employed as a chiral catalytic source for the borane (BH$_3$·SMe$_2$) mediated asymmetric reduction of prochiral phenacyl halides to provide the corresponding secondary alcohols up to 89% yield and 83%ee [92].

Table 4.14 Guanidine catalysed symmetric [1,3]-proton transfer of imines

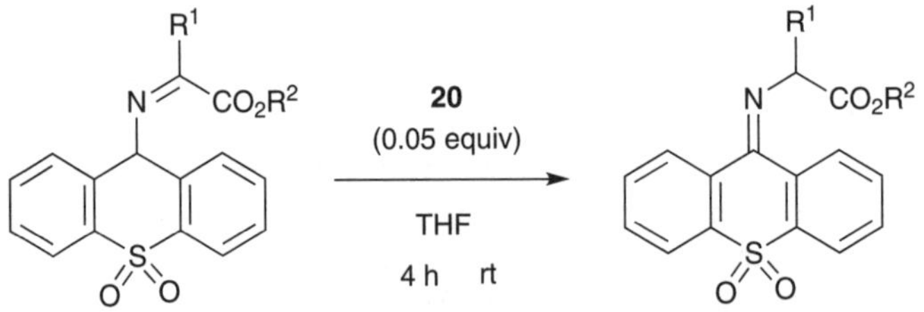

Run	R^1	R^2	Yield (%)	ee (%)
1	Ph	Me	63	43
2	^{i}Pr	Et	95	45
3	Me	Et	98	24
4	(CH$_2$)$_2$CO$_2$Me	Me	98	0

4.3.3.7 Tandem Reaction

Michael and Aldol Reactions

TMG (**1**) is used favourably for tandem reactions of Michael and aldol additions [93,94] (Scheme 4.37). Bicyclic furanopyran [93a] and a nucleotide derivative [93b] were synthesized from the corresponding α,β-unsaturated systems in one-pot reaction. A [3.3.1]bicyclic ring system, leading to huperzine A [94a], a candidate drug for Alzheimer decease, and its spirocyclopropyl derivative [94b], is efficiently constructed by Robinson-type annulation from β-keto ester and methacrolein in the presence of TMG (**1**) [94] (Scheme 4.37c).

Michael and Displacement Reactions

A bicyclo[3.1.0]hexane system was catalytically prepared by cyclopropanation of cyclopentenone through Michael addition followed by displacement using TMG (**1**) as a catalyst in high yield and high diastereoselectivity [95] (Scheme 4.38). TMG (**1**) used was quantitatively recovered as the hydrobromide salt by simple filtration. DBU was found to react less.

4.3.3.8 Vinyl Halide from Hydrazone

TMG (**1**) or BTMG (**2**) participate in the preparation of vinyl halides from hydrazones and halogen molecules [96]. Examples of the two-step synthesis of vinyl halides from ketones through hydrazones are shown in Table 4.15 [96a]. This procedure has been frequently employed in chemical synthesis and often provides alternative access to vinyl iodide difficult to prepare otherwise.

4.4 Guanidinium Salt

As summarized in Figure 4.2, a reactive guanidinium salt, which is produced from guanidine and an appropriate electrophile, plays an important role as an active intermediate for guanidine-participating reactions. In this section, the focus is on an alternative stable guanidinium salt as a possible synthetic tool, which is expected to be useful contribution to organic synthesis.

4.4.1 Guanidinium Ylide

Three-membered nitrogen heterocycles, aziridines, are very important molecules not only as key components of biologically active natural products, but also as reactive synthetic precursors for a wide variety of nitrogen-containing compounds [97]. Among them, aziridine-2-carboxylates are versatile precursors for the synthesis of amino acid derivatives, including unnatural type products [2]. Preparation of aziridine is basically classified into three types of reactions: intramolecular substitution of β-aminoalcohols by nucleophilic nitrogens; addition of carbenes to imines; and addition of nitrenes to olefins. Recently, a new synthetic method was reported for the preparation of 3-arylaziridine-2-carboxylates **25** from guanidinium salts **22** which carry a glycine unit and aryl (including heterocyclic) aldehydes [98]; guanidinium ylide **23** may be formed from **22** under basic conditions and acts as a nucleophile and urea **26**, useable as a

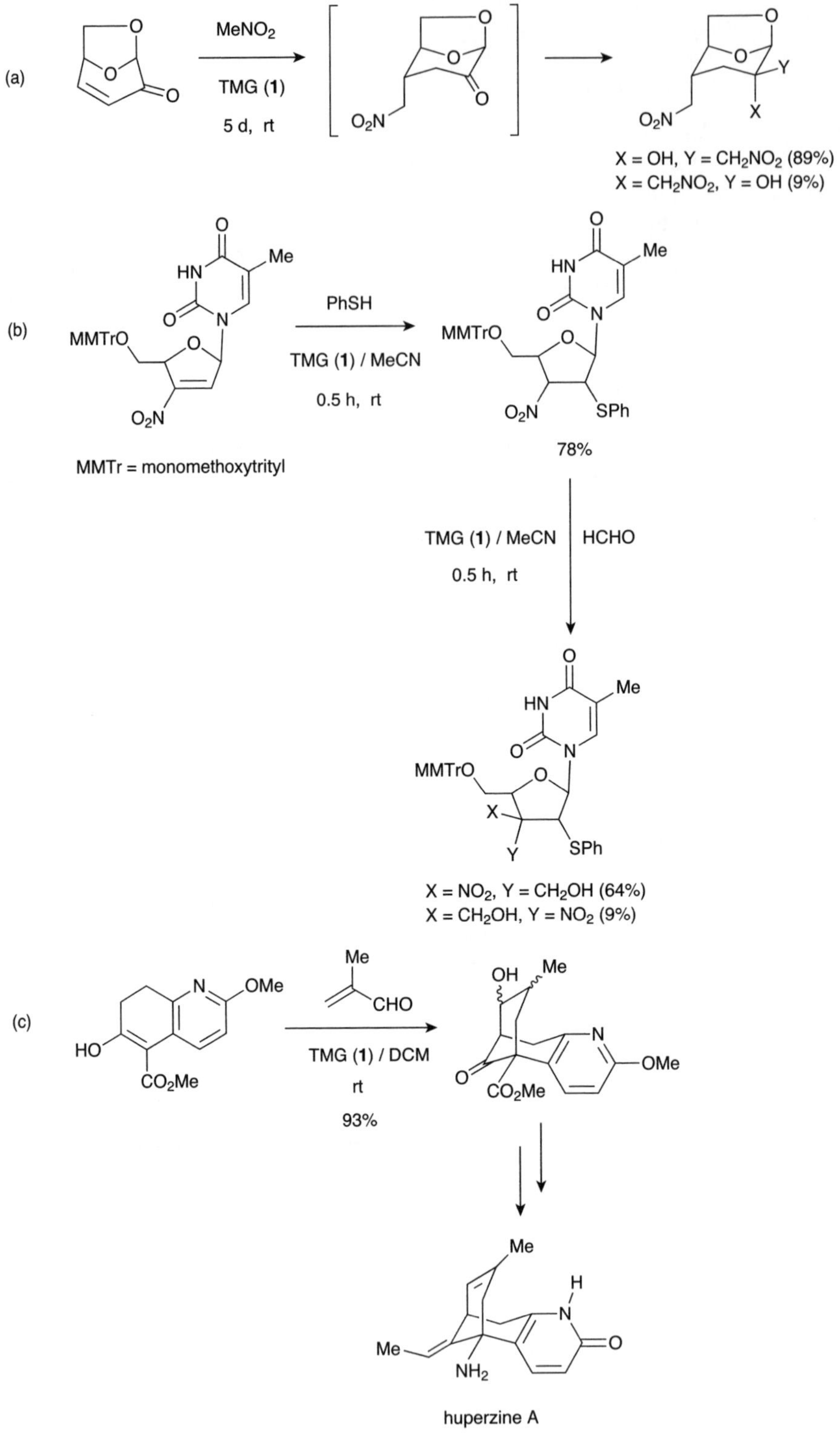

Scheme 4.37 Examples of sequential Michael and aldol reactions

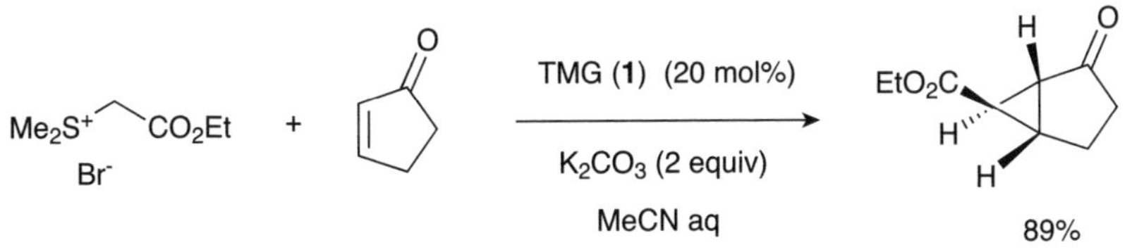

Scheme 4.38 TMG (1) catalysed cyclopropanation through tandem Michael-displacement reactions

Table 4.15 Two-step synthesis of vinyl halides from ketones through hydrazones

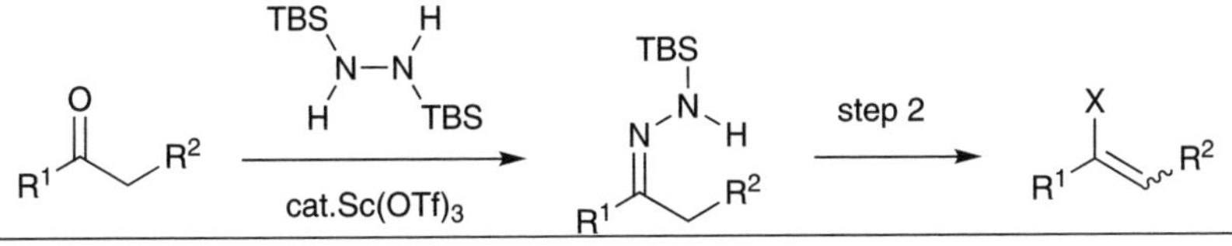

Run	R¹	R²	Step 2	Yield (%)
1	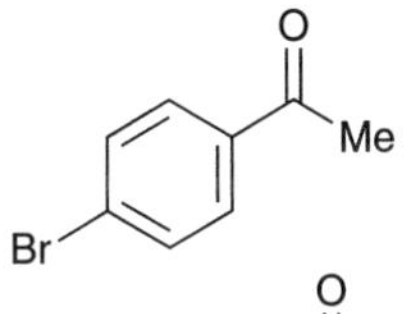 4-Br-C₆H₄-C(O)- (aryl ketone group)	Me	Br$_2$, BTMG (2) DCM, 23 °C	90 (X = Br)
2	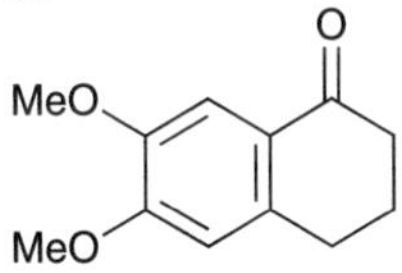 6,7-dimethoxy-tetralone (fused bicyclic)	—	I$_2$, TMG (1) THF, 0 °C	82 (X = I)
3	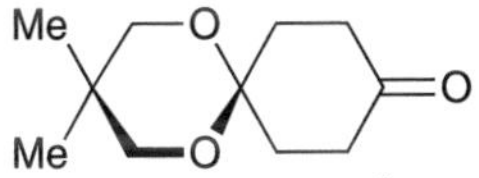 1,3-dioxane-protected cyclohexanone (spiroketal)	—	Br$_2$, BTMG (2) DCM, 23 °C	65 (X = Br)[a]
4	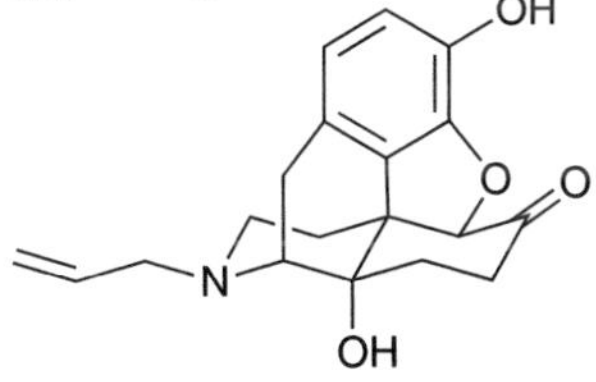 N-allylnoroxymorphone-type alkaloid	—	I$_2$, TMG (1) THF, 0 °C	84 (X = I)[b]
5	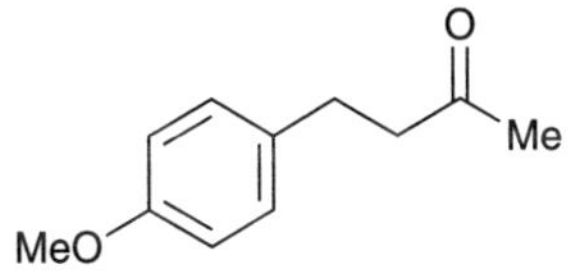 4-MeO-C₆H₄-CH₂-CH₂-	Me	I$_2$, TMG (1) THF, 0 °C	71 (X = I)[c]

[a] gem-Dibromide was formed in 7% yield.
[b] tetrasub.: trisub. = 57:43.
[c] terminal: inner-E: inner-Z = 62:21:17.

synthetic precursor of **22**, is produced as a co-product of aziridines (Scheme 4.39). Introduction of chiral centres into the guanidinium template (L = Ph) results in effective asymmetric induction in the aziridine formation. In this unique cycle, aziridines are generated with excellent to moderate stereoselectivity depending upon the choice of the aryl aldehydes. In general, *trans*-aziridines are efficiently obtained with satisfactory enantioselectivity when aryl aldehydes carrying an electron-donating group (EDG) such as piperonal are used as electrophiles. Based on stereochemical results a cycle mechanism for the asymmetric induction through spiro intermediates **24** has been postulated as shown in Scheme 4.39.

Mechanistic approaches to asymmetric aziridine synthesis have been carried out systematically using a variety of *p*-substituted benzaldehydes (Table 4.16). Two kinds of reaction mechanism, controlled by the nature of the *p*-substituent of aryl aldehydes, are proposed: an S_Ni-like mechanism, via cationic-like transition state for the fragmentation of intermediate adducts to aziridine products (step 2) by intramolecular nucleophilic substitution, when EDG-substituted benzaldehydes are used; and an $S_N 2$-like mechanism, where electron-withdrawing group (EWG) substituted benzaldehydes are used [99].

Reaction of chiral guanidinium ylides with α,β-unsaturated aldehydes also gives α,β-unsaturated aziridine-2-carboxylates in good to moderate yields with the chirality of the ylides effectively transferred to the 2 and 3 positions of the aziridine products (up to 93% de and 98% ee) [100]. The aziridines formed can, easily and stereoselectively, undergo ring-opening reaction with oxygen nucleophiles to afford α-amino-β-hydroxy esters. Thus, *D-erythro*-sphingosine was synthesized in good overall yield with high optical purity starting from both *trans*-(2*R*,3*S*) and *cis*-(2*R*,3*R*)-3-[(*E*)-pentadec-1-enyl]aziridine-2-carboxylates obtained from (*E*)-hexadec-2-enyl aldehyde (Scheme 4.40).

Furthermore, a spiro imidazolidine–oxazolidine intermediate (e.g. **24** in Scheme 4.39) was successfully isolated in the reaction of guanidinium ylide participating aziridination using α-bromocinnamaldehyde, and the stereochemical alignment of new stereogenic centres in the spiro system was unambiguously determined to be of *trans*-configuration by X-ray crystallographic analysis [101]. The proposed role of the spiro compound as an intermediate for the aziridination reaction was established by its smooth chemical conversion into aziridine product.

In ylide chemistry, the Wittig reaction is a well known process; a phosphonium ylide reacts with a carbonyl compound such as an aldehyde or a ketone to give an alkene by C−C bond formation together with phosphine oxide, in which the phosphorus atom of the ylide acts as oxygen acceptor from the carbonyl substrate (Scheme 4.41a). In the reaction of sulfonium ylide with a carbonyl substrate, the product is oxirane in which the oxygen atom comes from the carbonyl substrate (Scheme 4.41b). However, in the guanidinium ylides mechanism, the external nitrogen atom, among three ylide nitrogen atoms, is incorporated into the aziridine and the remaining amidine moiety acts as an oxygen acceptor to be converted to urea (Scheme 4.41c). Thus, these three types of ylide participating reaction with aryl aldehydes afford quite comparable profiles in the product formation.

4.4.2 Ionic Liquid

It is known that ionic liquid shows unique physico-chemical character and has been applied to a wide variety of fields in the chemical industry [102]. The guanidinium salts shown in

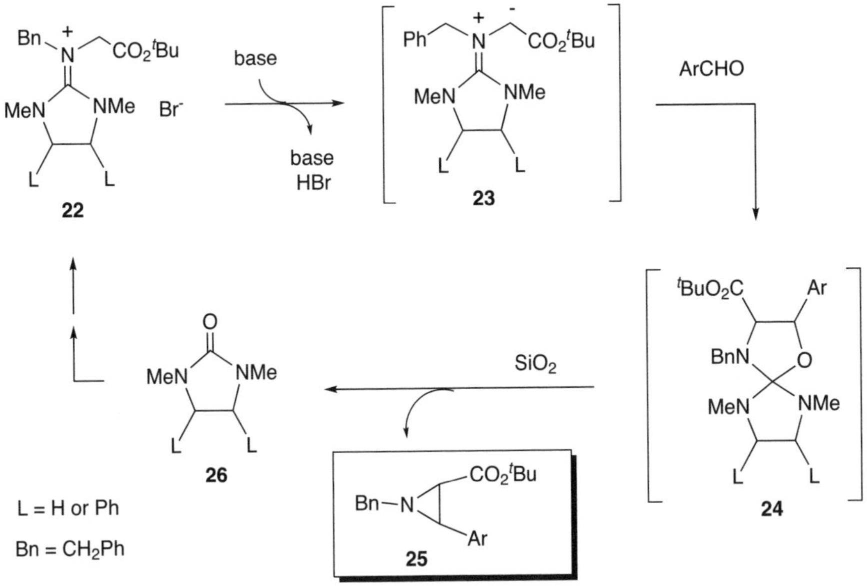

Selected data for asymmetric aziridination with **22** (L = Ph) in the presence of TMG (**1**) as a base

run	22	Ar	time (h)	yield (%) of 25			de (%)
				cis (ee)	trans (ee)	total	
1	(R,R)	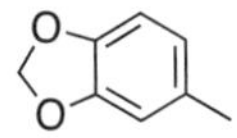 3,4-methylenedioxyphenyl	4	6	82 (97)	88	86
2	(S,S)	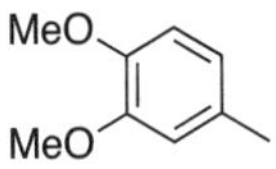 3,4-dimethoxyphenyl	4	8	73 (84)	81	80
3	(S,S)	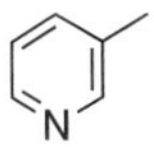 3-pyridyl	4	12	39	51	53
4	(S,S)	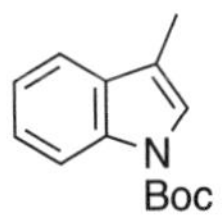 3-(N-Boc-indolyl)	7	6	70 (95)	76	84
5	(S,S)	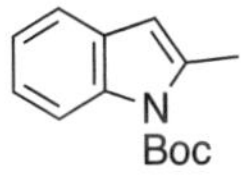 2-(N-Boc-indolyl)	5.5	9	87 (76)	96	82

Scheme 4.39 New synthesis of aziridine **25** from guanidinium salt **22** and aryl aldehydes through spiro intermediates under basic conditions

Table 4.16 Reactions of (S,S)-**22** and various p-substituted benzaldehydes using Ac_2O instead of SiO_2 in step 2

(S)-22 + X—C6H4—CHO →[TMG (1) (1.1 equiv), THF, 25 °C, 5 h (step 1)] [adduct] →[Ac_2O (2.2 equiv), $CHCl_3$, rt (step 2)] Bn–N⟨CO_2^tBu⟩—C6H4—X (**25**) + **26**

Run	X	Step 2, time (h)	25 (%) Cis (ee)	25 (%) Trans (ee)	Trans/cis	Total	26 (%)
1	OBu	0.5	3	64 (92)	96	67	65
2	OMe	0.5	4	77 (91)	95	81	65
3	Me	1.5	45 (90)	31 (93)	41	76	90
4	H	3	58 (86)	22 (88)	28	80	100
5	Cl	17	59 (86)	33 (84)	36	92	not purified
6	CO_2Me	17.5	52 (79)	28 (72)	35	80	69

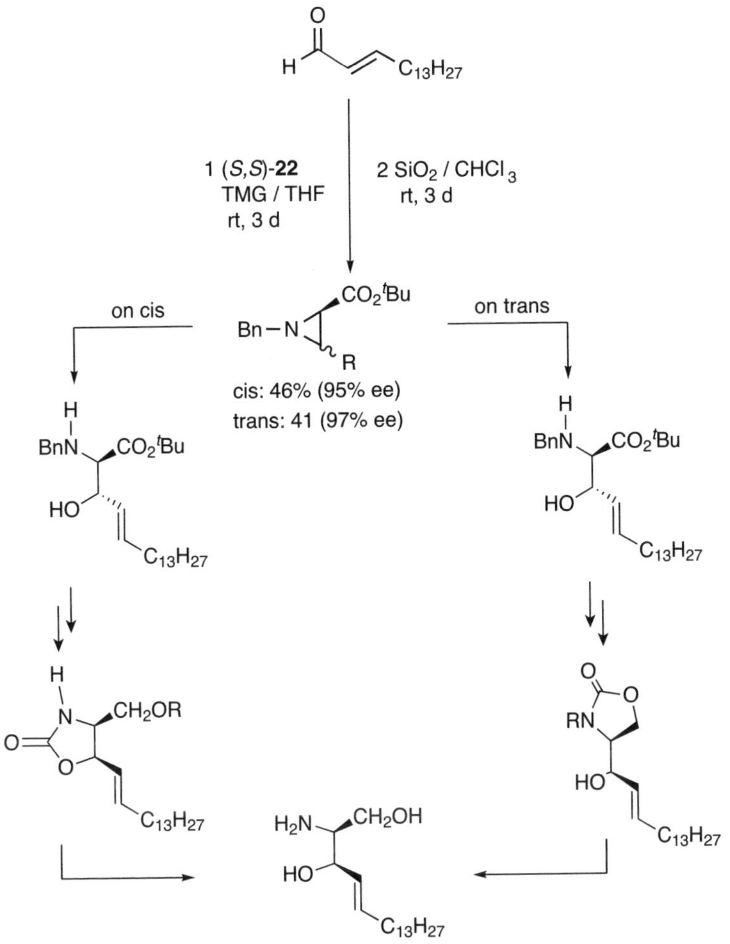

Scheme 4.40 *Asymmetric synthesis of sphingsine from cis- and trans-aziridines obtained by guanidinium ylide participating aziridination*

Figure 4.7 have been nominated as guanidine-type ionic liquids [103] and used in hydrogenation [103a,103d,103f], hydroformylation [103b], the aldol reaction [103c, 103e] and the palladium catalysed Heck reaction [103g]. In the last reaction, 2-butyl-1,1,3,3-tetremethylguanidinium acetate (**27**) plays multiple roles in the reaction, such as solvent, a strong base to facilitate β-elimination and a ligand to stabilize activated palladium species.

4.4.3 Tetramethylguanidinium Azide (TMGA)

In 1966 tetramethylguanidinium azide (TMGA) (**28**) was prepared as an hydroscopic colourless, but stable, crystal by treatment of TMG (**1**) with hydrogen azide and was introduced by Papa [104] as a reactive azidation reagent (Table 4.17).

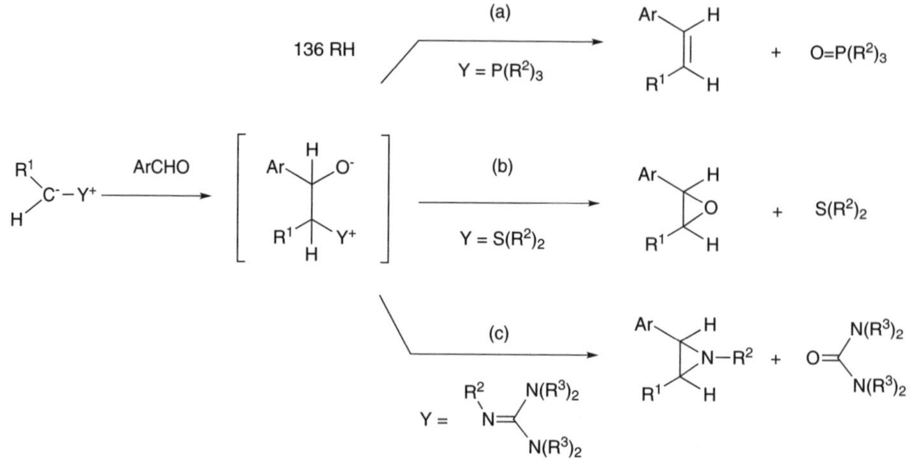

Scheme 4.41 *Schematic reaction profiles of phosphonium, sulfonium and guanidinium ylides with aryl aldehydes*

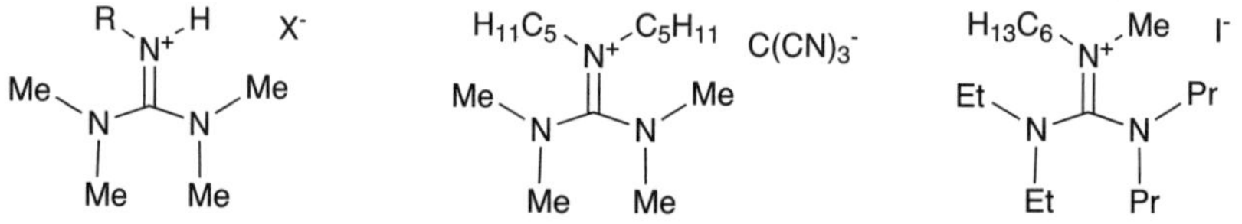

R = H, X = MeCH(OH)CH$_2$CO$_2$
R = H, X = MeCO$_2$
R = H, X = CF$_3$CO$_2$
R = Bu, X = MeCO$_2$ (**27**)

Figure 4.7 *Structures of guanidine-derived ionic liquids*

Table 4.17 *Azidation of alkyl halides with TMGA (**28**)*

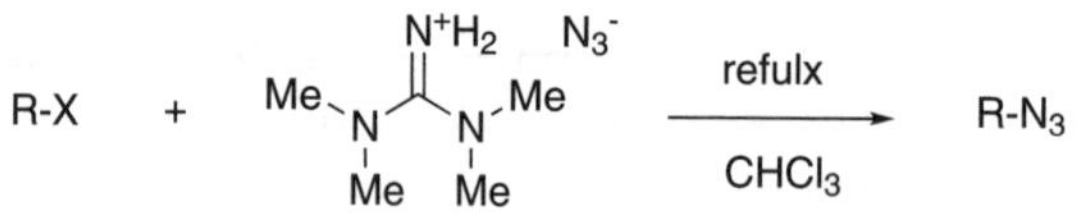

Run	R	X	Time (min)	Yield (%)
1	Ph(CH$_2$)$_2$	Br	90	100
2	PhCH$_2$	Cl	60	91
3	Me(CH$_2$)$_5$	Cl	240	60
4	EtO$_2$CCH$_2$	Cl	60	89
5	Ph$_2$CH	Br	60	81

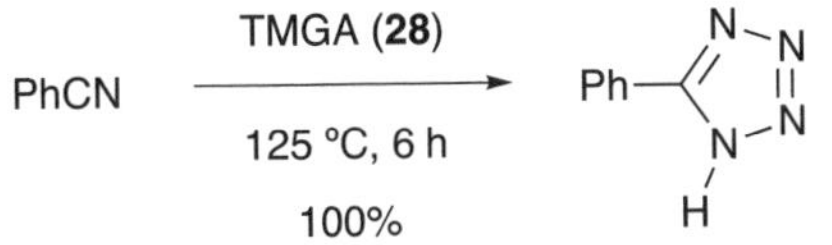

Scheme 4.42 Preparation of 5-phenyltetrazole

On the other hand, Papa reported that TMGA (**28**) could work as 1,3-dipolarophile in the tetrazole synthesis [104]. 5-Phenyltetrazole is quantitatively given by heating benzonitrile with **28** at 125 °C for six hours without solvent (Scheme 4.42).

The reaction of 1-arylsulfonylbicyclobutanes with TMGA (**28**) in N-methyl-2-pyrrolidone (NMP) or DMF at 80–90 °C regiospecifically produced cyclobutanes as a separable mixture of *cis* and *trans* isomers in high yield. Selected results are given in Table 4.18 [105].

Reactions using various azidation reagents were compared, among which TMGA (**28**) was found to be the most effective reagent for a desired displacement reaction [106] (Table 4.19).

TMGA (**28**) will convert to tetramethylguanidinium salt after completion of the azidation reaction, offering easy isolation of azide product from the reaction mixture by addition of diethyl ether. TMGA (**28**) is commercially available, nontoxic and safe to handle [107]. Thus, TMGA (**28**) participating azidation has been established as a widely applicable and simple operating method for the introduction of a nitrogen source in a variety of organic syntheses [108].

Glycosyl halides are reacted with **28** to afford the corresponding glycosyl azides in quantitative yields; complete inversion at the anomeric centre is observed in this reaction [107] (Scheme 4.43).

β-Functionalized vinyl azides are prepared through addition of TMGA (**28**) to acetylenic and allenic compounds [109]. Examples for the preparation of β-azido vinyl esters are given in Table 4.20.

Table 4.18 TMGA (**28**) mediated regiospecific addition of hydrazic acid to arylsulfonylbicyclobutanes

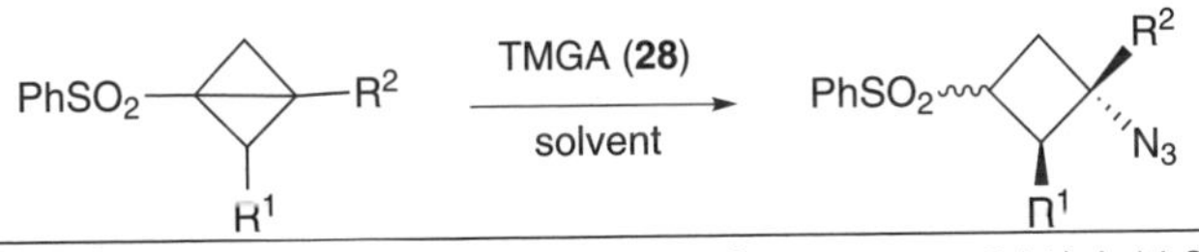

Run	R^1	R^2	Solvent	Time (h)	*trans*-S,N/*cis*-N,S	Yield (%)
1	H	H	NMP	1	1.2	96
2	H	Me	NMP	2	0.32	89
3	H	CO$_2$Me	DMF	2	1.5	86
4	H	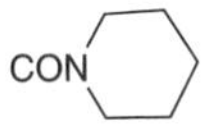 CON(C$_5$H$_{10}$)	NMP	3	5.6	95
5	Me	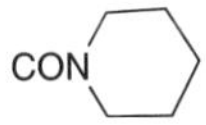 CON(C$_5$H$_{10}$)	NMP	20	2.0	93

Table 4.19 Azidation under various conditions

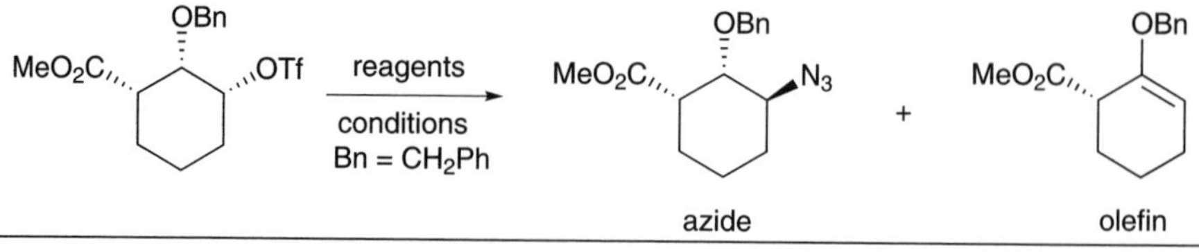

Run	Reagent/solvent	Yield (%) Azide	Yield (%) Olefin
1	NaN$_3$/DMF	56	14
2	NaN$_3$/DMF-MeCN	33	7
3	Zn(N$_3$)$_2$(pyridine)$_2$/DMF	no reaction	
4	Bn$_4$N$^+$N$_3^-$/PhH	0	60
5	TMSN$_3$/MeCN	complex mixture	
6	TMGA (**28**)/DCM	62	7.5

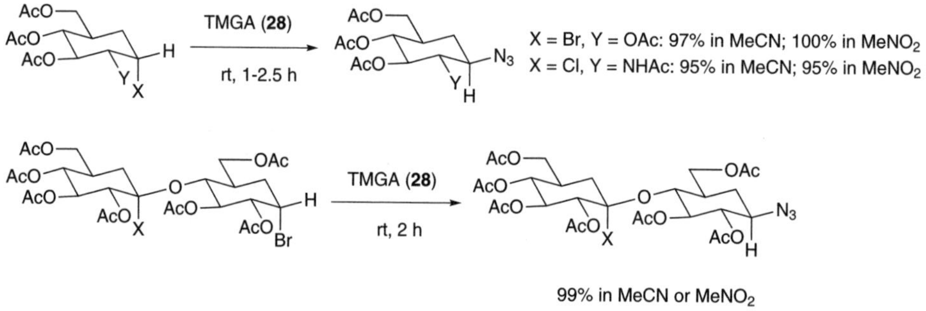

Scheme 4.43 TMG (**28**) mediated azidation of glycosyl halides

Table 4.20 TMGA (**28**) addition to acetylenic esters

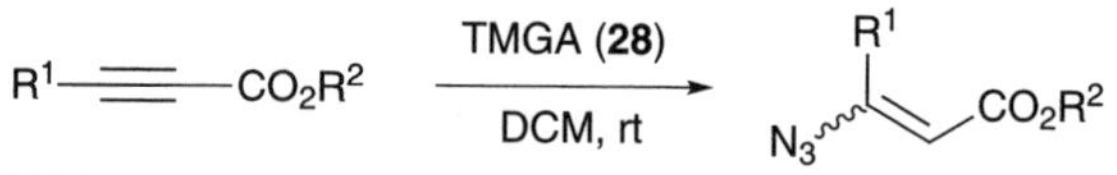

Run	R^1	R^2	Time (h)	E/Z	Yield (%)
1	H	Et	36	70^a/30	54
2	MeCO$_2$	Me	72	0/100	59
3	Ph	Et	72	100/0	85
4	Me	Me	24	70/30	46

^{a} (E)-Ethyl 3-[1-(4-ethoxycarbonyl)-1,2,3-triazoryl]acrylate was formed as by-product after the 1,3-dipolar cycloaddition of (E)-vinyl azide with ethyl propiolate.

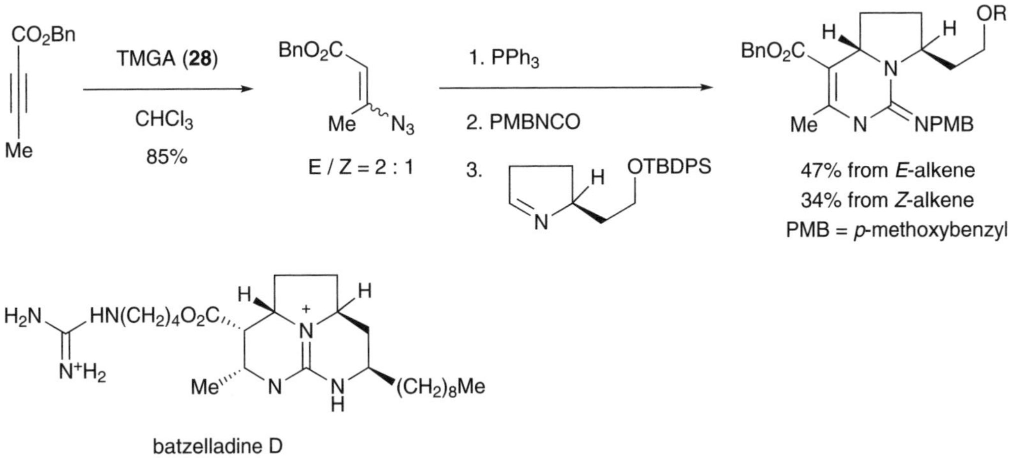

Scheme 4.44 Application of TMGA (28) addition to acetylenic ester to batzelladine synthesis

2-Butynoate was reacted with TMGA (28) to give β-azido acrylate as a mixture of geometric isomers. The 1,4-conjugate addition product served as the diene component in the [4 + 2] annulation reaction after conversion to vinyl carbodiimides by Staudinger-aza-Wittig condensation with benzylisocyanate [110]. These reactions are key steps in the total synthesis of batzelladine alkaloids (Scheme 4.44).

For the preparation of acceptor-substituted propargyl azide through substitution of the propargyl precursors bearing a leaving group, the reaction of halide with TMGA (28) resulted in the vinyl azide instead of a desired azide, because the presence of the acceptor substituent causes prototropic isomerization under basic conditions [111]. Expected displacement was observed when sulfur containing propargyl azide was subjected to the reaction, and the following oxidation with m-chloroperbenzoic acid (mCPBA) gave the sulfonyl substituted propargyl azide even in low yield (Scheme 4.45).

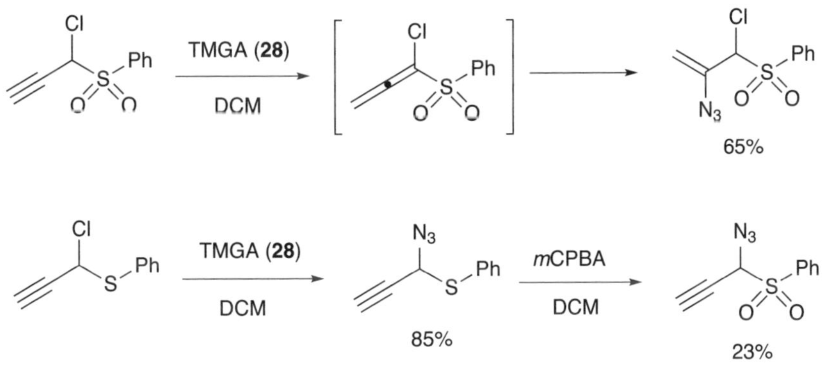

Scheme 4.45 Reaction of propargyl chloride with TMGA (28)

4.5 Concluding Remarks

Guanidines attract much attention as unique synthetic tools with multiple functions in organic synthesis, such as organosuperbase catalysts, metal ligands and templates for ionic liquids. The origin of their functionalities is the conjugated three-nitrogen system of the guanidinyl moiety, and the easy structural modification can allow the design of newly functionalized compounds with diverse potential. Thus, it can be reasonably expected that more extensive studies on guanidine chemistry in the future should result in progressing organic chemistry.

References

1. Costa, M., Chiusoli, G.P., Taffurelli, D. and Dalmonego, G. (1998) Superbase catalysis of oxazolidin-2-one ring formation from carbon dioxide and prop-2-yn-1-amines under homogeneous or heterogeneous conditions. *Journal of the Chemical Society – Perkin Transactions 1*, 1541–1546; Kovacevic, B. and Maksic, Z.B. (2001) Basicity of some organic superbases in acetonitrile. *Organic Letters*, **3**, 1523–1526.
2. Yamamoto, Y. and Kojima, S. (1991) Synthesis and chemistry of guanidine derivatives, in *The Chemistry of Amidines and Imidates*, **Vol. 2** (eds S. Patai and Z. Rappoport), John Wiley & Sons, Inc., Chichester.
3. Barton, D.H.R., Elliott, J.D. and Gero, S.D. (1981) The synthesis and properties of a series of strong but hindered organic bases. *Journal of the Chemical Society – Chemical Communications*, 1136–1137; Barton, D.H.R., Elliott, J.D. and Gero, S.D. (1982) Synthesis and properties of a series of sterically hindered guanidine bases. *Journal of the Chemical Society – Perkin Transactions 1*, 2085–2090.
4. Schwesinger, R. (1985) Extremely strong, non-ionic bases: syntheses and applications. *Chimia*, **39**, 269–272.
5. Ishikawa, T. and Isobe, T. (2002) Modified guanidine as chiral auxiliaries. *Chemistry – A European Journal*, **8**, 552–557.
6. Macquarrie, D.J., Modoe, J.E.G., Brunel, D. et al. (2001) Guanidine catalysts supported on silica and micelle templated silicas. New basic catalysts for organic chemistry. *Royal Society of Chemistry*, **266**, 196–202.
7. Berlinck, R.G.S., Kossuga, M.H. and Nascimento, A.M. (2005) Guanidine derivatives in *Science of Synthesis*, **Vol. 18**, George Thieme Verlag, pp. 1077–1116.
8. Schenck, M. (1912) Methylated guanidines. *Z. Physiol. Chem.*, **77**, 328–393.
9. Schenck, M. and v. Graevenitz, F. (1924) The preparation of the tetramethylguanidines. *Z. Physiol. Chem.*, **141**, 132–145.
10. Chinchilla, R., Najera, C. and Sanchez-Agullo, P. (1994) Enantiomerically pure guanidine-catalysed asymmetric nitroaldol reaction. *Tetrahedron: Asymmetry*, **5**, 1393–1402; Ma, D. and Cheng, K. (1999) Enatioselective synthesis of functionalized α-amino acids via a chiral guanidine-catalysed Michael addition reaction. *Tetrahedron: Asymmetry*, **10**, 713.
11. Kohn, U., Klopfleisch, M., Gorls, H. and Anders, E. (2006) Synthesis of hindered chiral guanidine bases starting from (S)-(N,N-dialkylaminomethyl)pyrrolidines and BrCN. *Tetrahedron: Asymmetry*, **17**, 811–818.
12. Cunha, S. and Rodrígues, M.T., Jr (2006) The first bithmuth(III)-catalysed guanylation of thioureas. *Tetrahedron Letters*, **47**, 6955–6956.
13. Zhang, W.-X., Nishiura, M. and Hou, Z. (2007) Catalytic addition of amine N–H bonds to carbodiimides by half-sandwich rare earth metal complexes: efficient synthesis of substituted guanidines through amine protonolysis of rare earth metal guanidinates. *Chemistry – A European Journal*, **13**, 4037–4051.
14. Iyer, M.S., Gigstad, K.M., Namdev, N.D. and Lipton, M. (1996) Asymmetrtic catalysis of the Strecker amino acid synthesis by a cyclic dipeptide. *Journal of the American Chemical Society*, **118**, 4910.

15. Gers, T., Kunce, D., Markowski, P. and Izdebski, J. (2004) Reagents for efficient conversion of amines to protected guanidines. *Synthesis*, 37–42.
16. Feichtinger, K., Zapf, C., Sings, H.L. and Goodman, M. (1998) Diprotected triflylguanidines: a new class of guanidinylation reagents. *The Journal of Organic Chemistry*, **63**, 3804–3805; Zapf, C.W. and Goodman, M. (2003) Synthesis of 2-amino-4-pyridinones from resin-bound guanidines prepared using bis(allyloxycarbonyl)-protected triflylguanidines. *The Journal of Organic Chemistry*, **68**, 10092–10097.
17. Zhang, Y. and Kennan, A.J. (2001) Efficient introduction of protected guanidines in Boc solid phase peptide synthesis. *Organic Letters*, **3**, 2341–2344; Castillo-Melendez, J.A. and Golding, B.T. (2004) Optimization of the synthesis of guanidines from amines via nitroguanidines using 3,5-dimethyl-*N*-nitro-1*H*-pyrazole-1-carboxamidine. *Synthesis*, 1655–1663.
18. Musiol, H.-J. and Moroder, L. (2001) *N,N*-Di-tert-butoxycarbonyl-1*H*-benzotriazole-1-carboxamidine derivatives are highly reactive guanidinylating reagents. *Organic Letters*, **3**, 3859–3861; Zahariev, S., Guarrnaccia, C., Lamba, D. *et al.* (2004) Solvent-free synthesis of azole carboximidamides. *Tetrahedron Letters*, **45**, 9423–9426.
19. Zapf, C.W., Creighton, C.J., Tomioka, M. and Goodman, M. (2001) A novel traceless resin-bound guanidinylating reagent for secondary amines to prepare *N,N*-disubstituted guanidines. *Organic Letters*, **3**, 1133–1136.
20. Feichtinger, K., Sings, H.L., Baker, T.J. *et al.* (1998) Triurethane-protected guanidines and triflylurethane-protected guanidines: new reagents for guanidinylation reactions. *The Journal of Organic Chemistry*, **63**, 8432–8439.
21. Davis, A.P. and Dempsey, K.J. (1995) Synthesis and investigation of a hindered, chiral, bicylic guanidine. *Tetrahedron: Asymmetry*, **6**, 2829–2840; Corey, E.J. and Grogan, M.J. (1999) Enantioselective synthesis of α-amino nitriles from *N*-benzhydryl imines and HCN with a chiral bicyclic guanidine as catalyst. *Organic Letters*, **1**, 157–160.
22. Echavarren, A., Galán, A., de Mendoza, J. *et al.* (1988) Anion-receptor molecules: synthesis of a chiral and functionalized binding subunit, a bicyclic guanidinium group derived from L- or D-asparagine. *Helvetica Chimica Acta*, **71**, 685–693; Echavarren, A., Galán, A., Lehn, J.-M. and de Mendoza, J. (1989) Chiral recognition of aromatic carboxylate anions by an optically active abiotic receptor containing a rigid guanidinium binding subunit. *Journal of the American Chemical Society*, **111**, 4994–4995; Kurzmeier, H. and Schmidtchen, F.P. (1990) Abiotic anion receptor functions. A facile and dependable access to chiral guanidinium anchor groups. *The Journal of Organic Chemistry*, **55**, 3749–3755; Münster, I., Rolle, U., Madder, A. and Clereq, P.J.D. (1995) Synthesis of a chiral di(hydroxyalkyl) substituted bicyclic guanidine. *Tetrahedron: Asymmetry*, **6**, 2673–2674.
23. Nagasawa, K., Georgieva, A., Takahashi, H. and Nakata, T. (2001) Acceleration of hetero-Michael reaction by symmetrical pentacyclic guanidines. *Tetrahedron*, **57**, 8959–8964.
24. Howard-Jones, A., Murphy, P.J., Thomas, D.A. and Caulkett, P.W.R. (1999) Synthesis of a novel C_2-symmetric guanidine base. *The Journal of Organic Chemistry*, **64**, 1039–1041; Kita, T., Georgieva, A., Hashimoto, Y. *et al.* (2002) C_2-Symmetric chiral pentacyclic guanidine: a phase-transfer catalyst for the asymmetric alkylation of *tert*-butyl glycinate Schiff base. *Angewandte Chemie – International Edition*, **41**, 2832–2834; Allingham, M.T., Howard-Jones, A., Murphy, P.J. *et al.* (2003) Synthesis and application of C_2-symmetric guanidine bases. *Tetrahedron Letters*, **44**, 8677–8680.
25. Ye, W., Leaw, D., Goh, S.L.M. *et al.* (2006) Chiral bicyclic guanidines: a concise and efficient aziridine-based synthesis. *Tetrahedron Letters*, **47**, 1007–1010.
26. Isobe, T. and Ishikawa, T. (1999) 2-Chloro-1,3-dimethylimidazolinium chloride. 3. Utility for chlorination, oxidation, reduction, and rearrangement reactions. *The Journal of Organic Chemistry*, **64**, 5832–5835; Isobe, T. and Ishikawa, T. (1999) 2-Chloro-1,3-dimethylimidazolinium chloride. 1. A powerful dehydrating equivalent to DCC. *The Journal of Organic Chemistry*, **64**, 6984–6988; Isobe, T. and Ishikawa, T. (1999) 2-Chloro-1,3-dimethylimidazolinium chloride. 2. Its application to the construction of heterocycles through dehydration reactions. *The Journal of Organic Chemistry*, **64**, 6989–6992.
27. Isobe, T., Fukuda, K. and Ishikawa, T. (2000) Modified guanidines as potential chiral superbases. 1. Preparation of 1,3-disubstituted 2-iminoimidazolidines and the related guanidines through chloroamidine derivatives. *The Journal of Organic Chemistry*, **65**, 7770–7773;

Isobe, T., Fukuda, K., Tokunaga, T. *et al.* (2000) Modified guanidines as potential chiral superbases. 2. Preparation of 1,3-unsubstituted and 1-substituted 2-iminoimidazolidine derivatives and a related guanidine by the 2-chloro-1,3-dimethylimidazolinium chloride-induced cyclization of thioureas. *The Journal of Organic Chemistry*, **65**, 7774–7778; Isobe, T., Fukuda, K., Yamaguchi, K. *et al.* (2000) Modified guanidines as potential chiral superbases. 3. Preparation of 1,4,6-triazabicyclooctene systems and 1,4-disubstitued 2-iminoimidazolidines by the 2-chloro-1,3-dimethylimidazolinium chloride-induced cyclization of guanidines with a hydroxyethyl substituent. *The Journal of Organic Chemistry*, **65**, 7779–7785.

28. Simoni, D., Invidiata, F.P., Manferdini, M. *et al.* (1998) Tetramethylguanidine (TMG)-catalysed addition of dialkyl phosphites to α,β-unsaturated carbonyl compounds, alkenenitriles, aldehydes, ketones, and imines. *Tetrahedron Letters*, **39**, 7615–7618.
29. Simoni, D., Rondanin, R., Morini, M. *et al.* (2000) 1,5,7-Triazabicyclo[4.4.0]dec-1-ene (TBD), 7-methyl-TBD (MTBD) and polymer-supported TBD (P-TBD): three efficient catalysts for the nitroaldol (Henry) reaction and for the addition of dialkyl phosphites to unsaturated systems. *Tetrahedron Letters*, **41**, 1607–1610.
30. Reddy, M.V.N., Kumar, B.S., Balakrishna, A. *et al.* (2007) One-pot synthesis of novel α-amino phosphonates using tetramethylguanidine as a catalyst. *Arkivoc*, 246–254.
31. Bernardi, L., Bonini, B.F., Capito, E. *et al.* (2004) Organocatalysed solvent-free aza-Henry reactions: a breakthrough in the one-pot syntehsis of 1,2-diamine. *The Journal of Organic Chemistry*, **69**, 8168–8171.
32. Pahadi, N.K., Ube, H. and Terada, M. (2007) Aza-Henry reaction of ketimines catalysed by guanidine and phosphazene bases. *Tetrahedron Letters*, **48**, 8700–8703.
33. Leadbeater, N.E. and van der Pol, C. (2001) Development of catalysts for the Baylis–Hillman reaction: the application of tetramethylguanidine and attempts to use a supported analogue. *Journal of the Chemical Society – Perkin Transactions 1*, 2831–2835; Graunger, R.S., Leadbeater, N.E. and Pamies, M.A. (2002) The tetramethylguanidine catalysed Baylis–Hillman reaction: effects of co-catalysts and alcohol solvents on reaction rate. *Catalysis Communications*, **3**, 449–452.
34. Cooper, D.M., Grigg, R., Hargreaves, S. *et al.* (1995) X = Y-ZH compounds as potential 1,3-dipoles. Part 44. Asymmetric 1,3-dipolar cycloaddition reaction of imines and chiral cyclic dipolarophiles. *Tetrahedron*, **51**, 7791–7808.
35. Shen, J., Nguyen, T.T., Goh, Y.-P. *et al.* (2006) Chiral bicyclic guanidine-catalysed enantioselective reactions of anthrones. *Journal of the American Chemical Society*, **128**, 13692–13693.
36. Wang, L., Huang, X., Jiang, J. *et al.* (2006) Catalytic cyanosilylation of ketones using organic catalyst 1,1,3,3-tetramethylguanidine. *Tetrahedron Letters*, **47**, 1581–1584.
37. Kitani, Y., Kumamoto, T., Isobe, T. *et al.* (2005) Guanidine-catalysed asymmetric trimethylsilylcyanation of carbonyl compounds. *Advanced Synthesis and Catalysis*, **347**, 1653–1658.
38. Ahmad, S.M., Braddock, D.C., Cansell, G. and Hermitage, S.A. (2007) Dimethylformamide, dimethylacetamide and tetramethylguanidine as nucleophilic organocatalysts for the transfer of electrophilic bromine from N-bromosuccinimide to alkenes. *Tetrahedron Letters*, **48**, 915–918.
39. Van Aken, E., Wynberg, H. and Van Bolhuis, F. (1992) Nitroalkanes in C–C bond forming reactions: a crystal structure of a complex of a guanidine catalyst and a nitroalkane substrate. *Journal of the Chemical Society – Chemical Communications*, 629–630; Van Aken, E., Wynberg, H. and Van Bolhuis, F. (1993) Nitroalkanes in C–C bond forming reactions: a crystal structure of a complex of a guanidine catalyst and a nitroalkane substrate. *Acta Chemica Scandinavica*, **47**, 122–124.
40. Simoni, D., Invidiata, F.P., Manfredini, S. *et al.* (1997) Facile synthesis of 2-nitroalkenols by tetramethylguanidines (TMG)-catalysed addition of primary nitroalkanes to aldehydes and acyclic ketones. *Tetrahedron Letters*, **38**, 2749–2752.
41. Hossain, N. and Herdewijn, P. (1998) Facile synthesis of $3'$-C-branched 1,5-anhydrohexitol nucleosides. *Nucleosides & Nucleotides*, **17**, 1781–1786; Hossain, N., van Halbeek, H., De Clercq, E. and Herdewijn, P. (1998) Synthesis of $3'$-C-branched $1',5'$-anhydromannitol nucleosides as new antiherpes agents. *Tetrahedron*, **54**, 2209–2226; Ishikawa, T., Shimizu,

Y., Kudoh, T. and Saito, S. (2003) Conversion of D-glucose to cyclitol with hydroxymethyl substituent via intramolecular silyl nitronate cycloaddition reaction: application to total synthesis of (+)-cyclophellitol. *Organic Letters*, **5**, 3879–3882.
42. Ma, D., Pan, Q. and Han, F. (2002) Diastereoselective Henry reactions of *N,N*-dibenzyl α-amino aldehydes with nitromethane catalysed by enantiopure guanidines. *Tetrahedron Letters*, **43**, 9401–9403.
43. Sohtome, Y., Takemure, N., Takada, K. *et al.* (2007) Organocatalytic asymmetric nitroaldol reaction: cooperative effects of guanidine and thiourea functional groups. *Chemistry, an Asian Journal*, **2**, 1150–1160; Sohtome, T., Hashimoto, Y. and Nagasawa, K. (2005) Guanidine-thiourea bifunctional organocatalyst for the asymmetric Henry (nitroaldol) reaction. *Advanced Synthesis and Catalysis*, **347**, 1643–1648; Sohtome, T., Takemura, N., Iguchi, T. *et al.* (2006) Diastereoselective Henry reaction catalysed by guanidine-thiourea bifunctional organocatalyst. *Synlett*, 144–146.
44. Terade, M., Nakano, M. and Ube, H. (2006) Axially chiral guanidine as highly active and enantioselelctive catalyst for electrophilic amination of unsymmetrically substituted 1,3-dicarbonyl compounds. *Journal of the American Chemical Society*, **128**, 16044–16045.
45. Nysted, L.N. and Burtner, R.R. (1962) Steroidal aldosterone antagonists. VI. *The Journal of Organic Chemistry*, **27**, 3175–3177.
46. Pollini, G.P., Barco, A. and De Giuli, G. (1972) Tetramethylguanidine-catalysed addition of nitromethane to α, β-unsaturated carboxylic acid esters. *Synthesis*, 44–45; Alvarez, F.S. and Wren, D. (1973) Synthesis of (±)-deoxyprostaglandin E_1, F_α and $F_{1\beta}$ and its 15β-epimers by conjugate addition of nitromethane to 2-(6′-carbomethoxyhexyl)-2-cyclopenten-1-one. *Tetrahedron Letters*, 569–572; Hewson, A.T. and MacPherson, D.T. (1983) Conjugate addition to the ethylene ketal of 2-carbomethoxy-2-cyclopentenone a synthesis of sarkomycin. *Tetrahedron Letters*, **24**, 647–648; Naito, T., Honda, Y., Bhavakul, V. *et al.* (1997) Radical cyclization in heterocycle synthesis. II. Total synthesis of (±)-anantine and (±)-isoanantine. *Chemical & Pharmaceutical Bulletin*, **45**, 1932–1939.
47. Barton, D.H.R., Swift, K.A.D. and Tachdjien, C. (1995) The synthesis of 1-hydroxy-3-phenylsulfonylpiperidine-2-thione derivatives utilizing methylthiocarbonylation as the key original step. *Tetrahedron*, **51**, 1887–1892.
48. Ye, W., Xu, J., Tan, C.-T. and Tan, C.-H. (2005) 1,5,7-Triazabicyclo[4.4.0]dec-5-ene (TBD) catalysed Michael reactions. *Tetrahedron Letters*, **46**, 6875–6878.
49. Henderson, D.A., Collier, P.N., Pave, G. *et al.* (2006) Studies on the total synthesis of lactonamycin: construction of model ABCD ring systems. *The Journal of Organic Chemistry*, **71**, 2434–2444.
50. Alcazar, V., Moran, J.R. and de Mendoza, J. (1995) Guanidinium catalysed conjugate addition of pyrrolidine to unsaturated lactones. *Tetrahedron Letters*, **36**, 3941–3944; Martin-Portugues, M., Alcazar, V., Prados, P. and de Mendoza, J. (2002) Guanidinium-catalysed addition of pyrrolidine to 2-(5*H*)-furanone. *Tetrahedron*, **58**, 2951–2955; For the preparation of the related guanidines, see: Echavarren, A., Galan, A., de Mendoza, J. and Salmeron, A. (1988) Anion-receptor molecules: synthesis of a chiral and functionalized binding subunit, a bicyclic guanidinium groups derived from L- or D-asparagine. *Helvetica Chimica Acta*, **71**, 685–693; Gleich, A. and Schmidtchen, F.P. (1990) Artificial molecular anion hosts. Synthesis of a chiral bicyclic guanidinium salt serving as a functionalized anchor group for oxo anions. *Chemische Berichte*, **123**, 907–915.
51. Ishikawa, T., Araki, Y., Kumamoto, T. *et al.* (2001) Modified guanidines as chiral supedrbases: application to asymmetric Michael reaction of glycine imine with acrylate or its related compounds. *Chemical Communications*, 245–246.
52. Ryuda, A., Yajima, N., Haga, T. *et al.* (2008) (±)-1,2-Bis(2-methylphenyl)ethylene-1,2-diamine as a chiral framework for 2-iminoimidazolidine with 2-methylphenyl pendant and the guanidine-catalysed asymmetric Michael reaction of *tert*-butyl diphenyliminoacetate and ethyl acrylate. *The Journal of Organic Chemistry*, **73**, 133–141.
53. Kumamoto, T., Ebine, K., Endo, M. *et al.* (2005) Guanidine-catalysed asymmetric addition reactions: Michael reaction of cyclopentenone with dibenzyl malonates and epoxidation of chalcone. *Heterocycles*, **66**, 347–359.

54. Terada, M., Ube, H. and Yaguchi, Y. (2006) Axially chiral guanidine as enantioselective base catalyst for 1,4-addition reaction of 1,3-dicarbonyl compounds with conjugated nitroalkenes. *Journal of the American Chemical Society*, **128**, 1454–1455.
55. Ye, W., Jiang, Z., Zhao, Y. *et al.* (2007) Chiral bicyclic guanidine as a versatile Brönsted base catalyst for the enantioselective Michael reactions of dithiomalonates and β-keto thioesters. *Advanced Synthesis and Catalysis*, **349**, 2454–2458.
56. DeOliveira, E., Torres, J.D., Silve, C.C. *et al.* (2006) Tetramethylguanidine covalently bonded onto silica gel as catalyst for the addition of nitromethane to cyclopentenone. *Journal of the Brazilian Chemical Society*, **17**, 994–999.
57. Saito, N., Ryoda, A., Nakanishi, W. *et al.* (2008) Guanidine-catalysed asymmetric synthesis of 2,2-disubstituted chroman skeletons by intramolecular oxa-Michael addition. *European Journal of Organic Chemistry*, 2759–2766.
58. Terada, M., Ikehara, T. and Ube, H. (2007) Enantioselective 1,4-addition reactions of diphenyl phosphite to nitroalkenes catalysed by an axially chiral guanidine. *Journal of the American Chemical Society*, **129**, 14112–14113.
59. Fu, X., Jiang, Z. and Tan, C.-H. (2007) Bicyclic guanidine-catalysed enantioselective phospha-Michael reaction: synthesis of chiral β-aminophosphine oxides and β-aminophosphines. *Chemical Communications*, 5058–5060.
60. Ono, N., Kamimura, A. and Kaji, A. (1984) A new synthesis of allylic alcohols or their derivatives via reductive elimination from γ-(phenylthio)-β-nitro alcohols with tributyltin hydride. *Tetrahedron Letters*, **25**, 5319–5322.
61. Menchaca, R., Martínez, V., Rodríguez, A. *et al.* (2003) Synthesis of natural ecteinascidin (ET-729, ET-745, ET-759B, ET-736, ET-637, ET-594) from cyanosafracin B. *The Journal of Organic Chemistry*, **68**, 8859–8866; Corey, E.J., Gin, D.Y. and Kania, R.S. (1996) Enantioselective total synthesis of ecteinascidin 743. *Journal of the American Chemical Society*, **118**, 9202–9203.
62. Blanc, A.C., Macquarrie, D.J., Valle, S. *et al.* (2000) The preparation and use of novel immobilized guanidine catalysts in base-catalysed epoxidation and condensation reactions. *Green Chemistry*, **2**, 283–288.
63. Li, J., Jiang, W.-Y., Han, K.-L. *et al.* (2003) Density functional study on the mechanism of bicyclic guanidine-catalysed Strecker reaction. *The Journal of Organic Chemistry*, **68**, 8786–8789.
64. Groeger, H. (2003) Catalytic enantioselective Strecker reactions and analogous syntheses. *Chemical Reviews*, **103**, 2795–2827.
65. Tanaka, K., Kamatani, M., Mori, H. *et al.* (1998) Synthesis of a new phospholipase A_2 inhibitor of an aldehyde terpenoid and its possible inhibitory mechanism. *Tetrahedron Letters*, **39**, 1185–1188; Kocienski, P.J., Brown, R.C.D., Pommier, A. *et al.* (1998) Synthesis of salinomycin. *Journal of the Chemical Society – Perkin Transactions 1*, 9–40.
66. Barton, D.H.R., Jaszbetenyi, J.C., Lin, W. and Shinada, T. (1996) Oxidation of hydrazones by hypervalent organoiodine reagents: regeneration of the carbonyl group and facile syntheses of α-acetoxy and α-alkoxy azo compounds. *Tetrahedron*, **52**, 14673–14688; (1997) *Tetrahedron*, **53**, 14821, (erratum); Barton, D.H.R., Jaszbetenyi, J.C., Lin, W. and Shinada, T. (1996) The invention of radical reactions. Part XXXVI. Synthetic studies related to 3-deoxy-D-manno-2-octulosonic acid (KDO) *Tetrahedron*, **52**, 2717–2726.
67. Isobe, T., Fukuda, K. and Ishikawa, T. (1998) Simple preparation of chiral 1,3-dimethyl-2-iminoimidazolidines (monocyclic guanidines) and applications to asymmetric alkylative esterification. *Tetrahedron: Asymmetry*, **9**, 1729–1735.
68. Isobe, T., Fukuda, K. and Ishikawa, T. unpublished results.
69. Chiappe, C., More, G.L. and Munforte, P. (1997) Lifetime of the glucosyl oxocarbenium ion and stereoselectivity in the glycosidation of phenols with 1,2-anhydro-3,4,6-tri-*O*-methyl-α-D-glucopyranose. *Tetrahedron*, **53**, 10471–10478.
70. Anastasia, M., Allevi, P., Ciuffeda, P. and Fiecchi, A. (1983) A convenient protection of the 3β-hydroxy or 3β-chloro substituent of 6-oxo-steroids. *Synthesis*, 123–124.
71. Kim, S. and Chang, H. (1984) 1,1,3,3-Tetramethylguanidine: an effective catalyst for the *tert*-butyldimethylsilylation of alcohols. *Synthetic Communications*, **14**, 899–904.

72. Isobe, T., Fukuda, K., Araki, Y. and Ishikawa, T. (2001) Modified guanidines as chiral supedrbases: the first example of asymmetric silylation of secondary alcohols. *Chemical Communications*, 243–244.
73. Wipf, P. and Lynch, S.M. (2003) Synthesis of highly oxygenated dinaphthyl ethers via S_NAr reactions promoted by Barton's base. *Organic Letters*, **5**, 1155–1158; Wipf, P., Lynch, S.M., Birmingham, A. *et al.* (2004) Natural product based inhibitors of the thioredoxin–thioredoxin reductase system. *Organic and Biomolecular Chemistry*, **2**, 1651–1658; Thommen, A.S.B., Raju, G.S., Blagg, J. *et al.* (2004) Double benzyne-furan cycloaddition and assembly of 1,1′-binaphthyl and 1,1′-dinaphthyl ether sytems. *Tetrahedron Letters*, **45**, 3181–3184.
74. Bera, S., Langley, G.J. and Pathak, T. (1998) Sugar-modified uridine bisvinyl sulfone: synthesis of a bifunctionalized nucleoside Michael acceptor and its use in stereoselective tandem cyclization. *The Journal of Organic Chemistry*, **63**, 1754–1760.
75. Williams, J.M., Brands, K.M.J., Skerlj, R.T. *et al.* (2005) Practical synthesis of the new carbapenem antibiotics ertapenem sodium. *The Journal of Organic Chemistry*, **70**, 7479–7487.
76. Evindar, G. and Batey, R.A. (2003) Copper- and palladium-catalysed intramolecular aryl guanidinylation: an efficient method for the synthesis of 2-aminobenzimidazoles. *Organic Letters*, **5**, 133–136.
77. Candiani, I., DeBernardinis, S., Cabri, W. *et al.* (1993) A facile one-pot synthesis of polyfunctionalized 2-unsubstituted benzo[*b*]furans. *Synlett*, 269–270.
78. Costa, M., Chiusoli, G.P. and Rizzardi, M. (1996) Base-catalysed direct introduction of carbon dioxide into acetylenic amines. *Chemical Communications*, 1699–1700.
79. Barrett, A.G.M., Cramp, S.M., Hennessy, A.J. *et al.* (2001) Oxazole synthesis with minimal purification: synthesis and application of a ROMPgel Tosmic reagent. *Organic Letters*, **3**, 271–273.
80. Barton, D.H.R., Kervagoret, J. and Zard, S.Z. (1990) A useful synthesis of pyrroles from nitroolefins. *Tetrahedron Letters*, **46**, 7587–7598.
81. Travins, J.M. and Etzkorn, F.A. (1997) Design and enantioselective synthesis of a peptidomimetic of the turn in the helix-turn-helix DNA-binding protein motif. *The Journal of Organic Chemistry*, **62**, 8387–8393; Debenham, S.D., Debenham, J.S., Burk, M.J. and Toone, E.J. (1997) Synthesis of carbon-linked glycopeptides through catalytic asymmetric hydrogenation. *Journal of the American Chemical Society*, **119**, 9897–9898; Varie, D.L., Brennan, J., Briggs, B. *et al.* (1998) Bioreduction of (*R*)-carvone and regioselective Baeyer–Villiger oxidations: application to the asymmetric synthesis of cryptophycin fragment A. *Tetrahedron Letters*, **39**, 8405–8408; Ritzen, A., Basu, B., Waellgerg, A. and Frejd, T. (1998) Phenyltrisalanine: a new, C_3-symmetric, trifunctional amino acid. *Tetrahedron: Asymmetry*, **9**, 3491–3496.
82. Barrett, A.G.M., Cramp, S.M., Roberts, R.S. and Zecri, F.J. (1999) Horner–Emmons synthesis with minimal purification using ROMPGEL: a novel high-loading matrix for supported reagents. *Organic Letters*, **1**, 579–582.
83. Fedorov, A., Combes, S. and Finet, J.-P. (1999) Influence of the steric hindrance of the aryl group of pentavalent triarylbismuth derivatives in ligand coupling reactions. *Tetrahedron*, **55**, 1341–1352; Matano, Y. and Imahori, H. (2004) A new, efficient method for direct α-alkenylation of β-dicarbonyl compounds and phenols using alkenyltriarylbismuthonium salts. *The Journal of Organic Chemistry*, **69**, 5505–5508.
84. Kozikowski, A.P., Campiani, G., Aagaard, P. and McKinney, M. (1993) An improved synthetic route to huperizine A: new analog and their inhibition of acetylcholinesterase. *Journal of the Chemical Society – Chemical Communications*, 860–862; McGhee, W.D., Riley, D.P., Christ, M.E. and Christ, K.M. (1993) Palladium-catalysed generation of *O*-allylic urethanes and carbonates from amines/alcohols, carbon dioxide, and allylic chlorides. *Organometallics*, **12**, 1429–1433; Jin, Z. and Fuchs, P.L. (1996) Syntheses via vinyl sulfones. 66. Palladium[0]-mediated aminospirocyclization of tertiary allylic sulfones. Stereospecific construction of the azabicyclic ring system of cephalotaxine. *Tetrahedron Letters*, **32**, 5253–5256; Luzzio, F.A. and Mayorov, A.V. (2003) Palldium(0)-mediated preparation of trans-4-substituted-1-(phthalimido)-2-cyclopentenes. *Synlett*, 532–536; Li, S., Xie, H., Zhang, S. *et al.* (2005) Tetramethylguanidine as an inexpensive and efficient ligand for the palladium-catalysed Heck reaction. *Synlett*, 1885–1888.

85. Trost, B.M. and Yasutaka, T. (2001) A catalytic asymmetric Wagner–Meerwein shift. *Journal of the American Chemical Society*, **123**, 7162–7163.
86. Li, S., Lin, Y., Cao, J. and Zhang, S. (2007) Guanidine/Pd(OAc)$_2$-catalysed room temperature Suzuki cross-coupling reaction in aqueous media under aerobic conditions. *The Journal of Organic Chemistry*, **72**, 4067–4072.
87. Ma, X., Zhou, Y., Zhang, J. *et al.* (2008) Solvent-free Heck reaction catalysed by a recyclable Pd catalyst supported on SBA-15 via an ionic liquid. *Green Chemistry*, **10**, 59–66.
88. Koehn, U., Guenther, W., Goerls, H. and Anders, E. (2004) Preparation of chiral thioureas, ureas and guanidines from (S)-2-(N,N-dialkylaminomethyl)pyrrolidines. *Tetrahedron: Asymmetry*, **15**, 1419–1426; Koehn, U., Schulz, M., Goerls, H. and Anders, E. (2005) Neutral zinc (II) and molybdenum(0) complexes with chiral guanidine ligands: synthesis, characterisation and applications. *Tetrahedron: Asymmetry*, **16**, 2125–2131.
89. Bunge, S.D., Lance, J.M. and Bertke, J.A. (2007) Synthesis, structure, and reactivity of alkylzinc complexes stabilized with 1,1,3,3-tetramethylguanidine. *Organometallics*, **26**, 6320–6328.
90. Barton, D.H.R., Finet, J.-P., Motherwell, W.B. and Pichon, C. (1987) The chemistry of pentavalent organobismuth reagents. Part 8. Phenylation and oxidation of alcohols by tetraphenylbismuth esters. *Journal of the Chemical Society – Perkin Transactions 1*, 251–259.
91. Hjelmencrantz, A. and Berg, U. (2002) New approach to biomimetic transamination using bifunctional [1,3]-proton transfer catalysis in thioxanthenyl dioxide imines. *The Journal of Organic Chemistry*, **67**, 3585–3594.
92. Basavaiah, D., Rao, K.V. and Reddy, B.S. (2006) (5S)-1,3-Diaza-2-imino-3-phenylbicyclo[3.3.0] octane: first example of guanidine based in situ recyclable chiral catalytic source for borane-mediated asymmetric reduction of prochiral ketones. *Tetrahedron: Asymmetry*, **17**, 1036–1040.
93. Forsyth, A.C., Paton, R.M. and Watt I. (1989) Highly selective base-catalysed addition of nitromethane to levoglucosenone. *Tetrahedron Letters*, **30**, 993–996; Forsyth, A.C., Gould, R.O., Paton, R.M. *et al.* (1993) Stereoselective addition of nitromethane to levoglucosenone; formation and structure of 2:1 and 1:2 adducts. *Journal of the Chemical Society – Perkin Transactions 1*, 2737–2741; Hossain, N., Garg, N. and Chattopadhyaya, J. (1993) New synthesis of 2′,3′-dideoxy-2′,3′-didehydro-3′-C-substituted thymidines. *Tetrahedron*, **49**, 10061–10068.
94. Xia, Y. and Kozikowski, A.P. (1989) A practical synthesis of the Chinese 'nootropic' agent huperzine A: a possible lead in the treatment of Alzheimer's disease. *Journal of the American Chemical Society*, **111**, 4116–4117; Kozikowski, A.P., Prakash, K.R.C., Saxena, A. and Doctor, B.P. (1998) Synthesis and biological activity of an optically pure 10-spirocyclopropyl analog of huperzine A. *Chemical Communications*, 1287–1288.
95. Zhang, F., Moher, E.D. and Zhang, T.Y. (2007) TMG catalysed cyclopropanation of cyclopentenone. Illustration by a simple synthesis of bicycle[3.1.0]hexan-2-one derivatives. *Tetrahedron Letters*, **48**, 3277–3279.
96. Barton, D.H.R., Bashiardes, G. and Fourrey, J.-L. (1983) An improved preparation of vinyl iodides. *Tetrahedron Letters*, **24**, 1605–1608; Barton, D.H.R., Chen, M., Jaszberenyi, J.C. and Taylor, D.K. (1997) Preparation and reaction of 2-*tert*-butyl-1,1,3,3-tetramethylguanidine: 2,2,6-trimethylcyclohexen-1-yl iodide. *Organic Syntheses*, **74**, 101–105; Furrow, M.E. and Myers, A.G. (2004) Practical procedures for the preparation of N-*tert*-butyldimethylsilylhydrazones and their use in modified Wolff-Kishner reductions and in the synthesis of vinyl halides and *gem*-dihalides. *Journal of the American Chemical Society*, **126**, 5436–5445.
97. Aube, J. (1991) *Comprehensive Organic Synthesis*, Vol. 1 (eds B.M. Trost and I. Fleming), Pergamon, Oxford, p. 835; Rosen, T. (1991) *Comprehensive Organic Synthesis*, Vol. 2 (eds B.M. Trost and I. Fleming), Pergamon, Oxford, p. 28; Kemp, J.E.G. (1991) *Comprehensive Organic Synthesis*, Vol. 7 (eds B.M. Trost and I. Fleming), Pergamon, Oxford, p. 469; Tanner, D. (1994) Chiral aziridines-their synthesis and use in stereoselective transformations. *Angewandte Chemie – International Edition*, **33**, 599–619; Osborn, H.M.I. and Sweeney, J. (1997) The asymmetric synthesis of aziridines. *Tetrahedron: Asymmetry*, **18**, 1693–1715; Atkinson, R.S. (1999) 3-Acetoxyaminoquinazolinones (QNHOAc) as aziridinating agents: ring-opening of N-(Q)-substituted aziridines. *Tetrahedron*, **55**, 1519–1559; Mitchinson, A. and

Nadin, A. (2000) Saturated nitrogen heterocycles. *Journal of the Chemical Society – Perkin Transactions 1*, 2862–2892.

98. Hada, K., Watanabe, T., Isobe, T. and Ishikawa, T. (2001) Guanidinium ylides as a new and recyclable source for aziridines and their roles as chiral auxiliaries. *Journal of the American Chemical Society*, **123**, 7705–7706; [*Chem. & Eng. News*, Aug. 13, 32 (2001)].

99. Haga, T. and Ishikawa, T. (2005) Mechanistic approaches to asymmetric synthesis of aziridines from guanidinium ylides and aryl aldehydes. *Tetrahedron*, **61**, 2857–2869.

100. Wannaporn, D. and Ishikawa, T. (2005) Chirality transfer from guanidinium ylides to 3-alkenyl (or 3-alkynyl) aziridine-2-carboxylates and application to the syntheses of (2R,3S)-3-hydroxyleucinate and D-*erythro*-sphingosine. *The Journal of Organic Chemistry*, **70**, 9399–9406.

101. Wannaporn, D., Ishikawa, T., Kawahata, M. and Yamaguchi, K. (2006) Guanidinium ylide mediated aziridination: identification of a spiro imidazolidine-oxazolidine intermediate. *The Journal of Organic Chemistry*, **71**, 6600–6603.

102. Plechkova, N.V. and Seddon, K.R. (2008) Application of ionic liquids in the chemical industry. *Chemical Society Reviews*, **37**, 123–150.

103. Huang, J., Jiang, T., Gao, H. *et al.* (2004) Pd nanoparticles immobilized on molecular sieves by ionic liquids: heterogeneous catalysts for solvent-free hydrogenation. *Angewandte Chemie – International Edition*, **43**, 1397–1399; Yang, Y., Deng, C. and Yuan, Y. (2005) Characterization and hydroformylation performance of mesoporous MCM-1-supported water soluble Rh complex dissolved in ionic liquids. *Journal of Catalysis*, **232**, 108–116; Zhu, A., Jiang, T., Wang, D. *et al.* (2005) Direct aldol reactions catalysed by 1,1,3,3-tetramethylguanidine lactate without solvent. *Green Chemistry*, **7**, 514–517; Huang, J., Jiang, T., Han, B. *et al.* (2005) A novel method to immobilize Ru nanoparticles in SBA-15 firmly by ionic liquid and hydrogenation of arenes. *Catalysis Letters*, **103**, 59–62; Zhu, A., Jiang, T., Han, B. *et al.* (2006) Study on guanidine-based task-specific ionic liquids as catalysts for direct aldol reactions without solvent. *New Journal of Chemistry*, **30**, 736–740; Miao, S., Liu, Z., Han, B. *et al.* (2006) Ru nanoparticles immobilized on montmorillonite by ionic liquids: a highly efficient heterogeneous catalyst for the hydrogenation of benzene. *Angewandte Chemie – International Edition*, **45**, 266–269; Li, S., Lin, Y., Xie, H. *et al.* (2006) Brønsted guanidine acid-base ionic liquids: novel reaction media for the palladium-catalysed Heck reaction. *Organic Letters*, **8**, 391–394; Song, J., Zhang, Z., Jiang, J. *et al.* (2008) Epoxidation of styrene to styrene oxide using carbon dioxide in ionic liquids. *Journal of Molecular Catalysis A-Chemical*, **279**, 235–238.

104. Papa, A.J. (1966) Synthesis and azidolysis of 2-chlorotetramethylguanidine. Synthetic utility of hexa- and tetramethylguanidinium azide. *The Journal of Organic Chemistry*, **31**, 1426–1430.

105. Gaoni, Y. (1988) Regiospecific additions of hydrazoic acid and benzylamine to 1-(arylsulfonyl) bicyclo[1.1.0]butanes. Application to the synthesis of *cis* and *trans* 2,7-methanoglutamic acids. *Tetrahedron Letters*, **29**, 1591–1594.

106. Xiao, D., Carroll, P.J., Mayer, S.C. *et al.* (1997) Stereoselective synthesis of a conformationally restricted β-hydroxy-λ-amino acid. *Tetrahedron: Asymmetry*, **8**, 3043–3046.

107. Li, C., Shih, T.-L., Jeong, J.U. *et al.* (1994) The use of tetramethylguanidinium azide in non-halogenated solvents avoids potential explosion hazards. *Tetrahedron Letters*, **35**, 2645–2646.

108. Blaszczyk, R. (2008) Tetramethylguanidinium azide (TMGA)-a versatile azidation agent. *Synlett*, 299–300.

109. Palacios, F., Aparicio, D., de los Santos, J.M. *et al.* (1995) One pot synthesis of β-functionalized vinyl azides through addition of tetramethylguanidinium azide to acetylenic and allenic compounds. *Organic Preparations and Procedures International*, **27**, 171–178.

110. Arnold, M.A., Day, K.A., Duron, S.G. and Gin, D.Y. (2006) Total synthesis of (+)-batzelladine A and (−)-batzelladine D. *Journal of the American Chemical Society*, **128**, 13255–13260.

111. Fotsing, J.R. and Banart, K. (2005) First propargyl azides bearing strong acceptor substituents and their effective conversion into allenyl azides: influence of the electronic effects of substituents on the reactivity of propargyl azides. *European Journal of Organic Chemistry*, 3704–3714.

5

Phosphazene: Preparation, Reaction and Catalytic Role

Yoshinori Kondo
Graduate School of Pharmaceutical Sciences, Tohoku University,
Aramaki Aza Aoba 6-3, Aoba-ku, Sendai 980-8578, Japan

5.1 Introduction

Selection of the appropriate base for the reaction to be performed has been one of the important matters for organic chemists. Although uncharged organic bases are usually weaker than their inorganic counterparts, such as alkali metal hydroxides, oxides and alkoxides, organic bases have become standard reagents widely used in organic synthesis (Figure 5.1).

The use of organic bases has some advantages over the use of ionic bases, such as milder reaction conditions, better solubility and absence of a coordinating metal ion. Recently, particular attention has been focussed on the design and preparation of organic superbases. The cation free strong neutral bases are strong nonionic systems, which allow the deprotonation of a wide range of weak acids, to form highly reactive 'naked' anions. In this respect, cyclic amidines, guanidines, phosphazenes and phosphatranes have been of special interest for general and synthetic organic chemistry.

The representative of a phosphazene base was first prepared in the 1970s [1] and, subsequently, simple, large scale syntheses of tris(dialkylamino)iminophophoranes and tris(dialkylamino)-*N*-alkyliminophosphoranes (the P1 phosphazene bases) were developed [2]. Furthermore, a (dma)$_3$P=N- unit on phosphorus for synthesizing highly nucleophilic tris[tris(dimethylamino)phosphinimino]phosphine was employed. The homologation

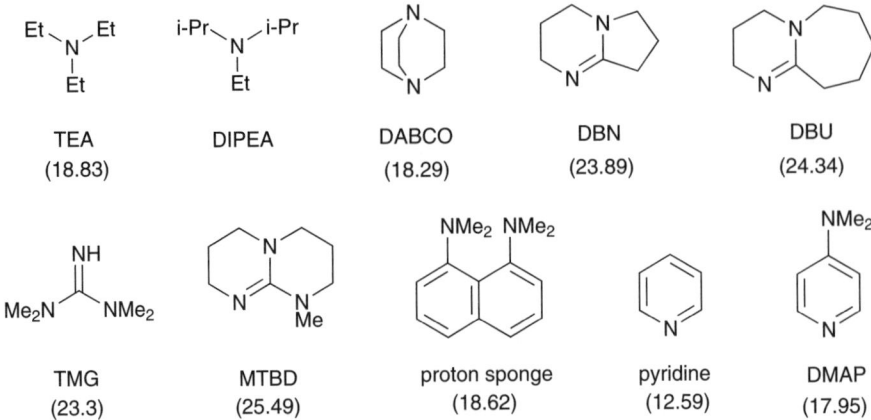

Figure 5.1 Organic bases for organic synthesis

concept was used for the construction of a family of extremely strong noncharged organic bases, P1–P7 phosphazene bases (iminophosphoranes), which are much stronger bases than the well known diazabicycloundecene (DBU) or triazabicyclodecene (TBD) bases [3] (Figure 5.2).

The hydrolytically stable phosphazene bases, especially ^{t}Bu-P4, were found to be useful in organic synthesis. Besides the highly enhanced basicity, phosphazene bases combine: high solubility in nonpolar organic solvents; easy handling and easier workup through cleaner reactions; low sensitivity to moisture and oxygen; and the possibility of operating at lower temperature and high selectivity. There are many examples which demonstrate the superiority of the phosphazene bases and polymer-supported phosphazene reagents over common inorganic and organic nitrogen bases in organic synthesis.

A simple high yield two-step protocol for the preparation of P1 is known. Reaction of phosphorus pentachloride with dimethylamine affords the peralkylated cation as tetrafluoroborate, which is then demethylated with thiolate. The high basicity of triaminoiminophosphoranes requires drastic conditions [4] (Scheme 5.1).

To further enhance basicity by the same formal homologation, replacement of the dimethylamino group by modified, potentially stronger electron-donating groups is considered to allow a tuning of the basicity. The nucleophilicity of the bases should be effectively controlled by proper choice of the alkyl group on the basic centre. The synthesis for P2, P3, P4 phosphazenes takes advantage of the P1 building block and HBF$_4$ salts are usually obtained in high yields by replacement of the chlorine atom on the phosphorus by the iminophosphorane. The most convenient method for the liberation of P4 phosphazene was found to be the reaction of the hydrogen tetrafluoroborate (HBF$_4$) salt with potassium amide (KNH$_2$) in liquid ammonia and separation of the base from inorganic material by extraction with hexane [5] (Scheme 5.2).

A density functional theory (B3LYP/6-311 + G**), *ab initio* (HF/3-21G*) and semi-empirical (PM3) study of intrinsic basicities, protonation energies or protonation enthalpies of phosphazene bases has been reported. The study shows that the organic superbases can reach the basicity level of the strongest inorganic superbases, such as alkali metal

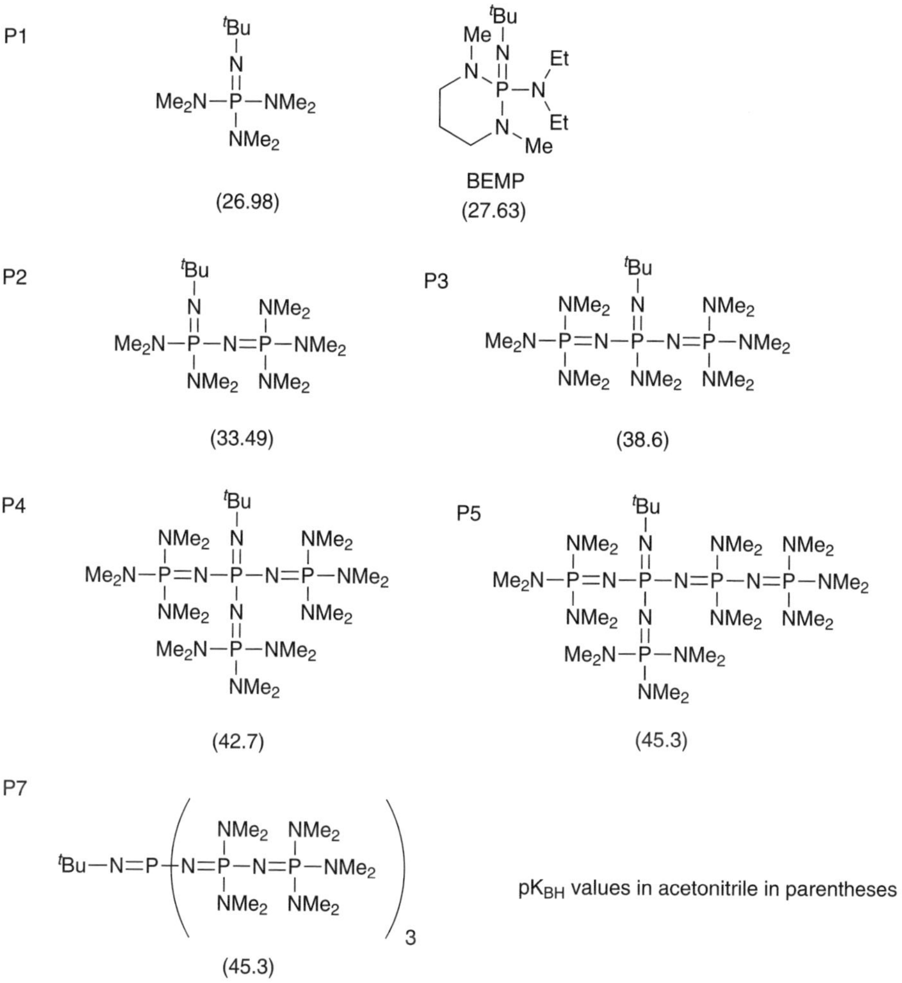

P1

'Bu
|
N
‖
Me₂N—P—NMe₂
|
NMe₂

(26.98)

BEMP

(27.63)

P2

'Bu
|
N   NMe₂
‖ |
Me₂N—P—N=P—NMe₂
| |
NMe₂   NMe₂

(33.49)

P3

'Bu
|
NMe₂ N   NMe₂
| ‖ |
Me₂N—P=N—P—N=P—NMe₂
| | |
NMe₂   NMe₂   NMe₂

(38.6)

P4

'Bu
|
NMe₂ N   NMe₂
| ‖ |
Me₂N—P=N—P—N=P—NMe₂
| | |
NMe₂ N   NMe₂
     ‖
Me₂N—P—NMe₂
     |
     NMe₂

(42.7)

P5

'Bu
|
NMe₂ N   NMe₂ NMe₂
| ‖ | |
Me₂N—P=N—P=N—P—N=P—NMe₂
| | | |
NMe₂   N   NMe₂ NMe₂
         ‖
    Me₂N—P—NMe₂
         |
         NMe₂

(45.3)

P7

            ⎛ NMe₂ NMe₂   ⎞
            | | | |
'Bu—N=P—(—N=P—N=P—NMe₂   )
            | | | |
            ⎝ NMe₂ NMe₂   ⎠₃

(45.3)

pK_BH values in acetonitrile in parentheses

Figure 5.2 Phosphazene bases

hydroxides, hydrides and oxides. The strongest organic phosphazene bases are predicted to reach the gas phase basicity level of ca. 300 kcal/mol (number of phosphorus atoms in the system $n > 7$) [6].

UV photoelectron spectroscopy was used to investigate the electronic structure of phosphazene bases. The spectral assignment was based on the band intensities, HeI/HeII

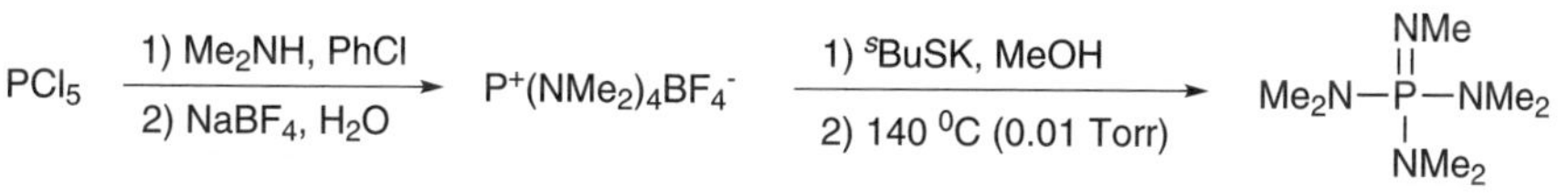

$$PCl_5 \xrightarrow[\text{2) NaBF}_4, H_2O]{\text{1) Me}_2NH, PhCl} P^+(NMe_2)_4 BF_4^- \xrightarrow[\text{2) 140 °C (0.01 Torr)}]{\text{1) }^sBuSK, MeOH} Me_2N-\underset{\underset{NMe_2}{|}}{\overset{\overset{NMe}{\|}}{P}}-NMe_2$$

Scheme 5.1 Synthesis of P1

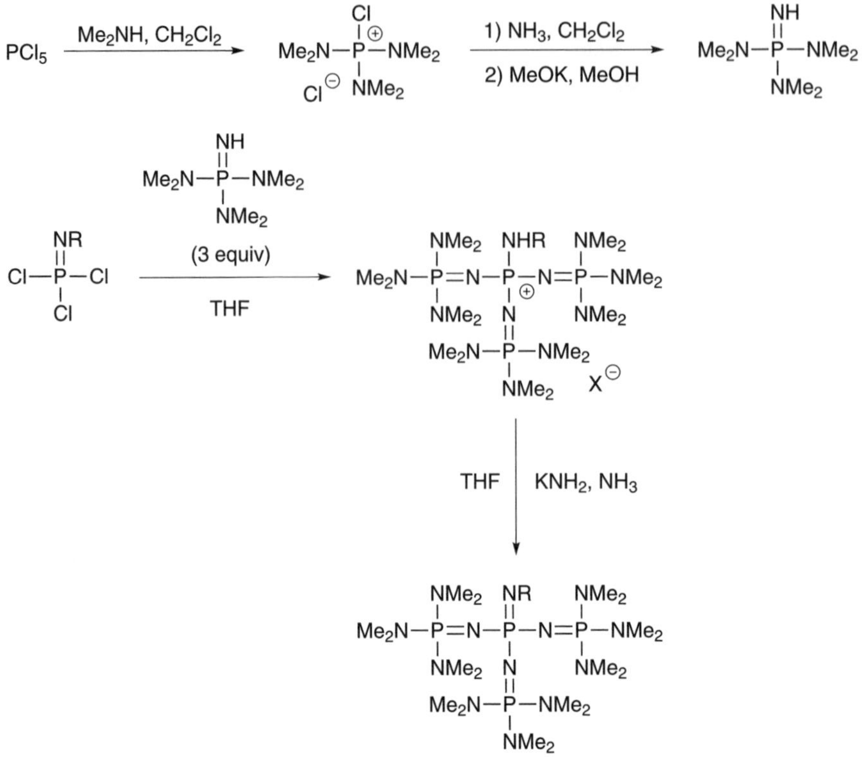

Scheme 5.2 *Synthesis of P4*

band intensity changes, comparison with the spectra of related compounds and MO calculations [7]. A UV–Visible spectrophotometric method for the measurement of relative acidities in heptane was developed and the phosphazene base ^{t}Bu-P4 was used as the deprotonating agent. Its protonated form is a good counter ion for the anions of the acids because it is bulky, has delocalized charge and, therefore, does not have specific interactions with the anions.

A self-consistent scale of relative acidities in heptane spanning for 3 pKa units was constructed [8]. A self-consistent spectrophotometric basicity scale of various organic bases including phosphazenes in acetonitrile was investigated. The span of the scale is almost 12 pKa units. The scale is anchored to the pKa value of pyridine of 12.33. Comparison of the basicity data of phenyliminophosphoranes and phenyltetramethylguanidines implies that the P=N bond in the (arylimino)tris(1-pyrrolidinyl)phosphatranes involves a contribution from the ylidic (zwitterionic) structure analogous to that found in phosphorus ylides [9]. A series of RN=P(Pyrr)$_3$ iminophosphoranes (P1 phosphazene), where R is amino, α-naphthyl- or substituted phenyl group, was prepared and characterized by other properties.

The pKa values of 12 different synthesized phosphazenes and C$_6$H$_5$N=P(NMe$_2$)$_3$ were determined in acetonitrile relative to the reference bases using ^{13}C NMR spectroscopy. The obtained pKa-values were compared with the corresponding values of RNH_2 amines. The pKa values of the synthesized phosphazenes in acetonitrile ranged from 14.6 to 26.8 pKa units [10]. The UV–Visible spectrophotometric titration method was used to establish the relative basicity of phenyl-substituted phosphazenes (P1, P3 and P4 bases), and to extend the ion pair basicity scale for tetrahydrofuran (THF) medium. These measurements gave a continuous basicity scale in THF ranging from 2.6 (2-MeO-pyridine) to 26.6 (2-Cl-$C_6H_4P_4$(pyrr)phosphazene) in pKa units, that is for 24 orders of magnitude and containing 58 compounds (pyridines, anilines, amines, guanidines, amidines, phosphazenes) [11]. The gas phase basicity (GB) values were determined for 19 strong bases, among them such well known bases as BEMP (1071.2 KJ/mol), Verkade's methyl-substituted base (1083.8 kJ/mol), Et–N=P$(NMe_2)_2$–N=P$(NMe_2)_3$ (Et-P2 phosphazene, 1106.9 KJ/mol) and t-Bu-N=P$(NMe_2)_3$ (t-Bu-P1 phosphazene, 1058.0 KJ/mol). The first experimental GB values were determined for P2 phosphazenes and an important region of the gas phase basicity scale is now covered with organic bases. The GB values for several superbases were calculated using density functional theory at the B3LYP/6-311 + G** level [12].

A new unique principle for creating novel nonionic superbases has been reported (Figure 5.3). It is based on attachment of tetraalkylguanidino, 1,3-dimethylimidazolidine-2-imino or bis(tetraalkylguanidino)carbimino groups to the phosphorus atom of the iminophosphorane group using tetramethylguanidine or the easily available 1,3-dimethylimidazolidine-2-imine. Their base strengths are established in THF solution by means of spectrophotometric titration and compared with reference superbases designed specially for this study, P2- and P4-iminophosphoranes. The gas phase basicities of several guanidine and N',N',N',N'-tetramethylguanidino(tmg) substituted phosphazenes, and of their cyclic analogues, have been calculated and the crystal structures of $(tmg)_3P=N-^tBu$ and $(tmg)_3P=N-^tBu$ HBF_4 determined. The enormous basicity-increasing effect of this principle is experimentally verified for the tetramethylguanidino groups in the THF medium, and the basicity increase when moving from $(dma)_3P=N-^tBu$ (pKa = 18.9) to $(tmg)_3P=N-^tBu$ (pKa = 29.1) is 10 orders of magnitude [13].

The gas phase basicities and pKa values of tris(phosphazeno) substituted azacalix[3](2,6) pyridine in acetonitrile and some related compounds were examined by the density functional theory (DFT) computational method. It was shown that the hexakis(phosphazeno) derivative of azacalx[3](2,6)pyridine is a hyperstrong neutral base, as evidenced by the absolute proton affinity of 314.6 kcal/mol and pKa (MeCN) of 37.3 units. It is a consequence of the very strong bifurcated hydrogen bond (32 kcal/mol) and substantial cationic resonance effect [14].

A combination of phosphazene base concept and the disubstituted 1,8-naphthalene spacer was shown and a new bisphosphazene 1,8-bis(hexamethyltriaminophosphazenyl) naphthalene (HMPN) represents the most basic representative of this class of 'proton sponge', as evidenced by the theoretically estimated proton affinity PA = 274 kcal/mol and the measured pK_{BH^+} (MeCN) 29.9. HMPN is by nearly 12 orders of magnitude more basic than Alder's classical 1,8-bis(dimethylamino)naphthalene (DMAN). The new bisphosphazene, HMPN, has been prepared and fully characterized. The spatial structure of HMPN and its conjugate acid have been determined by X-ray technique and theoretical DFT calculations. It is found that monoprotonated HMPN has an unsymmetrical intramolecular hydrogen bridge. This cooperative proton chelating effect renders the bisphosphazene more basic than P1 phosphazene bases. The density functional calculations are in

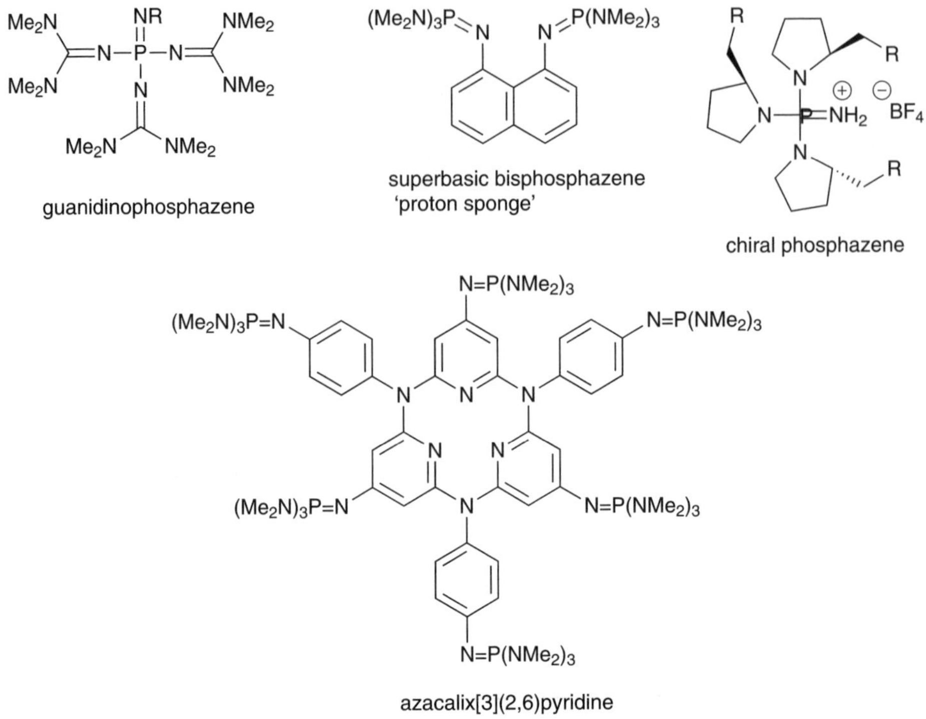

Figure 5.3 *Newly designed phosphazenes*

good accordance with the experimental results. They show that the high basicity of HMPN is a consequence of the high energy content of the base in its initial neutral state and the intramolecular hydrogen bonding in the resulting conjugate acid with contributions to proton affinity of 14.1 and 9.5 kcal/mol, respectively [15a]. It is shown by DFT calculation that HMPN and trisguanidylphosphazene are very powerful neutral organic superbases, as evidenced by the calculated proton affinities in the gas phase and the corresponding calculated pKa values in acetonitrile (given within parentheses): 305.4 kcal/mol (44.8) and 287.8 kcal/mol (37.8), respectively [15b].

A chiral example of phosphazene bases was synthesized by treatment of (S)-2-(dialkylaminomethyl)pyrrolidine derived from 5-oxo-(S)-proline, with phosphorus pentachloride and subsequent addition of gaseous ammonia. The phosphazenes were isolated as HBF$_4$ salts in high yields and fully characterized by ^{1}H, ^{13}C and ^{31}P NMR spectroscopy, various 1D and 2D NMR experiments and mass spectrometry (EI). The molecular structure and the absolute configuration of the HBF$_4$ salts were determined by X-ray analysis [16].

5.2 Deprotonative Transformations Using Stoichiometric Phosphazenes

In this section, various transformations promoted by the use of stoichiometric phosphazenes are discussed, classified by the types of the phosphazenes P1–P4.

5.2.1 Use of P1 Base

5.2.1.1 Alkylation of C-Nucleophile

The catalytic enantioselective alkylation of the benzophenone imine of glycine *tert*-butyl ester was realized by an efficient homogeneous reaction with alkyl halides, the phosphazene base (BEMP or BTTP) and chiral quaternary ammonium salts derived from the cinchona alkaloids. Wang-resin bound derivatives of the glycine Schiff base ester were alkylated in the presence of quaternary ammonium salts derived from cinchonidine or cinchonine using phosphazene bases to give either enantiomer of the product α-amino acid derivatives in 51–89% ee. The enantioselective conjugate addition of Schiff base ester derivatives to Michael acceptors either in solution (56–89% ee) or in the solid phase (34–82% ee) gave optically active unnatural α-amino acid derivatives. The reaction was conducted in the presence of chiral quaternary salts derived from the cinchona alkaloids using phosphazene bases. Reacting imine derivatives of resin-bound amino acids with α,ω-dihaloalkanes provides highly versatile intermediates to racemic α,α-disubstituted amino acids with wide variety of side-chain functionality. They allow the creation of amino acids with diverse functionalities placed at varying chain length from the α-center of the amino acid [17] (Scheme 5.3).

Iminic derivatives of (4R,5S)-1,5-dimethyl-4-phenylimidazolidin-2-one have been diastereoselectively alkylated with activated alkyl halides or electrophilic olefins either under phase transfer catalysis (PTC) conditions or in the presence of the phosphazene base BEMP at −20 °C in the presence of lithium chloride (LiCl). Hydrolysis of the alkylated imino imides gave (S)-α-amino acids with recovery of the imidazolidinone chiral auxiliary [18].

The phenylsulfinyl fluoroacetate can be alkylated with a wide range of alkyl halides and Michael acceptors. Subsequent thermal elimination of phenyl sulfinic acid leads to α-fluoro-α,β-unsaturated ethyl carboxylates, an important class of intermediates for fluorine containing biologically active compounds [19] (Scheme 5.4).

Using a phosphazene base allows unreactive nitroaromatic compounds to condense with ethyl isocyanoacetate to give C-annelated pyrroles. Stable 2H-isoindoles with electron-withdrawing groups have been prepared using the reaction of dinitrobenzene derivatives with isocyanoacetate in the presence of a phosphazene base (BTPP). The structure of an isoindole was confirmed by X-ray crystallographic analysis, and this substance existed in the solid phase only as the 2H-isomer. The reaction of 6-nitroquinoline gave a pyridine fused isoindole [20] (Scheme 5.5).

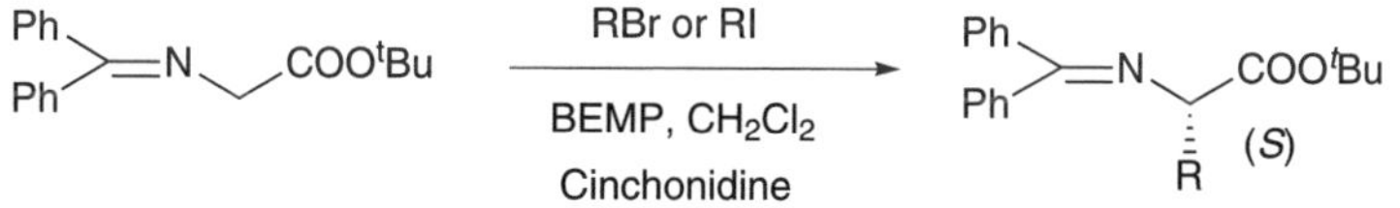

Scheme 5.3 Alkylation of glycinate in the presence of BEMP

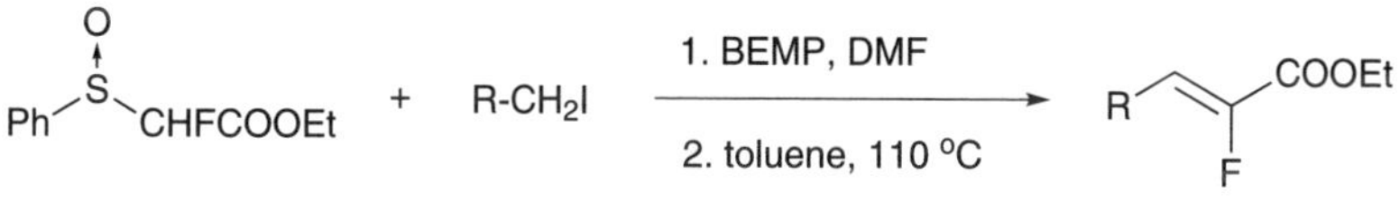

Scheme 5.4 Alkylation of sulfoxide

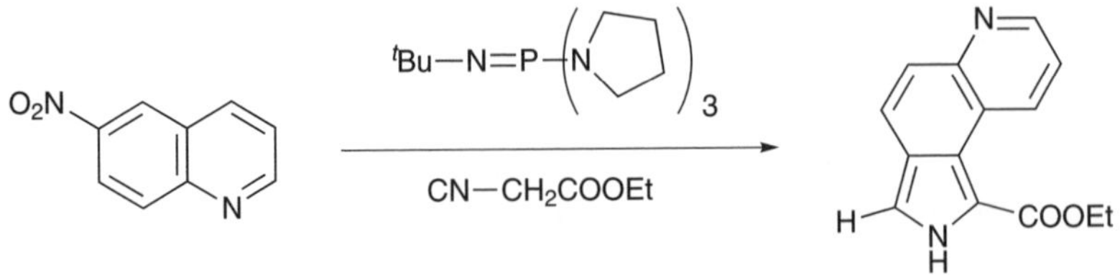

Scheme 5.5 Synthesis of pyrrole derivative from nitroarene

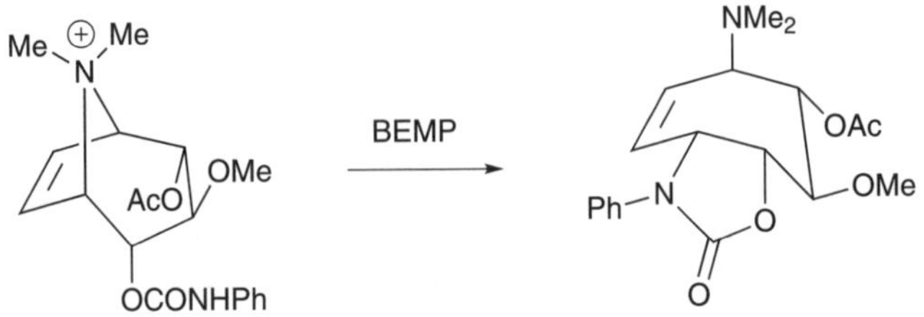

Scheme 5.6 Intramolecular alkylation for cyclization

5.2.1.2 Alkylation and Arylation of N-Nucleophile

The cyclization associated with the quasi-axial orientation of the carbamate residue involves complications, not unexpected because of the steric/stereoelectronic conditions. After a series of inadequately selective reactions with a variety of conventional bases (NaH, KH, LDA, ʹBuOK, etc.) the Schwesinger iminophosphorane bases eventually enabled a breakthrough and gave the desired product in high yield [21] (Scheme 5.6).

The amination of 3-bromoisoxazoles by a nucleophilic aromatic substitution reaction was facilitated by the use of phosphazene bases. 3-Bromoisoxazoles were found to be inert to substitution under thermal conditions. However, the use of phosphazene bases under microwave irradiation facilitates the amination process and allows the corresponding 3-aminoisoxazoles to be isolated in moderate yield [22] (Scheme 5.7).

5.2.1.3 Alkylation of S-Nucleophile

Disaccharides of 1-thioglycosides, an important class of glycomimics, can be synthesized by S-alkylation in exceptionally high yields when iminophosphorane bases are employed. The reaction conditions employed appear to be general and stereospecific. Axial and equatorial 4-triflates and primary tosylates of alkyl pyranosides provided excellent yields of

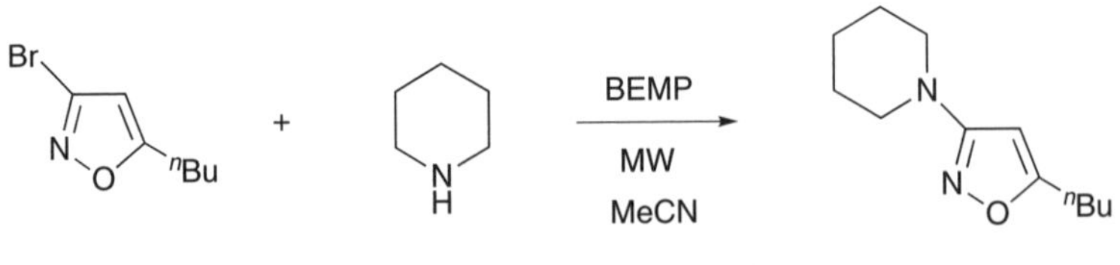

Scheme 5.7 Amination of 3-bromoisoxazole

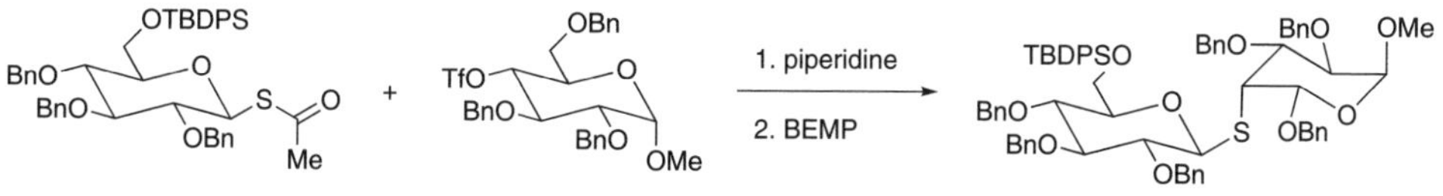

Scheme 5.8 Synthesis of 1-thioglycoside

thio-dissacharides without substantial elimination product. The iminophosphorane bases also proved to be useful in solid support-bound coupling of thioglycosides though with lower efficiency [23] (Scheme 5.8).

5.2.1.4 Amide Formation

α-Amino acids are soluble in acetonitrile when treated with phosphazene bases. As a result, the protection/deprotection events that are usually used for peptide coupling reactions can be minimized. This is illustrated in the synthesis of the important angiotensin-converting enzyme (ACE) inhibitor enalapril [24] (Scheme 5.9).

5.2.1.5 Heterocycle Formation

3,6-Dihydro-2H-1,4-oxazin-2-ones act as reactive chiral cyclic alanine equivalents and can be distereoselectively alkylated using the organic base BEMP when using unactivated alkyl halides. In most cases, the diastereoselectivity is excellent although the reactions are always carried out at room temperature. Hydrolysis of the alkylated oxazinones obtained allows the preparation of enantiomerically enriched (S)-α-methyl α-amino acids. The organic base methodology has also been applied to the synthesis of (R)-α-methyl α-amino acids starting from (R)-alanine. When dihalides are used as electrophiles in the presence of BEMP, a spontaneous N-alkylation also takes place giving bicyclic oxazinones, which can be hydrolyzed to an enantiomerically pure heterocyclic compound [25] (Scheme 5.10).

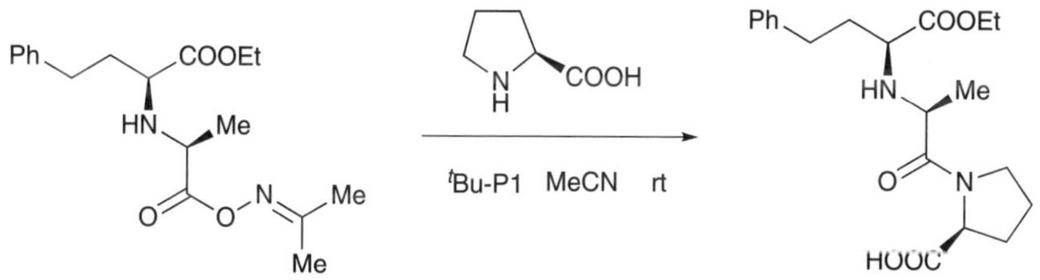

Scheme 5.9 Peptide coupling reaction

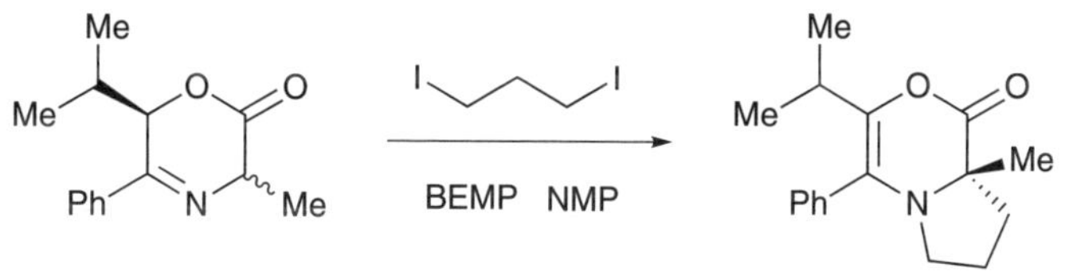

Scheme 5.10 Alkylation of chiral alanine equivalent

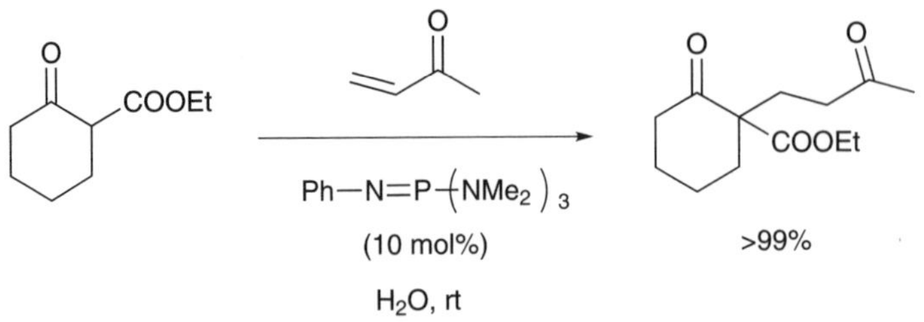

Scheme 5.11 1,4-Addition reaction to enone

5.2.1.6 Michael Addition

Michael addition reaction of various β-ketoesters with several Michael acceptors in water containing 10 mol% of N-phenyl-tris(dimethylamino)iminophosphorane results in high yield conversion to the corresponding adducts [26] (Scheme 5.11).

5.2.1.7 Polymer-Supported Reaction

The polymer bound indolecarboxylate was N-alkylated with 3-cyanobenzyl bromide in the presence of the strong base BEMP (Scheme 5.12). The replacement of the base by the weaker base DIPEA did not lead to any alkylation product [27].

Polymer bound acrylic ester is reacted in a Baylis–Hillman reaction with aldehydes to form 3-hydroxy-2-methylidenepropionic acids or with aldehydes and sulfonamides in a three-component reaction to form 2-methylidene-3-[(arylsulfonyl)amino]propionic acids. In order to show the possibility of Michael additions, the synthesis of pyrazolones was chosen. The Michael addition was carried out with ethyl acetoacetate and BEMP as base to form the resin bound β-keto ester. This was then transformed into the hydrazone with phenylhydrazine hydrochloride in the presence of TMOF and DIPEA [28]. The polymer bound phenol was readily coupled to a variety of allyl halides by using the P1-^{t}Bu to generate a reactive phenoxide [29].

Polymer-supported reagents have been applied to the synthesis of the natural product carpanone, resulting in a clean and efficient synthesis without the need for a conventional

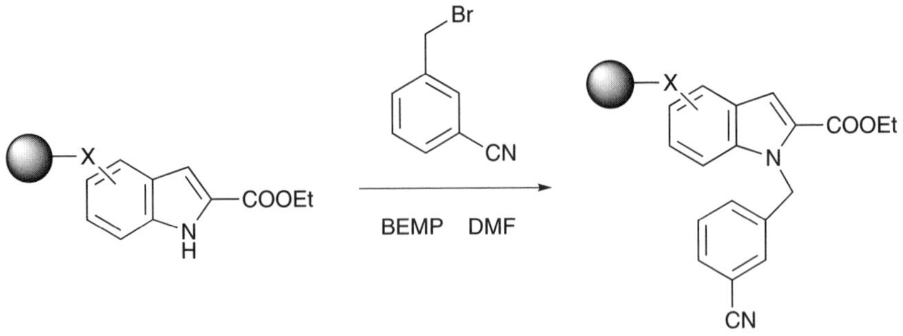

Scheme 5.12 Alkylation of immobilized indole

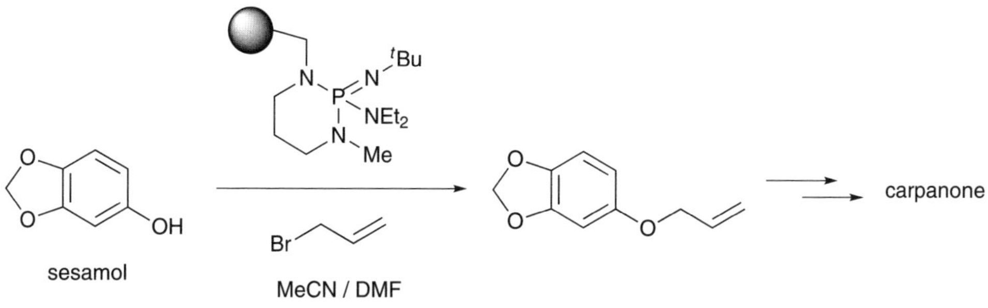

Scheme 5.13 Alkylation of phenol using P-BEMP

purification technique. The synthesis begins from commercially available sesamol, which is readily allylated in acetonitrile containing a small quantity of dimethylformamide using allyl bromide and a polymer-supported phosphazene base (P-BEMP) to give the aryl ether in 98% yield [30] (Scheme 5.13).

Polymer-supported reagents and other solid sequestering agents can be used to generate an array of 1,2,3,4-tetrasubstituted pyrrole derivatives without any chromatographic purification step [31] (Scheme 5.14).

Polymer-supported reagents and sequestering agents may be used to generate an array of variously substituted amino acid derivatives which were converted to hydroxamic acid derivatives as potential inhibitors of matrix metalloproteinases without any chromatographic purification step [32] (Scheme 5.15).

The preparation of a novel polystyrene-supported dehydrating agent and its application to the synthesis of 1,3,4-oxadiazoles under thermal and microwave conditions has been reported. An alternative procedure using tosyl chloride and P-BEMP has also been presented [33] (Scheme 5.16).

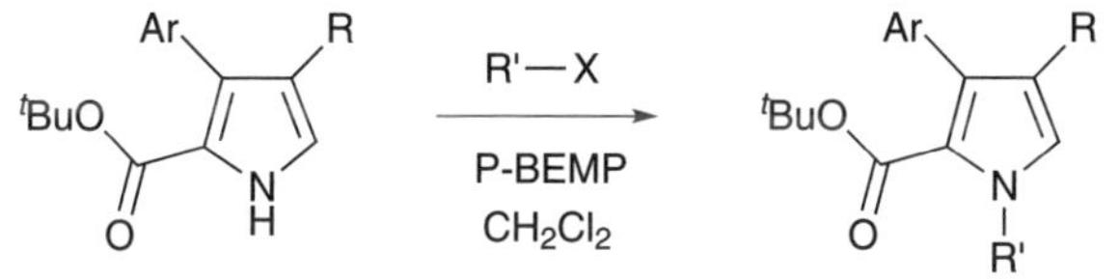

Scheme 5.14 Alkylation of pyrrole derivative using P-BEMP

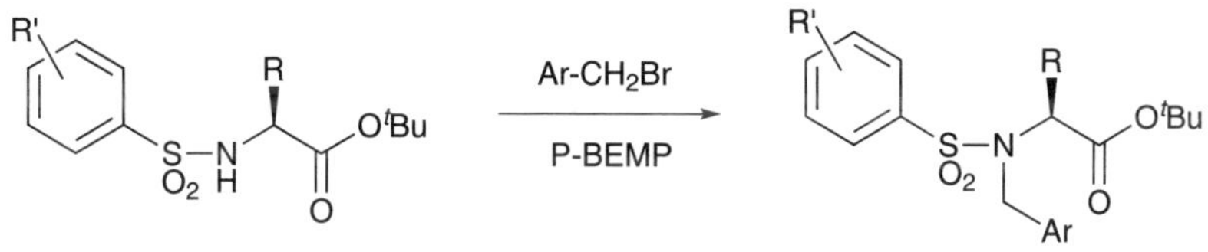

Scheme 5.15 Alkylation of sulfonamide using P-BEMP

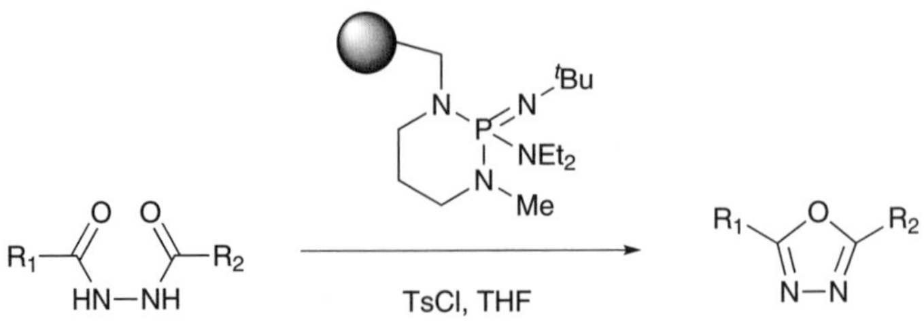

Scheme 5.16 Cyclization using P-BEMP

5.2.2 Use of P2 Base

5.2.2.1 Aldol Reaction of Iminoglycinate

It was found that a 'naked' enolate can undergo an aldol addition with full conversion provided that the ΔpKa between the aldol product and the conjugated acid of the base is large enough (more than six units stronger). Under such conditions the base being regenerated can be used in catalytic amounts. When the iminoglycinate is derived from hydroxypinanone, 83% of the threo-isomer having ~98% ee was obtained. When Et-P2 was used as a base, the threo/erythro selectivity was higher as well as the diastereoselectivity of the approach of the aldehyde on enolate. The high diastereoselectivity observed at C2 (97% R) is difficult to rationalize on the basis of product stability but similar R-diastereoselectivity has been observed during other catalytic Michael additions [34] (Scheme 5.17).

5.2.2.2 Deconjugation of Alkenyl Sulfone

Treatment of vinyl sulfones with 0.1–0.2 equiv. of phosphazene base, Et-P2 in THF at 25 °C affords the corresponding allyl sulfones in high yield. Et-P2 has shown clear and superiority to both DBU and KO^{t}Bu in these reactions [35] (Scheme 5.18).

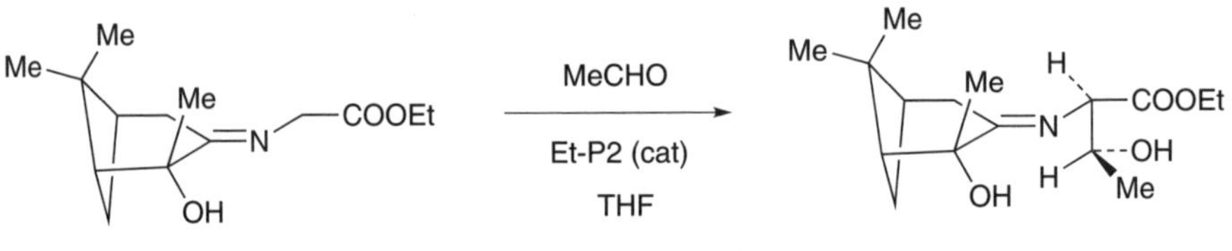

Scheme 5.17 Aldol reaction using P2

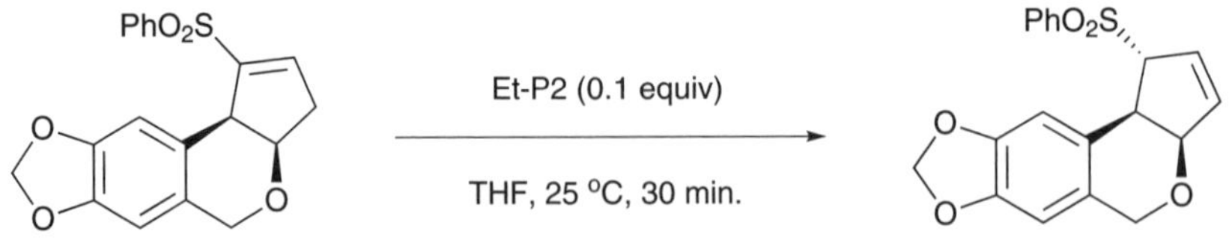

Scheme 5.18 Isomerization using P2

5.2.2.3 Generation of S-Ylide and Reaction

It was found that oxathiane is a very efficient chiral auxiliary that allows the preparation of various pure *trans*-diarylepoxides in high yields and with high enantiomeric purities. In an attempt to improve these results and shorten the reaction time, more reactive 'naked' carbanions were generated by using a phosphazene base, and Et-P2 is potentially able to provide a high concentration of ylide and could be used in dichloromethane (CH_2Cl_2). The percentage of the *cis* isomer was reduced to 6%. *Trans*-1-(2-pyridyl)-, *trans*-1-(3-pyridyl)-, *trans*-1-(2-furyl)-, and *trans*-1-(3-furyl)-2-phenyl epoxides with enantiomeric purities ranging from 96.8 to 99.8% are obtained in two steps from pure (*R,R,R*)-oxathiane. It was also found that *trans*-aryl-vinyl epoxides could be synthesized with 77–100% conversion from conjugated aldehydes and chiral sulfonium salts with ee's ranging from 95 to 100%. When a conjugated ketone such as methyl vinyl ketone was used under the same conditions, the reaction provided the corresponding cyclopropane in good yield and with high enantioselectivity [36] (Scheme 5.19).

Unknown diaryl and alkyl-phenyl *N*-tosyl aziridines have been successfully synthesized from pure (*R,R,R,S*)-(−)-sulfonium salt derived from Eliel's oxathiane, tosylimines, and using a phosphazene base Et-P2 to generate the ylide. Both *cis* and *trans* aziridines have exceptionally high enantiomeric purities (98.7–99.9% ee). The *R*-configuration found at C2 is consistent with the model and all previous results, therefore all *trans*-aziridines and *cis*-aziridines have been assigned the (2*R*,3*R*)- and the (2*R*,3*S*)-configurations, respectively. The chiral auxiliary is used in a stoichiometric amount but is recovered in high yield and reused [37] (Scheme 5.19).

5.2.2.4 Reaction of o-Halobenzyl Sulfone

The synthetic applications of *o*-halobenzyl sulfones as precursors of 1,3- and 1,5-zwitter ionic synthons, have been investigated and their sulfonyl carbanions, generated by means of the phosphazene base Et-P2 reacted with different electrophiles, such as alkyl halides and aldehydes. When the alkylation was performed with ethyl bromoacetate, *ortho*-substituted cinnamates were obtained after a subsequent β-elimination of sulfinic acid. When the aldol reaction was performed with paraform aldehyde, vinyl sulfones were obtained. The

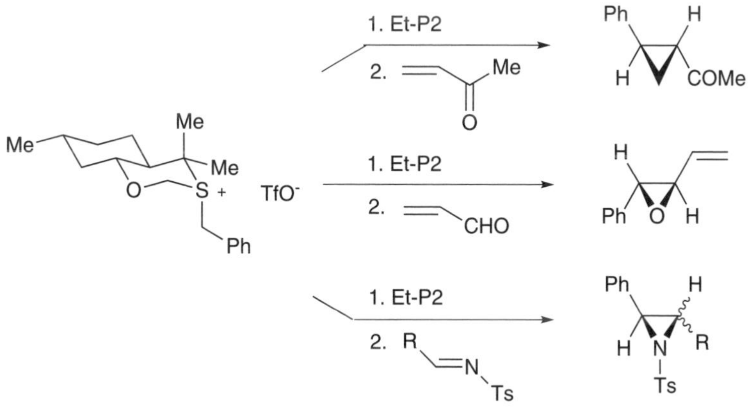

Scheme 5.19 *Generation of S-ylide in the presence of P2*

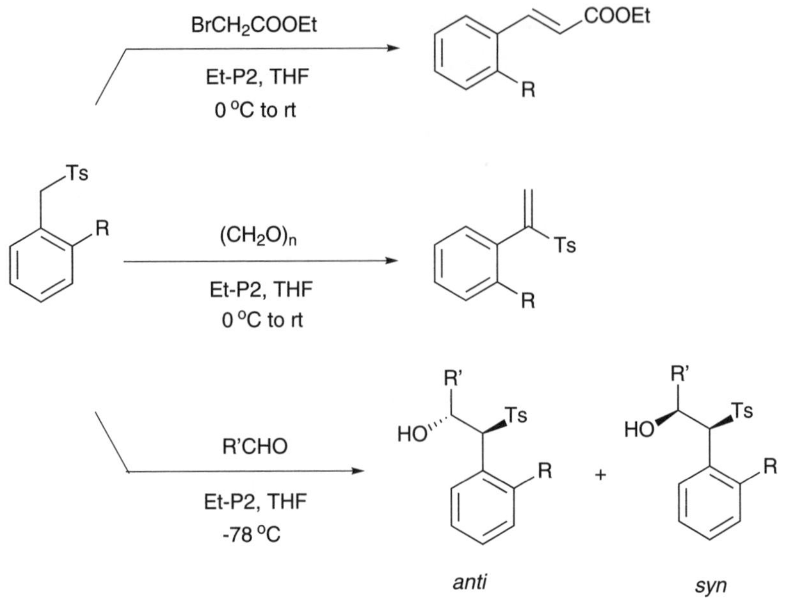

Scheme 5.20 Reaction of sulfones in the presence of P2

reaction of the *o*-halobenzyl sulfones with aliphatic aldehydes at −78 °C gave aldol-type products and the *anti* products are the favoured products [38] (Scheme 5.20).

5.2.2.5 Sigmatropic Rearrangement

The phosphazene bases BTPP and ′Bu-P2 mediate the rearrangement of unactivated *N*-alkyl-*O*-benzoyl hydroxamic acid derivatives to give 2-benzoyl amides. The rate of reaction was found to be dependent upon the steric nature of the *N*-alkyl substituent [39a]. Treatment of malonyl derived *O*-acylhydroxamic acid derivatives with the phosphazene superbase ′Bu-P2 gives 2,3-dihydro-4-isoxazole carboxylic ester derivatives. The rate and yield of the reaction depend upon the *O*-acyl substituent [36b] (Scheme 5.21).

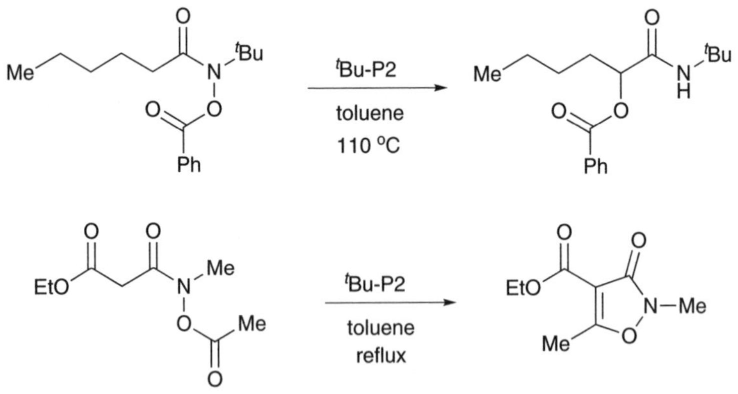

Scheme 5.21 P2 promoted rearrangement

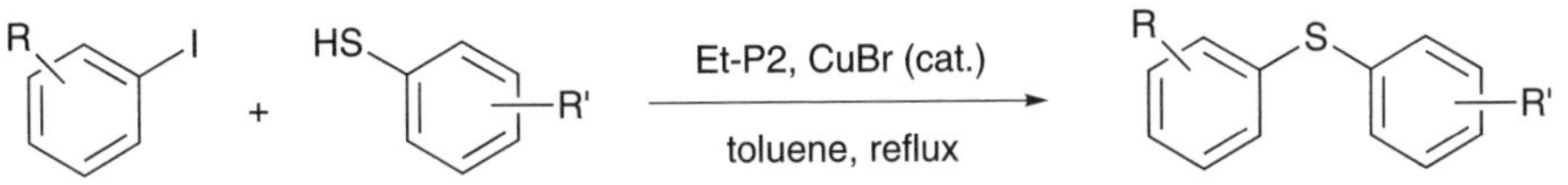

[Scheme 5.22: Ar-I + HS-Ar'-R' → Ar-S-Ar'-R', with Et-P2, CuBr (cat.), toluene, reflux]

Scheme 5.22 *Copper catalysed coupling in the presence of P2*

5.2.2.6 Sulfide Formation via S_NAr Reaction

In the presence of the phosphazene Et-P2 base as well as DBU the coupling of aryl iodides with arenethiols only requires catalytic amounts of copper(I) bromide (CuBr). Under these conditions, the reaction can be performed in refluxing toluene to give biaryl thioethers in excellent yields [40] (Scheme 5.22).

5.2.3 Use of P4 Base

5.2.3.1 Alkylation of C-Nucleophile

It was shown that monoalkylation of 8-phenylmenthyl phenylacetate using lithiated bases leads to poor or no diastereoselectivities (50/50 to 69/31) and high yields (75–98%) while alkylation using ^{t}Bu-P4 leads to high diastereoselectivities (92/8 to 98/2) and high yields (65–95%) [41] (Scheme 5.23).

It is postulated that, in the case of phenylacetates, the degree of aggregation of the lithium enolate is responsible of the poor diastereoselectivities. NMR studies revealed that methyl phenylacetate enolate generated with the ^{t}Bu-P4 phosphazene base was 'naked' or tightly associated with the ^{t}Bu-P4H$^+$ cation depending on very small variations in solvent composition. Both forms reacted more rapidly than the corresponding lithium enolate in a model alkylation experiment using dimethyl sulfate.

Supported ^{t}Bu-P4 enolate chemistry of phenylacetyloxymethyl polystyrene resin was investigated using high resolution magic angle spinning (HR-MAS) NMR spectroscopy. Direct analysis of the crude reaction suspension through the use of a diffusion filter allowed not only rapid selection of the optimal experimental conditions, but also the characterization of the enolate on the solid phase. Comparison with solution experiments and literature data indicated partially the structure of the enolate. HR-MAS NMR spectra of the enolate revealed also a tight interaction of ^{t}Bu-P4 base with the polymer matrix [41].

The structurally novel bicyclic oxazinone was prepared based on D-glucopyranose. The lithium enolates of these compounds undergo highly diastereoselective alkylation reactions with reactive alkyl halides, in modest yields. Use of the phosphazene P4 base enhanced the yields of these processes, suggesting that metal enolate aggregation is at least partly

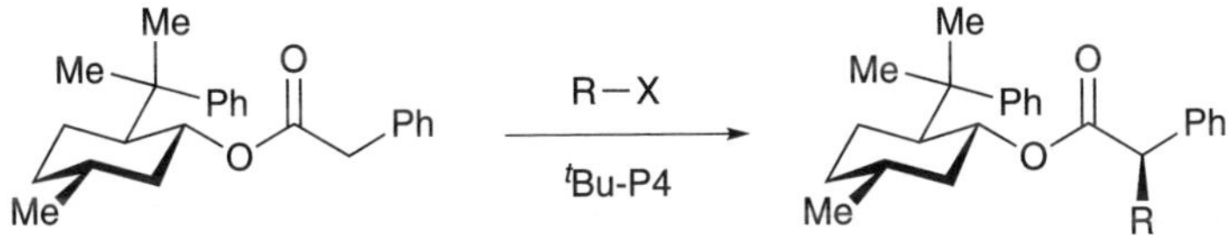

Scheme 5.23 *Diastereoselective alkylation in the presence of P4*

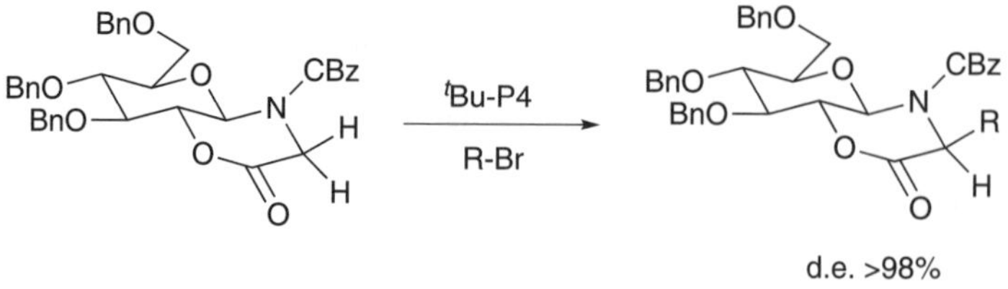

Scheme 5.24 *Diastereoselective alkylation of bicyclic oxazinone*

responsible for the depressed yields. The stereochemistry of the products has been unequivocally ascertained by nuclear Overhauser effect (nOe) measurements and *ab initio* calculations [42] (Scheme 5.24).

A synthesis of all four stereoisomers [(1*S*,2*S*)-, (1*R*,2*R*)-, (1*S*,2*R*)- and (1*R*,2*S*)-] of 1-amino-2-(hydroxymethyl)cyclobutanecarboxylic acid was presented. The synthesis is based on the chiral glycine equivalent, employed in both enantiomeric forms. The key step involves the cyclization of the silyl-protected iodohydrins to the corresponding spiro derivatives with the aid of the phosphazene base *t*Bu-P4. The final compounds were found to display moderate potency as ligands for the glycine binding of the *N*-methyl-D-aspartate (NMDA) receptor [43] (Scheme 5.25).

Three-membered cyclic sulfones (episulfones) undergo substitution on treatment with base–electrophile mixtures, such as *t*Bu-P4 phosphazene base–benzyl bromide, to give the corresponding alkenes following loss of sulfur dioxide (SO_2). The formation of the trisubstituted episulfone was observable but the compound proved unstable to the workup procedure and rapidly decomposed to the alkene as a mixture of stereoisomers [44] (Scheme 5.26).

Four cyclotetrapeptides containing one or two chiral amino acids have been *C*-alkylated or *C*-hydroxyalkylated through phosphazenium enolates. The reactions are completely diastereoselective with respect to the newly formed backbone stereogenic centres. With the *t*Bu-P4 base, all groups are first benzylated and *C*-benzylation then takes place at a sarcosine, rather than an *N*-benzylglycine residue. In contrast to open chain *N*-benzyl peptides, the *N*-benzylated cyclotetrapeptides could not be debenzylated under dissolving-metal conditions (Na/NH_3). Conformational analysis shows that the prevailing species have *cis*/*trans*/*cis*/*trans* peptide bonds [45] (Scheme 5.27).

Cyclosporin A can be regioselectively alkylated at the NH of Val-5 with reactive bromides in the presence of phosphazene base *t*Bu-P4 to yield alkylated products. These are devoid of immunosuppressive activity *in vitro* but they have binding affinity for cyclophilin A and represent a new class of cyclosporin antagonists. ^{1}H-NMR studies have shown that the compounds exist in a single, all *trans* conformation [45c].

5.2.3.2 Generation α-Sulfinyl Carbanion

The effect of *t*Bu-P4 on the yield and diastereoselectivity of additions of thus formed 'naked' α-sulfinyl carbanions to butyraldehyde has been studied. Condensation of ethyl benzylsulfone with butyraldehyde gave a mixture of two expected *syn* and *anti* diastereomers. It appeared that the *anti* isomer was favoured and the diastereoselectivity increased

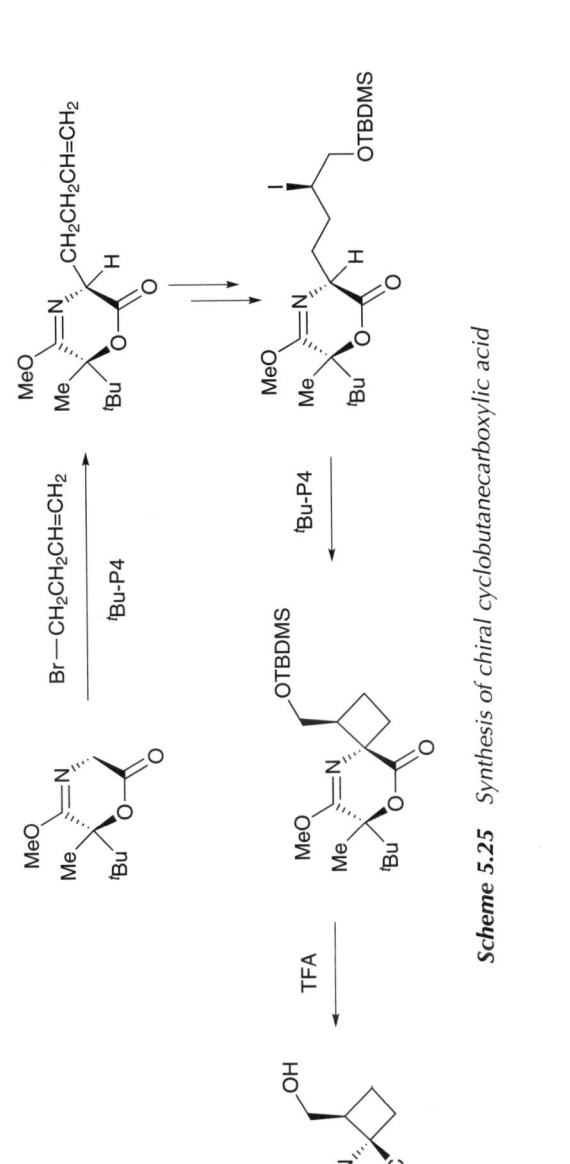

Scheme 5.25 Synthesis of chiral cyclobutanecarboxylic acid

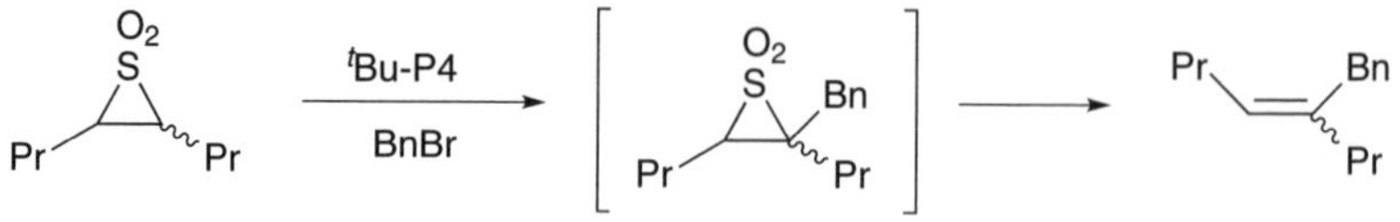

Scheme 5.26 Alkylation of episulfone in the presence of P4

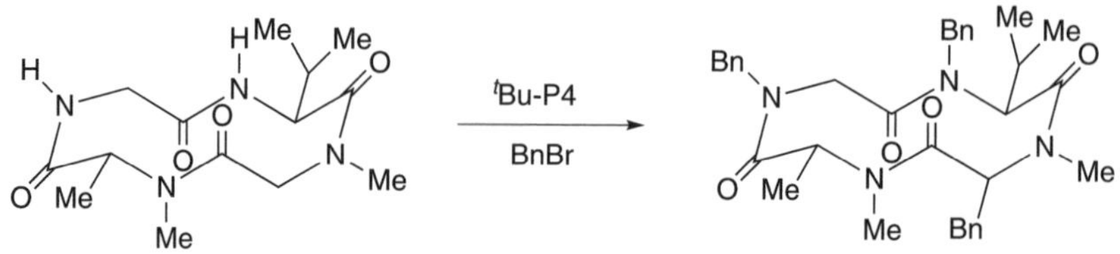

Scheme 5.27 Benzylation of cyclic peptide

when 'Bu-P4 was used as base. A further increase in the diastereoselectivity was observed when the alcoholate was quenched with trimethylchlorosilane [46] (Table 5.1).

5.2.3.3 Julia–Kocienski Olefination

The reaction between a carbanion derived from alkyl 3,5-bis(trifluoromethyl)phenyl sulfones and aldehydes affords, with good yields and stereoselectivities, the corresponding 1,2-disubstituted alkene through the Julia–Kocienski olefination reaction. This one-pot protocol can be performed using the phosphazene base at −78 °C and has been successfully used in a high yielding and stereoselective synthesis of various stilbenes such as resveratrol [47] (Scheme 5.28).

5.2.3.4 Benzofuran Cyclization

The hindered nonionic phosphazene base 'Bu-P4 efficiently deprotonates o-arylmethoxy benzaldehydes, leading to a direct synthesis of benzofuran derivatives. Strong ionic bases such as LDA, LTMP and KH failed for this cyclization [48] (Scheme 5.29).

Table 5.1 Diastereoselective addition of sulfone to aldehyde

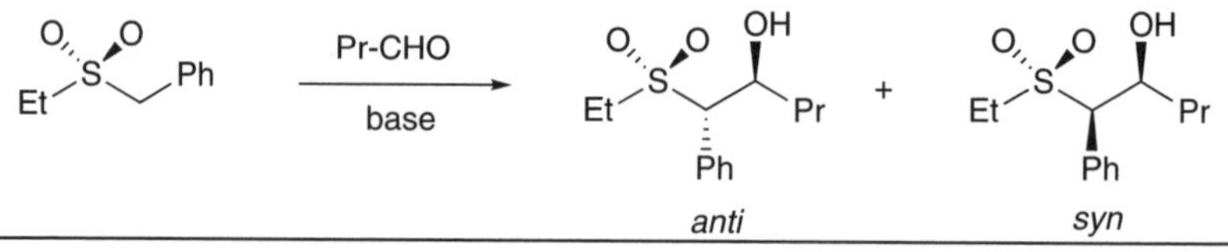

Base	anti	syn	Yield (%)
BuLi	59	41	100
'Bu-P4	72	28	56
'Bu-P4/TMSCl	82	18	72

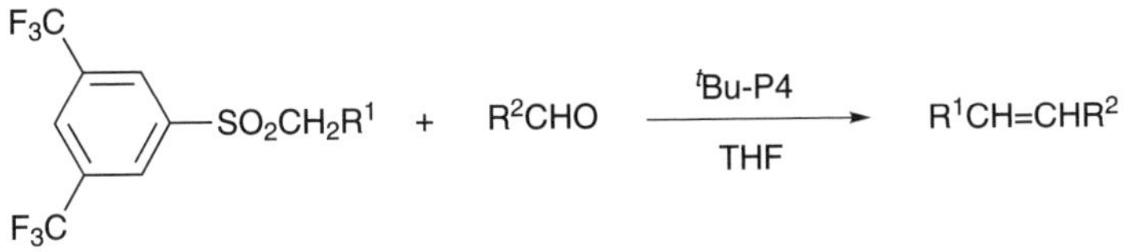

Scheme 5.28 Julia–Kocienski olefination

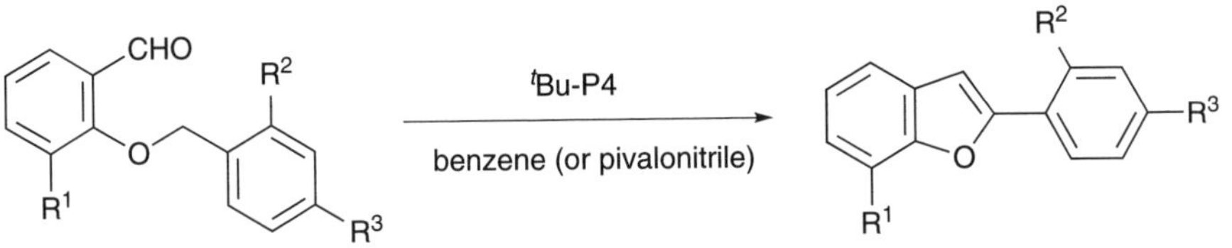

Scheme 5.29 Synthesis of benzofuran

5.2.3.5 Oxy-Cope Rearrangement

Compounds containing a 1,5-hexadien-3-ol system undergo anionic oxy-Cope rearrangement when treated with the phosphazene superbase ^{t}Bu-P4. The sigmatropic rearrangement occurs in hexane as well as in THF. The weaker phophazene base Et-P2 failed to induce rearrangement. This is the first example of the use of a metal-free base to induce anionic oxy-Cope rearrangement [49] (Scheme 5.30).

5.2.3.6 Ether Formation via S$_N$Ar Reaction

In the presence of ^{t}Bu-P4 base and copper(I) bromide, aryl halides couple with phenols to give biaryl ethers at *about* 100 °C, while the use of DBU as a base for this coupling reaction is not effective [50] (Scheme 5.31).

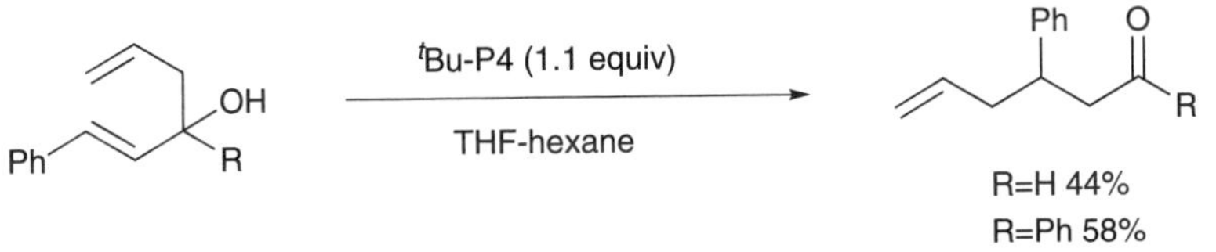

Scheme 5.30 Anionic oxy-Cope rearrangement

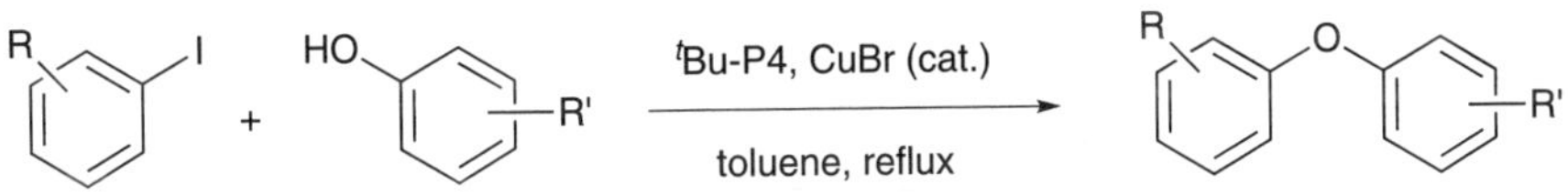

Scheme 5.31 Copper catalysed coupling in the presence of P4

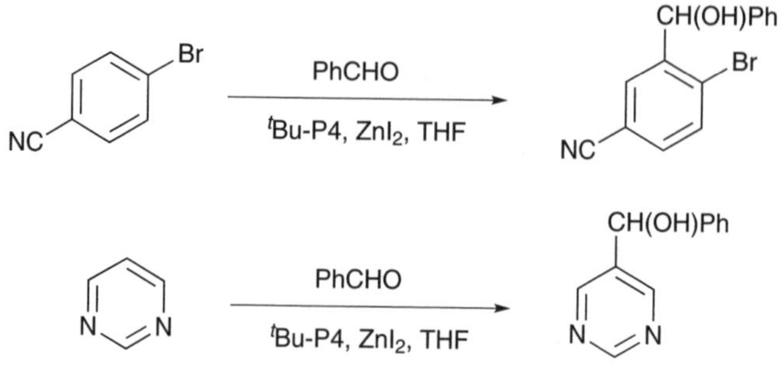

Scheme 5.32 Deprotonative functionalization of aromatic compounds

5.2.3.7 Reaction of Heteroaryl Carbanion

A novel type of deprotonative functionalization of aromatics was achieved with unique regioselectivities and excellent chemoselectivity using ^{t}Bu-P4 base and zinc iodide (ZnI$_2$) in the presence of an electrophile [51] (Scheme 5.32).

5.2.4 Use of P5 Base

Pyrido[1,2-a]azepinone was deprotonated from the pyridine unit by lithium diisopropylamide affording lithium salts, which were trapped by electrophiles like deuterium oxide (D$_2$O), benzoyl chloride and aldehyde. In the latter case, subsequent intramolecular attack of the intermediate alkoxide to the lactam moiety leads to [7] (2,6)pyridinophanes. On reaction with a lithium free phophazene base ^{t}Bu-P5 deprotonation takes place at the α-carbonyl position of the azepinone ring affording the enolate [52] (Scheme 5.33).

5.3 Transformation Using Phosphazene Catalyst

5.3.1 Addition of Nucleophiles to Alkyne

The addition of O- and N-nucleophiles to alkynes catalysed by a phosphazene base, ^{t}Bu-P4 base was investigated. Alkynes were easily transformed to enol ethers and enamines in dimethyl sulfoxide (DMSO) by the addition of nucleophiles. When phenylacetylene was

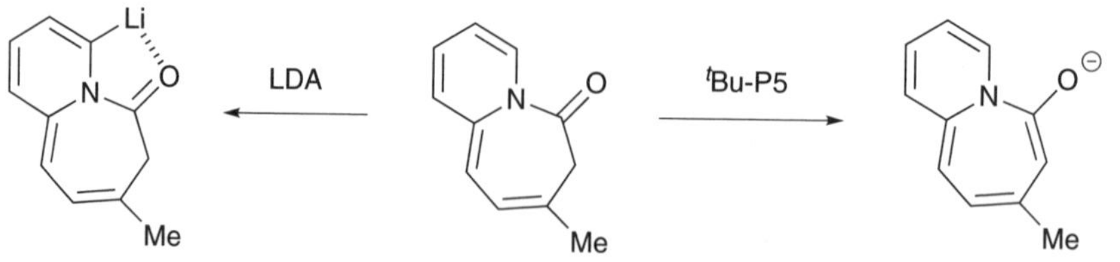

Scheme 5.33 Deprotonation using P5

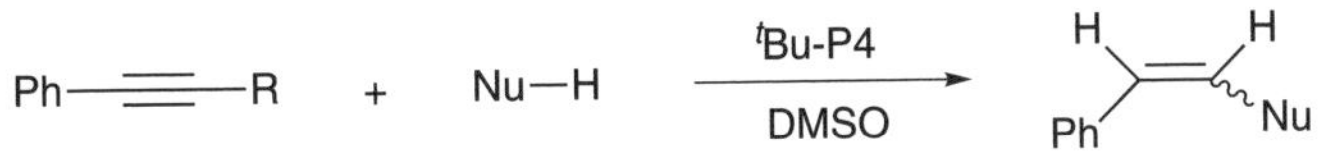

$$Ph\!\!=\!\!=\!\!-R + Nu\!-\!H \xrightarrow[\text{DMSO}]{^t\text{Bu-P4}} \underset{Ph}{\overset{H}{\diagdown}}\!\!=\!\!\underset{Nu}{\overset{H}{\diagup}}$$

Scheme 5.34 *P4 catalysed addition of nuclephiles to alkyne*

reacted with diisopropylamine, a unique head-to-head dimerization of phenylacetylene was observed to give the enyne derivative [53] (Scheme 5.34).

5.3.2 Catalytic Activation of Silylated Nucleophiles

Trialkylsilyl groups have played a very important role as effective protecting groups in organic synthesis. The activation of the nucleophile–silicon bond is important for the generation of a reactive anion to achieve a new bond formation. A novel catalytic activation of various *O*-, *N*-, and *C*-nucleophile-silicon bonds using a ^{t}Bu-P4 base has been investigated to perform reactions with various elecrophiles [54] (Figure 5.4).

5.3.2.1 Arylation of Silylated O-Nucleophile

The strong affinity of a ^{t}Bu-P4 base for a proton is regarded as synthetically useful. However, the ability of a ^{t}Bu-P4 base to activate silylated nucleophiles has not been known. Recently, many important aryloxylations of aryl halides have been reported using transition metal catalysed reactions; however, the transition metal free nucleophilic substitution reaction promoted by an organobase is also considered to be an attractive process. The reaction of TMS-OPh with 2-fluoronitrobenzene was carried out in the presence of 10 mol% ^{t}Bu-P4 base and the reaction proceeded smoothly at room temperature to give the biaryl ether in quantitatively yield (Table 5.2, run 2). When phenol was reacted under the same reaction conditions with 2-fluoronitrobenzene, the yield of the biaryl ether was only 1.6%. TBDMS-OPh also showed high reactivity in spite of the steric hindrance of the TBDMS group and the desired biaryl was obtained in 96% yield (run 3). TBDMS ethers of other phenol derivatives were used as substrates (runs 5–8), and 2-*t*-butyl, 2-bromo and 2-iodo derivatives gave excellent results in spite of the steric bulkiness of the nucleophile. The tolerance of the bromo or iodo group is very attractive for the subsequent transformation; the conventional metal catalysed reactions do not allow this kind of selectivity. When tetrabutylammonium fluoride (TBAF) was employed as a catalyst, the reaction was very sluggish and only a trace amount of product was obtained [54] (run 9).

Substituted aryl fluorides reacted with TBDMS-OC$_6$H$_4$OMe-*p* in the presence of 10 mol% ^{t}Bu-P4 base (Table 5.3). Ethyl *p*-fluorobenzoate reacted at 80 °C to give the biaryl ether in

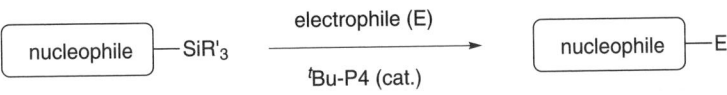

Figure 5.4 *Activation of silylated nucleophiles by ^{t}Bu-P4 base*

Table 5.2 P4 catalysed nucleophilic aromatic substitution

Run	Nu	R	Catalyst (mol%)	Solvent	Time (h)	Yield (%)
1	PhO	H	^{t}Bu-P4 base (10)	DMF	1	1.6
2	PhO	TMS	^{t}Bu-P4 base (10)	DMF	6	quant
3	PhO	TBDMS	^{t}Bu-P4 base (10)	DMF	1	96
4	PhO	TBDMS	^{t}Bu-P4 base (10)	DMSO	1	96
5	2-^{t}Bu-C$_6$H$_4$O	TBDMS	^{t}Bu-P4 base (10)	DMSO	1	99
6	2-Br-C$_6$H$_4$O	TBDMS	^{t}Bu-P4 base (10)	DMSO	6	95
7	2-I-C$_6$H$_4$O	TBDMS	^{t}Bu-P4 base (10)	DMSO	8	87
8	4-MeOC$_6$H$_4$O	TBDMS	^{t}Bu-P4 base (10)	DMSO	1	98
9	PhO	TBDMS	TBAF (10)	DMF	1	trace
10	n-HexO	TBDMS	^{t}Bu-P4 base (10)	DMSO	24	72

92% yield (run 1). Other bases, such as BEMP or DBU, were found almost inactive as a promoter (runs 2, 3). *p*-Fluorobenzonitrile reacted at 100 °C to give the desired product in 92% yield (run 4). *p*-Fluorobenzotrifluoride also reacted at 100 °C to give the desired product in 93% yield (run 5). When the reaction of *o*-fluorobromobenzene was carried out at 100 °C in DMF, the substitution occurred only at the fluorine substituted position to give the bromo biaryl ether exclusively in 85% yield (run 6). The reaction of *o*-iodofluorobenzene was also examined and the iodophenyl aryl ether was obtained in 43% yield (run 7). The reverse and complimentary regioselectivity to transition metal catalysed reactions is attractive for selective functionalization of aromatic compounds [54].

Table 5.3 Synthesis of diaryl ether

Run	X	Y	Catalyst	Solvent	Temp (°C)	Time (h)	Yield (%)
1	H	COOEt	^{t}Bu-P4 base	DMSO	80	2	92
2	H	COOEt	BEMP	DMSO	80	2	1
3	H	COOEt	DBU	DMSO	80	2	0
4	H	CN	^{t}Bu-P4 base	DMSO	100	4	92
5	H	CF$_3$	^{t}Bu-P4 base	DMSO	100	10	93
6	Br	H	^{t}Bu-P4 base	DMF	100	48	85
7	I	H	^{t}Bu-P4 base	DMF	100	48	43

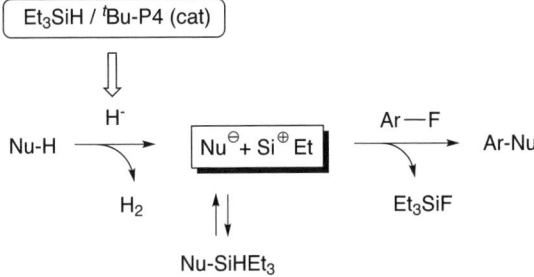

Figure 5.5 In situ silylation of nucleophiles

As above described, ^{t}Bu-P4 base was found to catalyse the coupling reaction of aryl fluorides with silyl ethers. The use of silyl ether was required for these reactions, and direct arylation of unsilylated alcohols using superbase catalysed coupling reactions has been a challenge. Therefore, the combination of triethylsilyl hydride (Et$_3$SiH) and catalytic ^{t}Bu-P4 is considered to represent an attractive hydride generating system with which to carry out sequential deprotonation and S$_N$Ar reaction [55] (Figure 5.5).

The reaction of 2-fluoronitrobenzene with n-hexanol was carried out in the presence of Et$_3$SiH and 10 mol% of ^{t}Bu-P4 at 100 °C for 2 h (Table 5.4, run 1). The arylation reaction proceeded to give the ether quantitatively. The reaction with n-butanol also proceeded smoothly. Secondary alcohols, such as 2-phenylethylalcohol and 2-butanol, were arylated under the same reaction conditions to give ethers in excellent yields (runs 3 and 4). Aryl fluorides with weaker electron-withdrawing groups were then examined. 4-Fluorobenzonitrile and 4-fluorobenzotrifluoride were found to be good substrates [55] (runs 5, 6).

p-Methoxyphenol was used as a representative phenol substrate and the S$_N$Ar reaction using Et$_3$SiH/cat. ^{t}Bu-P4 was examined. The reaction of 2-fluoronitrobenzene proceeded at room temperature to give the diaryl ether in 76% yield [55] (Scheme 5.35).

5.3.2.2 Benzofuran Natural Product

Dictyomedins A and B from Dictyostelium cellular slime moulds were recently discovered by Oshima and co-workers as physiologically active substances which inhibit their own

Table 5.4 Nucleophilic substitution of fluoroarenes with alcohols

Run	R	R'OH	Time (h)	Yield (%)
1	o-NO$_2$	^{n}hexanol	2	100
2	o-NO$_2$	^{n}BuOH	2	97
3	o-NO$_2$	PhMeCHOH	2	92
4	o-NO$_2$	EtMeCHOH	18	71
5	p-CN	^{n}BuOH	13	98
6	p-CF$_3$	^{n}BuOH	13	58

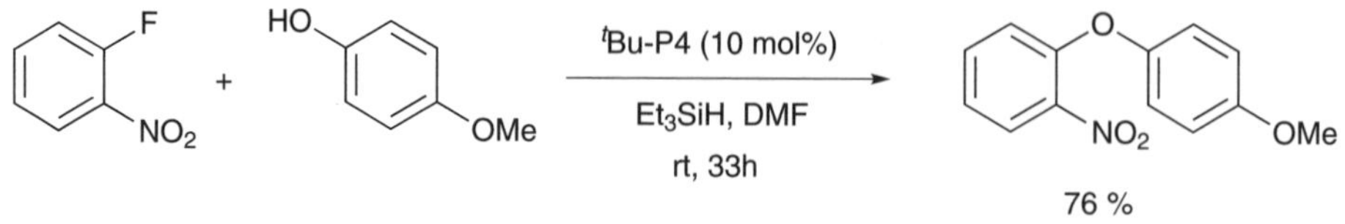

Scheme 5.35 *Nucleophilic substitution using phenol derivative*

development. Although they showed attractive biological activities in a preliminary biological evaluation, the isolated sample amount was not sufficient enough for various biological screening tests, because only a trace amount of dictyomedins can be isolated from a large amount of a dried fruit body extract of Dictyostelium. So the synthetic supply of these compounds was strongly desirable from the biological interest. Dictyomedins also have the unique structural feature of a 4-aryldibenzofuran structure. Diaryl ether synthesis is considered to be one of the most important reactions in the synthetic plans [56] (Figure 5.6).

The syntheses of dictyomedins A and B were started with the coupling of ethyl 3-bromo-4-fluorobenzoate and 3-benzyloxy-4-methoxyphenol. The reaction of the fluorobenzoate with the phenol derivative proceeded smoothly at room temperature in the presence of two equivalents of TMSNEt$_2$ as an additive and 10 mol% of ^{t}Bu-P4 base to give the diaryl ether in 94% yield. Subsequently, Suzuki–Miyaura coupling was carried out using two types of boronic acid derivatives. 4-Benzyloxy-3-methoxyphenylboronic acid pinacolate and 4-benzyloxy phenylboronic acid pinacolate were easily prepared from vanillin or 4-bromophenol, respectively. The Suzuki coupling of the bromide with boronates was carried out at 100 °C for 24 h using PdCl$_2$(dppf) and the desired arylated products were obtained in high yields. In order to form dibenzofuran ring system using palladium catalysed cyclization, the halogenation was first examined. The diaryl ethers were treated with NBS in DMF at room temperature and the desired monobrominated products were obtained. Then, palladium catalysed intramolecular biaryl formation was investigated and the bromides were treated with Pd(OAc)$_2$ and PCy$_3$ in the presence of potassium carbonate (K$_2$CO$_3$) at 100 °C for one hour under irradiation of microwave. The use of microwave dramatically accelerated the cyclization and the irradiation was critical to obtain the products in high yields in a shorter reaction time. The deprotection of benzyl groups was carried out by hydrogen (H$_2$), 10% Pd/C and the consequent hydrolysis of the ethoxycarbonyl group afforded dictyomedins A and B in 98 and 86% yields, respectively [56] (Scheme 5.36).

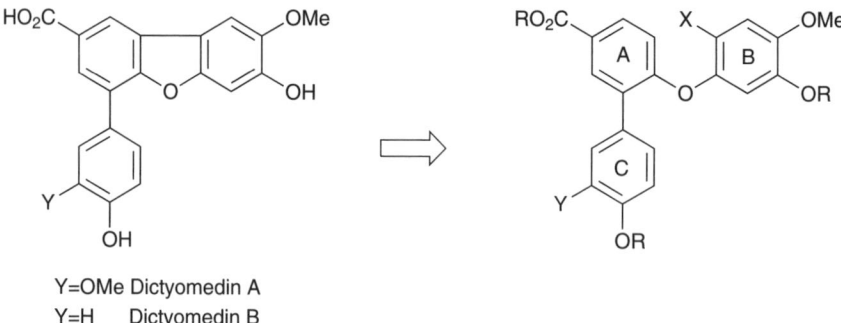

Y=OMe Dictyomedin A
Y=H   Dictyomedin B

Figure 5.6 *Diarylether intermediate for dictyomedin synthesis*

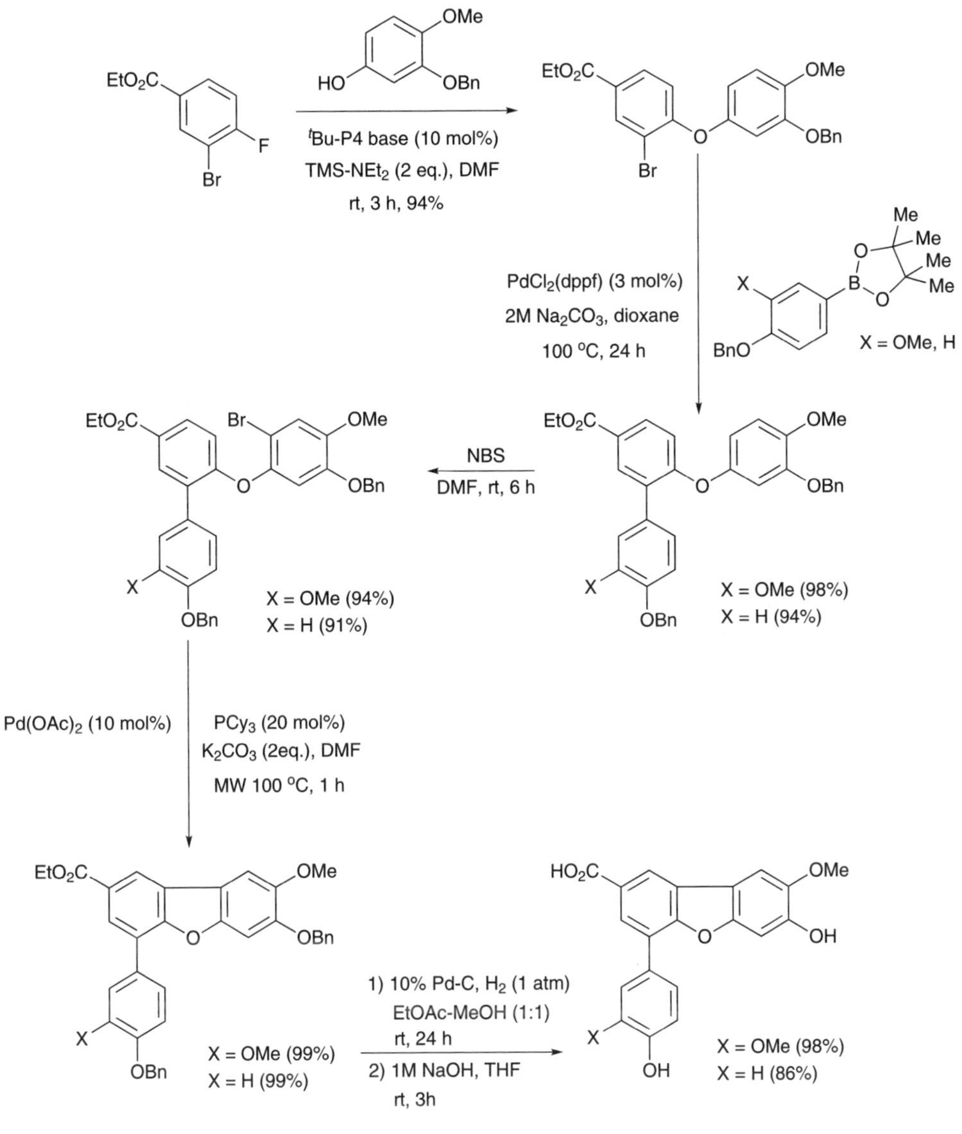

Scheme 5.36 Synthesis of dictyomedins

5.3.2.3 Generation of Aryl anion from Arylsilane

Aryltrimethylsilanes have been used as important synthons and various desilylative functionalizations have been investigated. Among these, anion mediated generation of aryl anions is one of the most important methods for selective bond formation. However, the reaction has been limited to aryltrimethylsilanes with strong electron-withdrawing groups on the aromatic rings. It was found that ^{t}Bu-P4 base could be used as an excellent catalyst to activate organosilicon compounds and the selective conversions of aryltrimethylsilanes catalysed by ^{t}Bu-P4 base have been investigated [57].

The reaction of 1-trimethylsilylnaphthalene with pivaldehyde in the presence of 20 mol% ^{t}Bu-P4 base proceeded smoothly at room temperature to give the alcohol in 91% yield. Other phosphazene bases with weaker basicities, such as ^{t}Bu-P2 base and BEMP, showed no catalytic activity. As one of the conventional strong organic bases, DBU was found to be inactive. Caesium fluoride (CsF) was then examined as a fluoride anion donor, but no carbon–silicon bond cleavage was observed. Reactions with other aldehydes have been examined; that with benzaldehyde was found to proceed somewhat slowly at room temperature. Other aryl aldehydes with electron-donating groups were also employed as electrophiles and the reactions proceeded smoothly at room temperature [57] (Table 5.5).

The reactions of other aryltrimethylsilanes have been examined. 2-Trimethylsilylnaphthalene reacted with pivalaldehyde in the presence of t-Bu-P4 base at room temperature to give the alcohol in 68% yield. 4-Fluorophenyltrimethylsilane and 4-bromophenyltrimethylsilane gave the corresponding alcohols in 64 and 73% yields respectively. Similarly, 4-trifluoromethylphenyltrimethylsilane and 2-trifluoromethylphenyltrimethylsilane reacted smoothly to give the alcohols in 69 and 73% yields respectively. 4-Methoxycarbonylphenyltrimethylsilane reacted to give the alcohol in 46% yield. The reactions of heteroaryltrimethylsilanes were then examined. 2-Pyridyltrimethylsilane and 3-pyridyltrimethylsilane reacted smoothly to give the alcohols in 67 and 80%. 2-Thienyltrimethylsilane also reacted to give the alcohol in 81% yield [57] (Table 5.6).

2-Trimethylsilylated benzamide was reacted with benzaldehyde to give the 1,2-adduct, which was treated with AcOH-toluene to give the phthalide in 76% yield. Phthalides have been used as precursors for the synthesis of anthraquinones [57] (Scheme 5.37).

5.3.2.4 Peterson Olefination

Condensation of α-silylalkyl compounds with carbonyl compounds to form alkenes has been known as the Peterson olefination reaction. Usually an α-silanyl carbanion generated *in situ*

Table 5.5 Selective functionalization of arylsilane

Run	Base	R	Temp. (°C)	Time (h)	Yield (%)
1	^{t}Bu-P4	^{t}Bu	rt	1	91
2	^{t}Bu-P2	^{t}Bu	rt	24	0
3	BEMP	^{t}Bu	rt	24	0
4	DBU	^{t}Bu	rt	24	0
5	CsF	^{t}Bu	rt	24	0
6	^{t}Bu-P4	Ph	80	6	61
7	^{t}Bu-P4	4-MeOC$_6$H$_4$	rt	1	78
8	^{t}Bu-P4	2-MeOC$_6$H$_4$	rt	1	68

Table 5.6 Transformation of functionalized arylsilane

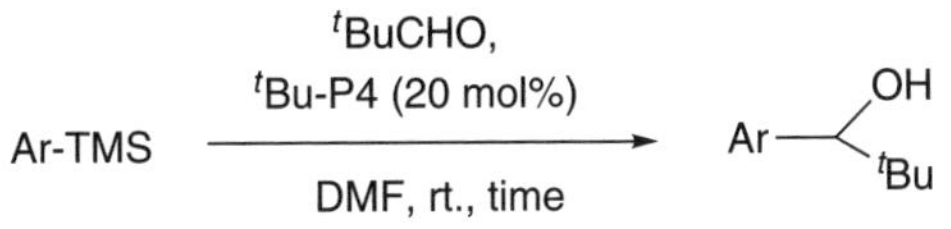

Ar-TMS $\xrightarrow[\text{DMF, rt., time}]{\substack{^{t}\text{BuCHO,} \\ ^{t}\text{Bu-P4 (20 mol\%)}}}$ Ar–CH(OH)(^{t}Bu)

Run	Ar	Time (h)	Yield (%)
1	2-Naphthyl	3	68
2	4-FC$_6$H$_4$	1	64
3	4-BrC$_6$H$_4$	1	73
4	4-CF$_3$C$_6$H$_4$	2	69
5	2-CF$_3$C$_6$H$_4$	1	73
6	4-MeOOCC$_6$H$_4$	1	46
7	2-Pyridyl	1	67
8	3-Pyridyl	1	80
9	2-Thienyl	1	81

using a stoichiometric metallic base such as LDA has been used for the reaction. The reaction has been widely employed as an important method for alkene synthesis, because of the simpler workup and purification procedure than the Wittig reaction. This condensation reaction is considered to be more valuable if the transformation can be accomplished with a catalytic base.

When the reaction of benzophenone and ethyl trimethylsilylacetate in THF in the presence of 10 mol% ^{t}Bu-P4 base was carried out at −78 °C, the condensation reaction proceeded smoothly and the unsaturated ester was isolated in 94% yield (Table 5.7, run 1). Other phophazene bases with weaker basicity, such as a ^{t}Bu-P2 base and BEMP, showed no effect on the condensation under the same reaction conditions (runs 2, 3). DBU also showed no catalytic effect on the condensation (run 4). When TBAF was used for the reaction, the 1,2-adduct was isolated in 38% yield, but no formation of alkene was observed [58] (run 5).

It was found that ^{t}Bu-P4 base is an excellent catalyst for the condensation, and the further scope and limitations were investigated (Table 5.8). As another trimethylsilylalkyl substrate, N,N-diethyltrimethylsilylacetamide was reacted with benzophenone to give the unsaturated amide in 87% yield (run 1) and the reaction of trimethylsilylacetonitrile with benzophenone was also successful, giving the arylacryronitorile in 78% yield (run 2). Benzaldehyde was then used as a carbonyl compound and ethyl cinnamate was obtained in 89% yield using ethyl trimehylsilylacetate (run 3). Condensation of other arylaldehydes with ethyl trimethylsilylacetate proceeded smoothly and the corresponding arylacrylates

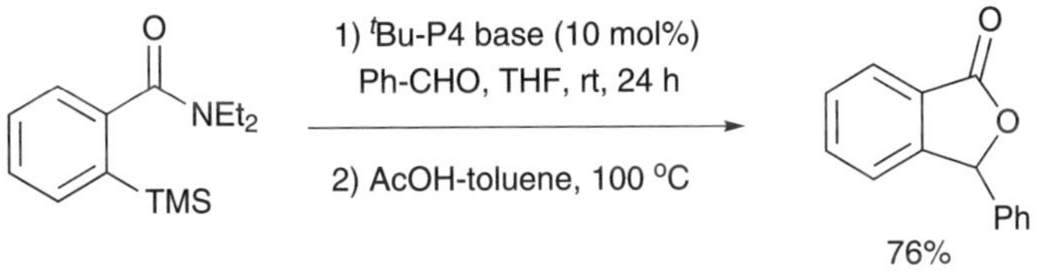

Scheme 5.37 Synthesis of phthalide

Table 5.7 Condensation of silylacetate with benzophenone

Run	Base	Time (h)	Yield (%)
1	^{t}Bu-P4 base	6	94
2	^{t}Bu-P2 base	24	0
3	BEMP	24	0
4	DBU	24	0
5	TBAF	6	0

were obtained in good yields (runs 4–6). The double bonds of the products from arylaldehydes were E geometry. When acetophenone was used as a carbonyl compound, the reaction with trimethylsilylacetonitrile proceeded smoothly to give the alkene in 63% yield as a mixture of E and Z isomers ($E:Z = 89:11$) (run 7). The reaction of chalcone with trimethylsilylacetate gave the diene in 81% yield ($E:Z = 100:0$) (run 8). Similarly, the condensation of chalcone with trimethylsilylacetonitrile also gave the corresponding diene in 80% yield ($E:Z = 50:50$) (run 9). As an example of aliphatic aldehyde, hexanal was used as a substrate and the desired unsaturated ester was obtained in 35% yield [58] (run 10).

Compared to ketones or aldehydes, the reactivities of formamides to nucleophiles were considered to be lower and the investigation was started with the reaction at room temperature (Table 5.9). When N-methylformanilide was reacted with ethyl trimethylsilylacetate in THF at room temperature, the condensation proceeded smoothly and the enaminoester was obtained in 90% yield (run 1). Similarly, trimethylsilylacetonitrile was

Table 5.8 Peterson reaction in the presence of P4

Run	R^1	R^2	EWG	Time (h)	Yield (%)
1	Ph	Ph	CONEt$_2$	12	87
2	Ph	Ph	CN	6	78
3	Ph	H	CO$_2$Et	6	89
4	4-MeC$_6$H$_4$	H	CO$_2$Et	6	91
5	4-MeOC$_6$H$_4$	H	CO$_2$Et	6	69
6	2-furyl	H	CO$_2$Et	6	85
7	Ph	Me	CN	13	63
8	Ph	CH=CHPh(E)	CO$_2$Et	6	81
9	Ph	CH=CHPh(E)	CN	13	80
10	n-Pentyl	H	CO$_2$Et	6	35

Table 5.9 Synthesis of enamine using the Peterson reaction

Run	X	R	EWG	Time (h)	Yield (%)
1	H	Me	CO_2Et	24	90
2	H	Me	CO_2Et	24	92[a]
3	H	Me	CN	24	78[a]
4	Me	Me	CO_2Et	48	74
5	OMe	Me	CO_2Et	48	47
6	CO_2Et	Me	CO_2Et	24	85
7	CO_2Et	Me	CN	48	87
8	CN	Me	CO_2Et	20	80
9	CN	Me	CN	48	80
10	H	$CH_2CH=CH_2$	CO_2Et	24	83[a]
11	H	$CH_2CH=CH_2$	CN	24	42[a]
12	H	CH_2Ph	CO_2Et	24	99[a]
13	H	CH_2Ph	CN	24	88[a]

[a] Without solvent.

condensed with N-methylformanilide to give the enaminonitrile in 78% (run 3). Various substituents on the aromatic ring of the formanilides were compatible and the tolerance of alkokycarbonyl and cyano groups during the condensation is considered to be synthetically important (runs 4–9). N-Allyl and N-benzyl formanilides were also reacted to give the corresponding enamines [58] (runs 10–13).

5.3.2.5 Promotion of Halogen–Zinc Exchange Reaction

The strong affinity of a ^{t}Bu-P4 base for protons is regarded as synthetically useful. However, the ability of a ^{t}Bu-P4 base to activate organometallic compounds is largely undocumented. Control of the reactivities of organometallic compounds is the key to the success of selective bond formation, and the catalytic promotion of organometallics by the ^{t}Bu-P4 base is an important subject. In recent years, organozinc compounds have been widely used in organic synthesis. One of the most powerful methods for the preparation of functionalized organozinc derivatives is the halogen–zinc exchange reaction, and the promotion of the exchange reaction by a catalytic ^{t}Bu-P4 base has been investigated [59] (Table 5.10).

When the reaction of 4-iodobenzoate and diethylzinc in THF in the presence of 30 mol % ^{t}Bu-P4 base was carried out at room temperature, the halogen–zinc exchange reaction proceeded smoothly. In the absence of ^{t}Bu-P4 base, the exchange reaction was very slow and only a trace of de-iodinated product was detected. Other phophazene bases with weaker basicity, such as ^{t}Bu-P2 base and ^{t}Bu-P1, showed less or no effect on the halogen–zinc exchange reaction. DBU also showed no promotive reactivity on the exchange reaction. When the amount of ^{t}Bu-P4 base was reduced to 10 mol%, the exchange reaction became significantly slower in THF. In order to optimize the reaction conditions further,

Table 5.10 Halogen–zinc exchange reaction in the presence of P4

I-C6H4-CO2Et → 1) ZnEt2, base, rt, solvent, time; 2) H3O+ → H-C6H4-CO2Et

Run	Base (mol%)	Solvent	ZnEt2 (mol%)	Time (h)	Yield (%)
1	^{t}Bu-P4 (30)	THF	240	11	quant.
2	—	THF	240	21	3[a]
3	P2 (30)	THF	240	12	33
4	P1 (30)	THF	240	12	0
5	DBU (30)	THF	240	11	0
6	^{t}Bu-P4 (10)	THF	200	11	56
7	^{t}Bu-P4 (10)	DMF	200	11	quant.
8	^{t}Bu-P4 (10)	NMP	200	11	quant.
9	^{t}Bu-P4 (1)	DMF	200	11	91

[a] The reaction was carried out at 60 °C.

the solvent was switched from THF to DMF. In DMF, 5 mol% of ^{t}Bu-P4 base was sufficient for activation of the exchange reaction. In the absence of ^{t}Bu-P4 base, the exchange reaction in DMF was quite slow and it was found that ^{t}Bu-P4 base apparently functions as an activator.

Using this exchange reaction, some functionalizations of aryl halides were examined. As an example of 1,2-addition to a carbonyl group, the arylzinc prepared from 4-iodobenzoate and diethylzinc in the presence of ^{t}Bu-P4 base in THF was reacted with benzaldehyde to give the benzhydrol derivative in 78% yield. As for the 1,4-addition reaction, the arylzinc prepared similarly in THF was reacted with chalcone and the 1,4-adduct was obtained in 71% yield under copper-free reaction conditions. Allylation was also carried out in the absence of copper additive, and allylbenzoate was obtained in 98% yield. It has been reported that arylzinc compounds are inert to 1,4-addition and allylation reaction in the absence of additives and conventionally the employment of copper species has been widely used. However, in this case the ^{t}Bu-P4 base is considered to promote the reactivity of arylzinc compounds toward electrophiles [59] (Scheme 5.38).

The arylzinc compounds prepared in DMF can also be used in the palladium catalysed Negishi coupling reaction and the reaction of the arylzinc with iodobenzene in the presence of palladium catalyst gave the corresponding biarylcarboxylated in 53% yield [59] (Scheme 5.39).

5.3.2.6 S$_N$Ar Reaction of C-Nucleophile

Introduction of O-nucleophiles to aromatics has been well investigated for S$_N$Ar reactions, but there are limited successful examples for the arylation of carbanions by S$_N$Ar reaction. Highly activated aryl fluorides have been the only successful substrates for the S$_N$Ar reactions using malonates as nucleophiles. Diethyl methylmalonate was reacted with

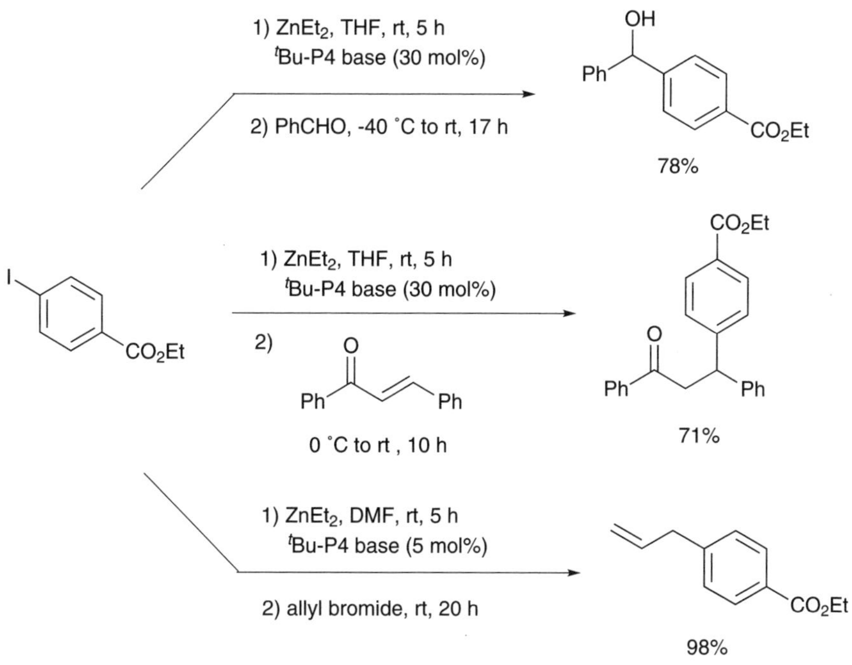

Scheme 5.38 Functionalization of arylzinc derivatives

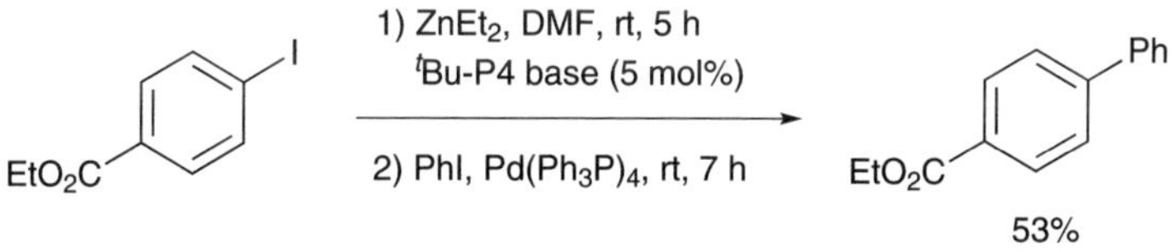

Scheme 5.39 Negishi coupling of arylzinc compound

2-fluoronitrobenzene using Et₃SiH/cat. ᵗBu-P4 at 80 °C for 1 h, and the arylation was found to proceed smoothly to give the arylmalonate in 99% yield (Table 5.11, run 1). The reaction with diethyl malonate was found to be slower and the arylmalonate was obtained in 56% yield after reacting for 22 h (run 2). Other α-substituted malonates also reacted with 2-fluoronitrobenzene to give arylmalonates (runs 3, 4). α-Substituted cyanoacetate and α-substituted malononitrile were also found to be excellent C-nucleophiles (runs 5, 6). 4-Fluoronitrobenzene reacted with methylmalonate to give the arylmalonate in 97% yield (run 7). The less reactive 2-fluoro- and 4-fluorobenzonitrile also reacted with methylmalonate to give the corresponding arylmalonates in 46% and 66% yield, respectively (runs 8, 9). Conventionally, aryl fluorides with weak electron-withdrawing groups have not been used for S_NAr reaction with these C-nucleophiles, and the method provides a new and highly effective S_NAr reaction protocol [55].

Table 5.11 Aromatic nucleophilic substitution using malonates

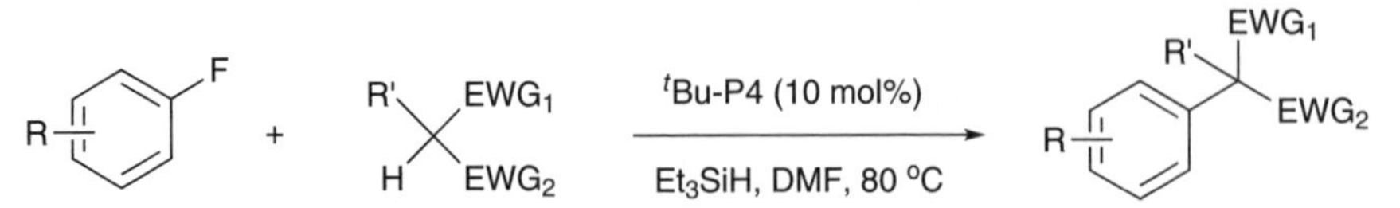

Run	R^1	R^2	EWG^1	EWG^2	Time (h)	Yield (%)
1	2-NO$_2$	Me	CO$_2$Et	CO$_2$Et	1	99
2	2-NO$_2$	H	CO$_2$Et	CO$_2$Et	22	56
3	2-NO$_2$	^{n}hexyl	CO$_2$Et	CO$_2$Et	24	76
4	2-NO$_2$	allyl	CO$_2$Et	CO$_2$Et	2	95
5	2-NO$_2$	allyl	CO$_2$Et	CN	3	89
6	2-NO$_2$	allyl	CN	CN	3	89
7	4-NO$_2$	Me	CO$_2$Et	CO$_2$Et	2	97
8	2-CN	Me	CO$_2$Et	CO$_2$Et	24	46
9	4-CN	Me	CO$_2$Et	CO$_2$Et	24	66

5.4 Proazaphosphatrane Base (Verkade's Base)

5.4.1 Properties of Proazaphosphatrane

Proazaphosphatranes are bicyclic, nonionic bases in which the phosphorus atom functions as the site of electron pair donation. In contrast to phosphazene bases, which are protonated on a nitrogen atom, proazaphosphatranes are protonated on the bridgehead phosphorus atom with a transannulation to form the corresponding azaphosphatranes [60] (Figure 5.7). The basicity of Verkade's superbase in acetonitrile solution is shown in that the corresponding pKa value is 29.0. Hence, its basicity is comparable or higher than that of some other P1 phosphazenes, but it is lower than the basicity of P2 phosphazenes. Structural characteristics of Verkade's superbase and its conjugate acid, as well as the origin of its basicity, have also been examined [61].

5.4.2 Synthesis Using Proazaphosphatrane

5.4.2.1 Activation of Allylsilane

Preparation of homoallylic alcohols was achieved by reacting aromatic aldehydes with allyltrimethylsilane in the presence of 20 mol% ^{i}Pr-proazaphosphatarene base. Lower

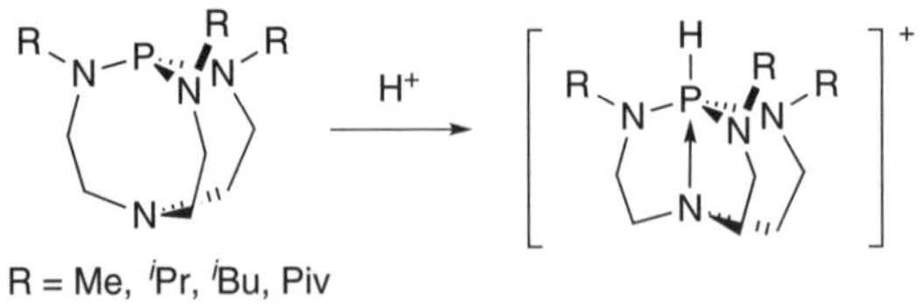

Figure 5.7 Protonation of proazaphosphatranes

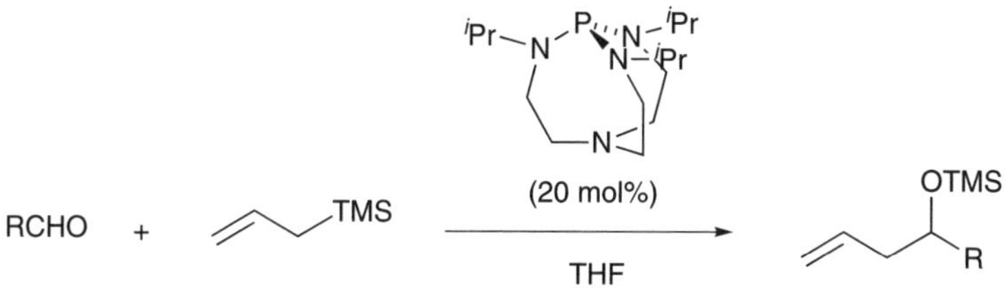

Scheme 5.40 1,2-Addition of allylsilane

yields were observed for aldehydes bearing electron-donating groups. For reasons not clear, the less sterically hindered CH$_3$-proazaphosphotrane proved to be ineffective for this transformation. The reaction is assumed to proceed via activation of the allylsilane by attack of the phosphorus atom of ^{i}Pr-proazaphosphatrane at the allylic silicon atom to form a phosphonium ion, with formation of an allylic anion that then adds to the aldehydes [62] (Scheme 5.40).

5.4.2.2 Activation of Trimethylsilyl Cyanide

The proazaphosphatrane base promotes the trimethylsilylcyanation of aryl and alkyl aldehydes and ketones in moderate to high yields at room temperature. ^{29}Si-NMR spectra suggested that a phosphorus–silicon adduct is formed as an intermediate [63] (Scheme 5.41).

5.4.2.3 1,4-Addition of Nucleophiles

The 1,4-addition of primary alcohols, higher nitroalkanes and a Schiff's base of an α-amino ester to α,β-unsaturated substrates produces the corresponding products in moderate to excellent yields in the presence of catalytic amounts of the nonionic strong bases P(RNCH$_2$CH$_2$)$_3$N in isobutyronitrile. Diasteroselectivity for the *anti* form of the product is higher than in the case of the Schiff's base in the absence of lithium ion [64] (Scheme 5.42).

5.4.2.4 α-Arylation of Alkanenitriles

A new catalyst system for the synthesis of α-aryl substituted nitriles has been reported. The ^{i}Bu-proazaphosphatrane serves as an efficient and versatile ligand for the palladium

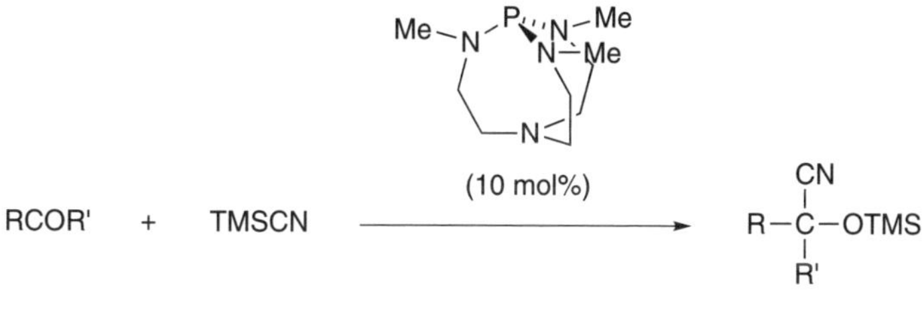

Scheme 5.41 1,2-Addition of trimethylsilylcyanide

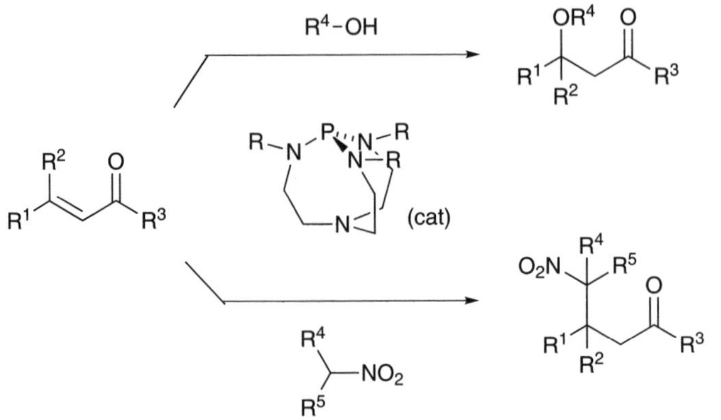

Scheme 5.42 *1,4-Addition catalysed by proazaphosphatrane*

catalysed direct α-arylation of nitriles with aryl bromides. Using the base, ethyl cyanoacetate and primary as well as secondary nitriles are efficiently coupled with a wide variety of aryl bromides possessing electron rich, electron poor, electron neutral and sterically hindered groups [65] (Scheme 5.43).

5.4.2.5 Dimerization of Allylsulfone

In the presence of CH$_3$-proazaphosphatrane catalyst, allyl phenyl sulfone readily dimerizes to give the product shown, for which only incomplete and inconclusive data exist in the literature. The dimer was shown from ^{1}H NMR spectral considerations to have the E configuration [66] (Scheme 5.44).

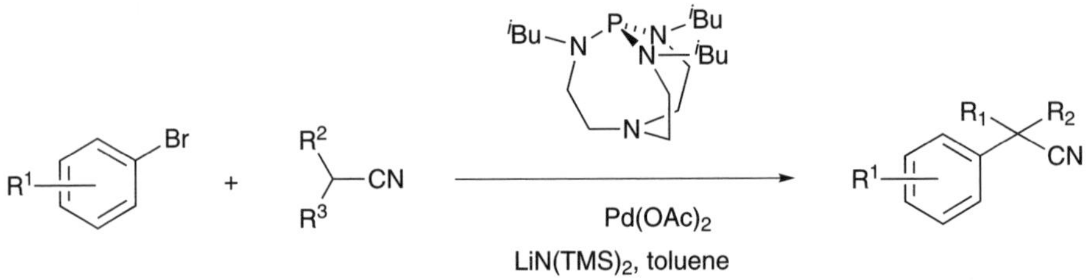

Scheme 5.43 *Arylation of substituted acetonitriles*

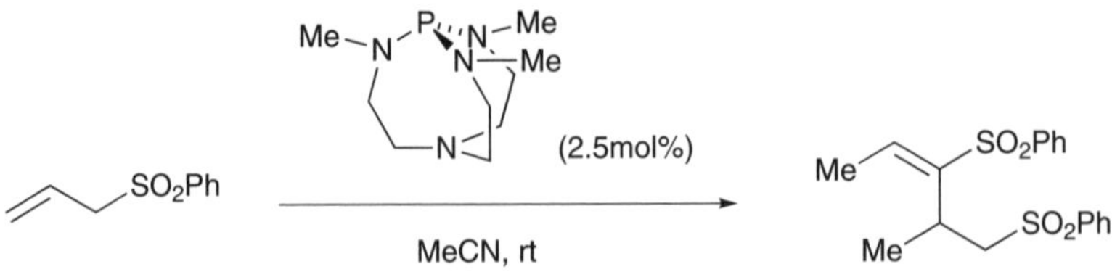

Scheme 5.44 *Dimerization of allyl phenyl sulfone*

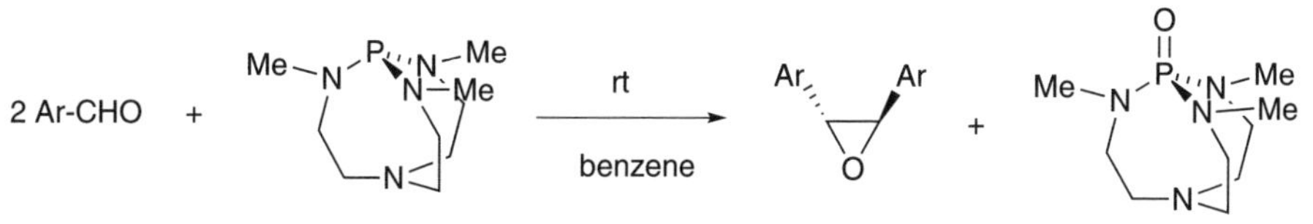

Scheme 5.45 Diaryl epoxide formation from aryl aldehyde

5.4.2.6 Formation of Diaryl Epoxide

In contrast to its acyclic analogue $P(NMe_2)_3$, which in benzene at room temperature reacts with two aryl aldehyde molecules bearing electron-withdrawing groups to give the corresponding diaryl epoxides as an isomeric mixture (*trans/cis* ratio 72/28 – 51/49), $P(MeNCH_2CH_2)_3N$ under the same reaction conditions is found to be a highly selective reagent that provides epoxides with *trans/cis* ratios as high as 99/1. These reactions are faster with the proazaphosphatrane because its phosphorus atom is apparently more nucleophilic than that in $P(NMe_2)_3$ [67] (Scheme 5.45).

5.4.2.7 β-Hydroxynitrile Synthesis

The successful synthesis of β-hydroxynitriles was reported in good to excellent yields from aldehydes and ketones in a simple reaction that is promoted by proazaphosphatrane bases. The reaction occurs in the presence of magnesium salts, which activate the carbonyl group and stabilize the enolate thus produced [68] (Scheme 5.46).

5.4.2.8 Stille Reaction

Proazaphosphatrane bases were used for palladium catalysed Stille reactions of aryl chlorides. These bases efficiently catalyse the coupling of electronically diverse aryl chlorides with an array of organotin reagents. The catalyst system based on benzyl (Bn)-proazaphosphatrane is active for the synthesis of sterically hindered biaryls. The use of the proazaphosphatrane allows room temperature coupling of aryl bromides and it also permits aryl triflates and vinyl chlorides to participate in Stille coupling [69] (Scheme 5.47).

5.4.2.9 Palladium Catalysed Amination of Aryl Halides

The proazaphosphatrane bases serve as an effective ligand for the palladium catalysed amination of a wide array of bromides and iodides. Other bicyclic or acyclic triaminophosphines, even those of similar basicity and/or bulk, were inferior. The palladium catalysed

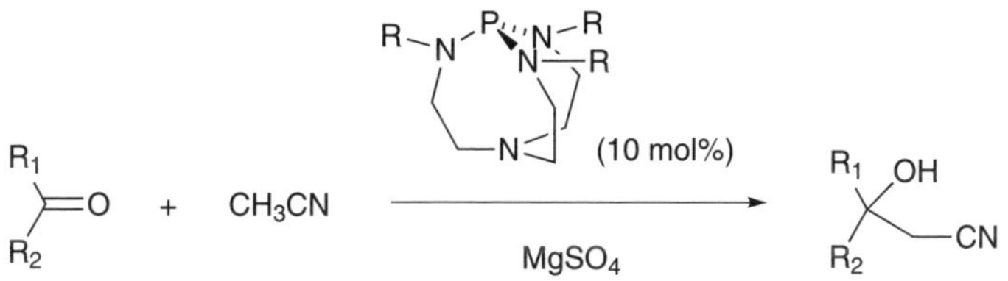

Scheme 5.46 1,2-Addition of acetonitrile

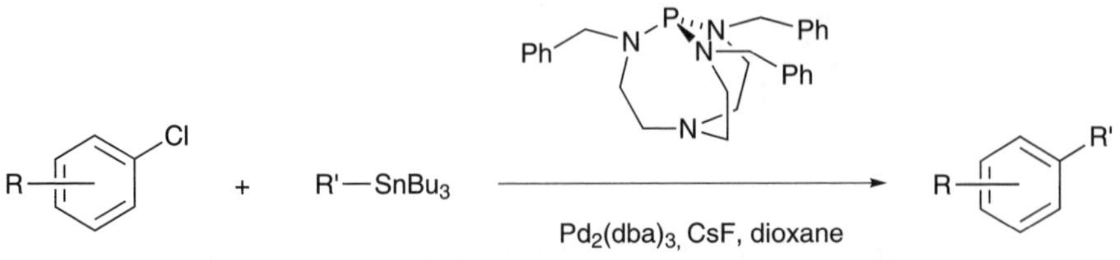

Scheme 5.47 Stille coupling using proazaphosphatrane ligand

amination reaction of aryl chlorides with amines also proceeded in the presence of the ^{i}Bu-proazaphosphatrane to afford the corresponding arylamines in good to excellent yields. Electron poor, electron neutral and electron rich aryl chlorides all participated with equal ease [70] (Scheme 5.48).

5.4.2.10 Reduction with Silane

The CH$_3$-proazaphosphatrane base can activate Si—H bonds in polymethylhydrosiloxane (PMHS) to reduce carbonyl compounds. Aldehydes and ketones are reduced under mild conditions by PMHS in the presence of the proazaphosphatrane, giving the corresponding alcohols in high yield. A variety of aromatic aldehydes were smoothly reduced to the corresponding alcohols in high yields with survival of the aromatic chloro, nitro, cyano and methoxy substituents. Conjugated as well as isolated double bonds also remained intact during regioselective reduction of the carbonyl groups [71] (Scheme 5.49).

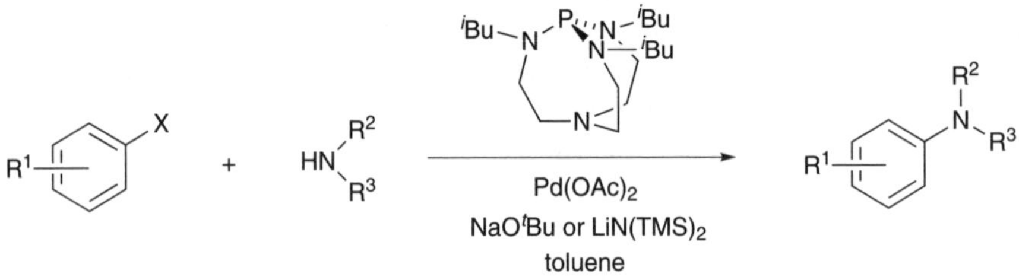

Scheme 5.48 Amination of aryl halides

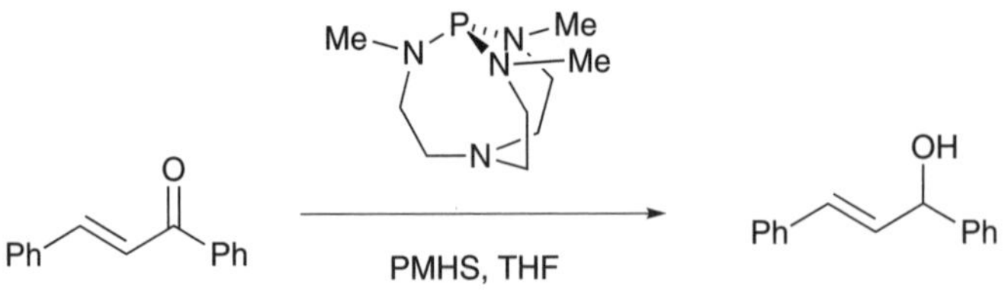

Scheme 5.49 Reduction with silane

5.5 Concluding Remarks

In the last two decades, nonionic nitrogen bases such as phosphazene bases and proazaphosphatrane bases have been found useful in organic synthesis. Various reactions previously restricted to ionic bases such as LDA and KO^{t}Bu have been carried out by employing these nonionic bases. These bases are now closing the gap between ionic and nonionic bases in stoichiometric applications because of the extraordinary basicity and weak nucleophilicity of nonionic bases. The protonated bulky cations are formed in the deprotonation process, and consequently produced anions are essentially close to naked and highly reactive. Many attractive catalytic transformations have also been developed and most of the reactions show excellent efficacy and selectivity. The history of these nonionic bases as organic catalysts is still short at present and further exploration is required to develop sophisticated catalytic transformations for organic synthesis using these nonionic bases.

References

1. Issleib, K. and Lischewski, M. (1973) Dimethylaminoiminophosphorane. *Synthesis and Reactivity in Inorganic and Metal-Organic Chemistry*, **3**, 255–266.
2. Marchenko, A.P., Koidan, G.N. and Kudryavtsev, A.A. (1980) Ammonolysis of triamidohalophosphonium halides. *Zhurnal Obshchei Khimii*, **50**, 679–680; Marchenko, A.P., Koidan, G. N., Pinchuk, A.M. and Kirsanov, A.V. (1984) Phosphorylated triamidoimidophosphates. *Zhurnal Obshchei Khimii*, **54**, 1774–1782.
3. Schwesinger, R. and Schlemper, H. (1987) Peralkylated polyaminophosphazenes – extremely strong, neutral nitrogen bases. *Angewandte Chemie – International Edition*, **26**, 1167–1169; Schwesinger, R., Hasenfratz, C., Schlemper, H. *et al.* (1993) How strong and how can uncharged phosphazene bases be? *Angewandte Chemie – International Edition*, **32**, 1361–1363.
4. Schwesinger, R., Willaredt, J., Schlemper, H. *et al.* (1994) Novel, very strong, uncharged auxiliary base; design and synthesis of monomeric and polymer-bound triaminoiminophosphorane bases of broadly varied steric demand. *Chemische Berichte*, **127**, 2435–2454.
5. Schwesinger, R., Schlemper, H., Hasenfratz, C. *et al.* (1996) Extremely strong, uncharged auxiliary bases; Monomeric and polymer-supported polyaminophophazenes (P_2-P_5). *Liebigs Annalen*, 1055–1081.
6. Koppel, I.A., Schwesinger, R., Breuer, T. *et al.* (2001) Intrinsic basicity of phosphorus imines and ylides: a theoretical study. *Journal of Physical Chemistry A*, **105**, 9575–9586; Maksic, Z.B. and Vianello, R. (2002) Quest for the origin of basicity: initial vs final effect in neutral nitrogen bases. *Journal of Physical Chemistry A*, **106**, 419–430.
7. Novak, I., Wei, X.M. and Chin, W.S. (2001) Electronic structures of very strong, neutral bases. *Journal of Physical Chemistry A*, **105**, 1783–1788.
8. Leito, I., Rodima, T., Koppel, I.A. *et al.* (1997) Acid-base equilibria in nonpolar media 1. A spectrophotomeric method for acidity measurements in heptane. *The Journal of Organic Chemistry*, **62**, 8479–8483; Rõõm, E., Kaljurand, I., Leito, I. *et al.* (2003) Acid-base equilibria in nonpolar media 3. Expanding the spectrophotomeric acidity scale in heptane. *The Journal of Organic Chemistry*, **68**, 7795–7799.
9. Kaljurand, I., Rodima, T., Leito, I. *et al.* (2000) Self-consistent spectrophotometric basicity scale in acetonitrile covering the range between pyridine and DBU. *The Journal of Organic Chemistry*, **65**, 6202–6208.
10. Rodima, T., Mäemets, V. and Koppel, I. (2000) Synthesis and N-aryl-substituted iminophosphoranes and NMR spectroscopic investigation of their acid-base properties in acetonitrile. *Journal of the Chemical Society – Perkin Transactions*, **1**, 2637–2644; Rodima, T., Kaljurand, I.,

Pihl, A. *et al.* (2002) Acid-base equilibria in non-polar media 2. Self-consistent basicity scale in THF solution ranging from 2-methoxypyridine to EtP1(pyrr) phosphazene. *The Journal of Organic Chemistry*, **67**, 1873–1881.
11. Kaljurand, I., Rodima, T., Pihl, A. *et al.* (2003) Acid-base equilibria in nonpolar media 4. Extension of the self-consistent basicity scale in THF medium. Gas-phase basicities of phosphazene. *The Journal of Organic Chemistry*, **68**, 9988–9993; Sooväli, L., Rodima, T., Kaljurand, I. *et al.* (2006) Basicity of some P1 phosphazenes in water and in aqueous surfactant solution. *Organic and Biomolecular Chemistry*, **4**, 2100–2105.
12. Kaljurand, I., Koppel, I.A., Kütt, A. *et al.* (2007) Experimental gas-phase basicity scale of superbasic phosphazenes. *Journal of Physical Chemistry A*, **111**, 1245–1250.
13. Kolomeitov, A.A., Koppel, I.A., Rodima, T. *et al.* (2005) Guanidinephosphazenes: design, synthesis, and basicity in THF and in the gas phase. *Journal of the American Chemical Society*, **127**, 17656–17666.
14. Despotović, I., Kovačević, B. and Maksić, Z.B. (2007) Hyperstrong neutral organic bases: phosphazeno azacalix[3](2,6)pyridine. *Organic Letters*, **9**, 4709–4712.
15. Raab, V., Gauchenova, E., Merkoulov, A. *et al.* (2005) 1,8-Bis(hexamethyltriaminophosphazenyl)naphthalene, HMPN: a superbasic bisphosphazene "proton sponge". *Journal of the American Chemical Society*, **127**, 15738–15743; Kovačević, B., Maksić, Z.B. (2006) High basicity of tris-(tetramethylguanidyl)-phosphine imide in the gas phase and acetonitrile – a DFT study. *Tetrahedron Letters*, **47**, 2553–2555.
16. Köhn, U., Schulz, M., Schramm, A. *et al.* (2006) A new class of chiral phosphazene bases: synthesis and characterization. *European Journal of Organic Chemistry*, 4128–4134.
17. O'Donnell, M.J., Delgado, F., Hostettler, C. and Schwesinger, R. (1998) An efficient homogeneous catalytic enantioselective synthesis of α-amino acid derivatives. *Tetrahedron Letters*, **39**, 8775–8778; O'Donnell, M.J., Delgado, F. and Pottorf, R.S. (1999) Enantioselective solid-phase synthesis of α-amino acid derivatives. *Tetrahedron*, **55**, 6347–6362;O'Donnell, M.J., Delgado, F., Domínguez, E. *et al.* (2001) Enantioselective solution- and solid-phase synthesis of glutamic acid derivatives via Michael addition reactions. *Tetrahedron: Asymmetry*, **12**, 821–828; Scott, W.L., O'Donnell, M.J., Delgado, F. and Alsina, J. (2002) A solid-phase synthetic route to unnatural amino acids with diverse side-chain substitutions. *The Journal of Organic Chemistry*, **67**, 2960–2969.
18. Guillena, G. and Nájera, C. (1998) PTC and organic bases-LiCl assisted alkylation of imidazolidinone-glycine iminic derivatives for the asymmetric synthesis of α-amino acids. *Tetrahedron: Asymmetry*, **9**, 3935–3938.
19. Allmendinger, T. (1991) Ethyl phenylsulfinyl fluoroacetate, a new and versatile reagent for the preparation of α-fluoro-α,β-unsaturated carboxylic acid ester. *Tetrahedron*, **47**, 4905–4914.
20. Lash, T.D., Thompson, M.L., Werner, T.M. and Spence, J.D. (2000) Synthesis of novel pyrrolic compounds from nitroarenes and isocyanoacetates using a phosphazene superbase. *Synlett*, 213–216; Murashima, T., Tamai, R., Nishi, K. *et al.* (2000) Synthesis and X-ray structure of stable 2*H*-isoindoles. *Journal of the Chemical Society – Perkin Transactions*, **1**, 995–998.
21. Falk-Heppner, M., Keller, M. and Prinzbach, H. (1989) Unsaturated 1,4-cis and 1,4-trans-Diamino-tetradeoxycycloheptites – enantiomerically pure, polyfunctionalized tropa derivatives by enzymatic hydrolysis. *Angewandte Chemie – International Edition*, **28**, 1253–1255.
22. Moore, J.E., Spinks, D. and Harrity, J.P.A. (2004) Microwave promoted amination of 3-bromoisoxazole. *Tetrahedron Letters*, **45**, 3189–3191.
23. Xu, W., Springfield, S.A. and Koh, J.T. (2000) Highly efficient synthesis of 1-thioglycosides in solution and solid phase using iminophosphorane bases. *Carbohydrate Research*, **325**, 169–176.
24. Palomo, C., Palomo, A.L., Palomo, F. and Mielgo, A. (2002) Soluble α-amino acid salts in acetonitrile: practical technology for the production of some dipeptides. *Organic Letters*, **4**, 4005–4008.
25. Chinchilla, R., Galindo, N. and Nájera, C. (1999) Chiral 3,6-dihydro-2*H*-1,4-oxazin-2-ones as alanine equivalents for the asymmetric synthesis of α-methyl α-amino acids (AMAAs) under mild reaction conditions. *Synthesis*, 704–717.
26. Bensa, D., Brunel, J.-M., Buono, G. and Rodoriguez, J. (2001) Highly efficient phosphazene base-catalysed addition of β-ketoesters in water. *Synlett*, 715–717.

27. Heinelt, U., Herok, S., Matter, H. and Wildgoose, P. (2001) Solid-phase optimisation of achiral amidinobenzyl indoles as potent and selective factor Xa inhibitors. *Bioorganic & Medicinal Chemistry Letters*, **11**, 227–230.
28. Richter, H., Walk, T., Holtzel, A. and Jung, G. (1999) Polymer bound 3-hydroxy-2-methylidenepropionic acids. A template for multiple core structure libraries. *The Journal of Organic Chemistry*, **64**, 1362–1365.
29. Du, X. and Armstrong, R.W. (1997) Synthesis of benzofuran derivatives on solid support via SmI_2-mediated radical cyclization. *The Journal of Organic Chemistry*, **62**, 5678–5679.
30. Baxendale, I.R., Lee, A.-L. and Ley, S.V. (2001) A concise synthesis of the natural product carpanone using solid-supported reagents and scavenger. *Synlett*, 1482–1484.
31. Caldarelli, M., Habermann, J. and Ley, S.V. (1999) Clean five-step synthesis of an array of 1,2,3,4-tetra-substituted pyrroles using polymer-supported reagents. *Journal of the Chemical Society – Perkin Transactions*, **1**, 107–110.
32. Xu, W., Mohan, R. and Morrissey, M.M. (1998) Polymer supported bases in solution-phase synthesis. 2. A convenient method for *N*-alkylation reaction of weakly acidic heterocycles. *Bioorganic & Medicinal Chemistry Letters*, **8**, 1089–1092.
33. Brain, C.T. and Brunton, S.A. (2001) Synthesis of 1,3,4-oxadiazole using polymer-supported reagents. *Synlett*, 382–384.
34. Solladié-Cavallo, A. and Crescenzi, B. (2000) Full conversion in diastereoselective aldol additions using 'naked' enolates under catalytic amount of phosphazene base: the hydroxypinanone as chiral auxiliary. *Synlett*, 327–330; Solladié-Cavallo, A., Koessler, J.-L., Isarno, T. *et al.* (1997). The hydroxypinanone as chiral auxiliary in Michael additions: an inversion of diastereoselectivity at low concentration of enolate, a substrate-directed approach. *Synlett*, 217–218.
35. Jin, Z., Kim, S.H. and Fuchs, P.L. (1996) Phosphazene base P2-Et mediated isomerization of vinyl sulfones to allyl sulfones. *Tetrahedron Letters*, **37**, 5247–5248; Jin, Z. and Fuchs, P.L. (1996) Use of phosphazene base and phase-transfer conditions for regiospecific alkylative isomerization of vinyl capable of undergoing β-elimination reactions. *Tetrahedron Letters*, **37**, 5249–5252; Kim, S.H., Figueroa, I. and Fuchs, P.L. (1997) Application of the Grubbs ring-closing metathesis for the construction of a macrocyclic ansa-bridge. Synthesis of the tricyclic core of roseophilin. *Tetrahedron Letters*, **38**, 2601–2604.
36. Solladié-Cavallo, A., Diep-Vohuule, A. and Isarno, T. (1998) Two-step synthesis of trans-2-arylcyclopropane carboxylates with 98–100ee by the use of a phosphazene base. *Angewandte Chemie – International Edition*, **37**, 1689–1691; Solladié-Cavallo, A., Roje, M., Isarno, T. *et al.* (2000) Pyridyl and furyl epoxides of more than 99% enantiomeric purities: use of a phosphazene base. *European Journal of Organic Chemistry*, 1077–1080; Solladié-Cavallo, A., Bouérat, L. and Roje, M. (2000) Asymmetric synthesis of trans-disubstituted aryl-vinyl epoxides: a p-methoxy effect. *Tetrahedron Letters*, **41**, 7309–7312.
37. Solladié-Cavallo, A., Roje, M., Welter, R. and Sunjić, V. (2004) Two-step asymmetric synthesis of disubstituted *N*-tosyl aziridines having 98–100% ee: use of a phosphazene base. *The Journal of Organic Chemistry*, **69**, 1409–1412.
38. Costa, A., Nájera, C. and Sansano, J.M. (2002) Synthetic applications of o-and p-halobenzyl sulfones as zwitterionic synthons: preparation of orhto-substituted cinnamates and biarylacetic acids. *The Journal of Organic Chemistry*, **67**, 5216–5225.
39. Clark, A.J., Al-Faiyz, Y.S.S., Patel, D. and Broadhurst, M.J. (2001) Rearrangement of unactivated *N*-alkyl-*O*-benzoyl hydroxamic acid derivatives with phosphazene bases. *Tetrahedron Letters*, **42**, 2007–2009; Clark, A.J., Patel, D. and Broadhurst, M.J. (2003) Base-mediated reaction of *N*-alkyl-*O*-acyl hydroxamic acids: synthesis of 3-oxo-2,3-dihydro-4-isoxazole carboxylic ester derivatives. *Tetrahedron Letters*, **44**, 7763–7765.
40. Palomo, C., Oiarbide, M., Lopez, R. and Gomez-Bengoa, E. (2000) Phosphazene bases for the preparation of biaryl thioethers from aryl iodides and arenethiols. *Tetrahedron Letters*, **41**, 1283–1286.
41. Solladié-Cavallo, A., Liptaj, T., Schmitt, M. and Solgadi, A. (2002) iso-Propylacetate: formation of a single enolate with tBuP4 as base. *Tetrahedron Letters*, **43**, 415–418; Solladié-Cavallo, A., Csaky, A.G., Gantz, I. and Suffert, J. (1994) Diastereoselective Alkylation of

8-phenylmenthyl phenylacetate: aggregated lithium enolate versus 'naked' enolate. *The Journal of Organic Chemistry*, **59**, 5343–5346; Fruchart, J.-S., Gras-Masse, H. and Melnyk, O. (2001) Methyl phenylacetate enolate generated with the P4-tBu Schwesinger base: 'naked' or not? *Tetrahedron Letters*, **42**, 9153–9155; Fruchart, J.-S., Lippens, G., Kuhn, C. *et al.* (2002) Solid-phase enolate chemistry investigated using HR-MAS NMR spectroscopy. *The Journal of Organic Chemistry*, **67**, 526–532.
42. Keynes, M.N., Earle, M.A., Sudharshan, M. and Hultin, P.G. (1996) Synthesis and asymmetric alkylation of glucose-derived bicyclic oxazinones: an evaluation of their use as "chiral glycines". *Tetrahedron*, **52**, 8685–8702.
43. Koch, C.-J., Höfner, G., Polborn, K. and Wanner, K.T. (2003) Synthesis of the four stereoisomers of 1-amino-2-(hydroxymethyl)cyclobutanecarboxylic acid and their biological evaluation as ligands for the glycine binding site of the NMDA receptor. *European Journal of Organic Chemistry*, 2233–2242; Koch, C.-J., Simonyiová, S., Pabel, J. *et al.* (2003) Asymmetric synthesis with 6-*tert*-butyl-6-methyl-3,6-dihydro-2*H*-1,4-oxazin-2-one as a new chiral glycine equivalent: preparation of enantiomerically pure α-tertiary and α-quaternary α-amino acids. *European Journal of Organic Chemistry*, 1244–1263.
44. Muccioli, A.B. and Simpkins, N.S. (1994) First examples of episulfone substitution reaction via α-sulfonyl carbanion intermediates. *The Journal of Organic Chemistry*, **59**, 5141–5143; Dishington, A.P., Douthwaite, R.E., Mortlock, A. *et al.* (1997) Episulfone substitution and ring-opening reactions via α-sulfonyl carbanion intermediates. *Journal of the Chemical Society – Perkin Transactions*, **1**, 323–337.
45. Pietzonka, T. and Seebach, D. (1992). *N*-Perbenzylation of oligopeptides with P4-phosphazene base; a new protecting group technique for modification and solubilization of peptides in apolar organic solvents. *Angewandte Chemie – International Edition*, **31**, 1481; Seebach, D., Bezencon, O., Jaun, B. *et al.* (1996) Further C-alkylation of cyclotetrapeptides via lithium and phosphazenium (P4) enolates: discovery of a new conformation. *Helvetica Chimica Acta*, **79**, 588–608; Papageorgiou, C., Kallen, J., France, J. and French, R. (1997) Conformational control of cyclosporin through substitution of the N-5 position. A new class of cyclosporin antagonists. *Bioorganic and Medicinal Chemistry*, **5**, 187–192.
46. Solladie-Cavallo, A., Roche, D., Fischer, J. and De Cian, A. (1996) Effect of a phosphazene base on the diastereoselectivity of addition of α-sulfonyl carbanions to butyraldehyde and isopropylideneglyceraldehyde. *The Journal of Organic Chemistry*, **61**, 2690–2694.
47. Alonso, D.A., Nájera, C. and Varea, M. (2004) 3,5-Bis(trifluoromethyl)phenyl sulfones in the modified Julia olefination: application to the synthesis of resveratrol. *Tetrahedron Letters*, **45**, 573–577.
48. Kraus, G.A., Zhang, N., Verkade, J.G. *et al.* (2000) Deprotonation of benzylic ethers using a hindered phosphazene base. A synthesis of benzofurans from *ortho*-substituted benzaldehydes. *Organic Letters*, **2**, 2409–2410.
49. Mamdani, H.T. and Hartley, R.C. (2000) Phosphazene bases and the anionic oxy-Cope rearrangement. *Tetrahedron Letters*, **41**, 747–749.
50. Palomo, C., Oiarbide, M., Lopez, R. and Gomez-Bengoa, E. (1998) Phosphazene P4-But base for the Ullmann biaryl ether synthesis. *Chemical Communications*, 2091–2092.
51. Imahori, T. and Kondo, Y. (2003) A new strategy for deprotonative functionalization of aromatics: transformations with excellent chemoselectivity and unique regioselectivities using *t*Bu-P4 base. *Journal of the American Chemical Society*, **125**, 8082–8083.
52. Maier, W., Keller, M. and Eberbach, W. (1993) Formation of [7](2,6)pyridinophanes by ring enlargement of a pyrido[1,2-a]azepinone. *Heterocycles*, **35**, 817–820.
53. Imahori, T., Hori, C. and Kondo, Y. (2004) Functionalization of alkynes catalysed by t-Bu-P4 base. *Advanced Synthesis and Catalysis*, **346**, 1090–1092.
54. Ueno, M., Hori, C., Suzawa, K. *et al.* (2005) Catalytic activation of silylated nucleophiles using *t*Bu-P4 as a base. *European Journal of Organic Chemistry*, 1965–1968.
55. Ueno, M., Yonemoto, M., Hashimoto, M. *et al.* (2007) Nucleophilic aromatic substitution using Et_3SiH/cat. t-Bu-P4 as a system for nucleophile activation. *Chemical Communications*, 2264–2266.
56. Ebisawa, M., Ueno, M., Oshima, Y. and Kondo, Y. (2007) Synthesis of dictyomedins using phosphazene base catalysed diaryl ether formation. *Tetrahedron Letters*, **48**, 8918–8921.

57. Suzawa, K., Ueno, M., Wheatley, A.E.H. and Kondo, Y. (2006) Phosphazene base-promoted functionalization of aryltrimethylsilanes. *Chemical Communications*, 4850–4852.
58. Kobayashi, K., Ueno, M. and Kondo, Y. (2006) Phosphazene base-catalysed condensation of trimethylsilylacetate with carbonyl compounds. *Chemical Communications*, 3128–3130.
59. Ueno, M., Wheatley, A.E.H. and Kondo, Y. (2006) Phosphazene base-promoted halogen-zinc exchange reaction of aryl iodides using diethylzinc. *Chemical Communications*, 3549–3550.
60. Verkade, J.G. and Kisanga, P.B. (2004) Recent applications of proazaphosphatranes in organic synthesis. *Aldlichimica Acta*, **37** (1), 3; Verkade, J.G. and Kisanga, P.B. (2003) Proazaphosphatranes: a synthesis methodology trip from their discovery to vitamin A. *Tetrahedron*, **59**, 7819–7858.
61. Kovačević, B., Barić, D. and Maksić, Z.B. (2004) Basicity of exceedingly strong non-ionic organic bases in acetonitrile – Verkade's superbase and some related phosphazenes. *New Journal of Chemistry*, **28**, 284–288.
62. Wang, Z., Kisanga, P. and Verkade, J.G. (1999) P(^{i}PrNCH$_2$CH$_2$)$_3$N: an effective Lewis base promoter for the allylation of aromatic aldehydes with allyltrimethylsilane. *The Journal of Organic Chemistry*, **64**, 6459–6461.
63. Wang, Z., Fetterly, B. and Verkade, J.G. (2002) P(MeNCH$_2$CH$_2$)$_3$N: an effective catalyst for trimethylsilylcyanation of aldehydes and ketones. *Journal of Organometallic Chemistry*, **646**, 161.
64. Kisanga, P.B., Ilankumaran, P., Fetterly, B.M. and Verkade, J.G. (2002) P(RNCH$_2$CH$_2$)$_3$N: efficient 1,4-addition catalysts. *The Journal of Organic Chemistry*, **67**, 3555–3560.
65. You, J. and Verkade, J.G. (2003) P(^{i}BuNCH$_2$CH$_2$)$_3$N: an efficient ligand for the direct α-arylation of nitriles with aryl bromides. *The Journal of Organic Chemistry*, **68**, 8003–8007.
66. Yu, Z. and Verkade, J.G. (2004) Catalytic dimerization of allyl phenyl sulfone in the presence of a proazaphosphatrane catalyst. *Advanced Synthesis and Catalysis*, **346**, 539–541.
67. Liu, X. and Verkade, J.G. (2000) P(MeNCH$_2$CH$_2$)$_3$N: a highly selective reagent for synthesizing trans-epoxides from aryl aldehydes. *The Journal of Organic Chemistry*, **65**, 4560–4564.
68. Kisanga, P., McLeod, D., D'Sa, B. and Verkade, J. (1999) P(RNCH$_2$CH$_2$)$_3$N-catalysed synthesis of β-hydroxy nitriles. *The Journal of Organic Chemistry*, **64**, 3090–3094.
69. Su, W., Urgaonkar, S., McLaughlin, P.A. and Verkade, J.G. (2004) Highly active palladium catalysts supported by bulky proazaphosphatrane ligands for Stille cross-coupling: coupling of aryl and vinyl chlorides, room temperature coupling of aryl bromides, coupling of aryl triflates, and synthesis of sterically hindered biaryls. *Journal of the American Chemical Society*, **126**, 16433–16439.
70. Urgaonkar, S., Nagarajan, M. and Verkade, J.G. (2003) P(^{i}BuNCH$_2$CH$_2$)$_3$N: an effective ligand in the palladium-catalysed amination of aryl bromides and iodides. *The Journal of Organic Chemistry*, **68**, 452–459; Urgaonkar, S., Xu, J.-H. and Verkade, J.G. (2003) Application of a new bicyclic triaminophosphine ligand in Pd-catalysed Buchwald–Hartwig amination reactions of aryl chlorides, bromides, and iodides. *The Journal of Organic Chemistry*, **68**, 8416–8423; Urgaonkar, S., Nagarajan, M. and Verkade, J.G. (2003) P[N(^{i}Bu)CH$_2$CH$_2$]$_3$N: A versatile ligand for the Pd-catalysed amination of aryl chlorides. *Organic Letters*, **5**, 815–818.
71. Wang, Z., Wroblewski, A.E. and Verkade, J.G. (1999) P(MeNCH$_2$CH$_2$)$_3$N: An efficient promoter for the reduction of aldehydes and ketones with poly(methylhydrosilane). *The Journal of Organic Chemistry*, **64**, 8021–8023.

6
Polymer-Supported Organosuperbases

Hiyoshizo Kotsuki
Faculty of Sciences, Kochi University, 2-5-1 Akebono-cho, Kochi 780-8520, Japan

6.1 Introduction

The immobilization of reagents or molecular catalysts in a polymeric form generally offers several advantages in organic synthesis. These include: the ease of handling without the need for special equipment or an inert atmosphere; the simplicity of separating the products from complex reaction mixtures; the ease of recovering and recycling the reagents; and the adaptability to continuous-flow synthetic processes. Accordingly, there have been a wide variety of investigations to discover new efficient polymer-supported reagents, and great achievements have been made in recent years, especially in the fields of pharmaceutical and agrochemical research, and combinatorial chemistry [1–9].

In many cases, polymer-supported organic base reagents can be synthesized by the direct attachment of a suitable strong base core on a polymeric backbone through a covalent bond linkage. Typically, 1,1,3,3-tetramethylguanidine (TMG, pK_a = 23.7) [10], 1,5,7-triazabicyclo[4.4.0]dec-5-ene (TBD, pK_a = 26.2) [10], 1,8-diazabicyclo[5.4.0]undec-7-ene (DBU, pK_a = 23.9) [10], the Schwesinger's phosphazene base, that is, 2-*tert*-butylimino-2-diethylamino-1,3-dimethylperhydro-1,3,2-diazaphosphorine (BEMP, pK_a = 27.6) [11], and the Verkade's proazaphosphatrane base (PAPT, pK_a = 32.9) [12], can be used as nonionic strong bases (Figure 6.1).

The synthetic study of polymer-supported superbase reagents has grown from the pioneering work from Tomoi's laboratory [13–15]. There it was showed that *N*-alkylation

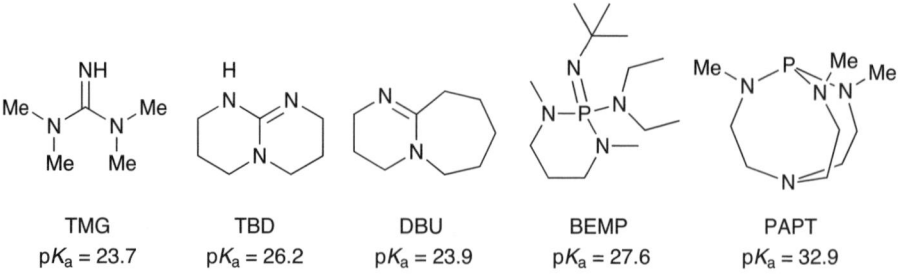

Figure 6.1 The pK_a values (in MeCN) of conjugate acids of typical organic superbases

of DBU upon treatment with chloromethylated or ω-bromoalkylated polystyrene resins was highly reliable and gave the desired family of molecules. The synthesis of other reagents of this type is normally based on this strategy [11,16,17], and some of these polymer-supported superbase reagents are now available commercially (Figure 6.2).

In this chapter, recent advances in polymer-supported superbase reagents or catalysts in several organic transformations will be outlined.

6.2 Acylation Reactions

Acylation of hydroxyl or amino compounds is important in protective chemistry, and hence several useful methods have been developed using polymer-supported superbases, relying primarily on their nature as acid scavengers. In an early stage (1996) of this chemistry,

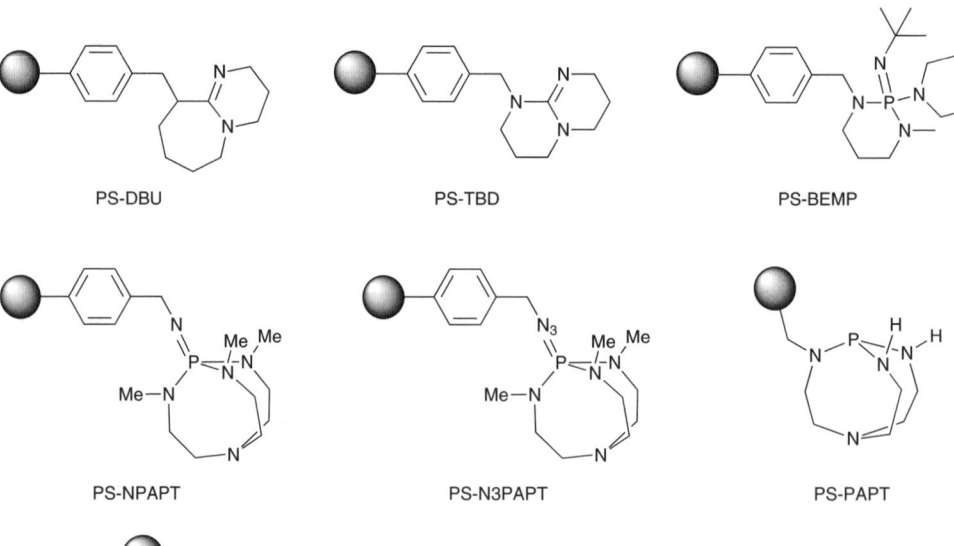

Figure 6.2 Representative examples of polymer-supported organic bases

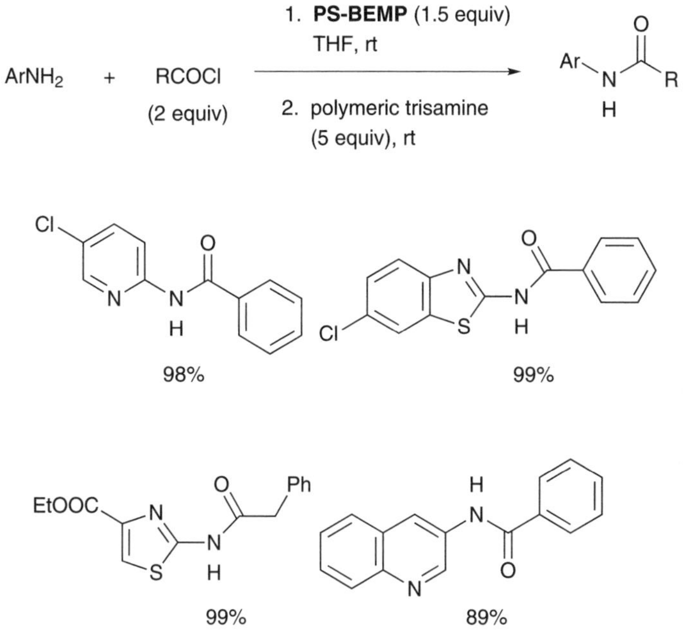

Scheme 6.1 Acylation of less reactive heterocyclic amines

alkylguanidine attached to polystyrene was used in the transesterification of triglycerides [18]. Thereafter, similar works have appeared in the literature [19–23], and TBD anchored on the mesoporous molecular sieve, MCM-41, has also been developed for the transesterification of β-ketoesters [24].

On the other hand, Kim and Le reported that the use of PS-BEMP was very convenient for the N-acylation of several weakly nucleophilic heterocyclic amines with acyl chloride (Scheme 6.1) [25]. In this example, treatment with polymer-based trisamine after completion of the reaction with PS-BEMP was convenient for removing excess acyl chloride. Thus, the corresponding N-acylated compounds could be obtained in good yields and high purities by simple filtration. The utility of PS-BEMP as an acid scavenger has also been established in other types of transformations to make free amines from their salts [26].

Similarly, Ilankumaran and Verkade reported that PS-NPAPT worked well as a catalyst (10 mol%) for the acetylation of a variety of primary alcohols using vinyl acetate as an effective acetylating agent even in the presence of acid labile groups within the molecule [27]. This catalyst could be used three times without a significant loss of reactivity, but after that it tended to decompose to a fine powder. Interestingly, this chemistry was recently extended to the green chemical production of biodiesel from soybean oil via transesterification in methanol using PS-N3PAPT as a catalyst at room temperature [28]. It was shown that the catalyst could be recycled 11 times, although its activity decreased gradually with time. Despite the extensive efforts to find clean and selective methods for the synthesis of carbamates and unsymmetrical alkyl carbonates from diethyl carbonate, there are few reports on the use of the mesoporous silica-supported TBD as a basic catalyst [29].

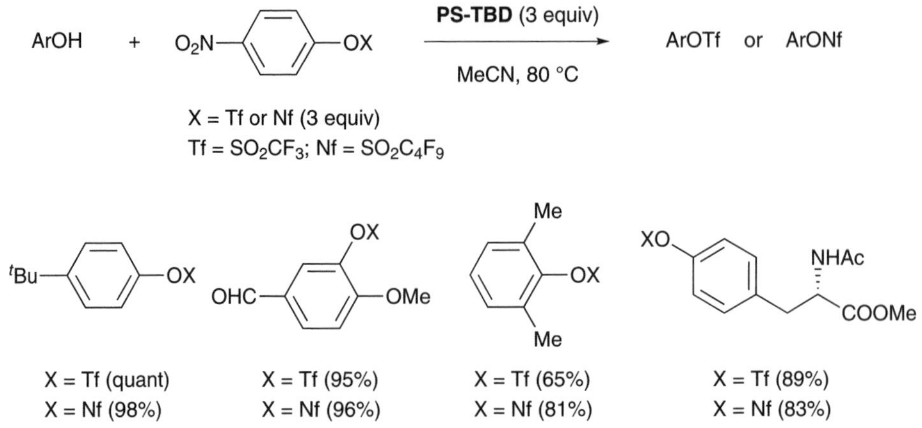

Scheme 6.2 Triflation or nonaflation of phenols via sulfonate transfer reactions

Due to the importance of aryl triflates and related compounds in modern organic chemistry, including metal catalysed cross-coupling reactions, an efficient procedure for the triflation or nonaflation of phenols has been developed by the combination of PS-TBD (3 equiv.) and a perfluoroalkanesulfonyl transfer reagent (3 equiv.) such as *p*-nitrophenyl triflate or nonaflate [30] (Scheme 6.2). In this case, the free PS-TBD reagent could be recovered by washing successively with diluted acid, base, water and organic solvents, and almost comparable results were obtained with use of the recycled catalyst.

6.3 Alkylation Reactions

The solid base character of polymer-supported superbase reagents can be conveniently applied to important carbon–carbon bond forming reactions. For example, PS-TBD has been shown to be an efficient promoter to realize the addition reaction of nitroalkanes to aldehydes and alicyclic ketones, that is, the Henry reaction [31] (Scheme 6.3).

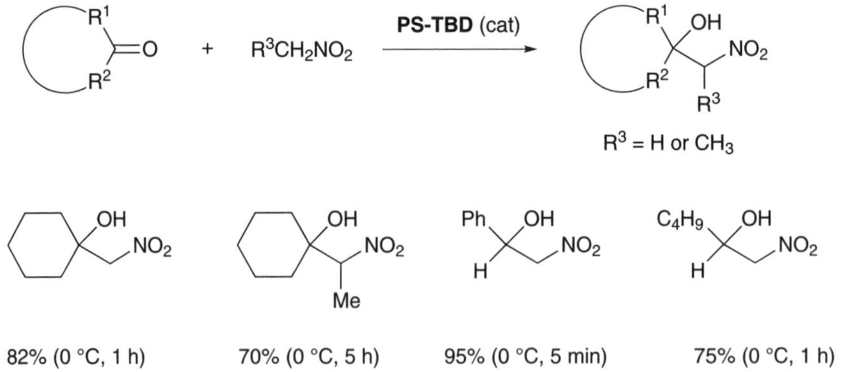

Scheme 6.3 Henry reactions of nitroalkanes with aldehydes or ketones

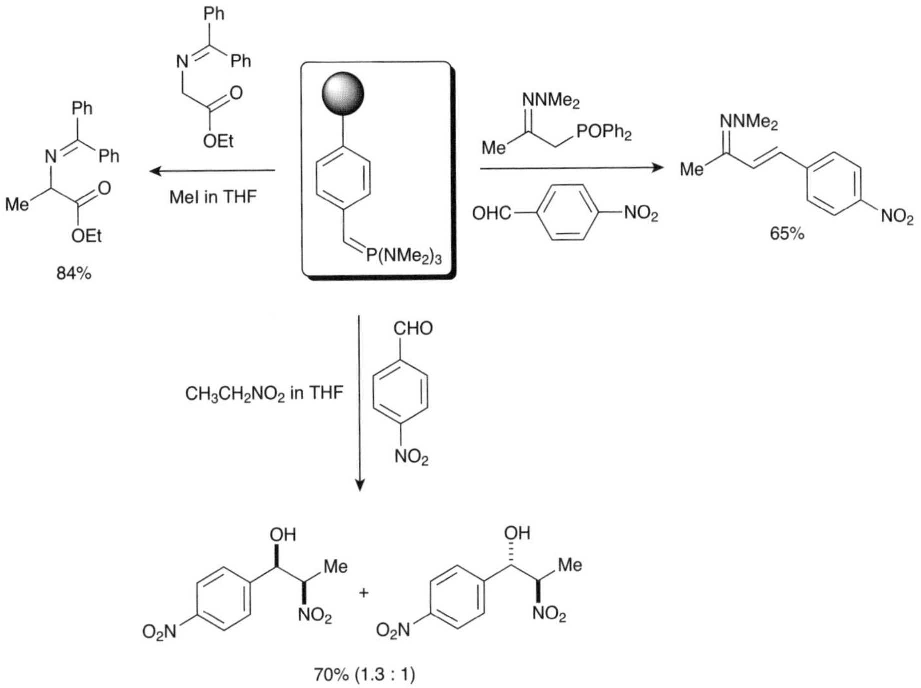

Scheme 6.4 *C–C bond forming reactions using polymer-supported phosphorus ylide*

This method is of great value in terms of its simplicity for purification as well as mild reaction conditions and fast reactions. As a related example, a new type of polymer-supported ylide base was found to be effective for the Henry reaction, the C-alkylation of α-amino acid derivatives, Wittig-olefination, and the N-alkylation of amide derivative [32,33] (Scheme 6.4).

Fetterley et al. have been very active in determining the utility of strongly basic PAPT-type reagents in organic synthesis, and have reported that α-aminonitrile compounds could be efficiently synthesized by the Strecker condensation of aldehydes, amines, and trimethylsilyl cyanide (TMSCN) in the presence of a nitrate salt of PS-PAPT as a catalyst [34] (Scheme 6.5). In this work, the reusability of the catalyst was also established, and the catalyst activity was ascribed to the participation of a nitrate ion of the catalyst acted as a 'proton shuttle.'

The Michael addition reaction is one of the most important carbon–carbon bond forming reactions in organic synthesis, and several examples using polymer-supported reagents have been reported. For example, Bensa et al. found that the Michael addition reaction of 1,3-dicarbonyl compounds with activated olefins as a Michael acceptor could be efficiently catalysed by PS-BEMP [35] (Scheme 6.6). The procedure does not require dry solvents or an inert atmosphere, and filtration of the catalyst gives substantially pure products. It is also

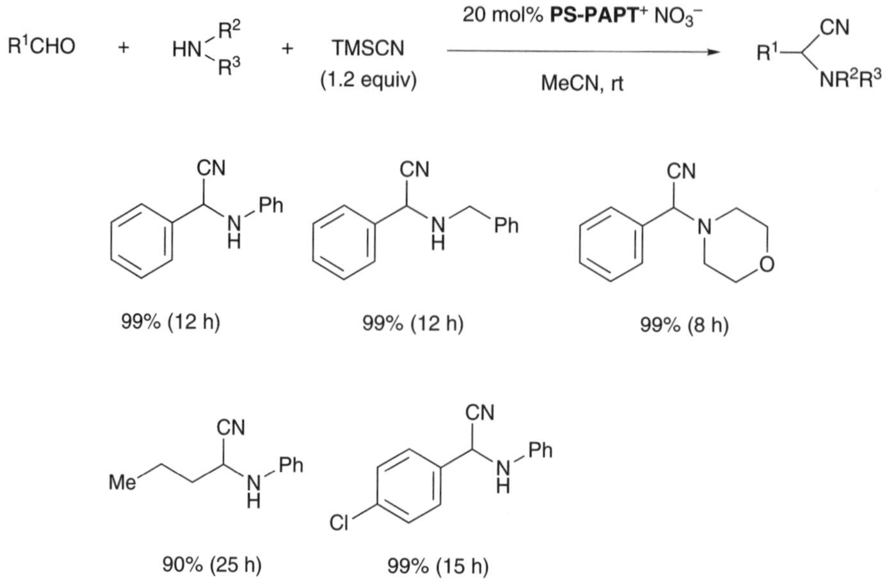

Scheme 6.5 Strecker-type condensation of aldehydes with amines and TMSCN

known that mesoporous silica materials or polymer-supported TBD could catalyse the similar type of Michael addition reactions under remarkably mild conditions [36–38].

Surprisingly, there have been only few synthetic studies on polymer-supported 'asymmetric' superbase reagents. Recently, Wannaporn and Ishikawa prepared a new chiral guanidine based polymer catalyst and applied it to the asymmetric Michael addition reaction of iminoacetate with methyl vinyl ketone [39] (Scheme 6.7). Although the catalyst shows only moderate levels of reactivity and enantioselectivity, the result demonstrates the possibility of expanding an exciting field of asymmetric synthesis using polymer-supported chiral superbase catalysts.

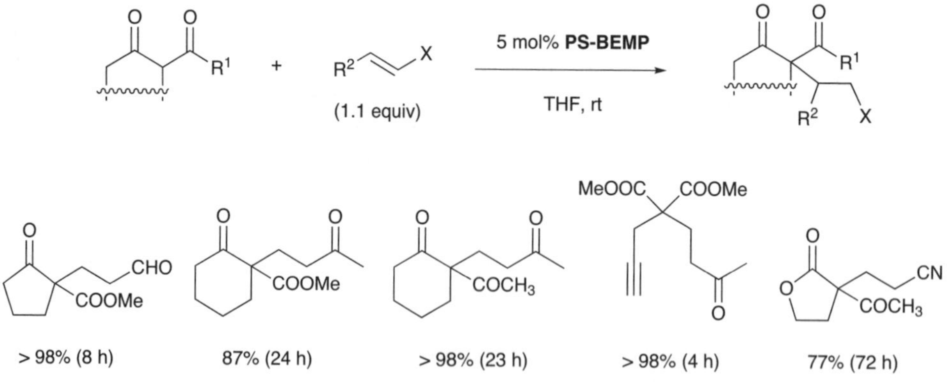

Scheme 6.6 Michael addition reactions of 1,3-dicarbonyl compounds

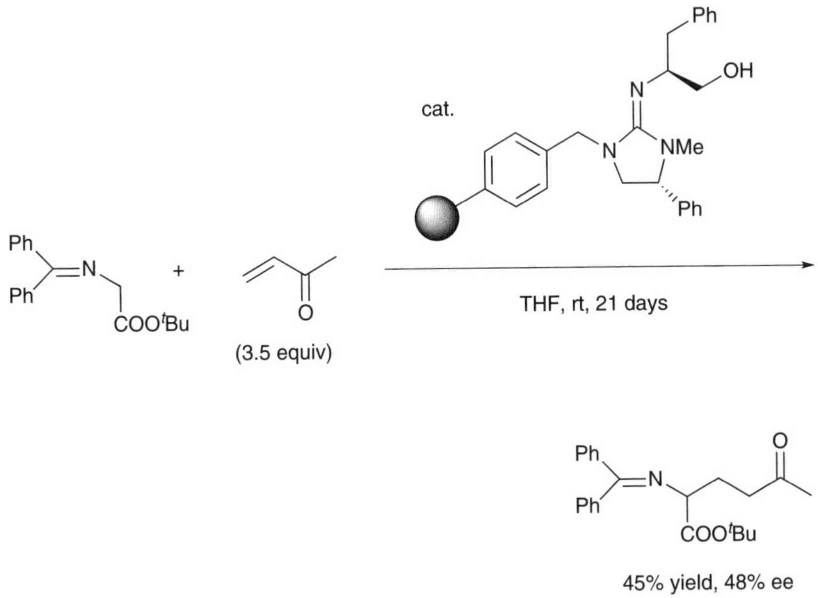

Scheme 6.7 *Chiral guanidine catalysed asymmetric Michael addition reaction of t-butyl diphenyliminoacetate with methyl vinyl ketone*

Very recently, Pilling et al. explored a new methodology using a combination of PS-BEMP (10%) and Amberlyst A15 (200%) to facilitate the Michael-initiated N-acyl iminium ion cyclization cascade starting from β-ketoamide substrates, thus providing a unique entry to a variety of complex heterocyclic molecules [40] (Scheme 6.8).

In addition, it has been shown that the catalytic use of PS-N3PAPT (10 mol%) was very effective for Michael addition reactions of a variety of Michael donors such as β-ketoesters, α-nitroketones and nitroalkanes with activated olefins such as methyl acrylate and methyl vinyl ketone [41] (Scheme 6.9). The recyclability of the catalyst was established after it had been used 12 times. When diethyl malonate or ethyl cyanoacetate was used as the Michael donor substrate, the expected disubstituted adducts were obtained in moderate to good yields.

In their extensive efforts to devise a new strong nonionic base, Verkade and coworkers found that a highly basic dendrimer containing a PAPT base fragment could act as an efficient catalyst for Michael addition reactions, nitroaldol (Henry) reactions and aryl isocyanate trimerization reactions [42] (Figure 6.3). In view of the characteristic nature of this dendrimer, which has sixteen catalytic sites per molecule, the attachment of other superbase functionalities might also be attractive.

The utility of polymer-supported superbase catalysts has been established in the Michael-type conjugate addition using hetero-atom nucleophiles such as amines and thiols [34,37,41]. For example, a variety of primary/secondary/aliphatic/aromatic amines and aliphatic/aromatic thiols could smoothly react with activated olefins in the presence of 10 mol% of a nitrate salt of PS-PAPT or PS-N3PAPT [34] (Scheme 6.10).

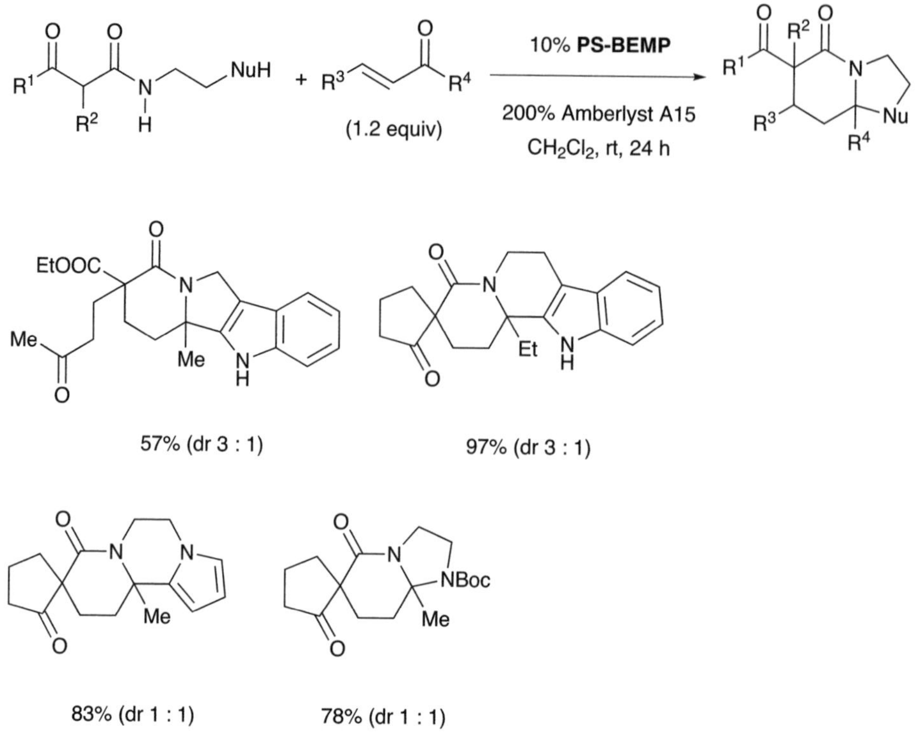

Scheme 6.8 Tandem Michael cyclization of β-ketoamide derivatives

Polymer- or mesoporous silica-supported TBD reagents have also been shown to catalyse efficiently Knoevenagel condensations and, in some cases, be feasible for continuous-flow synthesis in a microreactor [36,37,43,44].

Due to the strong basicity, high steric hindrance and low quaternizability of polymer-supported superbase reagents, several types of *O*- and *N*-alkylation reactions have been reported, with the goal of contributing to combinatorial chemistry [45–58]. For example,

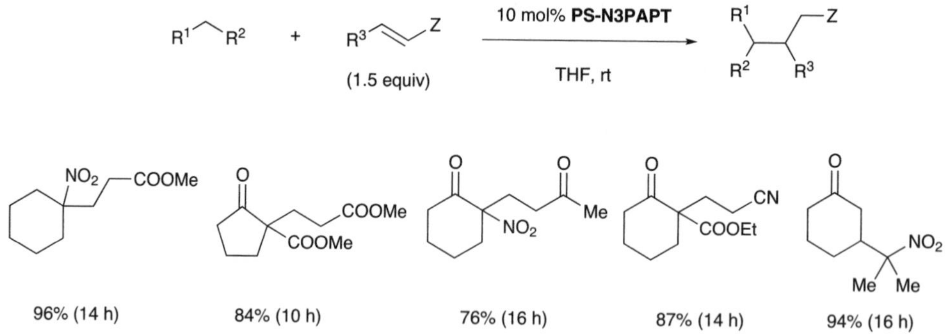

Scheme 6.9 Michael addition reaction of active methine or methylene compounds

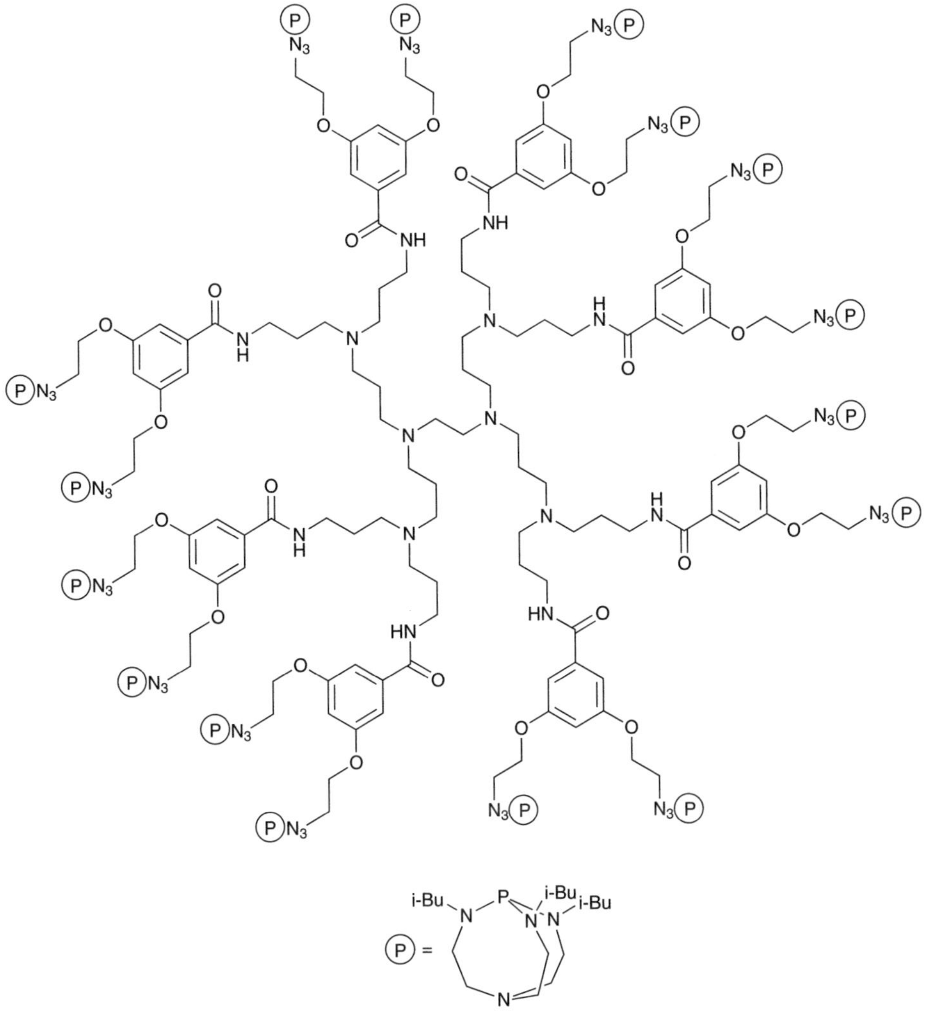

Figure 6.3 Azaphosphatrane incorporated dendrimer as a strong base catalyst

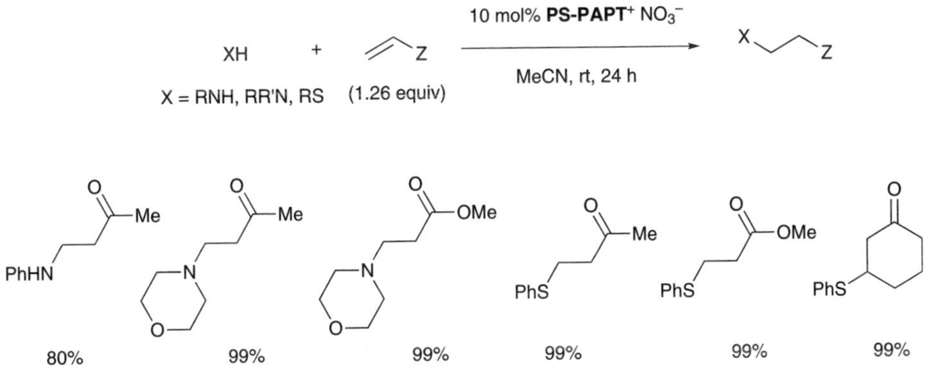

Scheme 6.10 Aza- or thia-Michael addition reactions of amines or thiols

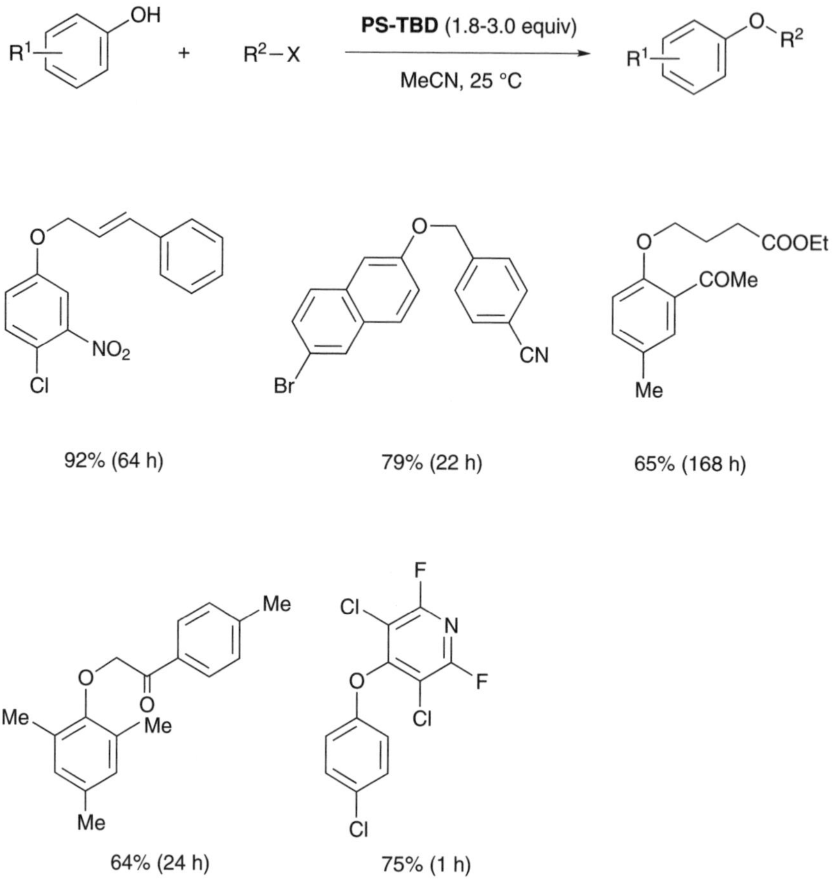

Scheme 6.11 Alkylation of phenols with alkyl halides

Xu et al. reported that a variety of phenols could be smoothly alkylated with alkyl halides or activated aromatic fluorides in the presence of 1.8–3.0 equiv. of PS-TBD [45] (Scheme 6.11).

The role of PS-TBD can be easily understood if it is considered that it works not only as a base to deprotonate phenols, but also as a scavenger to trap unreacted excess starting phenols. In addition, also established was a convenient strategy for performing multi-step conversions in one-pot operation as represented by Scheme 6.12 [47].

Thus, the procedure consists of the following sequential treatments to derive the desired double alkylated compounds: the first alkylation on a piperidine ring using R^1X in the presence of weakly basic PS-NMe$_2$ followed by addition of PS-NH$_2$ to sequester the remaining excess R^1X; the second alkylation on a pyrazole ring using R^2X in the presence of strongly basic PS-BEMP followed by addition of PS-NH$_2$; and, finally, filtration and concentration.

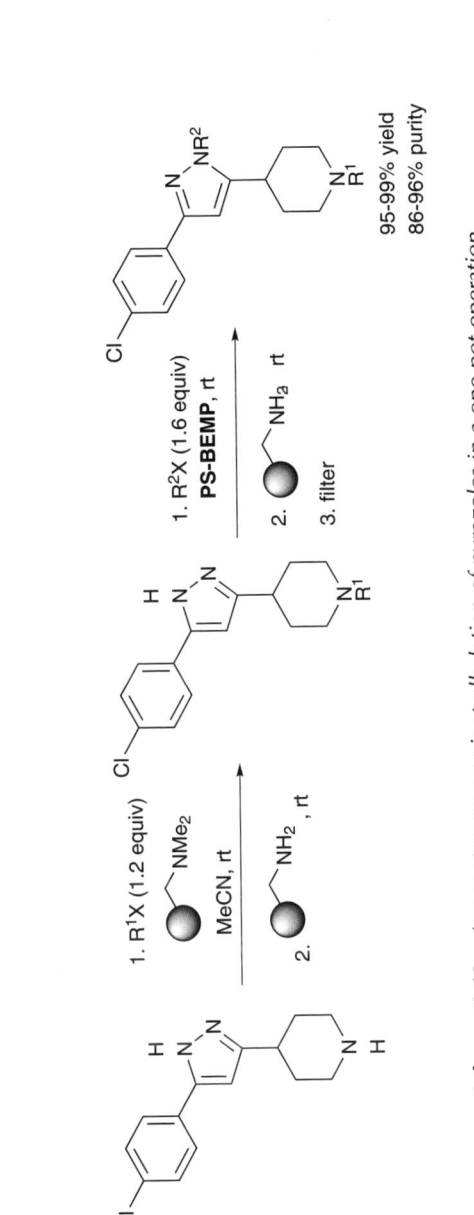

Scheme 6.12 A two-step convenient alkylation of pyrazoles in a one-pot operation

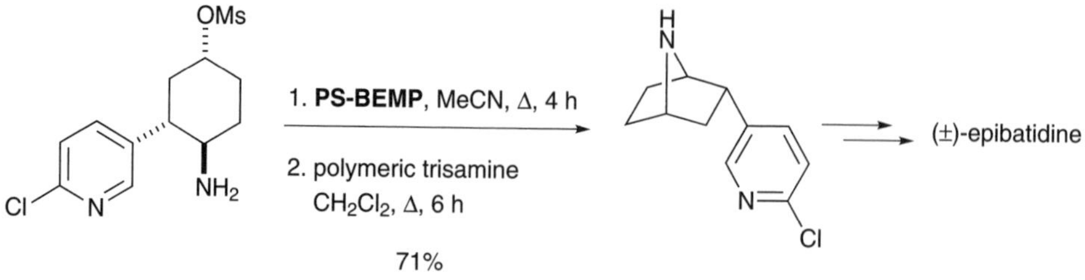

Scheme 6.13 Synthesis of a 7-aza-bicyclo[2.2.1]heptane system of epibatidine

An interesting application in this field is the work reported by Habermann *et al.*: a bicyclic 7-aza[2.2.1]heptane system, which has a mother skeleton of epibatidine, was efficiently constructed via PS-BEMP promoted intramolecular *N*-alkylation [59] (Scheme 6.13).

The synthetic utility of PS-TBD as a strong base has also been confirmed by other types of aromatic amination [60] or intramolecular alkoxyalkylation based on the nucleophilic aromatic substitution (S_NAr) of aryl fluorides [61].

As a relatively unusual reaction, it has been reported that the ring opening reaction of carbolinium salts with a variety of thiols was facilitated by the use of PS-TBD as a strong base [62] (Scheme 6.14).

6.4 Heterocyclization

With the increasing importance of the need to discover new drugs, considerable effort has been directed to identify a convenient method for rapidly preparing a large number of diverse molecules. To satisfy such keen demand, polymer-supported synthesis offers tremendous utility, and several methods have been developed to derive heterocyclic compounds based on a heterocyclization strategy. For example, Habermann *et al.* described a convenient method for preparing 2-amino-1,3-thiazole derivatives from α-bromoketone precursors by direct treatment with thiourea in refluxing THF in the presence of PS-TBD [63] (Scheme 6.15).

Similarly, for the construction of a central imidazolinone core skeleton of rhopaladin D, a marine alkaloid, the PS-BEMP promoted aza-Wittig reaction followed by intramolecular cyclization was used [64], and a tetrasubstituted pyrazole ring of sildenafil

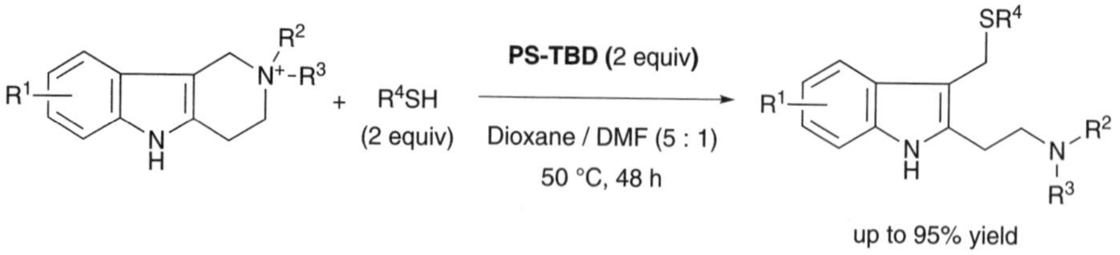

Scheme 6.14 Nucleophilic ring opening reactions of carbolinium salts with thiols

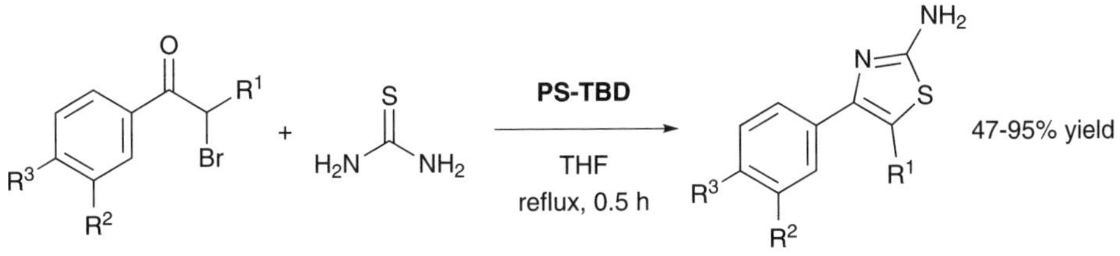

Scheme 6.15 Preparation of 2-amino-4-aryl-1,3-thiazoles via condensation of α-bromo-acetophenones with thiourea

(Viagra) was built up by the PS-BEMP promoted intramolecular alkylation of nitriles [49].

Graybill et al. discovered a new expeditious method for preparing a variety of 3-thio-1,2,4-triazole derivatives starting from acyl hydrazides and isothiocyanates via PS-BEMP mediated multi-step transformations: the so-called 'catch-cyclize-release' methodology [65] (Scheme 6.16). In this sequence, PS-BEMP was successfully used for both the deprotonation of acyl hydrazides and N-alkylation of weakly acidic triazoles. Using the conceptually analogous cyclodehydration strategy, a variety of 2-amino-1,3,4-oxadiazole derivatives, an important class of heterocycles in medicinal chemistry, were also prepared by the PS-BEMP promoted cyclization of diacyl hydrazides and related compounds [66–68]. In some cases, the use of microwave irradiation was extremely useful for facilitating rapid access to the desired products [68].

A convenient 'catch and release' methodology based on the PS-BEMP mediated sequestration was explored for the synthesis of 2-alkylthio-pyrimidinone derivatives

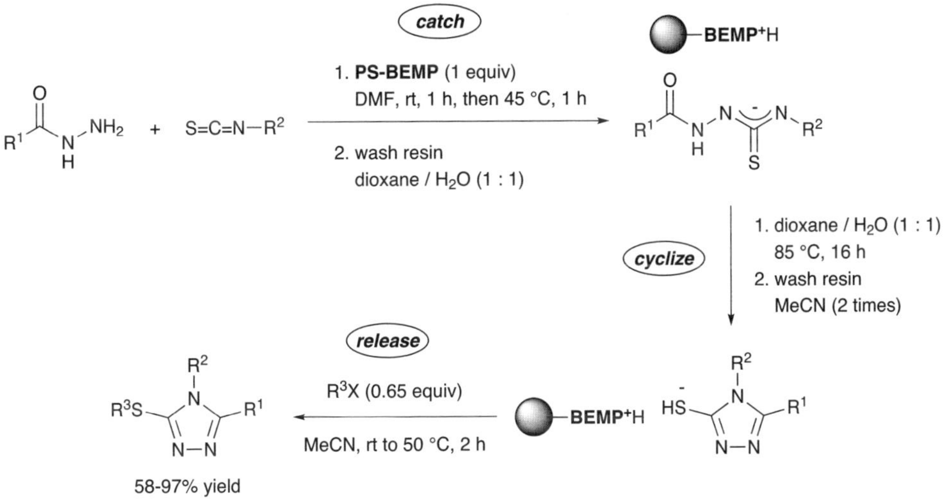

Scheme 6.16 PS-BEMP mediated 'catch-cyclize-release' preparation of 3-thio-1,2,4-triazole derivatives

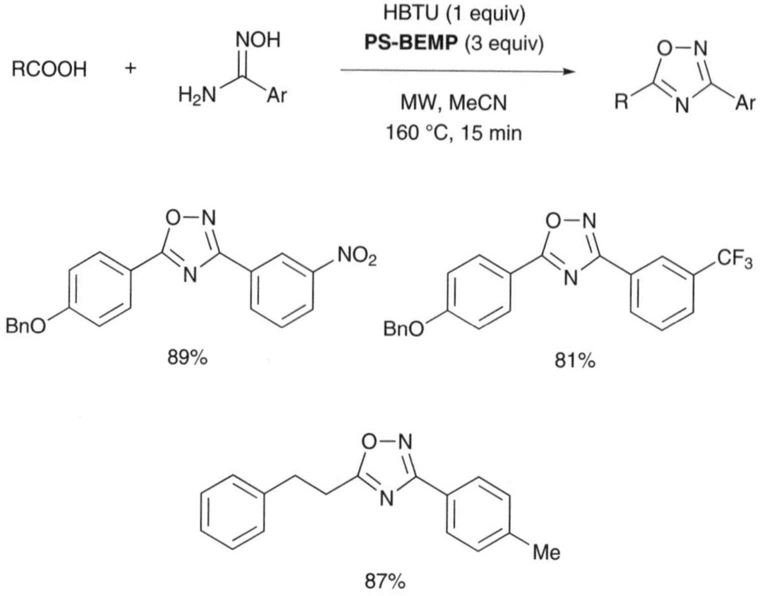

Scheme 6.17 Preparation of 1,2,4-oxadiazoles via condensation of amidoximes with carboxylic acids under the irradiation of microwave

[69]. On the other hand, 1,2,4-oxadiazole derivatives could be prepared efficiently by the condensation of amidoximes with carboxylic acids in the presence of one equivalent of O-(benzotriazol-1-yl)-N,N,N′,N′-tetramethyluronium hexafluorophosphate (HBTU) and three equivalents of PS-BEMP under microwave irradiation [70] (Scheme 6.17). This method is highly desirable in an automated format where most of the reagents can be added through liquid-delivery lines, and the PS-BEMP reagent can be easily removed by filtration upon completion of the reaction.

For the rapid on-demand synthesis of 4,5-disubstituted oxazole libraries, a flow reaction system equipped with two columns containing PS-BEMP and QP-BZA (QuadraPure benzylamine, a primary amine functionalized resin), respectively, was developed by Ley et al. [71,72] (Scheme 6.18). This system is very valuable from the viewpoint of rapid access to 25 different oxazoles in yields between 83 and 99%, the ease with which the reactions can be scaled-up to 10 g and the ability to recycle the PS-BEMP containing columns between runs.

6.5 Miscellaneous

In the molecular engineering of carbohydrates, glycosyl trichloroacetimidates play an important role as effective glycosyl donor molecules; they are generally prepared by trapping anomeric alcohols by treatment with trichloroacetonitrile in the presence of a strong base [73]. However, purification is sometimes diminished by the acid lability of trichloroacetimidate functionalities. To reduce such an inconvenience to remove trace

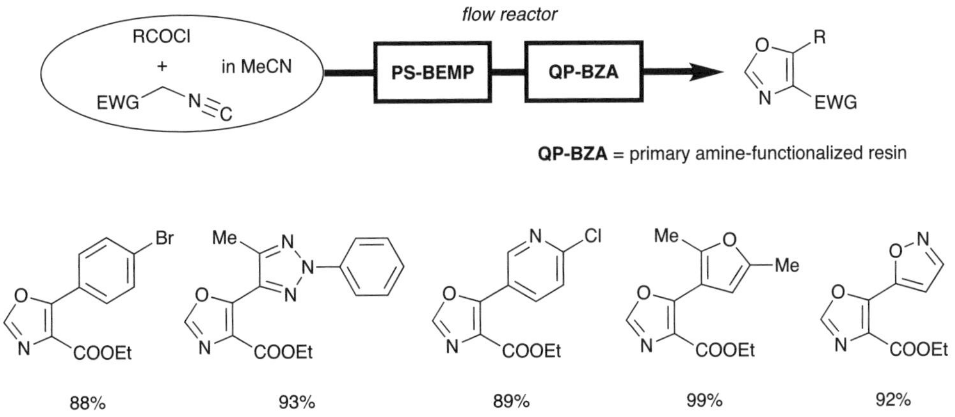

Scheme 6.18 Automated flow synthesis of 4,5-disubstituted oxazoles

amounts of base, polymer-supported superbase reagents such as PS-TBD, PS-DBU and PS-BEMP have been developed. Clearly, this constitutes one of the most powerful glycosylation methods in synthetic glycochemistry [74–76] (Scheme 6.19). In addition, an advantage of the method is that the polymer-supported reagents can be regenerated and reused.

Dondoni et al. further extended this chemistry to the highly sophisticated synthesis of trisaccharides by iterative glycosidation using a standard trichloroacetyl isocyanate-based sequestering technique [77] (Scheme 6.NaN). In this sequence, PS-BEMP played a key role in removing impurities arising from excess unreacted glycosyl acceptors in its trichloroacetyl urethane form, to provide a chromatography-free purification technique for obtaining a variety of oligosaccharides with high purities.

The PS-TBD catalyst has been shown to be effective for epoxide ring opening reactions with several nucleophiles such as thiols under solvent free conditions [37,78] (Scheme 6.21). In this case, the reusability of the catalyst was also established without a significant loss of reactivity and selectivity. As a related work, the utility of mesoporous silica-supported TBD catalysts was demonstrated in the reaction of propylene oxide with carbon dioxide to prepare the corresponding carbonate derivative under the ultrasonic activation [79].

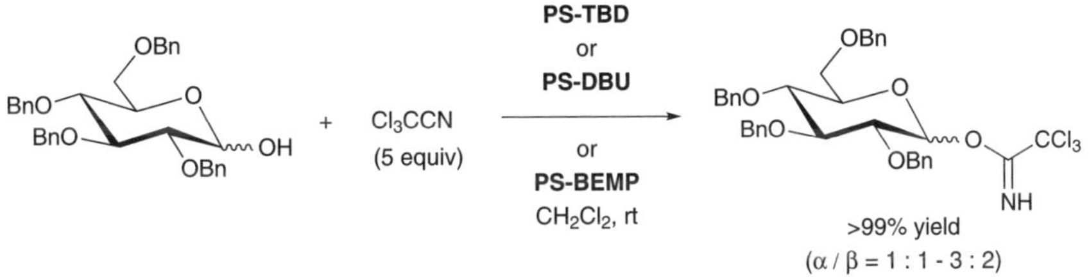

Scheme 6.19 Preparation of glycosyl trichloroacetimidates

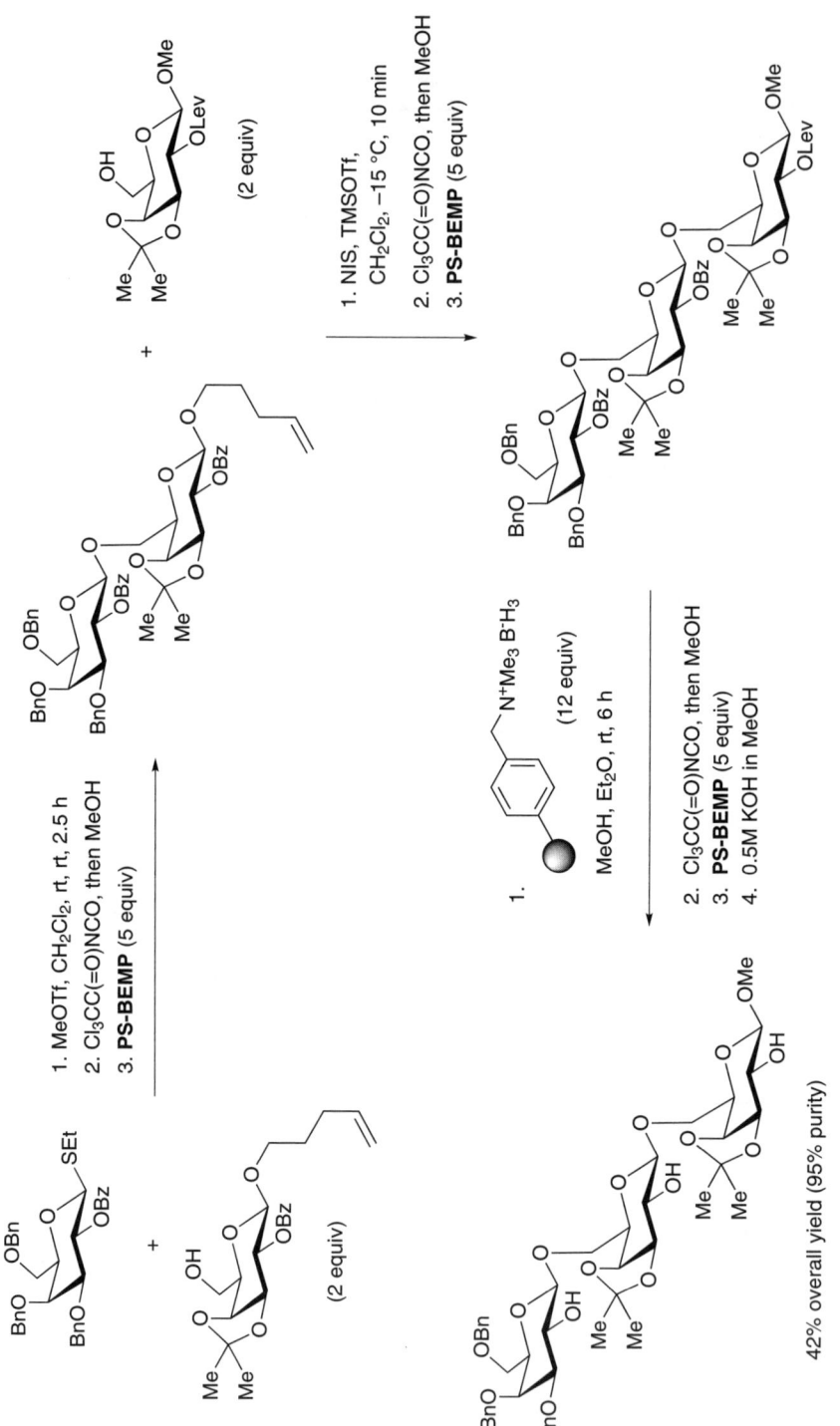

Scheme 6.20 *Synthesis of a trisaccharide without using chromatographic purification*

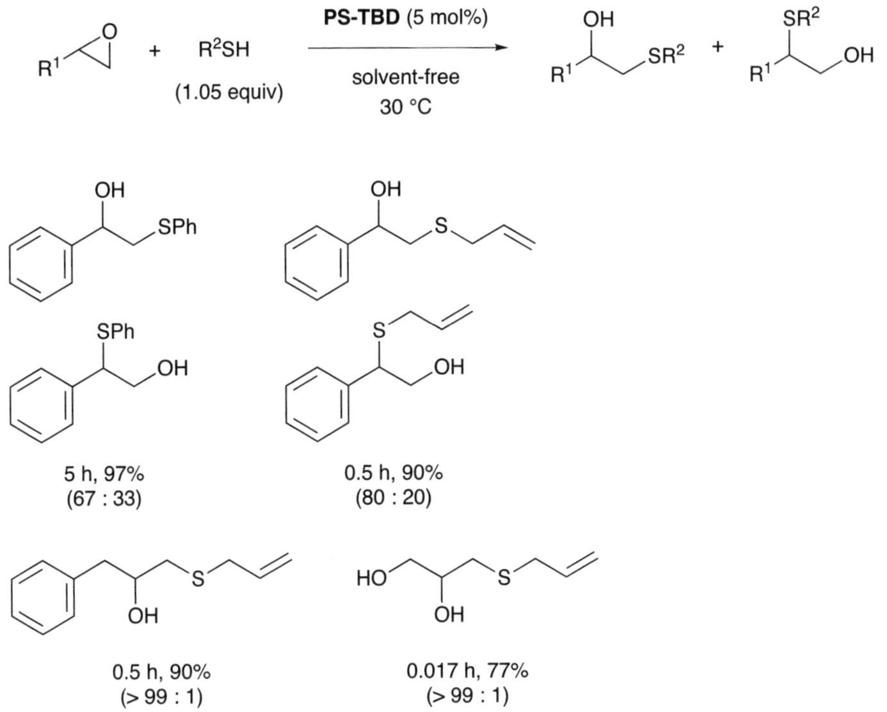

Scheme 6.21 PS-TBD catalysed thiolysis of 1,2-epoxides

In general, polymer-supported superbases are quite convenient as scavengers for removing acidic by-products by simple filtration [80–83]. Storer *et al.* are actively working in this field, and in their own multi-step organic synthesis they introduced a 'catch and release' technique that required neither aqueous work-up nor chromatographic purification [82] (Scheme 6.22).

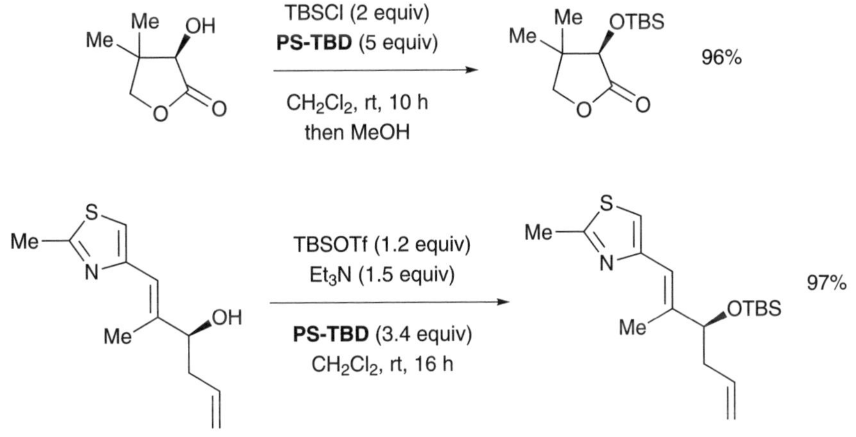

Scheme 6.22 TBS protection of alcohols with t-butyldimethylsilyl chloride in conjunction with PS-TBD

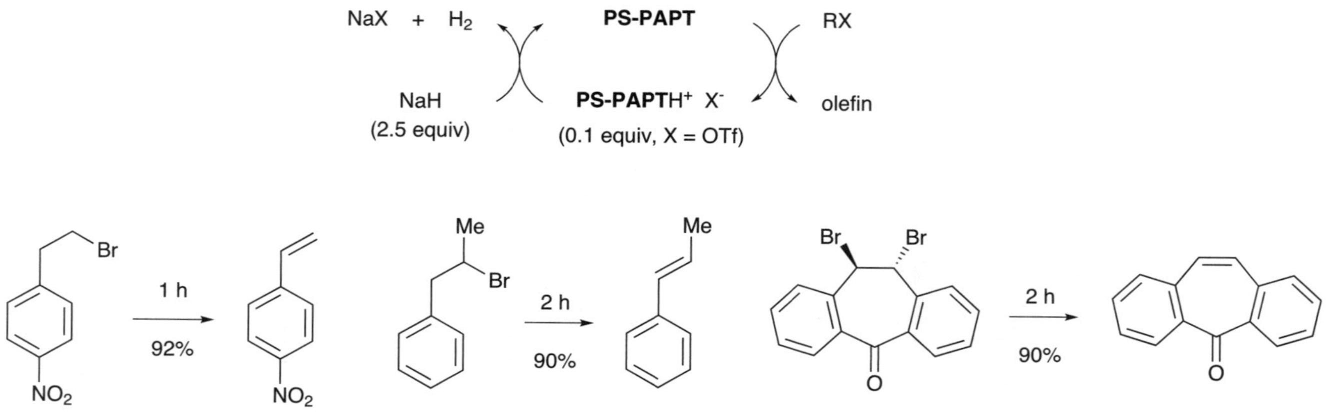

Scheme 6.23 Dehydrobromination and debromination of alkyl halides

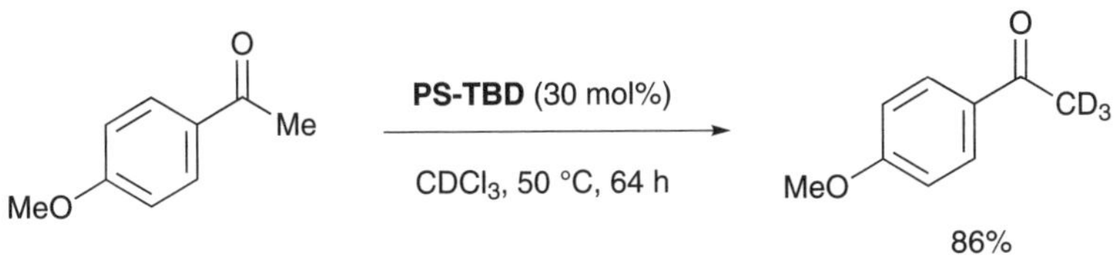

Scheme 6.24 *Effective H–D exchange reaction in CDCl$_3$*

It was found that the use of a triflate salt of PS-PAPT (0.1 equiv.) in conjunction with sodium hydride (2.5 equiv.) could be a remarkably effective method for achieving dehydrohalogenation of RX or debromination of vicinal dibromides to the corresponding olefins [84] (Scheme 6.23). In this case, deprotonation of a triflate salt of PS-PAPT with sodium hydride may have generated the actual strong base species, since sodium hydride itself did not work well or at all under the same reaction conditions.

Recently, PS-TBD was found to be basic enough to cause hydrogen/deuterium (H/D) exchange reactions with acidic substrates such as methyl ketones and terminal alkynes in CDCl$_3$ as a deuterium source as well as a solvent [85] (Scheme 6.24). An advantage of this method is that the conditions are compatible with other sensitive functional groups and aqueous work-up can be avoided, providing a convenient technique for incorporating deuterium into the acidic substrates.

It has also been reported that the use of PS-TBD as a sequestering agent in periodinate mediated oxidations was convenient for removing by-products and excess starting reagents [86].

For other works on the use of poly(aminophosphazene) catalysts or mesoporous silica-supported TBD reagents, only references are shown for convenience [87,88].

6.6 Concluding Remarks

The research area of organic synthesis using polymer-supported superbase reagents or catalysts has grown rapidly after reports of combinatorial chemistry appeared in the literature. Accordingly, this method provides a new frontier in the rapid production of a large number of chemical libraries that consist of structurally diverse molecules, and particularly with the use of automation or parallel flow systems in multi-step sequence. Furthermore, this method may be very useful in synthetic chemistry because of its significant ability to cleanly separate undesired by-products, unreacted starting materials and excess reagents from the desired products.

Despite these fascinating properties, there have been very few studies on the development of asymmetric organobase catalysts [31,39,89], compared with the dramatic progress in polymer-supported chiral lithium amide based asymmetric transformations [90]. It can be expected that new effective polymer-supported chiral superbase reagents will be discovered in the near future.

References

1. Shuttleworth, S.J., Allin, S.M. and Sharma, P.K. (1997) Functionalised polymers: recent developments and new applications in synthetic organic chemistry. *Synthesis*, 1217–1239.

2. Booth, R.J. and Hodges, J.C. (1999) Solid-supported reagent strategies for rapid purification of combinatorial synthesis products. *Accounts of Chemical Research*, **32**, 18–26.
3. Ley, S.V., Baxendale, I.R., Bream, R.N. *et al.* (2000) Multi-step organic synthesis using solid-supported reagents and scavengers: a new paradigm in chemical library generation. *Journal of the Chemical Society – Perkin Transactions 1*, 3815–4195.
4. Kirschning, A., Monenschein, H. and Wittenberg, R. (2001) Functionalized polymers – emerging versatile tools for solution-phase chemistry and automated parallel synthesis. *Angewandte Chemie – International Edition*, **40**, 650–679.
5. Clapham, B., Reger, T.S. and Janda, K.D. (2001) Polymer-supported catalysis in synthetic organic chemistry. *Tetrahedron*, **57**, 4637–4662.
6. Ley, S.V. and Baxendale, I.R. (2002) Organic synthesis in a changing world. *Chemical Record*, **2**, 377–388.
7. Baxendale, I.R., Storer, R.I. and Ley, S.V. (2003) Supported reagents and scavengers in multi-step organic synthesis, in *Polymeric Materials in Organic Synthesis and Catalysis* (ed. M.R. Buchmeiser) Wiley-VCH Verlag GmbH, Weinheim.
8. Ley, S.V., Baxendale, I.R. and Myers, R.M. (2006) Polymer-supported reagents and scavengers in synthesis, in *Comprehensive Medicinal Chemistry II*, **Vol. 3** (eds D. Triggle and J. Taylor), Elsevier, London, pp. 791–836.
9. Solinas, A. and Taddei, M. (2007) Solid-supported reagents and catch-and-release techniques in organic synthesis. *Synthesis*, 2409–2453.
10. Kovacevic, B. and Maksic, Z.B. (2001) Basicity of some organic superbases in acetonitrile. *Organic Letters*, **3**, 1523–1526.
11. Schwesinger, R., Willaredt, J., Schlemper, H. *et al.* (1994) Novel, very strong, uncharged auxiliary bases; design and synthesis of monomeric and polymer-bound triaminoiminophosphorane bases of broadly varied steric demand. *Chemische Berichte*, **127**, 2435–2454.
12. Kisanga, P.B., Verkade, J.G. and Schwesinger, R. (2000) pK_a measurements of P(RNCH$_2$CH$_3$)$_3$N. *The Journal of Organic Chemistry*, **65**, 5431–5432.
13. Tomoi, M., Kato, Y. and Kakiuchi, H. (1984) Polymer-supported bases, 2. Polystyrene-supported 1,8-bicyclo[5.4.0]undec-7-ene as reagent in organic syntheses. *Makromolekulare Chemie – Macromolecular Chemistry and Physics*, **185**, 2117–2124.
14. Iijima, K., Fukuda, W. and Tomoi, M. (1992) Polymer-supported bases. XI. Esterification and alkylation in the presence of polymer-supported bicyclic amidine or guanidine moieties. *Journal of Macromolecular Science – Pure and Applied Chemistry*, **A29**, 249–261.
15. Tamura, Y., Fukuda, W., Tomoi, M. and Tokuyama, S. (1994) Polymer-supported bases. XII. Regioselective synthesis of lysophospholipids using polymer-supported bicyclic amidines or guanidines. *Synthetic Communications*, **24**, 2907–2914.
16. Schwesinger, R. (1985) Extremely strong, non-ionic bases: syntheses and applications. *Chimia*, **39**, 269–272.
17. Schwesinger, R., Schlemper, H., Hasenfratz, C. *et al.* (1996) Extremely strong, uncharged auxiliary bases; monomeric and polymer-supported polyaminophosphazenes (P$_2$–P$_5$). *Liebigs Annalen*, 1055–1081.
18. Schuchardt, U., Vargas, R.M. and Gelbard, G. (1996) Transesterification of soybean oil catalysed by alkylguanidines heterogenized on different substituted polystyrenes. *Journal of Molecular Catalysis A-Chemical*, **109**, 37–44.
19. Gelbard, G. and Vielfaure-Joly, F. (1998) Polynitrogen strong bases: 1- New syntheses of biguanides and their catalytic properties in transesterification reactions. *Tetrahedron Letters*, **39**, 2743–2746.
20. Gelbard, G. and Vielfaure-Joly, F. (2000) Polynitrogen strong bases as immobilized catalysts for the transesterification of vegetable oils. *Comptes Rendus de l'Academie des Sciences Series IIC: Chemistry*, **3**, 563–567.
21. Gelbard, G. and Vielfaure-Joly, F. (2001) Polynitrogen strong bases as immobilized catalysts. *Reactive & Functional Polymers*, **48**, 65–74.
22. Lin, X., Chuah, G.K. and Jaenicke, S. (1999) Base-functionalized MCM-41 as catalysts for the synthesis of monoglycerides. *Journal of Molecular Catalysis A-Chemical*, **150**, 287–294.

23. Jerome, F., Kharchafi, G., Adam, I. and Barrault, J. (2004) "One pot" and selective synthesis of monoglycerides over homogeneous and heterogeneous guanidine catalysts. *Green Chemistry*, **6**, 72–74.
24. Kantam, M.L. and Sreekanth, P. (2001) Transesterification of β-keto esters catalysed by basic porous material. *Catalysis Letters*, **77**, 241–243.
25. Kim, K. and Le, K. (1999) Two efficient N-acylation methods mediated by solid-supported reagents for weakly nucleophilic heterocyclic amines. *Synlett*, 1957–1959.
26. Baxendale, I.R., Davidson, T.D., Ley, S.V. and Perni, R.H. (2003) Enantioselective synthesis of the tetrahydrobenzylisoquinoline alkaloid (−)-norarmepavine using polymer supported reagents. *Heterocycles*, **60**, 2707–2715.
27. Ilankumaran, P. and Verkade, J.G. (1999) Highly selective acylation of alcohols using enol esters catalysed by iminophosphoranes. *The Journal of Organic Chemistry*, **64**, 9063–9066.
28. Venkat Reddy, C.R., Fetterly, B.M. and Verkade, J.G. (2007) Polymer-supported azidoproazaphosphatrane: a recyclable catalyst for the room-temperature transformation of triglycerides to biodiesel. *Energy & Fuels*, **21**, 2466–2472.
29. Carloni, S., De Vos, D.E., Jacobs, P.A. et al. (2002) Catalytic activity of MCM-41-TBD in the selective preparation of carbamates and unsymmetrical alkyl carbonates from diethyl carbonate. *Journal of Catalysis*, **205**, 199–204.
30. Boisnard, S., Chastanet, J. and Zhu, J. (1999) A high throughput synthesis of aryl triflate and aryl nonaflate promoted by a polymer supported base (PTBD). *Tetrahedron Letters*, **40**, 7469–7472.
31. Simoni, D., Rondanin, R., Morini, M. et al. (2000) 1,5,7-Triazabicyclo[4.4.0]dec-1-ene (TBD), 7-methyl-TBD (MTBD) and the polymer-supported TBD (P-TBD): three efficient catalysts for the nitroaldol (Henry) reaction and for the addition of dialkyl phosphites to unsaturated systems. *Tetrahedron Letters*, **41**, 1607–1610.
32. Goumri-Magnet, S., Guerret, O., Gornitzka, H. et al. (1999) Free and supported phosphorus ylides as strong neutral Brønsted bases. *The Journal of Organic Chemistry*, **64**, 3741–3744.
33. Palacios, F., Aparicio, D., de los Santos, J.M. et al. (2000) Easy and efficient generation of reactive anions with free and supported ylides as neutral Brønsted bases. *Tetrahedron*, **56**, 663–669.
34. Fetterly, B.M., Jana, N.K. and Verkade, J.G. (2006) [HP(HNCH$_2$CH$_2$)$_3$N]NO$_3$: an efficient homogeneous and solid-supported promoter for aza and thia-Michael reactions and for Strecker reactions. *Tetrahedron*, **62**, 440–456.
35. Bensa, D., Constantieux, T. and Rodriguez, J. (2004) P-BEMP: a new efficient and commercially available user-friendly and recyclable heterogeneous organocatalyst for the Michael addition of 1,3-dicarbonyl compounds. *Synthesis*, 923–927.
36. Subba Rao, Y.V., De Vos, D.E. and Jacobs, P.A. (1997) 1,5,7-Triazabicyclo[4.4.0]dec-5-ene immobilized in MCM-41: a strongly basic porous catalyst. *Angewandte Chemie – International Edition*, **36**, 2661–2663.
37. Fringuelli, F., Pizzo, F., Vittoriani, C. and Vaccaro, L. (2004) Polystyryl-supported TBD as an efficient and reusable catalyst under solvent-free conditions. *Chemical Communications*, 2756–2757.
38. Srivastava, R. (2007) An efficient, eco-friendly process for aldol and Michael reactions of trimethylsilyl enolate over organic base-functionalized SBA-15 catalysts. *Journal of Molecular Catalysis A-Chemical*, **264**, 146–152.
39. Wannaporn, D. and Ishikawa, T. (2005) Polymer-supported and polymeric chiral guanidines: preparation and application to the asymmetric Michael reaction of iminoacetate with methyl vinyl ketone. *Molecular Diversity*, **9**, 321–331.
40. Pilling, A.W., Boehmer, J. and Dixon, D.J. (2007) Site-isolated base- and acid-mediated Michael-initiated cyclization cascades. *Angewandte Chemie – International Edition*, **46**, 5428–5430.
41. Reddy, C.R.V. and Verkade, J.G. (2007) Merrifield Resin-C$_6$H$_4$CH$_2$N$_3$P(MeNCH$_2$CH$_2$)$_3$N: an efficient reusable catalyst for room-temperature 1,4 addition reactions and a more convenient synthesis of its precursor P(MeNCH$_2$CH$_2$)$_3$N. *The Journal of Organic Chemistry*, **72**, 3093–3096.

42. Sarkar, A., Ilankumaran, P., Kisanga, P. and Verkade, J.G. (2004) First synthesis of a highly basic dendrimer and its catalytic application in organic methodology. *Advanced Synthesis and Catalysis*, **346**, 1093–1096.
43. Wiles, C., Watts, P. and Haswell, S.J. (2004) An investigation into the use of silica-supported bases within EOF-based flow reactors. *Tetrahedron*, **60**, 8421–8427.
44. Bogdan, A.R., Mason, B.P., Sylvester, K.T. and McQuade, D.T. (2007) Improving solid-supported catalyst productivity by using simplified packed-bed microreactors. *Angewandte Chemie – International Edition*, **46**, 1698–1701.
45. Xu, W., Mohan, R. and Morrissey, M.M. (1997) Polymer supported bases in combinatorial chemistry: synthesis of aryl ethers from phenols and alkyl halides and aryl halides. *Tetrahedron Letters*, **38**, 7337–7340.
46. Habermann, J., Ley, S.V. and Smits, R. (1999) Three-step synthesis of an array of substituted benzofurans using polymer-supported reagents. *Journal of the Chemical Society – Perkin Transactions 1*, 2421–2423.
47. Xu, W., Mohan, R. and Morrissey, M.M. (1998) Polymer supported bases in solution-phase synthesis. 2. A convenient method for *N*-alkylation reactions of weakly acidic heterocycles. *Bioorganic & Medicinal Chemistry Letters*, **8**, 1089–1092.
48. Caldarelli, M., Habermann, J. and Ley, S.V. (1999) Clean five-step synthesis of an array of 1,2,3,4-tetra-substituted pyrroles using polymer-supported reagents. *Journal of the Chemical Society – Perkin Transactions 1*, 107–110.
49. Baxendale, I.R. and Ley, S.V. (2000) Polymer-supported reagents for multi-step organic synthesis: application to the synthesis of sildenafil. *Bioorganic & Medicinal Chemistry Letters*, **10**, 1983–1986.
50. McComas, W., Chen, L. and Kim, K. (2000) Convenient synthesis of 2,9-disubstituted guanines mediated by solid-supported reagents. *Tetrahedron Letters*, **41**, 3573–3576.
51. Organ, M.G. and Dixon, C.E. (2000) The preparation of amino-substituted biaryl libraries: the application of solid-supported reagents to streamline solution-phase synthesis. *Biotechnology and Bioengineering*, **71**, 71–77.
52. Caldarelli, M., Habermann, J. and Ley, S.V. (1999) Synthesis of an array of potential matrix metalloproteinase inhibitors using a sequence of polymer-supported reagents. *Bioorganic & Medicinal Chemistry Letters*, **9**, 2049–2052.
53. Griffiths-Jones, C.M., Hopkin, M.D., Jönsson, D. *et al.* (2007) Fully automated flow-through synthesis of secondary sulfonamides in a binary reactor system. *Journal of Combinatorial Chemistry*, **9**, 422–430.
54. Baxendale, I.R., Lee, A.-L. and Ley, S.V. (2001) A concise synthesis of the natural product carpanone using solid-supported reagents and scavengers. *Synlett*, 1482–1484.
55. Baxendale, I.R., Lee, A.-L. and Ley, S.V. (2002) A concise synthesis of carpanone using solid-supported reagents and scavengers. *Journal of the Chemical Society-Perkin Transactions 1*, 1850–1857.
56. Lee, A.-L. and Ley, S.V. (2003) The synthesis of the anti-malarial natural product polysphorin and analogues using polymer-supported reagents and scavengers. *Organic and Biomolecular Chemistry*, **1**, 3957–3966.
57. Li, X., Abell, C., Warrington, B.H. and Ladlow, M. (2003) Polymer-assisted solution phase synthesis of the antihyperglycemic agent Rosiglitazone (Avandia™). *Organic and Biomolecular Chemistry*, **1**, 4392–4395.
58. Lizarzaburu, M.E. and Shuttleworth, S.J. (2003) Convenient preparation of aryl ether derivatives using a sequence of functionalized polymers. *Tetrahedron Letters*, **44**, 4873–4876.
59. Habermann, J., Ley, S.V. and Scott, J.S. (1999) Synthesis of the potent analgestic compound (±)-epibatidine using an orchestrated multi-step sequence of polymer supported reagents. *Journal of the Chemical Society – Perkin Transactions 1*, 1253–1255.
60. Hilty, P., Hubschwerlen, C. and Thomas, A.W. (2001) Expeditious solution phase synthesis of fluoroquinolone antibacterial agents using polymer supported reagents. *Tetrahedron Letters*, **42**, 1645–1646.
61. Tempest, P., Ma, V., Kelly, M.G. *et al.* (2001) MCC/S$_N$Ar methodology. Part 1: novel access to a range of heterocyclic cores. *Tetrahedron Letters*, **42**, 4963–4968.

62. Lizarzaburu, M.E. and Shuttleworth, S.J. (2004) 1,2,3,4-Tetrahydro-γ-carbolinium salts: novel reactions with thiols, mediated by polymer-supported reagents. *Tetrahedron Letters*, **45**, 4781–4783.
63. Habermann, J., Ley, S.V., Scicinski, J.J. *et al.* (1999) Clean synthesis of α-bromo ketones and their utilization in the synthesis of 2-alkoxy-2,3-dihydro-2-aryl-1,4-benzodioxanes, 2-amino-4-aryl-1,3-thiazoles and piperidino-2-amino-1,3-thiazoles using polymer-supported reagents. *Journal of the Chemical Society – Perkin Transactions 1*, 2425–2427.
64. Fresneda, P.M., Molina, P. and Sanz, M.A. (2000) The first synthesis of the bis(indole) marine alkaloid rhopaladin D. *Synlett*, 1190–1192.
65. Graybill, T.L., Thomas, S. and Wang, M.A. (2002) A convenient 'catch, cyclize, and release' preparation of 3-thio-1,2,4-triazoles mediated by polymer-bound BEMP. *Tetrahedron Letters*, **43**, 5305–5309.
66. Brain, C.T. and Brunton, S.A. (2001) Synthesis of 1,3,4-oxadiazoles using polymer-supported reagents. *Synlett*, 382–384.
67. Coppo, F.T., Evans, K.A., Graybill, T.L. and Burton, G. (2004) Efficient one-pot preparation of 5-substituted-2-amino-1,3,4-oxadiazoles using resin-bound reagents. *Tetrahedron Letters*, **45**, 3257–3260.
68. Baxendale, I.R., Ley, S.V. and Martinelli, M. (2005) The rapid preparation of 2-aminosulfonamide-1,3,4-oxadiazoles using polymer-supported reagents and microwave heating. *Tetrahedron*, **61**, 5323–5349.
69. Adams, G.L., Graybill, T.L., Sanchez, R.M. *et al.* (2003) A convenient 'catch and release' synthesis of fused 2-alkylthio-pyrimidinones mediated by polymer-bound BEMP. *Tetrahedron Letters*, **44**, 5041–5045.
70. Wang, Y., Miller, R.L., Sauer, D.R. and Djuric, S.W. (2005) Rapid and efficient synthesis of 1,2,4-oxadiazoles utilizing polymer-supported reagents under microwave heating. *Organic Letters*, **7**, 925–928.
71. Baumann, M., Baxendale, I.R., Ley, S.V. *et al.* (2006) Fully automated continuous flow synthesis of 4,5-disubstituted oxazoles. *Organic Letters*, **8**, 5231–5234.
72. Hornung, C.H., Mackley, M.R., Baxendale, I.R. and Ley, S.V. (2007) A microcapillary flow disc reactor for organic synthesis. *Organic Process Research & Development*, **11**, 399–405.
73. Schmidt, R.R. and Jung, K.-H. (2000) Trichloroacetimidates. *Carbohydrate Chemistry & Biochemistry*, **1**, 5–59.
74. Oikawa, M., Tanaka, T., Fukada, N. and Kusumoto, S. (2004) One-pot preparation and activation of glycosyl trichloroacetimidates: operationally simple glycosylation induced by combined use of solid-supported, reactivity-opposing reagents. *Tetrahedron Letters*, **45**, 4039–4042.
75. Ohashi, I., Lear, M.J., Yoshimura, F. and Hirama, M. (2004) Use of polystyrene-supported DBU in the synthesis and α-selective glycosylation study of the unstable Schmidt donor of L-kedarosamine. *Organic Letters*, **6**, 719–722.
76. Chiara, J.L., Encinas, L. and Díaz, B. (2005) A study of polymer-supported bases for the solution phase synthesis of glycosyl trichloroacetimidates. *Tetrahedron Letters*, **46**, 2445–2448.
77. Dondoni, A., Marra, A. and Massi, A. (2005) Hybrid solution/solid-phase synthesis of oligosaccharides by using trichloroacetyl isocyanate as sequestration-enabling reagent of sugar alcohols. *Angewandte Chemie – International Edition*, **44**, 1672–1676.
78. Fringuelli, F., Pizzo, F., Vittoriani, C. and Vaccaro, L. (2006) Polystyrene-supported 1,5,7-triazabicyclo[4.4.0]dec-5-ene as an efficient and reusable catalyst for the thiolysis of 1,2-epoxides under solvent-free conditions. *European Journal of Organic Chemistry*, 1231–1236.
79. Zhang, X., Zhao, N., Wei, W. and Sun, Y. (2006) Chemical fixation of carbon dioxide to propylene carbonate over amine-functionalized silica catalysts. *Catalysis Today*, **115**, 102–106.
80. Souers, A.J., Virgilio, A.A., Schürer, S.S. *et al.* (1998) Novel inhibitors of α4β1 integrin receptor interactions through library synthesis and screening. *Bioorganic & Medicinal Chemistry Letters*, **8**, 2297–2302.
81. Sheppeck, J.E., II, Kar, H. and Hong, H. (2000) A convenient and scaleable procedure for removing the Fmoc group in solution. *Tetrahedron Letters*, **41**, 5329–5333.

82. Storer, R.I., Takemoto, T., Jackson, P.S. *et al.* (2004) Multi-step application of immobilized reagents and scavengers: a total synthesis of epothilone C. *Chemistry – A European Journal*, **10**, 2529–2547.
83. Gulín, O.P., Rabanal, F. and Giralt, E. (2006) Efficient preparation of proline *N*-carboxyanhydride using polymer-supported bases. *Organic Letters*, **8**, 5385–5388.
84. Liu, X., Yu, Z. and Verkade, J.G. (1999) Free and polymer-bound tricyclic azaphosphatranes HP(RNCH$_2$CH$_2$)$_3$N$^+$: procatalysts in dehydrohalogenations and debrominations with NaH. *The Journal of Organic Chemistry*, **64**, 4840–4843.
85. Sabot, C., Kumar, K.A., Antheaume, C. and Mioskowski, C. (2007) Triazabicyclodecene: an effective isotope exchange catalyst in CDCl$_3$. *The Journal of Organic Chemistry*, **72**, 5001–5004.
86. Parlow, J.J., Case, B.L. and South, M.S. (1999) High-throughput purification of solution-phase periodinane mediated oxidation reactions utilizing a novel thiosulfate resin. *Tetrahedron*, **55**, 6785–6796.
87. Memeger, W., Jr, Campbell, G.C. and Davidson, F. (1996) Poly(aminophosphazene)s and protophosphatranes mimic classical strong anionic base catalysts in the anionic ring-opening polymerization of lactams. *Macromolecules*, **29**, 6475–6480.
88. Tajima, T. and Fuchigami, T. (2005) An electrolytic system that uses solid-supported bases for *in situ* generation of a supporting electrolyte from acetic acid solvent. *Angewandte Chemie – International Edition*, **44**, 4760–4763.
89. Brunel, J.M., Legrand, O., Reymond, S. and Buono, G. (1999) First iminodiazaphospholidines with a stereogenic phosphorus center. Application to asymmetric copper-catalysed cyclopropanation. *Journal of the American Chemical Society*, **121**, 5807–5808.
90. Ma, L. and Williard, P.G. (2006) Synthesis of polymer-supported chiral lithium amide bases and application in asymmetric deprotonation of prochiral cyclic ketones. *Tetrahedron: Asymmetry*, **17**, 3021–3029.

7

Application of Organosuperbases to Total Synthesis

Kazuo Nagasawa
Tokyo University of Agriculture and Technology, 2-24-16 Naka-cho,
Koganei, Tokyo 184-8588, Japan

7.1 Introduction

Organic amidine, guanidine and phosphazene type superbases are typical strong 'nonionic' bases and show good solubility in most organic solvents, even at low temperature. These bases have been applied to a variety of critical reaction steps in the total syntheses of natural products. Moreover, they have many characteristic reactivities and reaction enhancing effects. In this chapter, applications of amidine, guanidine and phosphazene superbases to natural product synthesis are discussed. The structures of the superbases described in this chapter are summarized in Figure 7.1.

7.2 Carbon–Carbon Bond Forming Reactions

7.2.1 Aldol Reaction

Many examples of guanidine base-promoted nitro aldol reactions [1] and their application to the synthesis of natural products have been reported. Ishikawa *et al.* synthesized (+)-cyclophellitol (**14**), an α-glucosidase inhibitor and also a potential inhibitor of HIV, via the intramolecular 1,3-dipolar cycloaddition reaction of silylnitronate **13**. In this synthesis, nitroalcohol **12** was prepared by the reaction of aldehyde **11** with nitromethane in the presence of TMG (**3**) as a 2 : 1 diastereomeric mixture [2] (Scheme 7.1).

Superbases for Organic Synthesis: Guanidines, Amidines, Phosphazenes and Related Organocatalysts
Edited by Tsutomu Ishikawa
© 2009 John Wiley & Sons, Ltd

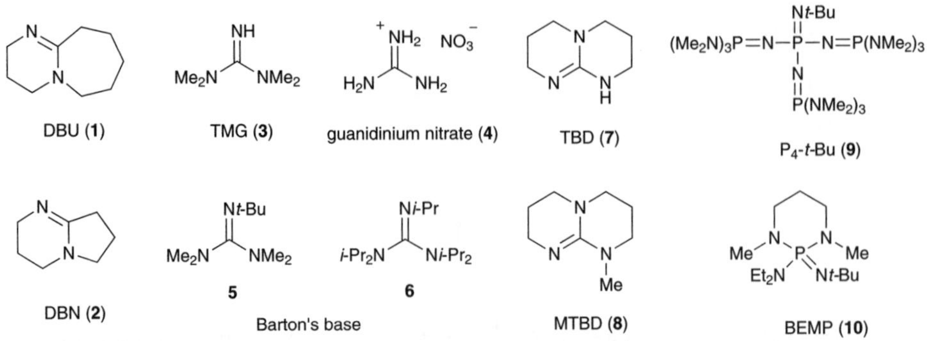

Figure 7.1 Structures of superbases

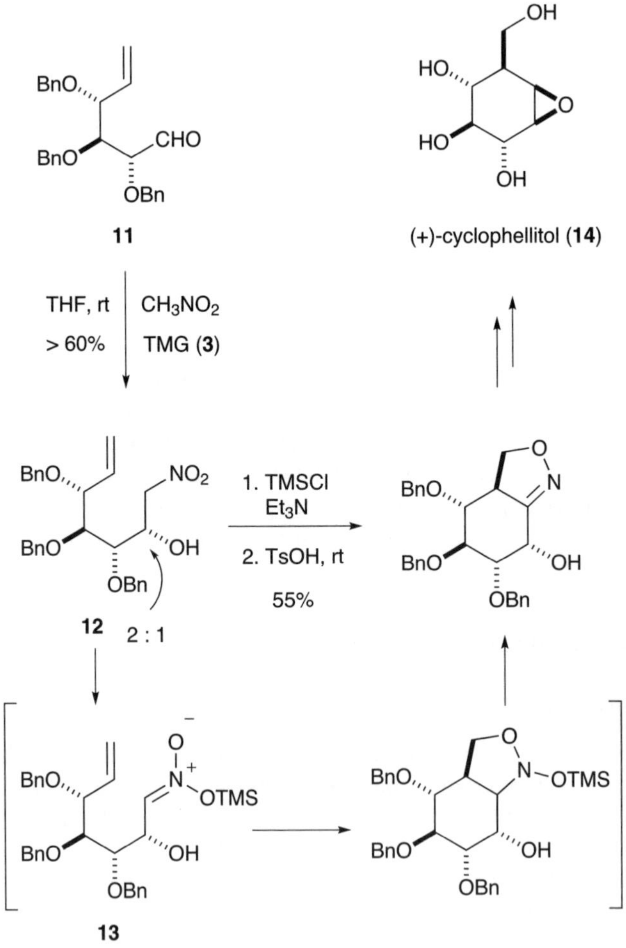

Scheme 7.1 Synthesis of (+)-cyclophellitol (**14**)

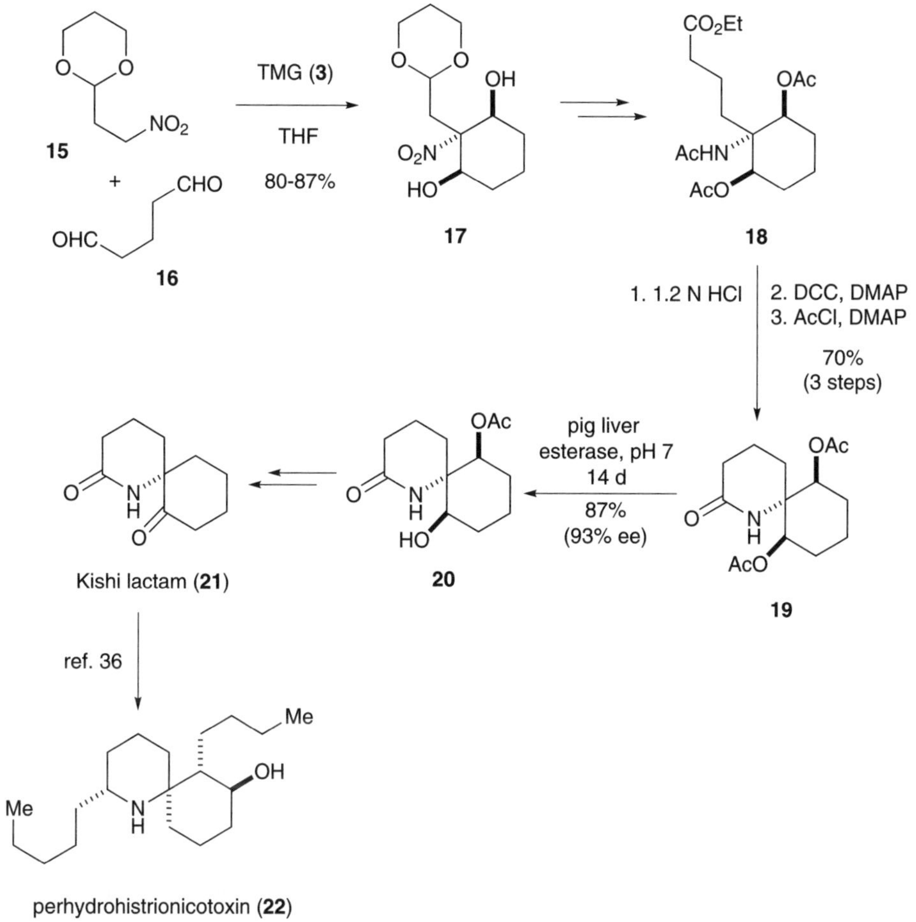

Scheme 7.2 Synthesis of a key intermediate **21** of perhydrohistrionicotoxin (**22**)

The double Henry reaction was examined by Luzzio and Fitch for the synthesis of a key intermediate **21** of perhydrohistrionicotoxin (**22**) [3]. Reaction of nitroalkane **15** and glutaraldehyde **16** using TMG (**3**) in dry tetrahydrofuran (THF) proceeded via a double nitroaldol reaction to give meso-**17** in 80–87% yield. After conversion of the diol to meso lactamdiacetate **19**, esterase mediated hydrolysis gave optically active **20** in 87% yield with 93% ee (Scheme 7.2). This hydroxyacetate was successfully led to the Kishi lactam (**21**) [4], a key intermediate for **22**.

Madin et al. reported a synthesis of gelsemine (**28**) [5] in which a DBU promoted skeletal rearrangement reaction was applied. Treatment of **23** with DBU in refluxing toluene proceeded through a retro aldol–aldol reaction process with epimerization at C3, and the resulting axial alcohol **25** cyclized at the nitrile group to generate lactone **27** after hydrolysis of the imidate **26**. The lactone **27** was successfully converted to gelsemine (**28**) (Scheme 7.3).

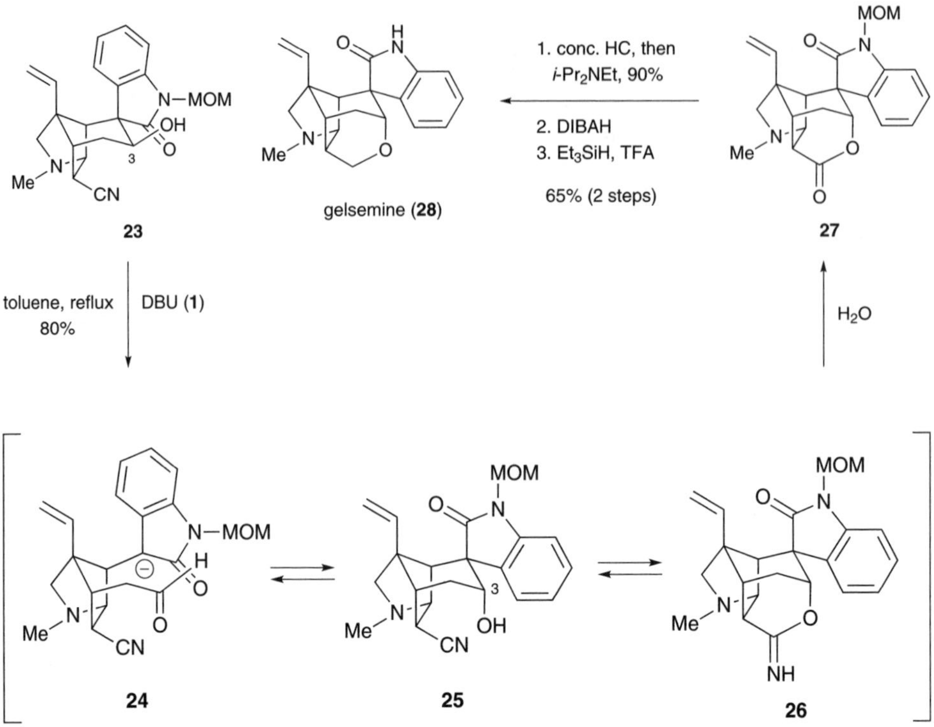

Scheme 7.3 Synthesis of gelsemine (**28**)

The benzofuran skeleton is common in natural products. A direct synthesis from *o*-arylmethoxybenzaldehyde by base promoted condensation reaction was reported by Kraus *et al.* [6] The reaction of *o*-arylmethoxybenzaldehyde **29** with 1.1 equiv. of phosphazene **9** in benzene or pivalonitrile at 90–100 °C gave **30** in moderate yield (Scheme 7.4). Strong ionic bases, such as LDA, LiTMP and KH, were ineffective for this cyclization reaction.

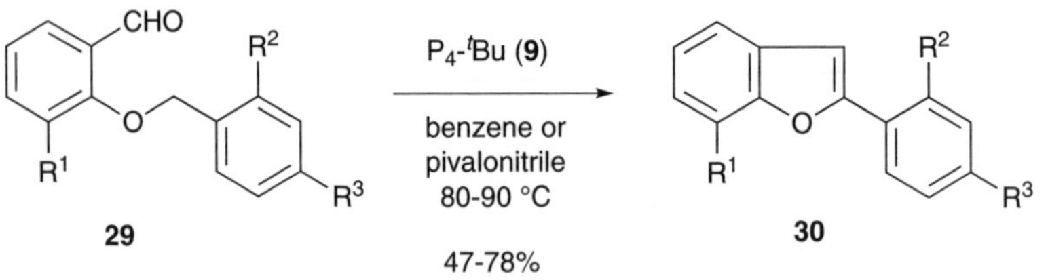

$R^1 = R^2 = R^3 = H$
$R^1 = R^2 = OMe, R^3 = H$
$R^1 = R^3 = H, R^2 = OMe$
$R^1 = R^3 = H, R^2 = OMOM$
$R^1 = R^2 = H, R^3 = NO_2$

Scheme 7.4 Synthesis of benzofuran skeleton **30**

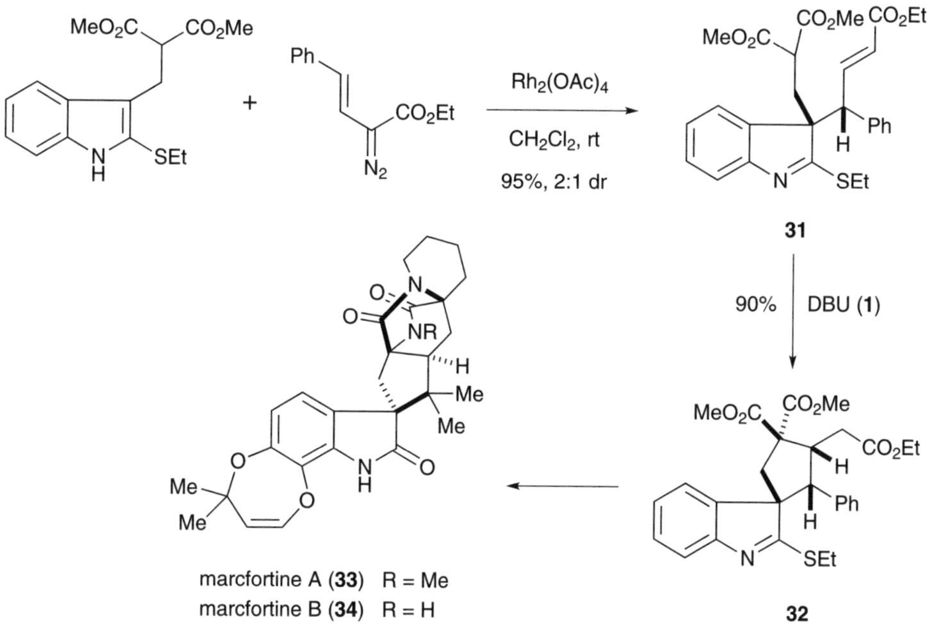

Scheme 7.5 Synthesis of spirocyclic indole skeleton **32**

7.2.2 Michael Reaction

Conjugate addition reaction is also promoted by superbases. Novikov *et al.* reported that the intramolecular Michael addition of α,β-unsaturated ester **31**, which was obtained from vinyl diazoacetate in the presence of rhodium acetate [Rh$_2$(OAc)$_4$], proceeded with DBU to give the spirocyclic indole skeleton **32** in 90% yield [7] (Scheme 7.5). This structure can be seen in intriguing, complex natural products, such as marcfortines **33** and **34**.

The intramolecular Michael addition reaction of ketone enolate to β-alkoxyacrylate proceeded selectively (in a ratio of 5.9 : 1) with Barton's base (**5**) to give tetrahydrofuran **36** [8]. In a model study, LDA was much less effective than Barton's base (28–45% versus 96% yield). The diester **36** was converted into **38**, which corresponds to the A-D ring system for lactonamycin (**39**) (Scheme 7.6).

In a synthesis of the tricyclic skeleton of FR901483 (**45**), Bonjoch and Solé reported a TMG (**3**) promoted conjugate addition reaction of nitroalkane to methyl acrylate. Reaction of methyl acrylate (**40**) and nitro acetal **41** [9,10], obtained from Diels–Alder reaction between nitroethylene and 2-(trimethylsilyloxy)-1,3-butadiene, gave nitro ester **42** in 71% yield. The ester **42** was further converted to the spiro compound **43**, and a palladium promoted cyclization reaction led to the azatricyclic skeleton **44** (Scheme 7.7).

Conjugate addition–elimination reaction of nitromethane with β-trifloxy acrylate in the presence of TMG (**3**) was reported by Chung *et al.* [11]. Treatment of enol triflate **46** with nitromethane and DMPU as a co-solvent in the presence of TMG (**3**) gave allyl nitro compound **47** in 60–70% yield via a conjugate addition–elimination process. In this

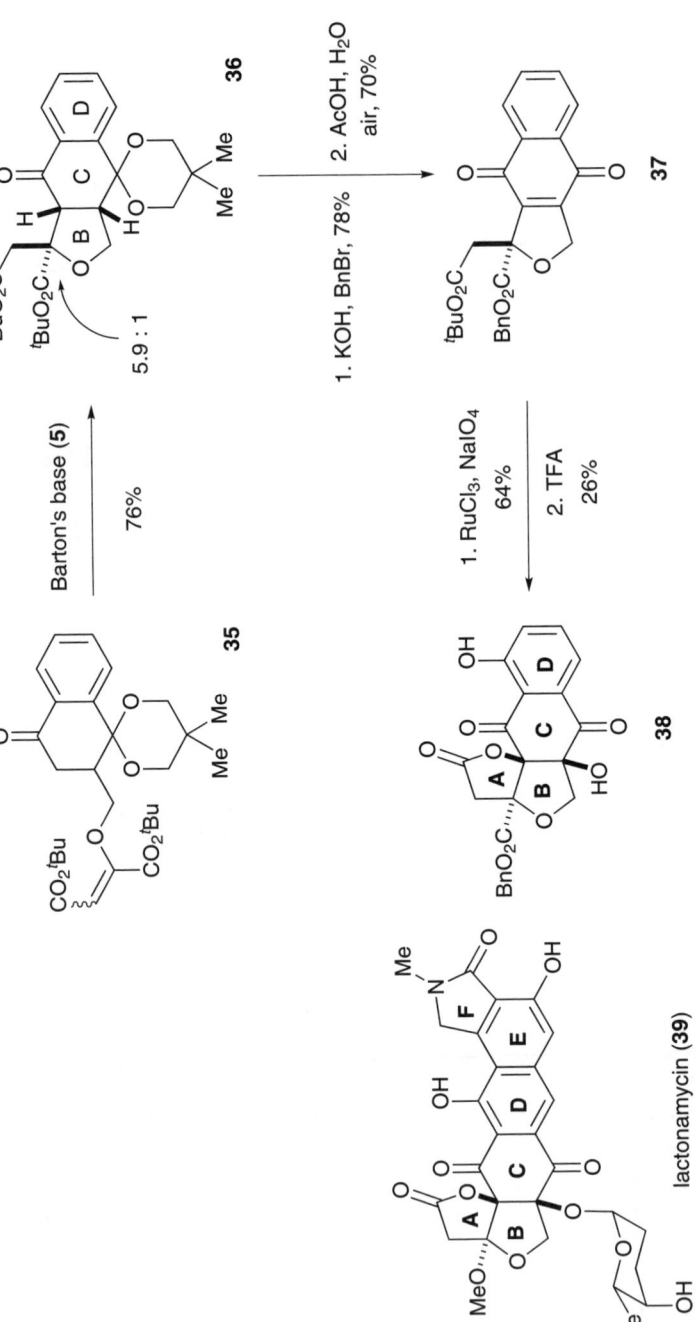

Scheme 7.6 Synthesis of A-D ring system for lactonamycin (**39**)

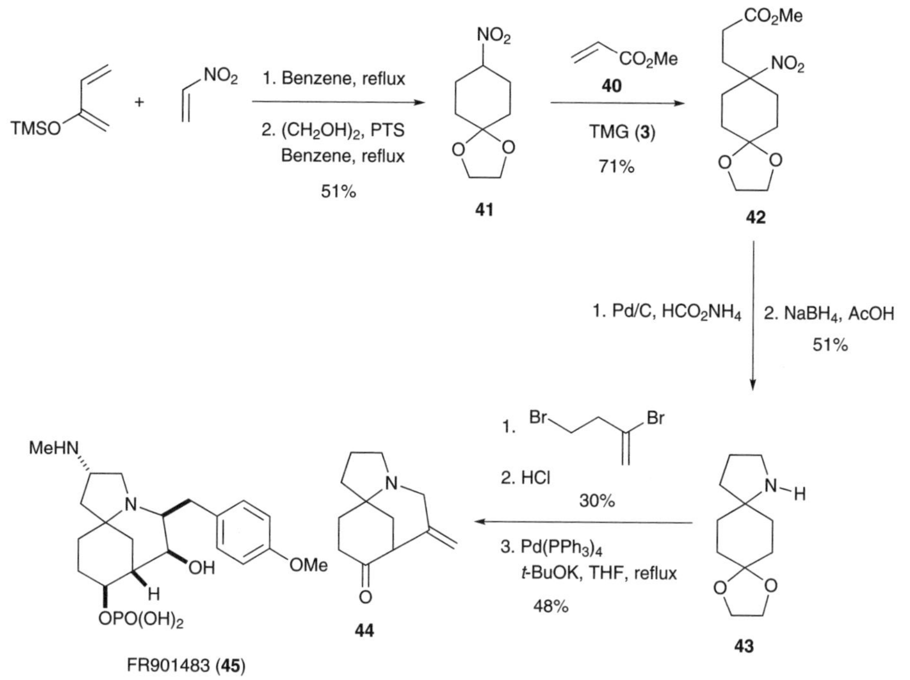

Scheme 7.7 Synthesis of tricyclic skeleton of FR901483 (**45**)

reaction, low temperature is required for a good yield. Triethylamine is not effective as a base. palladium catalysed substitution reaction of **47** with naphthosultam **48** took place to give naphthosultammethyl carbapenem **49**, a key intermediate for the anti-MRS agent L-786,392 (**50**) (Scheme 7.8).

7.2.3 Pericyclic Reaction

The oxy-Cope rearrangement reaction provides useful synthetic intermediates for natural products. Since the oxy-Cope rearrangement is known to be greatly accelerated when the alkoxide is used instead of the corresponding alcohol, metal-free superbases were applied to the reaction by Mamdani and Hartley [12]. Reaction of the alcohol **51** with phosphazene **9** at room temperature gave the rearranged product **52** in 53% yield (Scheme 7.9).

Corey and Kania reported an enantioselective Claisen rearrangement reaction of macrocyclic lactone for the synthesis of (+)-dollabellatrienone (**56**) [13]. Reaction of the lactone **53** with chiral (*S*,*S*)-diazaborolidine L$_2$BBr **54** and Barton's base (**6**) resulted in Claisen rearrangement to give carboxylic acid **55** in 86% yield with >98% ee (diastereoselectivity; >98:2) (Scheme 7.10). In this reaction, rapid deprotonation by sterically hindered guanidine base is the key to suppress side reactions.

[2,3]-Sigmatropic rearrangement of allylic ammonium ylides mediated by the combination of Lewis acid–nonionic phosphazene base was reported by Blid and Somfai [14]. The

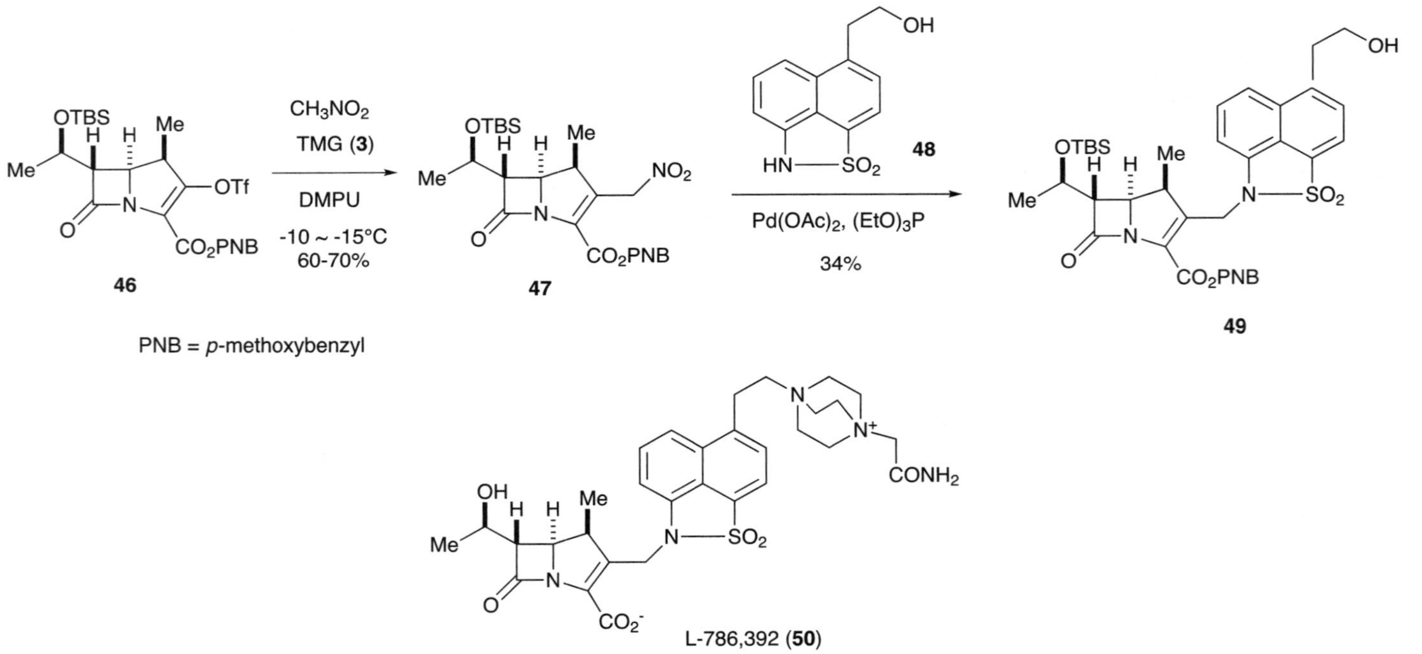

Scheme 7.8 Synthesis of naphthosultammethyl carbapenam 49

Scheme 7.9 Oxy-Cope rearrangement reaction promoted by phosphazene 9

Scheme 7.10 Synthesis of (+)-dollabellatrienone (56)

allyl amines **57** were reacted with born tribromide (BBr_3) and phosphazene base **9** in toluene at low temperature (-78 to $-20\,°C$), and the rearrangement reaction proceeded to give **58** (Scheme 7.11). In this reaction, neither LDA nor KHMDS was effective. The (E)-substrate of allyl amines gave excellent diastereoselectivity, whereas that of the (Z)-substrate was low (6:5).

Scheme 7.11 [2,3]-Sigmatropic rearrangement of allylic ammonium ylides

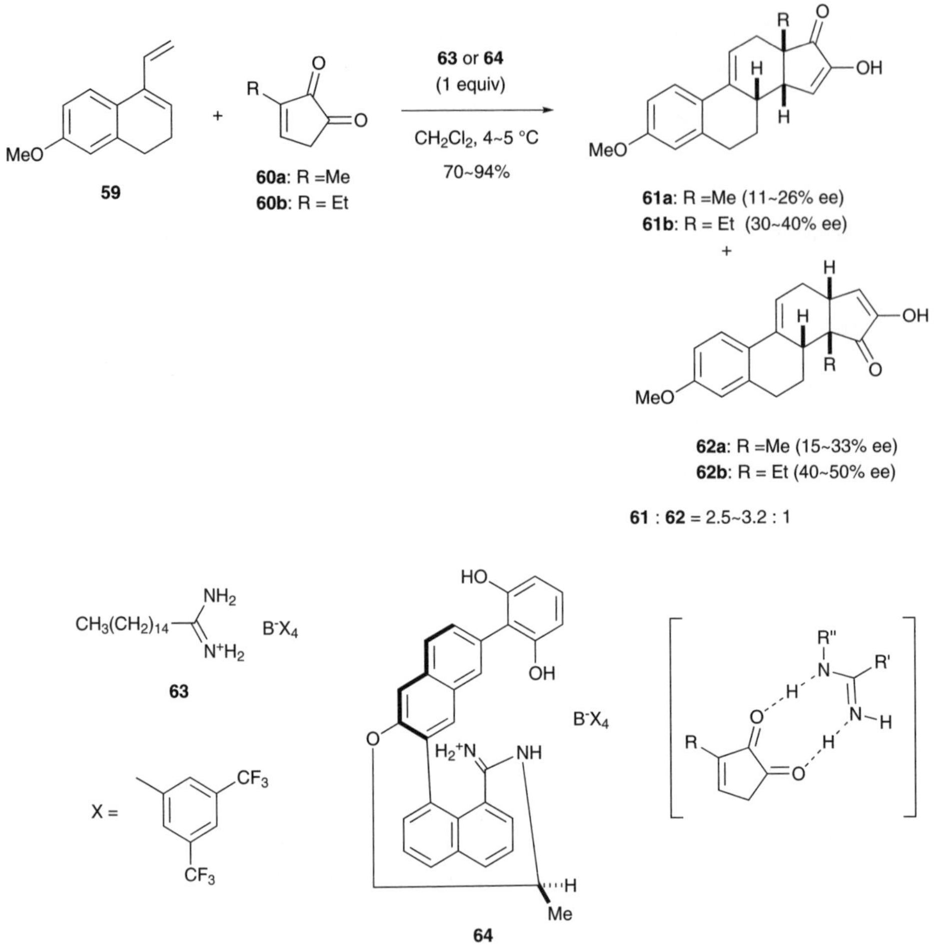

Scheme 7.12 Synthesis of (−) norgestrel intermediate

Schuster et al. reported the accelerating effect of the amidinium ion on the Diels–Alder reaction [15]. Reaction of the diene **59** and diketone **60a** or **60b** in the presence of lipophilic amidinium ion **63** (1 equiv.) gave **61** and **62** (2.5 ∼ 3.2 : 1), with a 100-fold rate increase compared to the uncatalysed conditions. When the reaction was run in the presence of chiral amidinium compound **64**, **61** and **62** were obtained in 70 ∼ 94% yield (ca 3 : 1) with 11 ∼ 50% ee. The Diels–Alder adduct **61b** is a key intermediate for synthesis of (−)-norgestrel (Scheme 7.12). The reaction enhancement effect of amidinium ion can be explained in terms of the hydrogen bond mediated interaction with diketone.

7.2.4 Wittig Reaction

Organic superbases effectively generate carbonyl-stabilized ylides to promote Wittig reactions. In the synthesis of (−)-mycalolide A (**68**) by Panek and Liu, construction of the C19-C20 olefin was troublesome under the regular Wittig reaction conditions, that is, n-Bu$_3$P-LDA, n-Bu$_3$P-LHMDS, Et$_3$P-^{t}BuOK, nor was the Horner–Wadsworth–Emmons (HWE) olefination protocol effective. Since the bromide **66** has an electron-deficient heteroaromatic moiety, Et$_3$P-DBU in DMF at 0 °C was employed, and **67** obtained as a single olefin isomer in 86% yield [16] (Scheme 7.13). This advanced intermediate was converted to **68** through the Yamaguchi–Yonemitsu macrolactonization.

The intramolecular HWE reaction in the presence of DBU was employed for the synthesis of (+)-rhizoxin D (**71**) by Jiang *et al.* [17]. Reaction of the aldehyde **69** with DBU-LiCl (Masamune–Roush conditions) [18] in acetonitrile at room temperature under high dilution conditions constructed the C2−C3 bond to form the macrolactone **70** (Scheme 7.14). Further elongation of the C20−C21 bond achieved the total synthesis of **71**.

A guanidine base has also been used for intramolecular HWE reaction. Nicolaou *et al.* reported a synthetic study of the originally proposed structure of diazonamide A (**74**) [19], employing the modified Masamune–Roush conditions [18]. Thus, reaction of aldehyde **72** with TMG (**3**)-LiCl in DMF at 70 °C generated **73** as a single atropisomer in 55–60% yield. Under other reaction conditions, for example, LHMDS in THF or DBU-LiCl in acetonitrile, **73** was obtained in only 0 ~ 35% yield (Scheme 7.15). Unfortunately, this advanced intermediate could not be transformed to the final product **74**, because the C29–30 olefin was resistant to oxidation.

Enamide ester, which is a useful synthetic intermediate for a variety of α-amino acids, can be prepared by means of the HWE reaction in the presence of TMG (**3**) or DBU [20,21]. In the synthesis of teicoplanin aglycon (**80**) reported by Evans *et al.* [22], one of the phenylalanine derivatives **79** was synthesized from the aldehyde **75**. HWE reaction of aldehyde **75** with phosphonate **76** using TMG (**3**) in THF gave (Z)-enamide ester **77** in 99% yield. Asymmetric hydrogenation of **77** catalysed by rhodium(I) complex **78** (1 mol%) gave the phenylalanine ester **79** in 96% with 94% ee (Scheme 7.16).

Since the above methodology provides easy access to a variety of α-amino acid derivatives, many applications for the synthesis of natural products have been reported [23–25]. The HWE reaction of the sterically hindered aldehyde **81** with phosphonate **82** using TMG (**3**) proceeded to give (Z)-enamide **83** in 80% yield from the alcohol (2-step yield) [26]. The resulting enamide **83** was submitted to the asymmetric hydrogenation reaction using Burk's rhodium(I) catalyst [27] to give **84** in 85% yield as the sole product (Scheme 7.17). The α-amino acid ester **84** was successfully converted to neodysiherbaine A (**85**).

Davis *et al.* reported synthetic studies of martefragin A (**91**) [28]. For the construction of the asymmetric centre next to the oxazole, HWE reaction of the aldehyde and subsequent asymmetric hydrogenation were applied. The HWE reaction of chiral aldehyde **86** with phosphonate **87** in the presence of DBU gave (Z)-enamide ester **88**, although epimerization was observed (75% ee). When the reaction was conducted using TMG (**3**), epimerization was suppressed (95% ee). Enamide ester was converted to the potential precursor **90** for **91** through the α-amino acid ester **89** via asymmetric hydrogenation (Scheme 7.18).

Enamide ester synthesis was applied for the synthesis of a complex indole natural product by Okano *et al.* [29]. (Z)-Enamide ester **94** was obtained by HWE reaction of aldehyde **92** with phosphonate **93** in the presence of TMG (**3**), in 84% yield. Treatment of the resulting ester with copper iodide (CuI) and CsOAc provided dihydropyrroloindole **95**, which was efficiently converted to yatakemycin (**96**) (Scheme 7.NaN).

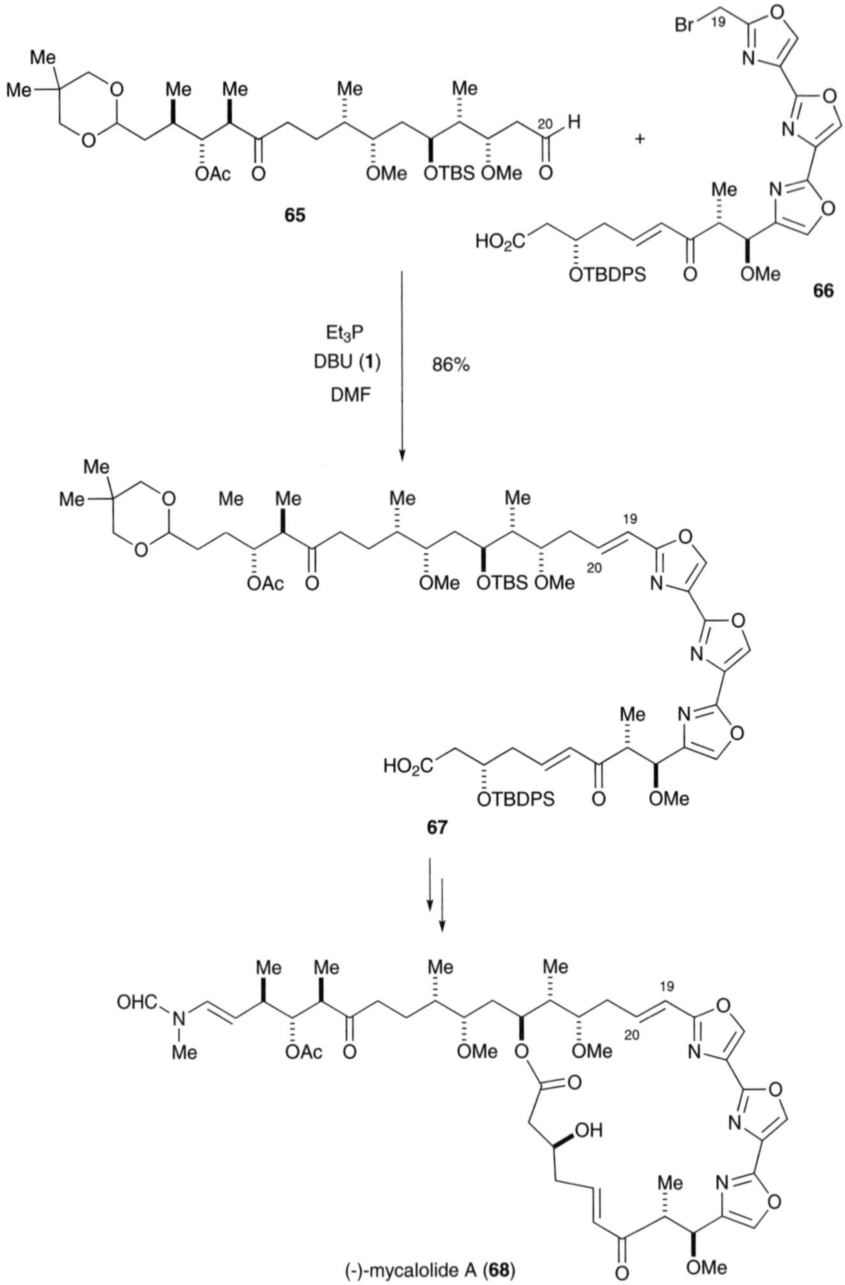

Scheme 7.13 Synthesis of (−) mycalolide A (**68**)

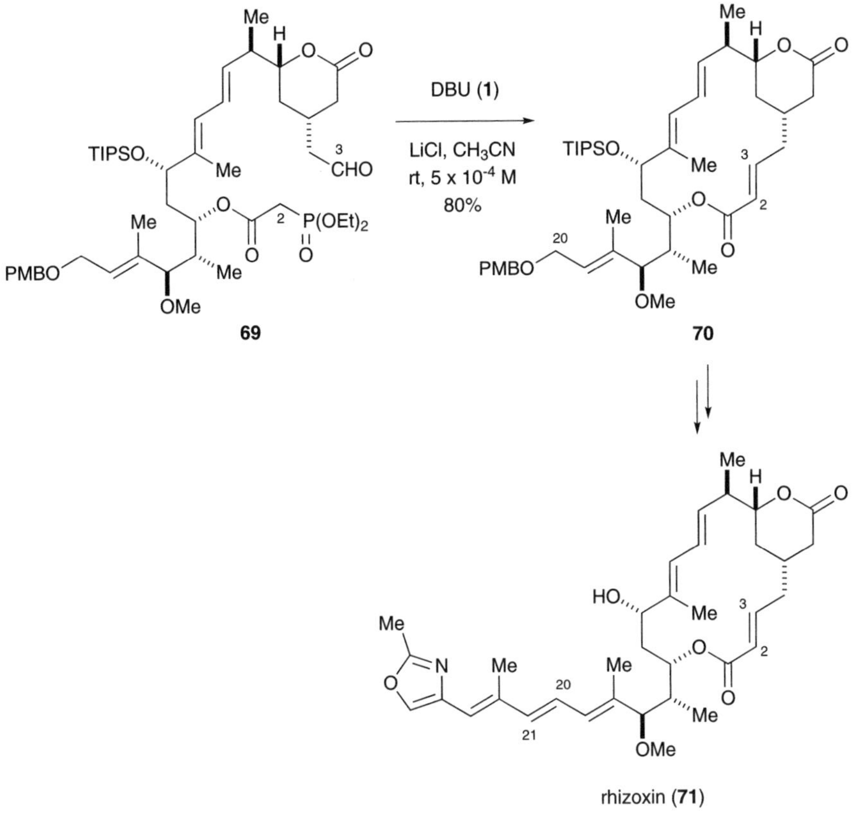

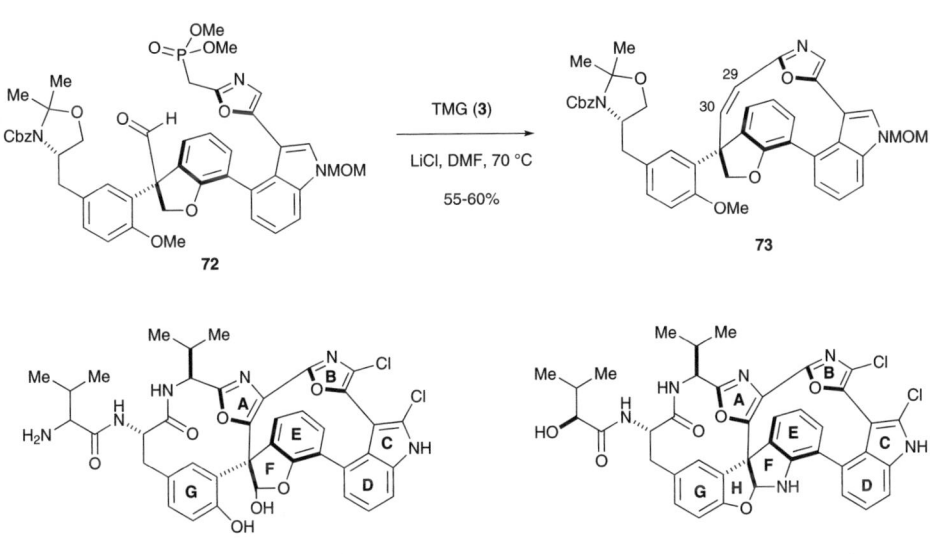

Scheme 7.14 Synthesis of (+)-rhizoxin D (**71**)

Scheme 7.15 Synthetic studies on diazonamide A (**74**)

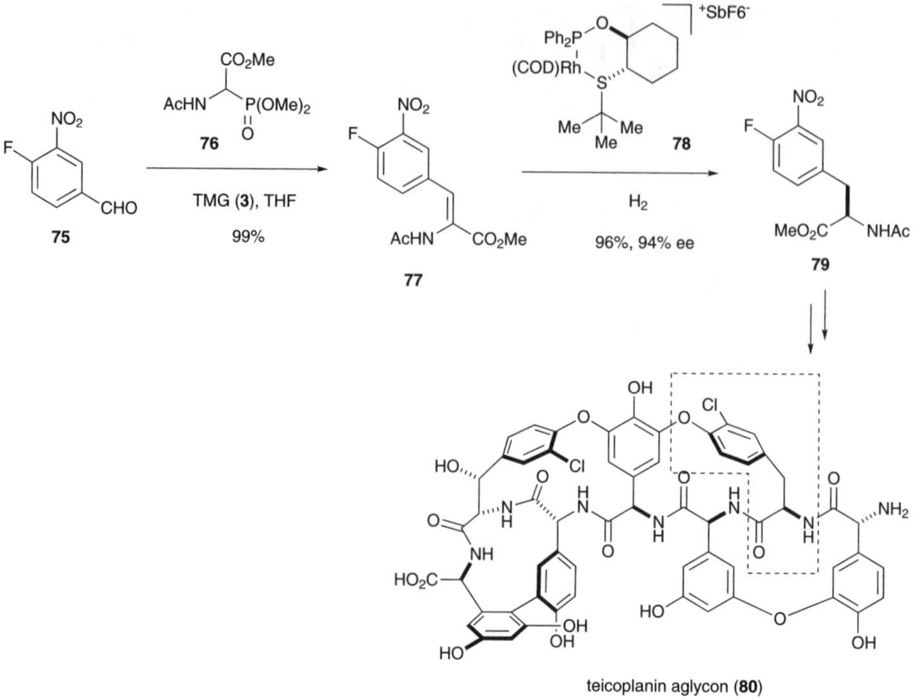

Scheme 7.16 Synthesis of teicoplanin aglycon (**80**)

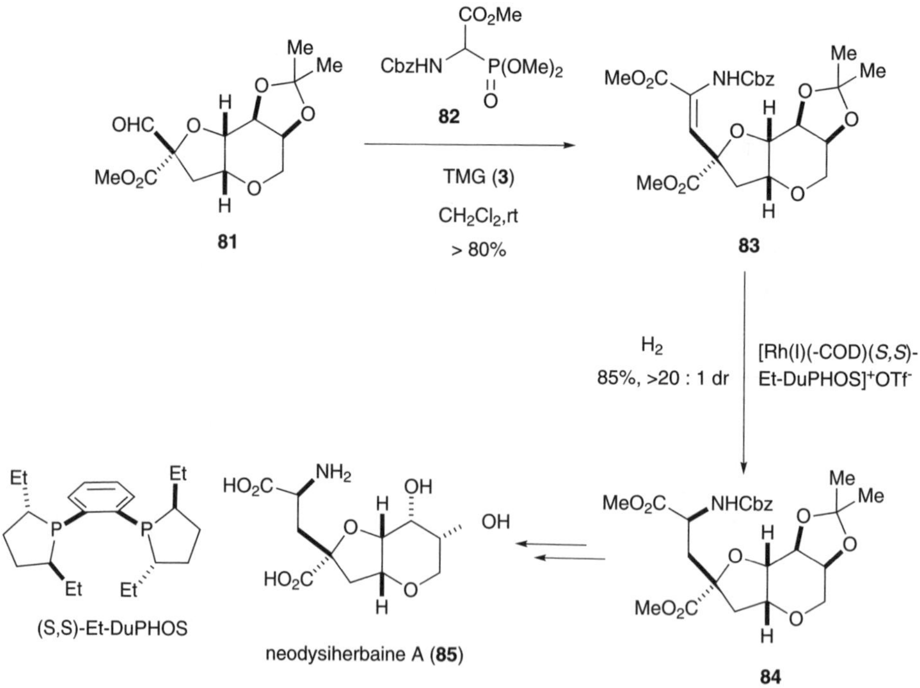

Scheme 7.17 Synthesis of neodysiherbaine A (**85**)

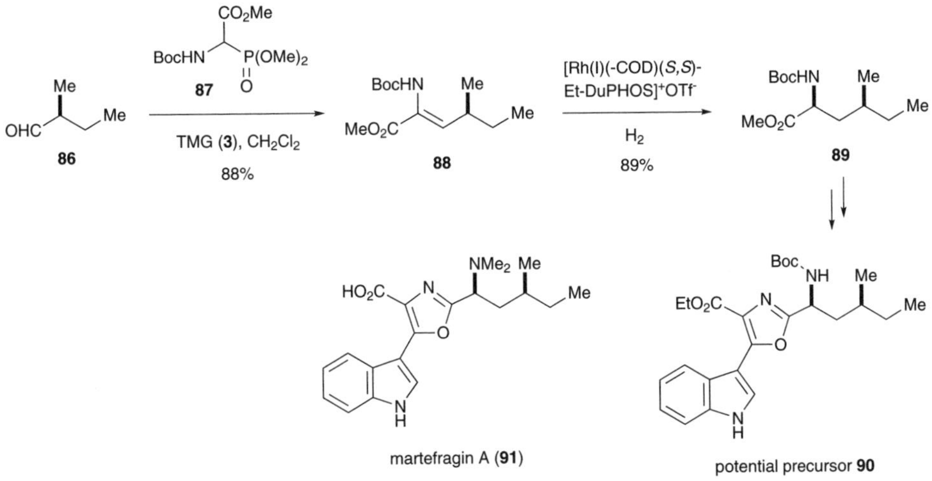

Scheme 7.18 Synthetic studies on martefragin A (**91**)

7.3 Deprotection

Guanidine/guanidinium nitrate is a selective *O*-deacetylation reagent in the presence of base labile protective groups [30]. Endo *et al.* reported the use of selective *O*-deacetylation in the synthesis of ecteinascidin 743 (**99**) [24] (Scheme 7.20). The two *O*-acetates in **97** were cleanly deprotected in the presence of the *N*-Troc group by using guanidinium nitrate in methanol.

In the synthesis of the C1–25 fragment of amphidinol **102**, selective *O*-deacetylation was performed in the presence of carbonate under the same conditions [31] (Scheme 7.21).

7.4 Elimination

The β-elimination reaction of carbonyl compounds having a leaving group at the β-position is promoted by superbases via an E1cB process. Allin *et al.* described a synthesis of deplacheine (**107**) [32], in which the β-methanesulfonyloxycarbonyl compound **105** obtained from the aldol reaction of **104** and acetaldehyde was selectively converted into the desired *E*-isomer **106** by the use of DBN in THF. This enone **106** was successfully led to the natural product **107** (Scheme 7.22).

Tanino and Kuwajima used a guanidine base promoted elimination reaction in the synthesis of ingenol (**110**) [33]. Reaction of ketone **108** with MTBD resulted in an elimination reaction and subsequent isomerization to give conjugated diene **109**, which was efficiently led to ingenol (**110**) (Scheme 7.23).

The DBU promoted epoxide opening reaction through an E1cB process has been applied to natural product synthesis. Trudeau and Morken reported a synthesis of fraxinellone (**113**) [34] (Scheme 7.24). Treatment of epoxide **111** with DBU in benzene gave allylic alcohol **112**, which was led to the natural product by oxidation of the resulting alcohol with TPAP.

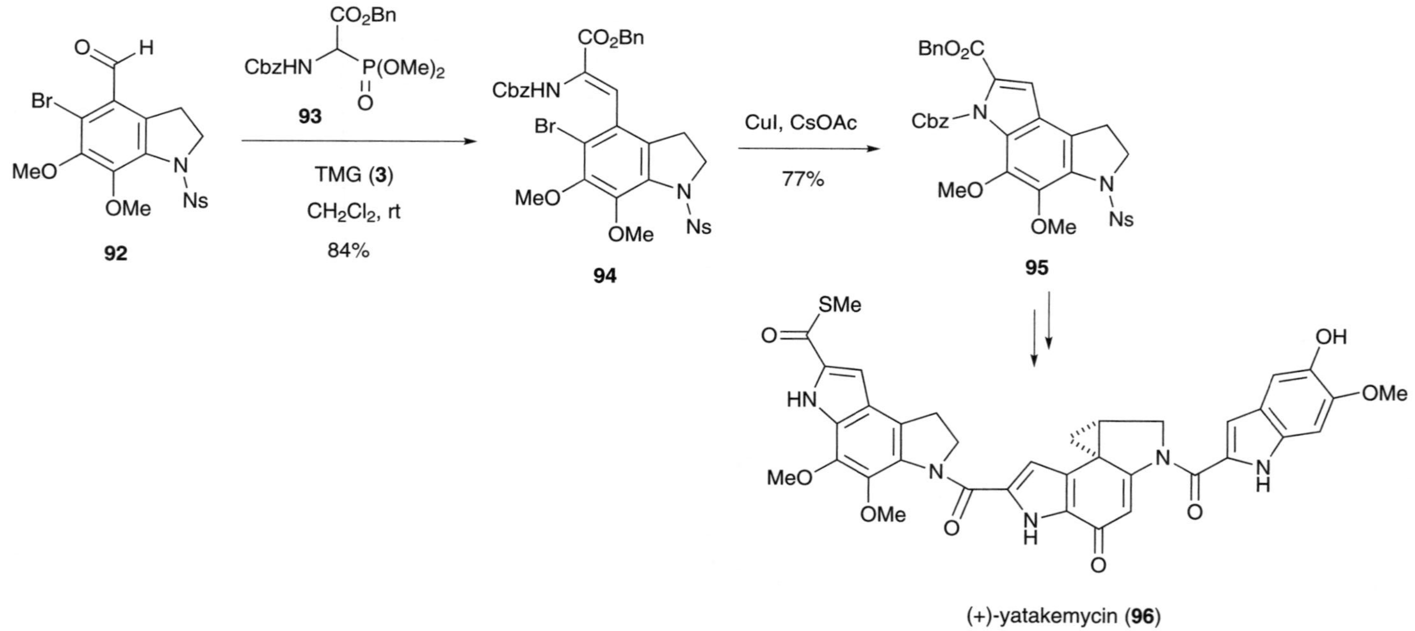

Scheme 7.19 Synthesis of (+)-yatakemycin (**96**)

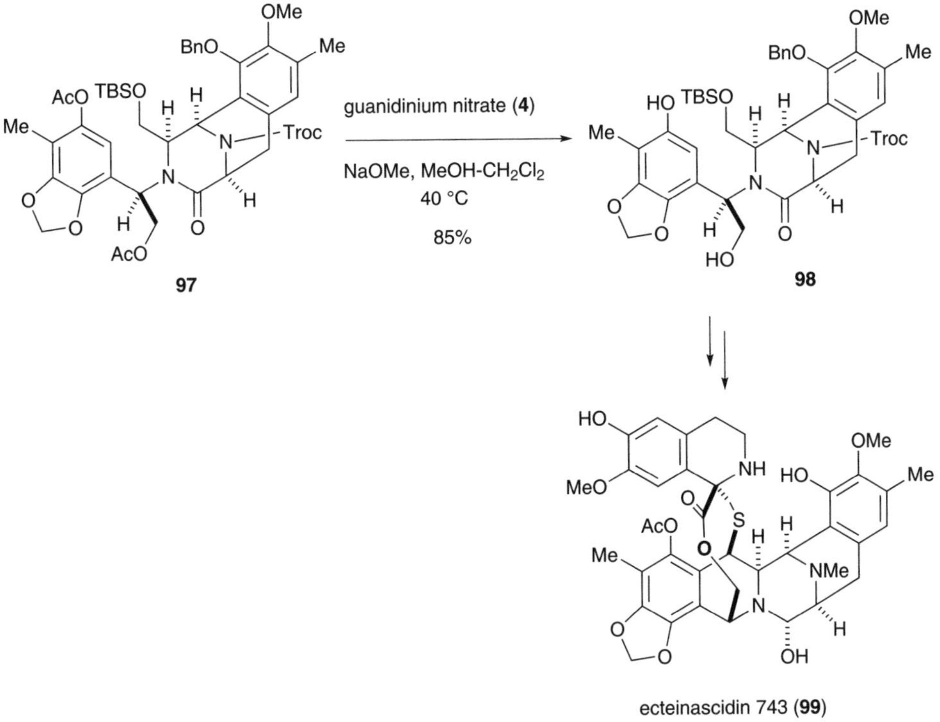

Scheme 7.20 Synthesis of ecteinascidin 743 (**99**)

Futagami *et al.* synthesized *ent*-ravidomycin (**115**) [35]. In the final stage of the synthesis, DBN was used for the simultaneous removal of acetate protection of a phenolic hydroxyl group and elimination of mesylate to give **115** (Scheme 7.25).

Hughes *et al.* achieved a synthesis of amythiamicin D (**121**) using a biosynthesis inspired hetero Diels–Alder reaction as a key step [36]. Synthesis of the key 1-alkoxy-2-azadiene **118** was conducted by coupling of the imidate **116** and amine hydrochloride **117** and subsequent elimination of the acetate with DBU [37]. The hetero Diels–Alder reaction of the azadiene **118** and enamide **119** proceeded under microwave conditions to give the pyridine **120**, which was effectively converted into the natural product **121** (Scheme 7.26).

Synthesis of oxazoles and thiazoles from corresponding oxazolines and thiazolines by dehydrogenation using the bromotrichloromethane ($BrCCl_3$)-DBU system was reported by Williams *et al.* [38]. This transformation was applied to synthesize many natural products. Doi and Takahashi reported the synthesis of telomestatin (**124**) [39], which has a macrocyclic polyoxazole structure. Reaction of oxazoline **122** with $BrCCl_3$-DBU gave trisoxazole **123** (Scheme 7.27). This left-hand segment was employed for the synthesis of telomestatin **124**.

Vinyl halide is a useful synthetic intermediate for natural products. Knapp *et al.* reported the synthesis of griseolic acid B (**130**) via intramolecular radical cyclization of vinyl iodide [40]. Aldehyde **125** was reacted with hydrazine to give hydrazone **126**, which was further

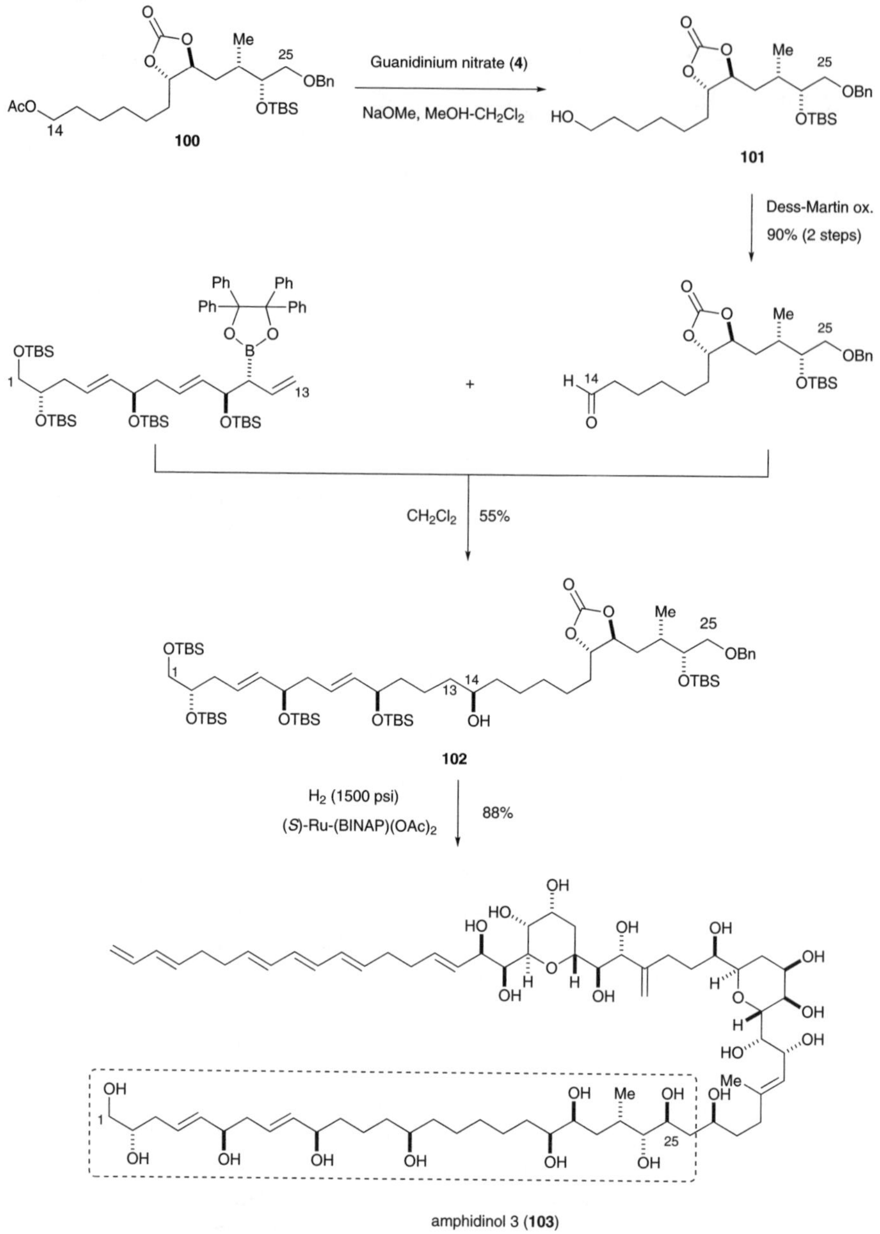

Scheme 7.21 Synthetic studies on amphidinol 3 (**103**)

reacted with iodine in the presence of TMG (**3**) to give gem-diiodide. This diiodide was treated with DBU to afford the vinyl iodide **127** [41]. After conversion of **127** to the maleate derivative **128**, radical cyclization afforded **129**, which was led to the natural product griseolic acid B **130** (Scheme 7.28).

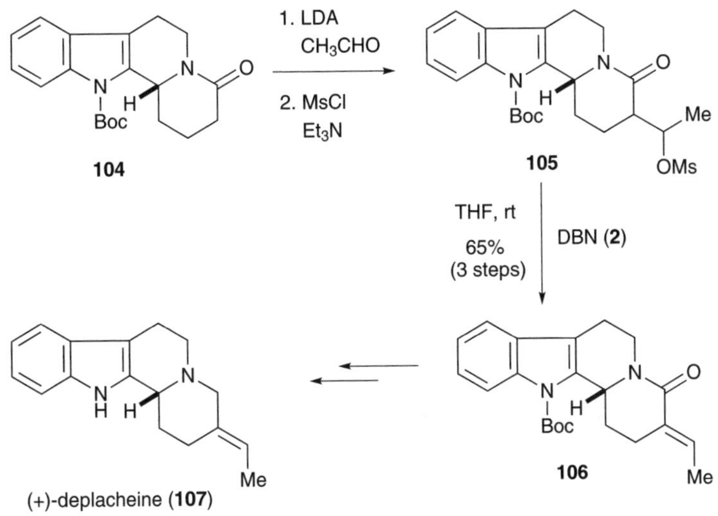

Elimination 229

Scheme 7.22 Synthesis of deplacheine (**107**)

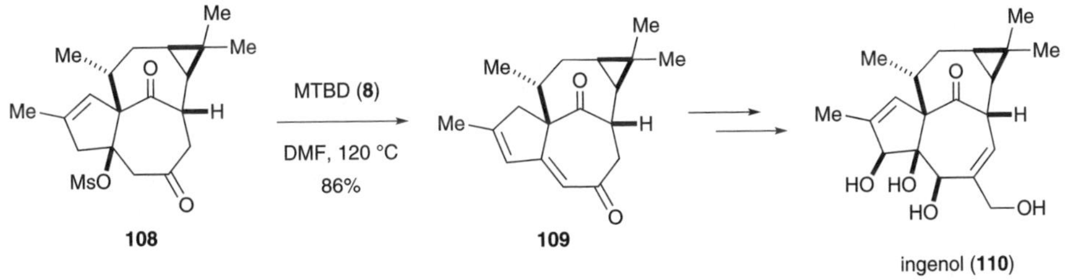

Scheme 7.23 Synthesis of ingenol (**110**)

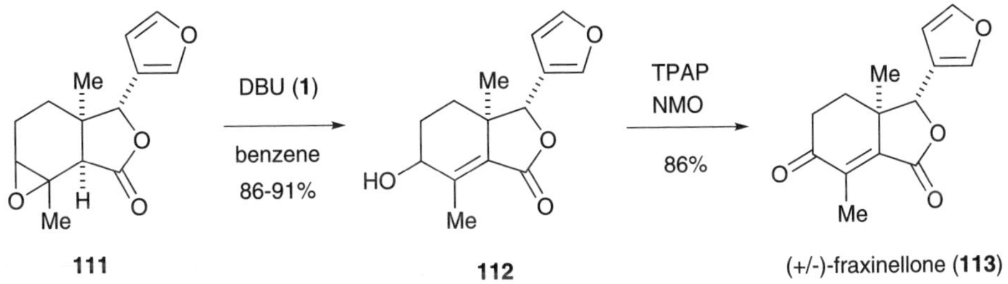

Scheme 7.24 Synthesis of (±)-fraxinellone (**113**)

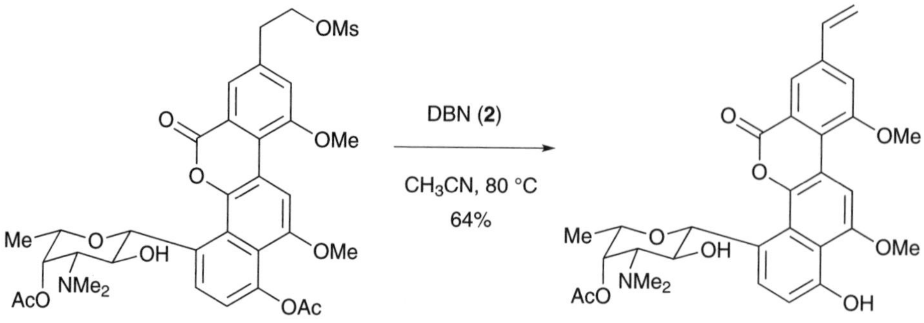

Scheme 7.25 Synthesis of ent-ravidomycin (**115**)

Similar vinyl iodide formation (**131** to **132**) was employed by Paquette et al. in synthetic studies on dumsin (**133**) [42]. In this synthesis, palladium catalysed Stille coupling was conducted using the vinyl iodide **132** (Scheme 7.29).

Synthetically useful alkynes can be obtained from 1,2-dibromoalkanes using a variety of bases. Ohgiya et al. developed the DBU promoted elimination reaction of 1,2-dibromoalkanes having an oxygen functional group at C3 [43]. Reaction of 1,2-dibromide **134** having PMB ether at C3 with DBU in DMF gave alkyne **135** in 73% yield. This intermediate was employed for the synthesis of sapinofuranone B (**136**) (Scheme 7.30).

Wender et al. achieved a synthesis of phorbol (**140**) [44], in which the BC-ring system **139** was efficiently constructed by transannular [5 + 2] cycloaddition reaction. Thus, reaction of acetoxypyranone **137** with DBU in acetonitrile generated the oxidopyrylium intermediate **138** by elimination of the acetoxy group and enolization, and this intermediate smoothly reacted with alkene to give **139** in 79% yield. This BC-ring system was successfully converted to phorbol (**140**) (Scheme 7.31).

Biosynthesis of the polycyclic diterpene intricarene (**144**) may occur from the natural product bipinnatin J (**141**) through transannular [5 + 2] cycloaddition reaction. Based upon this proposed biosynthetic route, Tang et al. examined a synthesis of **144** [45]. Synthetic **141** was treated with VO(acac)$_2$ and tert-butyl hydroperoxide, followed by acetic anhydride to give acetoxypyranone **142**, which was subsequently heated in acetonitrile in the presence of DBU to give intricarene (**144**) (Scheme 7.32).

7.5 Ether Synthesis

Ether synthesis by alkylation of alcohol using an inorganic base is sometimes troublesome. In such cases, soluble organic superbases are often effective. Knapp et al. performed the alkylation of **145** with isopropyl bromoacetate using the strong soluble base BEMP, achieving 95% yield [46]. Intramolecular ester enolate alkylation of **147** with LDA took place to give **148**, which was effectively led to the natural product octosyl acid A (**149**) (Scheme 7.33).

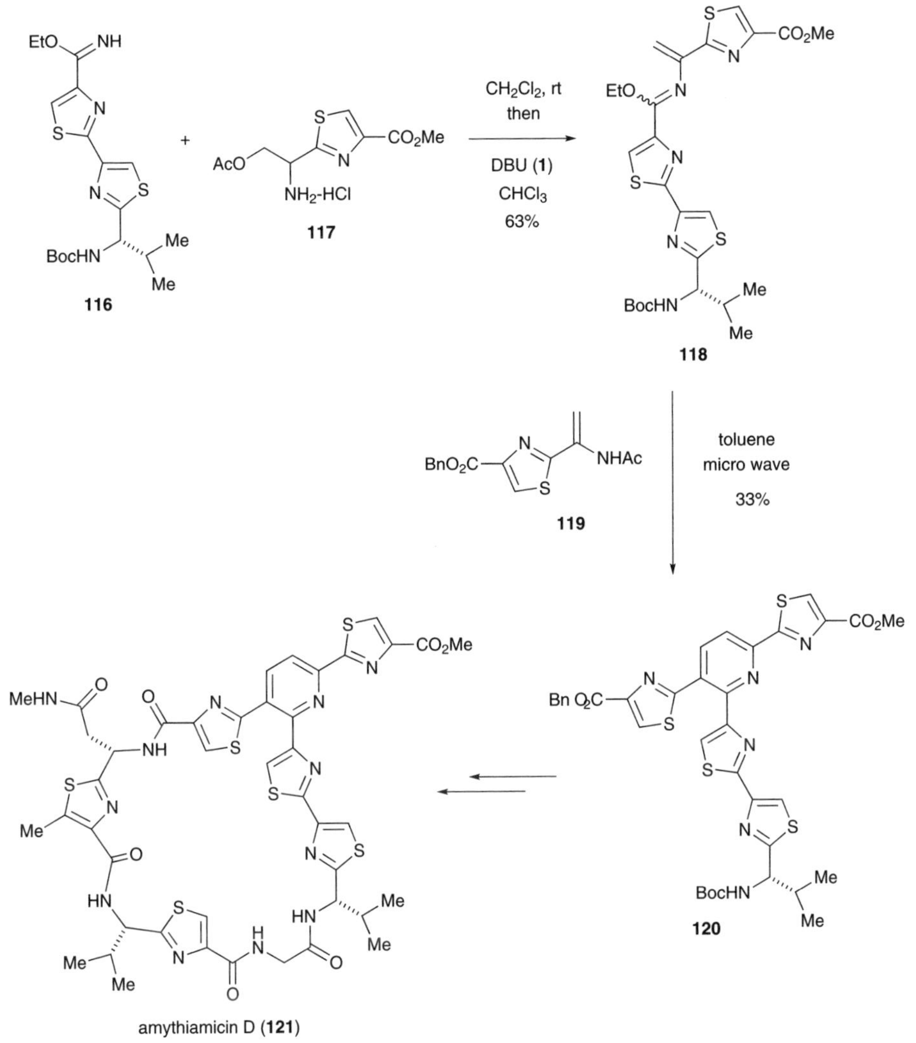

Scheme 7.26 Synthesis of amythiamicin D (**121**)

Hanessian *et al.* reported a synthesis of manassantins [47]. In this synthesis, the phenolic hydroxyl groups of **150** were alkylated with triflate ester **151** using BEMP to give **152** with complete inversion of the stereochemistry. The combination of caesium phenolate and mesylate of ethyl lactate induced partial racemization. The ester **152** was efficiently converted into manassantin A (**154**) (Scheme 7.34).

Intramolecular alkylation of alcohol is also feasible with superbases. Kusama *et al.* synthesized the oxetane **156**, containing the D-ring of taxol (**157**), from hydroxychloride **155** using DBU in toluene at reflux in 86% yield [48] (Scheme 7.35).

Copper mediated Ullman reaction, usually requires harsh conditions. Wipf and Lynch reported a relatively mild biaryl ether synthesis via an S_NAr (nucleophilic aromatic substitution)

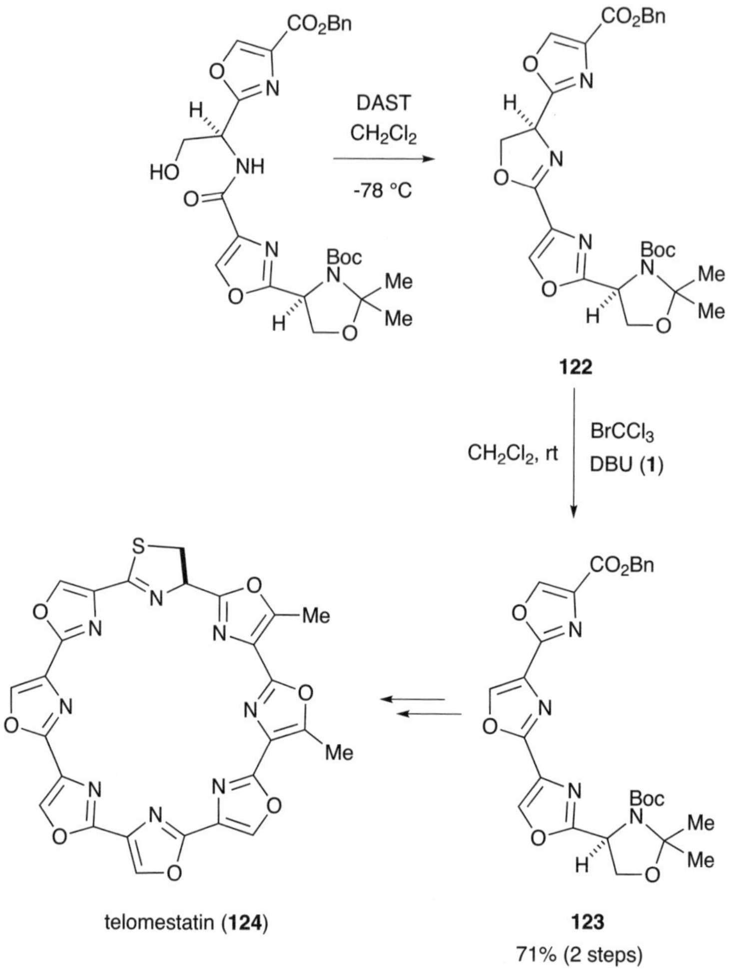

Scheme 7.27 Synthesis of telomestatin (**124**)

reaction promoted by guanidine base. Reaction of activated aryl fluoride **158** and naphthol **159** using Barton's base **5** provided dinaphthyl ether **160** in 85% yield [49]. The aryl ether underwent oxidative spirocyclization with PhI(OAc)$_2$ to give **161**, which represents the mother skeleton of palmarumycin CP$_1$ (**162**) and diepoxin σ (**163**) (Scheme 7.36).

A phosphazene base catalysed S$_N$Ar reaction for biaryl ether synthesis was reported by Ebisawa et al. [50]. Reaction of activated fluorobenzoate **164** and functionalized phenol **165** with TMSNEt$_2$ and a catalytic amount of phosphazene of P4-^{t}Bu (10 mol %) gave biaryl ether **166** in 94% yield. The biaryl ether was efficiently led to dictyomedin A (**167**) and B (**168**) (Scheme 7.37).

The aziridine ring opening reaction with phenol derivatives using copper(I) acetate (CuOAc)-DBU was reported by Li et al. [51]. Reaction of ethynyl nosyl-aziridine **170** and β-hydroxytyrosine derivative **169** in the presence of DBU (2 equiv.) and a catalytic amount

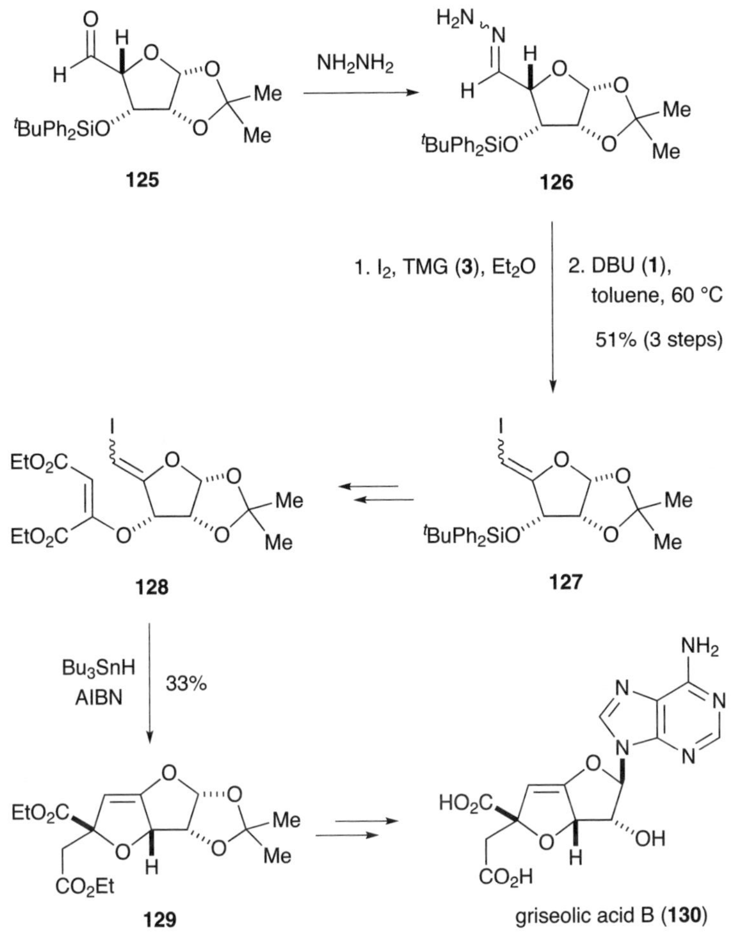

Scheme 7.28 *Synthesis of griseolic acid B (**130**)*

of copper(I) acetate (2.5 mol%) gave alkyl-aryl ether **171** in 90% yield. In this reaction, a quaternary β-substituted α-amino function was stereoselectively generated. The aziridine ring opening product **171** was efficiently led to ustiloxin D (**172**) (Scheme 7.38).

Recently, Forbeck *et al.* improved the aziridine ring opening reaction [52]. With the new protocol, using the guanidine base of TBD (2 equiv.), a range of functionalized phenol derivatives can be used to generate the aziridine ring opened product **175** in high yield (Table 7.1).

7.6 Heteroatom Conjugate Addition

Heteroatom conjugate addition reactions are promoted by superbases. Draper *et al.* described the synthesis of the estrogen antagonist Sch 57050 (**179**) [53]. Knoevenagel condensation reaction with ketone **176** and aldehyde **177** in the presence of piperidine

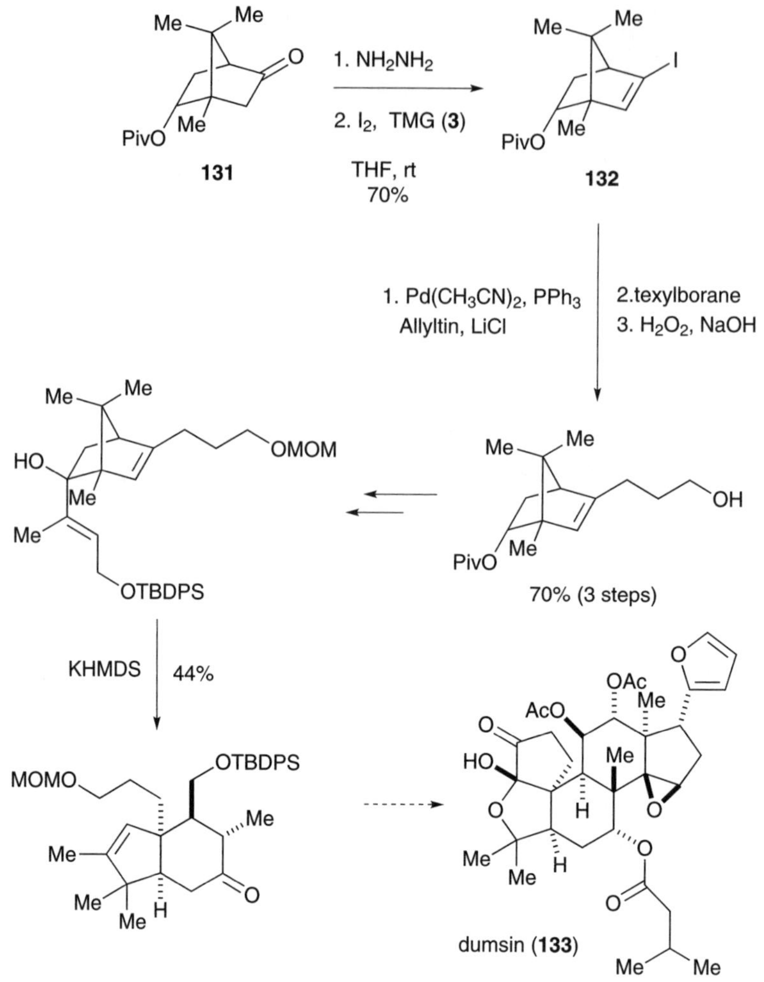

Scheme 7.29 Synthetic studies on dumsin (**133**)

followed by DBU treatment promoted oxy-Michael reaction to give chromanone **178**, which was further transformed into Sch 57 050 (**179**) (Scheme 7.39).

Conjugate addition reaction of a thiol group to unsaturated ketone was efficiently applied to a synthesis of ecteinascidin 743 (**99**) by Corey et al. [54]. Conjugate addition of the

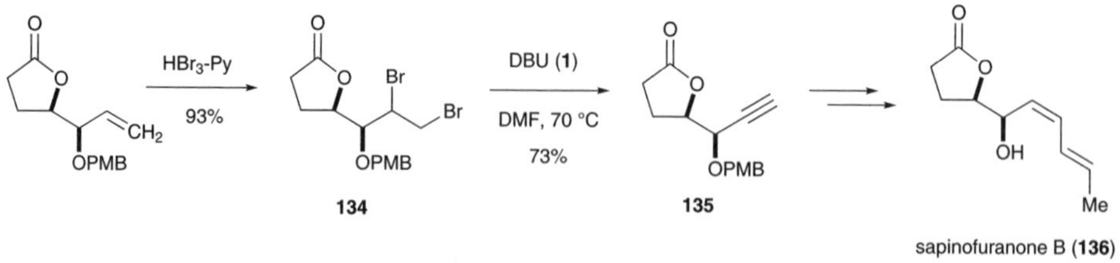

Scheme 7.30 Synthesis of sapinofuranone B (**136**)

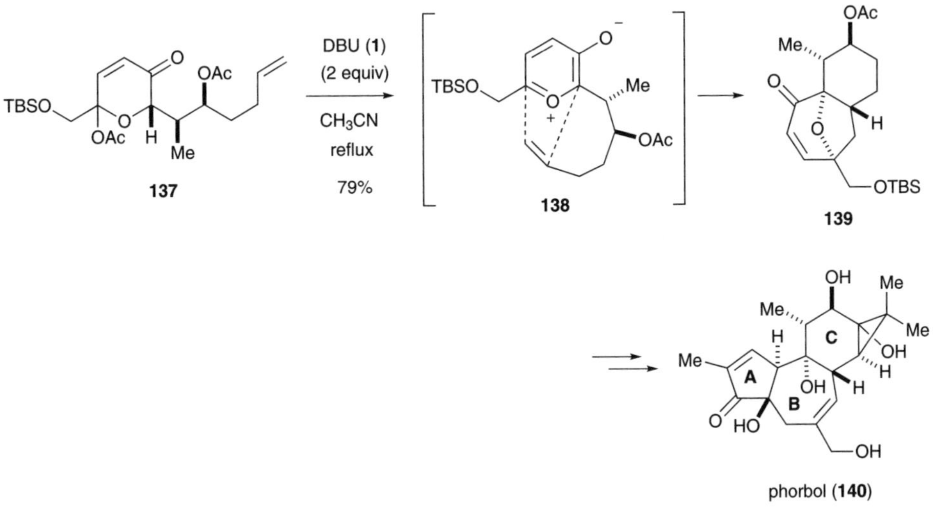

Scheme 7.31 Synthesis of phorbol (**140**)

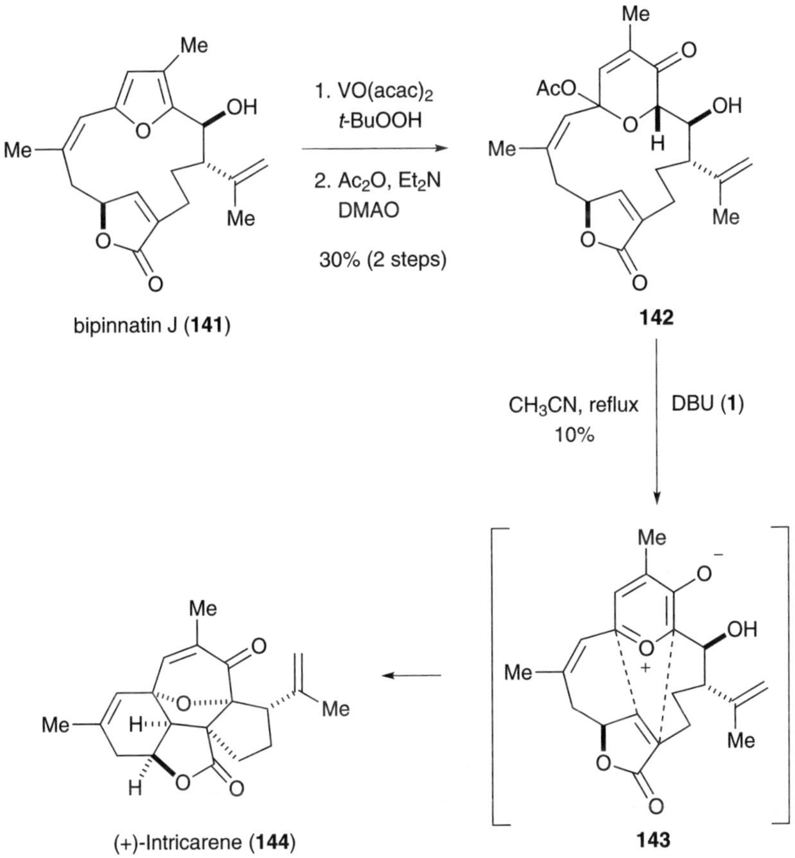

Scheme 7.32 Synthesis of (+)-intricarene (**144**)

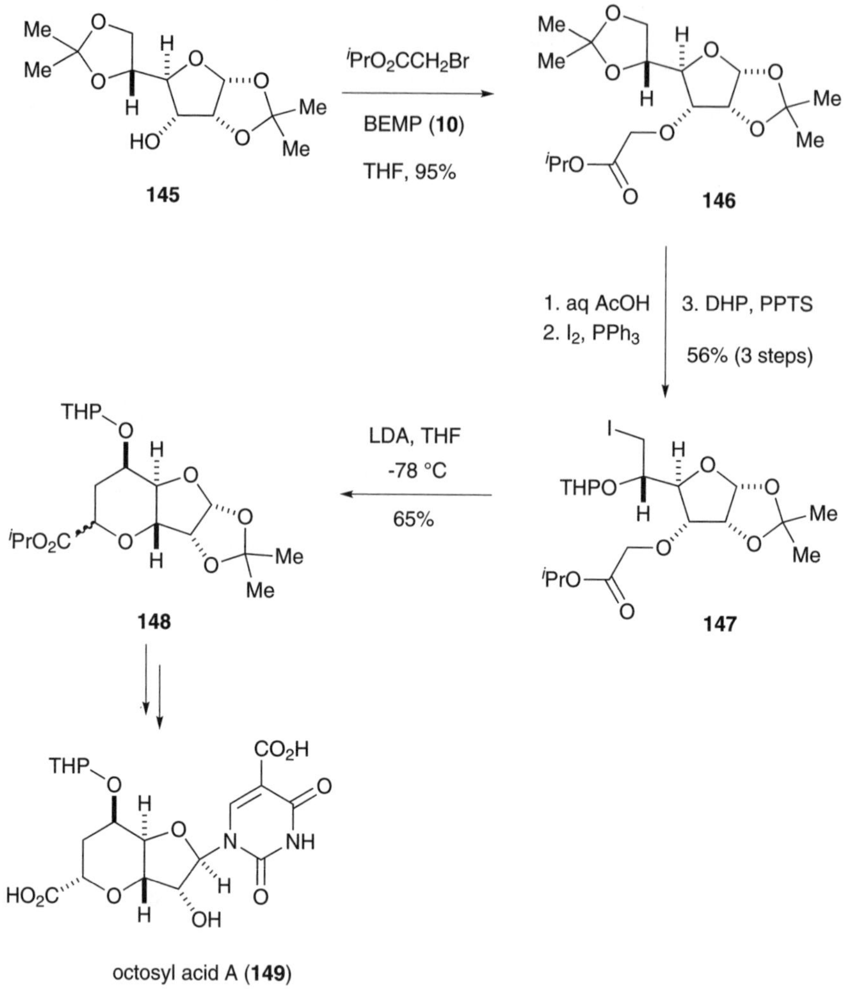

Scheme 7.33 Synthesis of octosyl acid A (**149**)

cysteine thiol group to the orthoquinone methide **181**, which was generated from the α-hydroxy ketone **180**, took place to give the 10-membered lactone **182** by the use of Barton's base (**5**) (Scheme 7.40).

Conjugate addition of hydroperoxide to unsaturated carbonyl compounds generates epoxide in the presence of superbases. Wood et al. conducted the epoxidation reaction of **183** by treatment with *tert*-butyl-hydroperoxide, using a catalytic amount of DBU [55]. The resulting epoxide **184** was converted to (±)-epoxysorbicillinol (**185**) (Scheme 7.41).

Genki et al. reported the epoxidation reaction of dienone using a combination of *tert*-butyl-hydroperoxide and a guanidine base [56]. Applying this methodology, they achieved a synthesis of (±)-preussomerin L (**188**) [56]. Reaction of dienone **186** with *tert*-butyl hydroperoxide and TBD in toluene gave bis-α,β-epoxyketone **187** in 91% yield. The ketone **187** was efficiently led to the natural product **188** (Scheme 7.42).

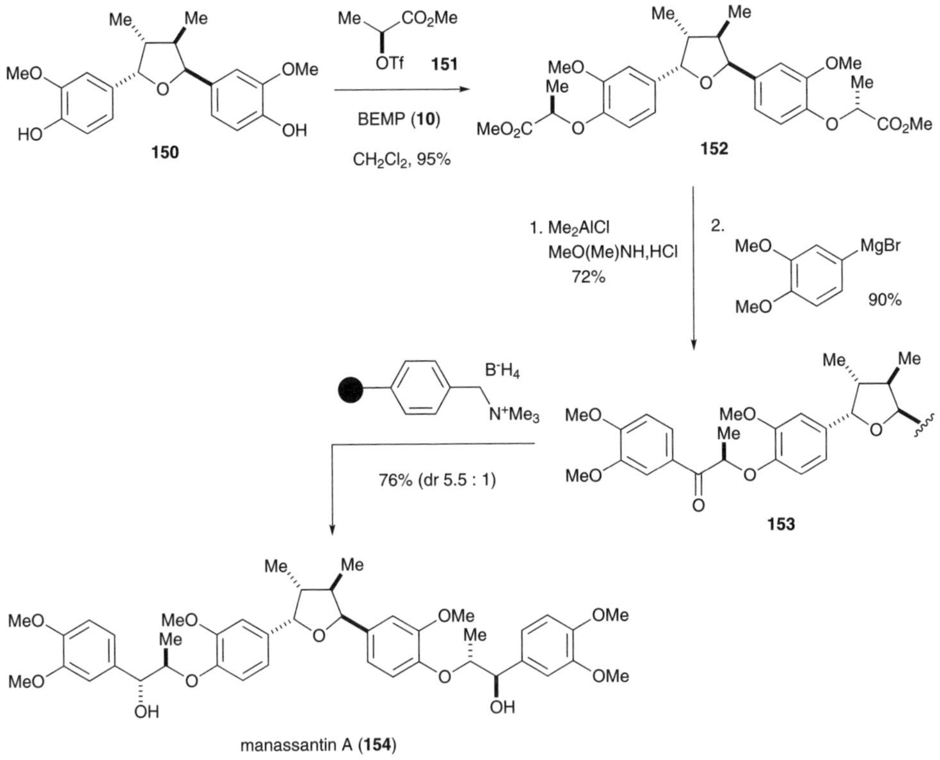

Scheme 7.34 Synthesis of manassantin A (**154**)

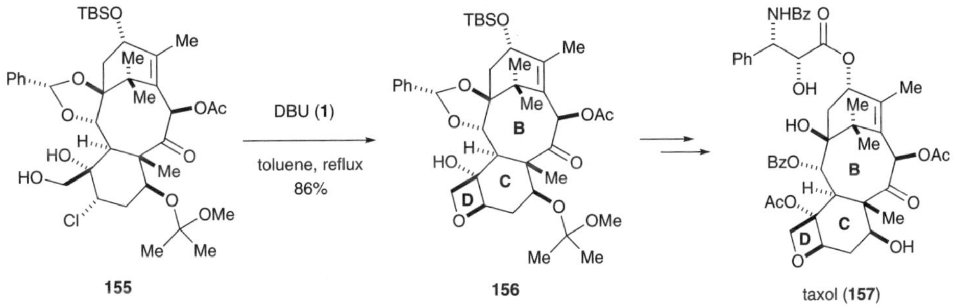

Scheme 7.35 Synthesis of taxol (**157**)

7.7 Isomerization

Baros *et al.* reported a quantitative isomerization of *exo*-olefin **189** to the thermodynamically stable α,β-unsaturated ketone **190** in the presence a catalytic amount of DBN at room temperature in quantitative yield [57]. The two TBS groups were smoothly deprotected to give (−)-eutypoxide B (**191**) (Scheme 7.43).

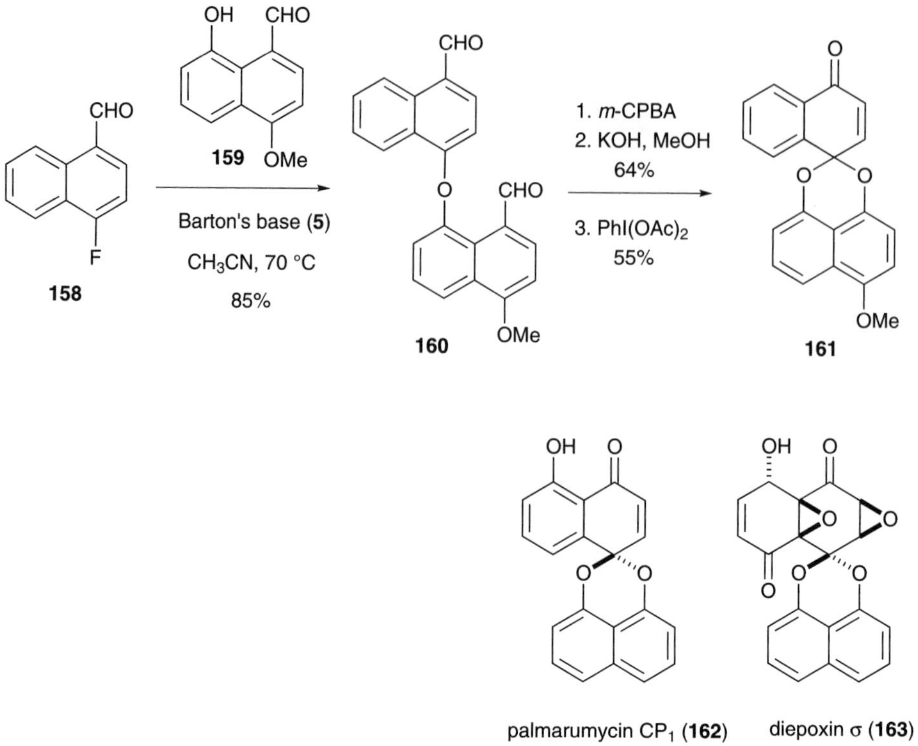

Scheme 7.36 Synthesis of mother skeleton of palmarumycin CP1 (**162**) and diepoxin σ (**163**)

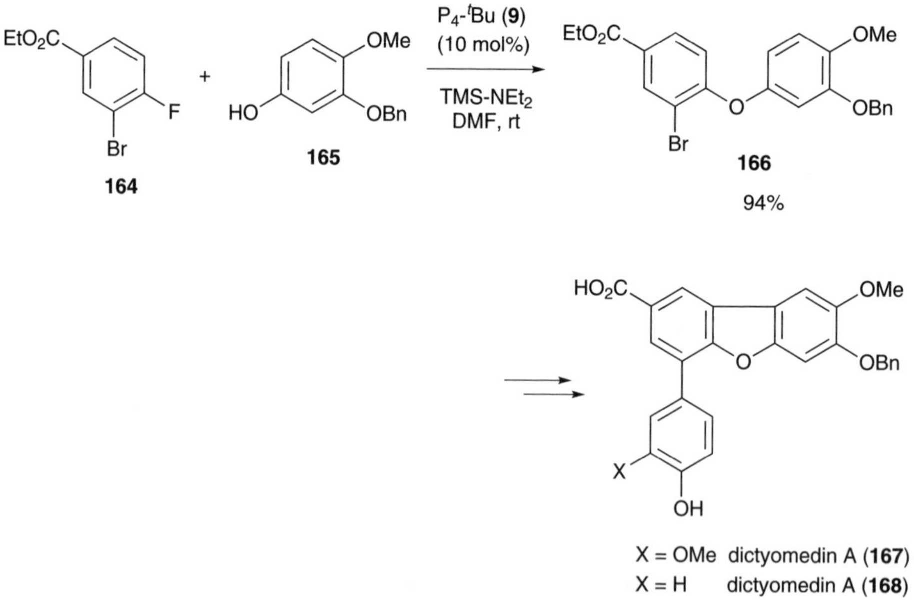

palmarumycin CP$_1$ (**162**)   diepoxin σ (**163**)

Scheme 7.37 Synthesis of dictyomedin A (**167**) and B (**168**)

X = OMe dictyomedin A (**167**)
X = H   dictyomedin A (**168**)

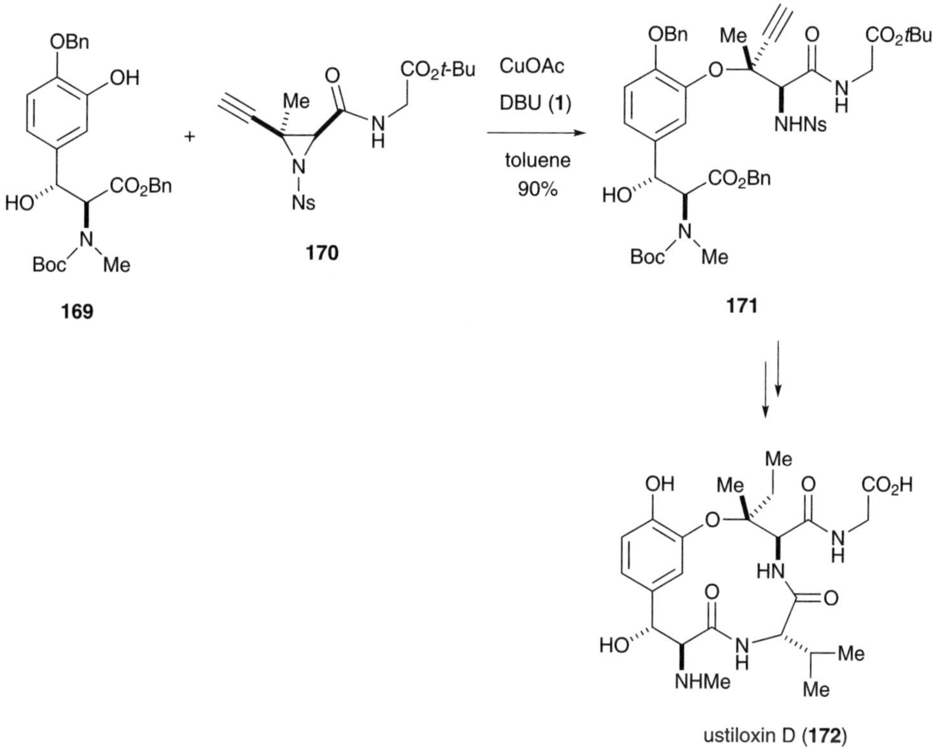

Scheme 7.38 Synthesis of ustiloxin D (**172**)

Table 7.1 Aziridine ring opening reaction with phenols in the presence of TBD (**7**)

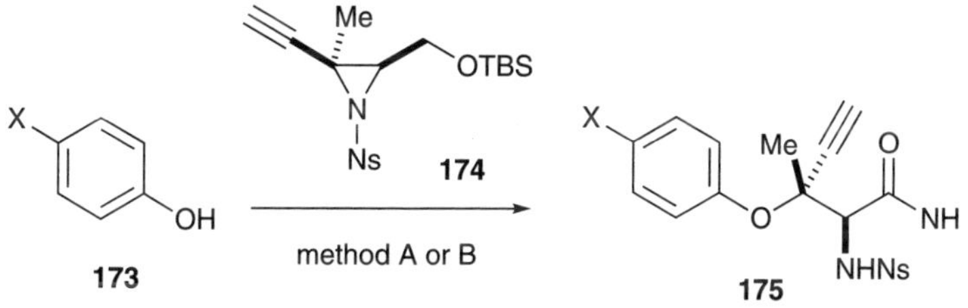

X	Yield of **175** (%)	
	method A	method B
H	39	78
Br	22	73
Me	31	71
CN	—	84
CHO	—	69

method A: Ar-OH (2.2 equiv.), DBU (**1**) (2.2 equiv.), CuOAc (2.5 mol%), toluene.
method B: Ar-OH (2 equiv.), TBD (**7**) (2 equiv.), toluene.

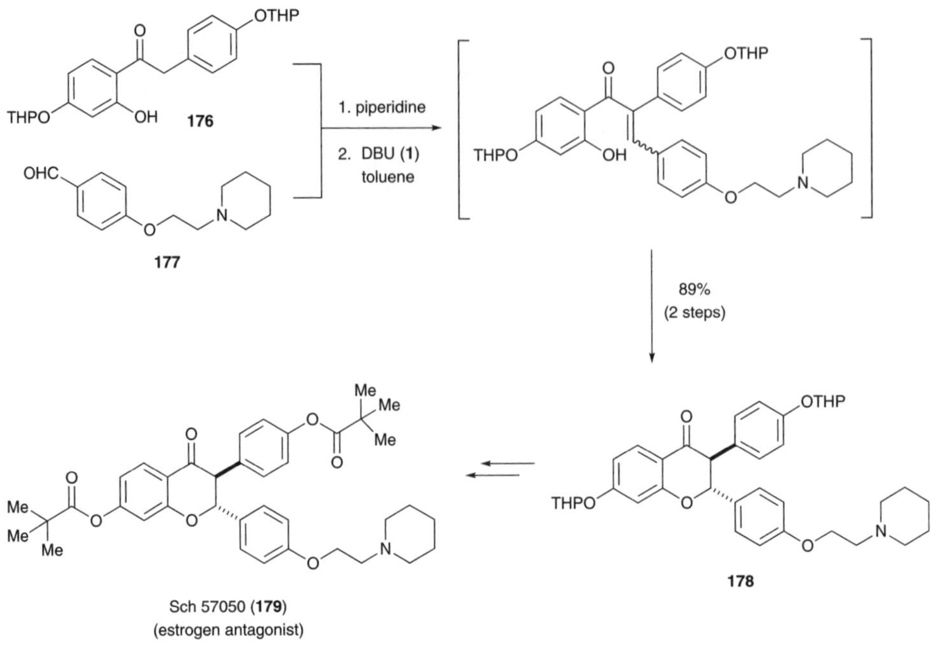

Scheme 7.39 Synthesis of estrogen antagonist Sch 57 050 (**179**)

Corey and Kania reported a synthesis of (+)-dolabellatrienone (**56**) [13], in which spontaneous cyclization of **192** occurred via the acid chloride to give a mixture of **193** and **194**. This mixture was subsequently converted to **56** exclusively by treatment with DBU (Scheme 7.44).

The cyathane diterpenoids have a characteristic 5-6-7 membered tricyclic skeleton. Piers et al. reported the synthesis of one of the cyathane diterpenoids, (±)-sarcodonin G (**198**) [58]. In this synthesis, a mixture of **196** and **197** (30:1) obtained from selenide **195** by oxidation, cleanly isomerized to give thermodynamically favoured **197** in the reaction with DBN in refluxing benzene (Scheme 7.45).

Similar conditions, that is, DBU in benzene at 75 °C, were effectively applied to the synthesis of scabronine G **201** by Waters et al. [59] (Scheme 7.46).

Corey et al. reported the reaction of singlet oxygen with conjugated diene **203**, which was obtained from the isomerization reaction of α,β-unsaturated lactone **202** using a catalytic amount of DBN. The resulting peroxide intermediate **204** was converted to (±)-folskolin (**205**) [60] (Scheme 7.47).

Renneberg et al. reported a synthesis of (+)-coraxeniolide A (**208**) [61]. In this synthesis, isomerization at the α-position (C4) was achieved by the use of TBD. Thus, alkylation of **206** at C4 was conducted with LDA and 1-bromo-4-methylpent-2-ene to give a mixture of coraxeniolide A (**208**) and its C4 epimer **207** in a ratio of 1:5.7 (Scheme 7.48). TBD treatment of the mixture caused epimerization at C4, changing the product ratio to 3:1.

An isomerization reaction at an α-position with superbase was also applied in the synthesis of taxol (**157**) by Kusama et al. [48]. C10α-Acetate **210**, obtained from the enol

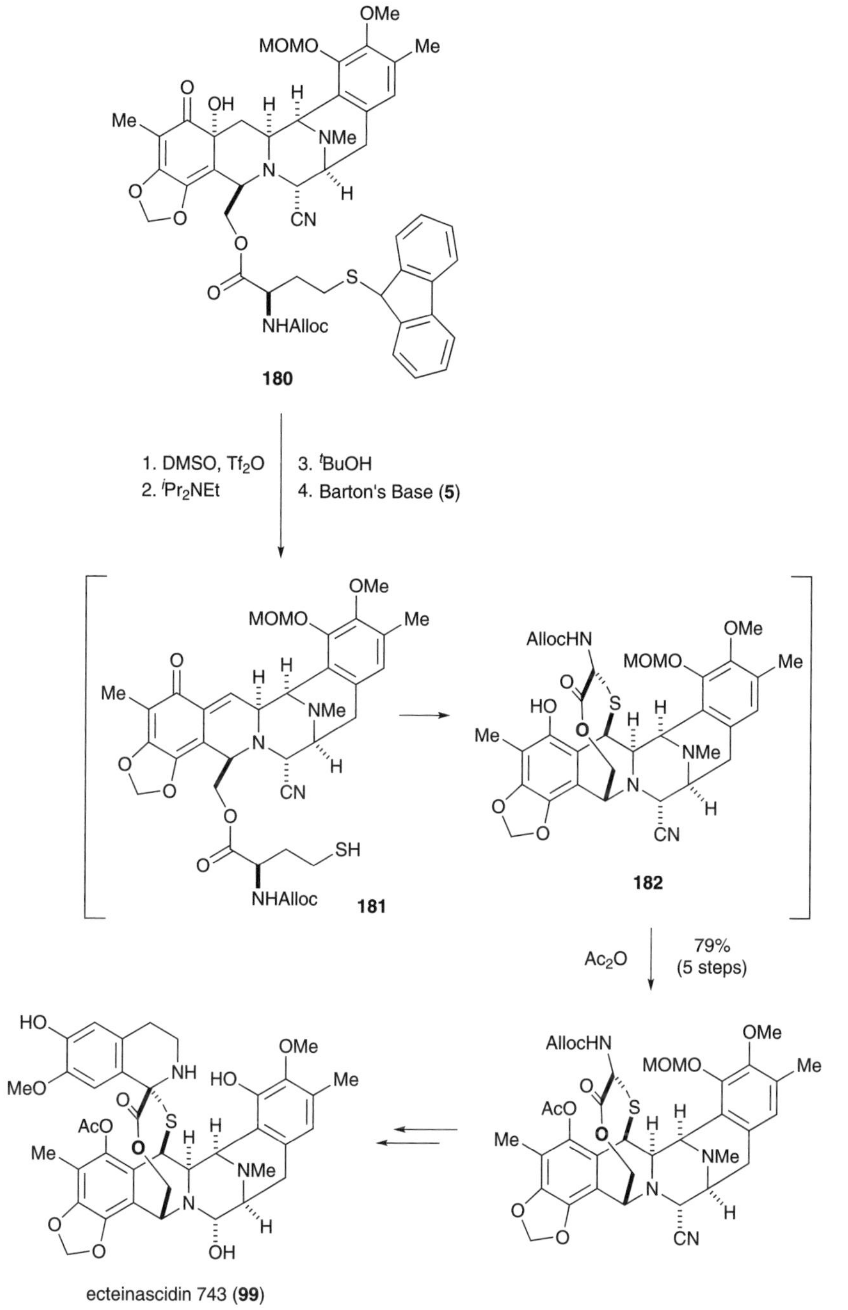

Isomerization 241

180

1. DMSO, Tf$_2$O
2. ^{i}Pr$_2$NEt
3. ^{t}BuOH
4. Barton's Base (**5**)

181

182

Ac$_2$O | 79% (5 steps)

ecteinascidin 743 (**99**)

Scheme 7.40 Synthesis of ecteinascidin 743 (**99**)

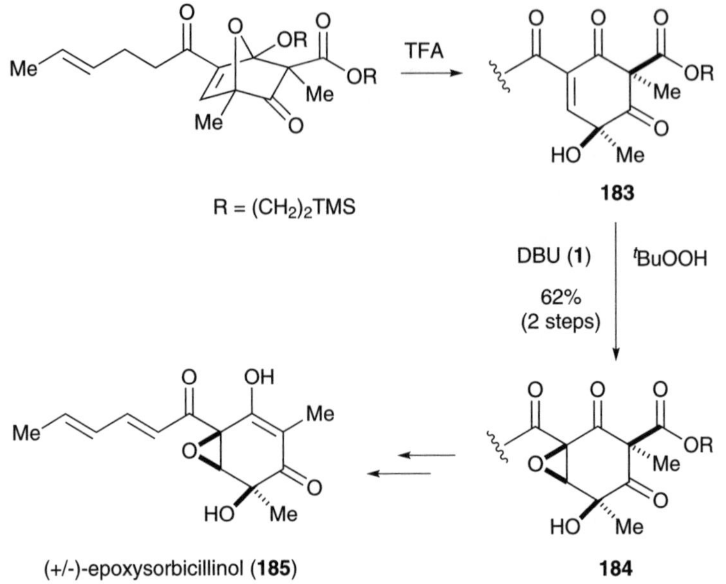

Scheme 7.41 Synthesis of (±)-epoxysorbicillinol (**185**)

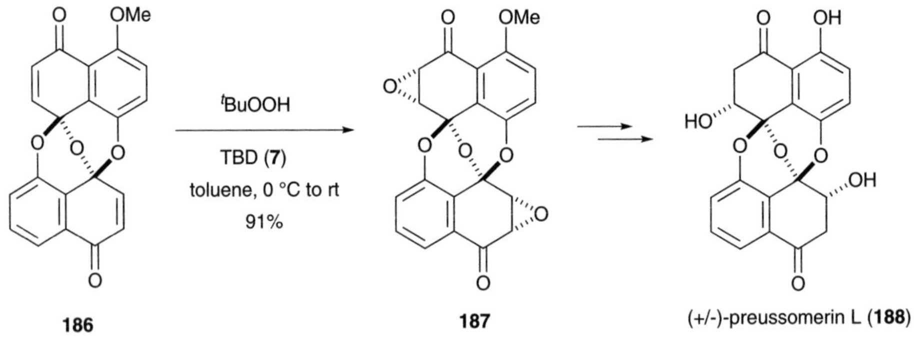

Scheme 7.42 Synthesis of (±)-preussomerin L (**188**)

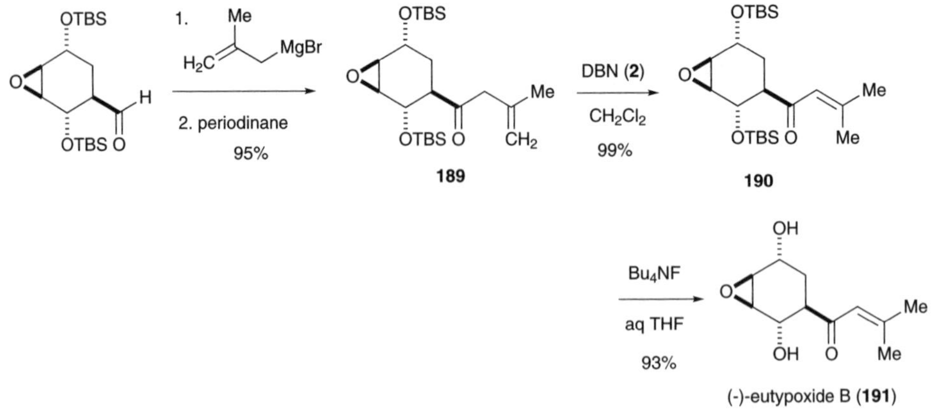

Scheme 7.43 Synthesis of (−)-eutypoxide B (**191**)

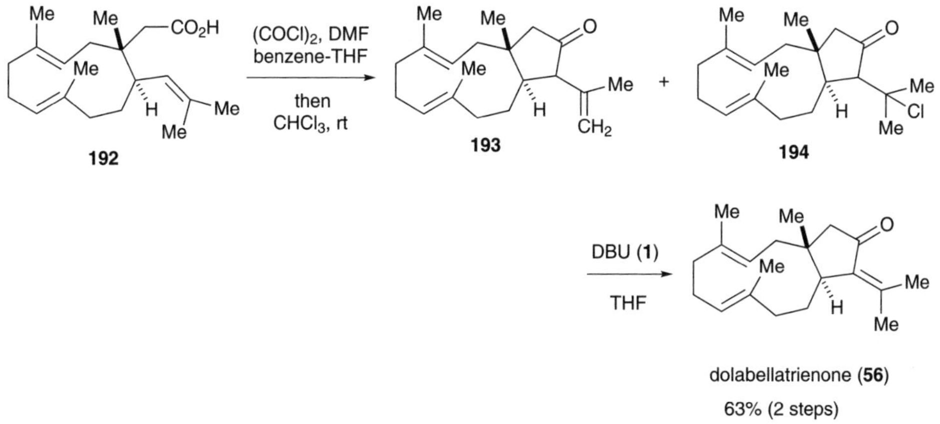

Scheme 7.44 *Synthesis of (+)-dolabellatrienone (**56**)*

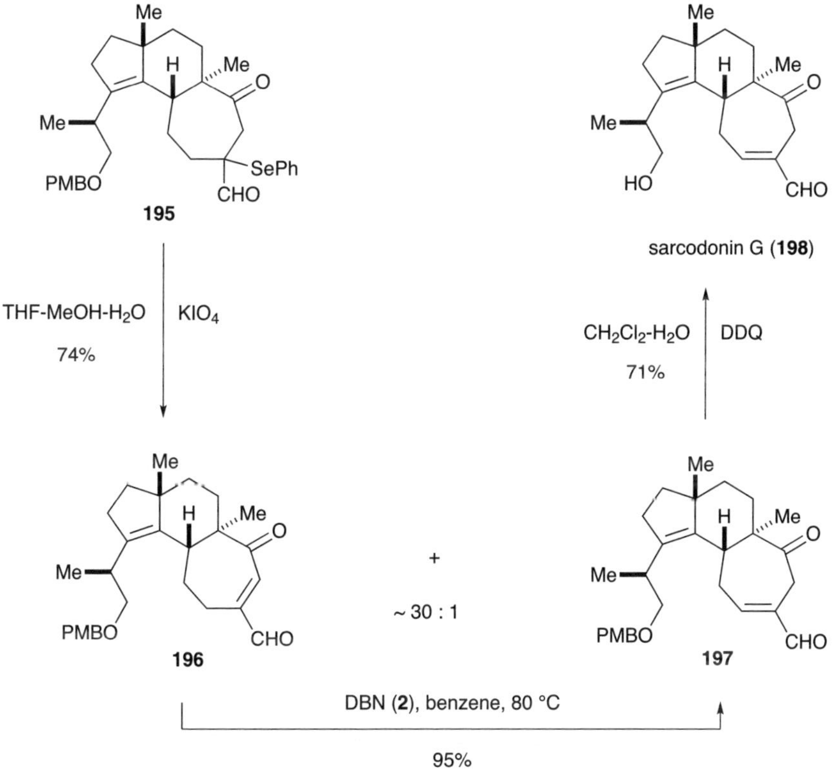

Scheme 7.45 *Synthesis of sarcodonin G (**198**)*

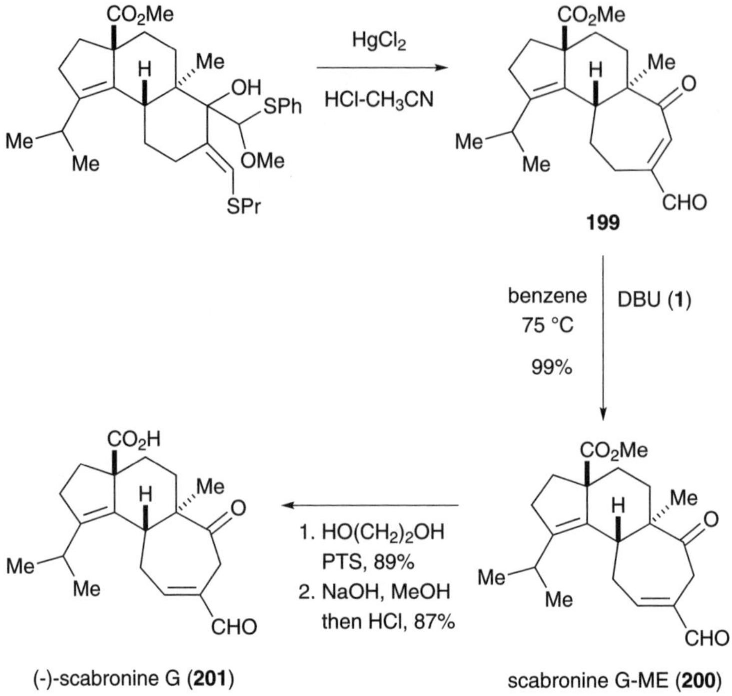

Scheme 7.46 Synthesis of (−)-scabronine G (**201**)

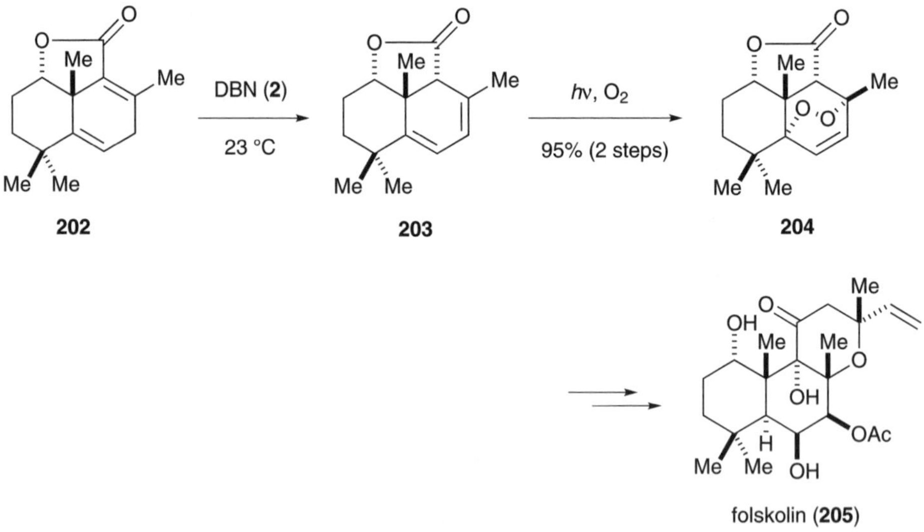

Scheme 7.47 Synthesis of folskolin (**205**)

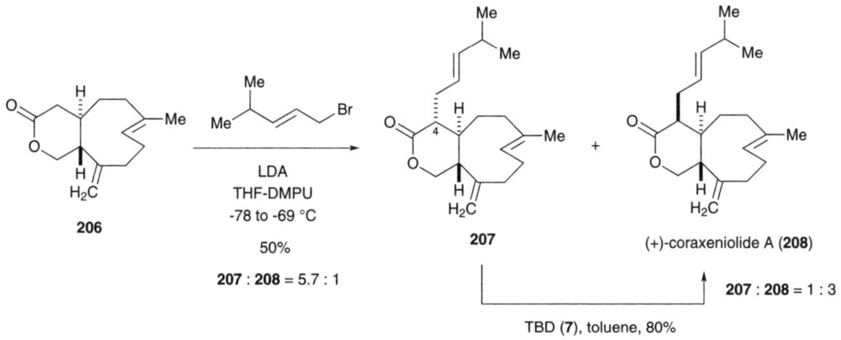

Scheme 7.48 Synthesis of (+)-coraxeniolide A (**208**)

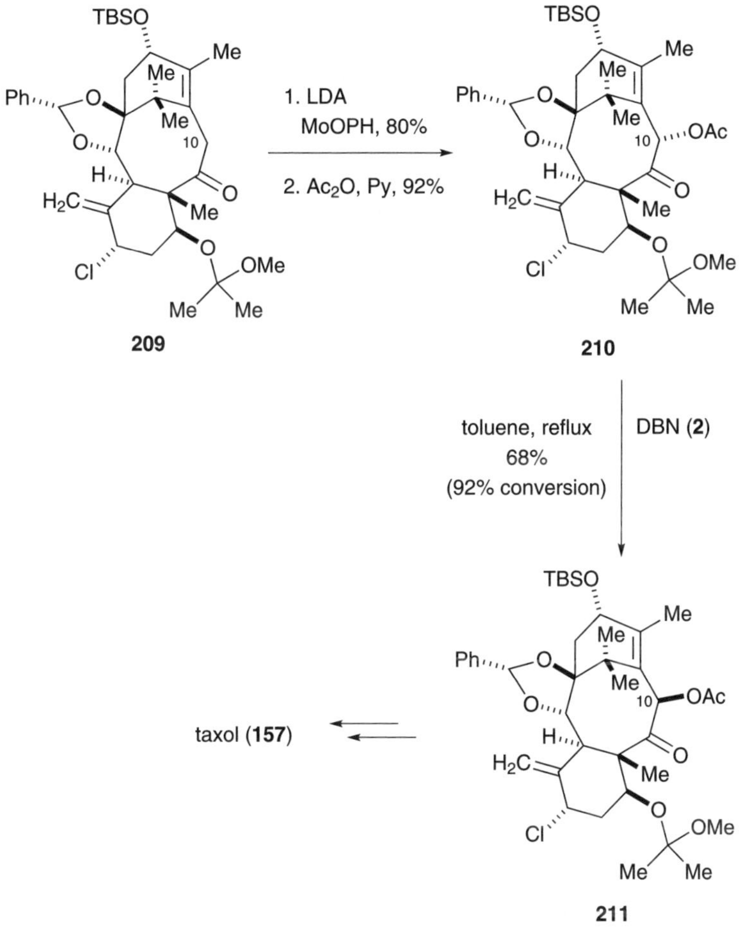

Scheme 7.49 Synthesis of taxol (**157**)

formation of **209** with LDA, followed by oxidation with MoOPH and successive acetylation, was cleanly epimerized to the desired C10β-acetate **211** by treatment with DBN in refluxing toluene in 68% yield with 92% conversion (Scheme 7.49).

Clark *et al.* reported an efficient synthesis of the A-E fragment of ciguatoxin CTX3C (**216**) [62]. Based upon a two-directional and iterative ring closing metathesis (RCM) strategy, the A-D ring system **212** of CTX3C was obtained. Allylation of the ketone **212** proceeded to give a mixture of **213** and **214** in a ratio of 4 : 1. The desired **214** was generated as the major product (1 : 4) by an isomerization reaction with Barton's base **5**. This intermediate **214** was successfully converted into the A-E fragment **216** of ciguatoxin CTX3C via RCM reaction of **215** (Scheme 7.50).

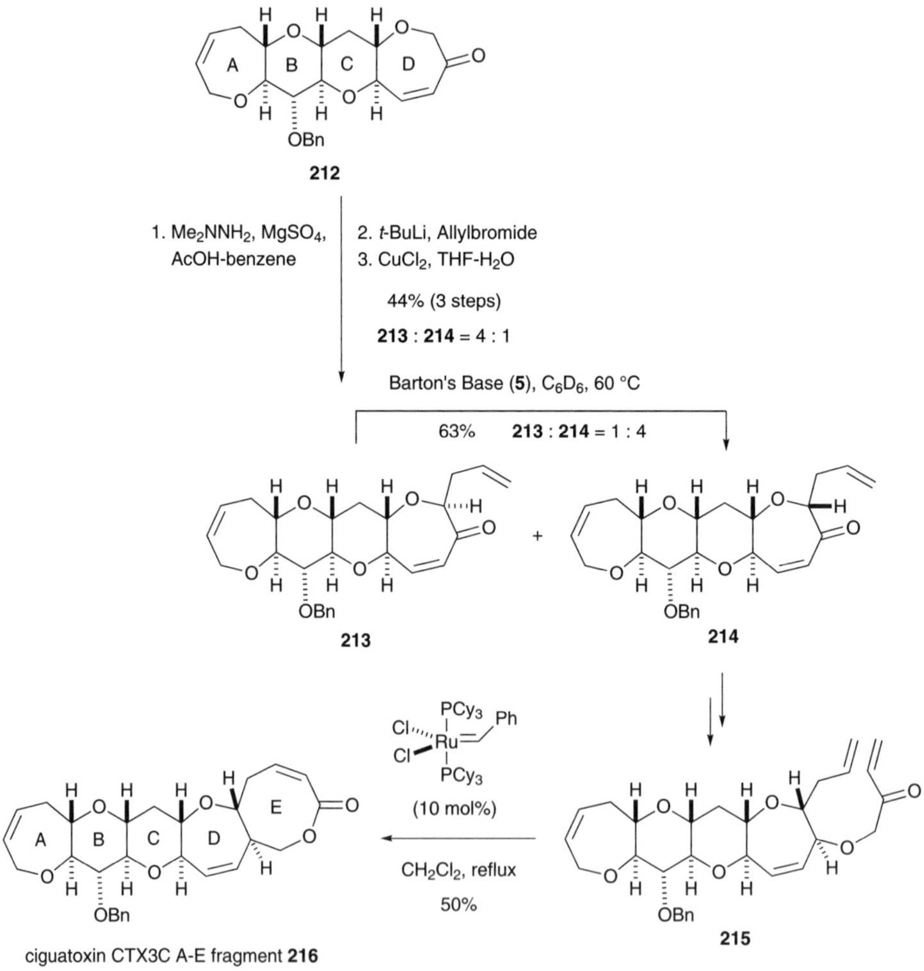

Scheme 7.50 *Synthesis of CTX3C A-E fragment* **216**

7.8 Concluding Remarks

In this chapter, applications of amidine, guanidine and phosphazene superbases to the synthesis of natural products have been discussed. Many structurally complex natural products have been synthesized efficiently and elegantly by making use of the reactions described. Currently, much attention is focussed on the development of chiral superbases and their application to asymmetric reactions. Such catalytic asymmetric reactions are expected to offer exciting and efficient new approaches to the synthesis of natural products and biologically active compounds.

References

1. Simoni, D., Invidiata, F.P., Manfredini, S. *et al.* (1997) Facile synthesis of 2-nitroalkanols by tetramethylguanidine (TMG)-catalysed addition of primary nitroalkanes to aldehydes and alicyclic ketones. *Tetrahedron Letters*, **38**, 2749–2752.
2. Ishikawa, T., Shimizu, Y., Kudoh, T. and Saito, S. (2003) Conversion of D-glucose to cyclitol with hydroxymethyl substituent via intramolecular silyl nitronate cycloaddition reaction: application to total synthesis of (+)-cyclophellitol. *Organic Letters*, **5**, 3879–3882.
3. Luzzio, F.A. and Fitch, R.W. (1999) Formal synthesis of (+)- and (−)-perhydrohistrionicotoxin: a 'double Henry'/enzymatic desymmetrization route to the Kishi lactam. *The Journal of Organic Chemistry*, **64**, 5485–5493.
4. Aratani, M., Dunkerton, L.U., Fukuyama, T. *et al.* (1975) Synthetic studies on histrionicotoxins. I. Stereocontrolled synthesis of (±)-perhydrohistrionicotoxin. *The Journal of Organic Chemistry*, **40**, 2009–2001.
5. Madin, A., O'Donnell, C.J., Oh, T. *et al.* (1999) Total synthesis of (±)-gelsemine. *Angewandte Chemie – International Edition*, **38**, 2934–2936.
6. Kraus, G.A., Zhang, N., Verkade, J.G. *et al.* (2000) Deprotonation of benzylic ethers using a hindered phosphazene base. A synthesis of benzofurans from *ortho*-substituted benzaldehydes. *Organic Letters*, **2**, 2409–2410.
7. Novikov, A.V., Kennedy, A.R. and Rainier, J.D. (2003) Sulfur ylide-initiated thio-Claisen rearrangements. The synthesis of highly substituted indolines. *The Journal of Organic Chemistry*, **68**, 993–996.
8. Henderson, D.A., Collier, P.N., Pavé, G. *et al.* (2006) Studies on the total synthesis of lactonamycin: construction of model ABCD ring systems. *The Journal of Organic Chemistry*, **71**, 2434–2444.
9. Bonjoch, J., Diaba, F., Puigbo, G. *et al.* (2003) A new synthetic entry to the tricyclic skeleton of FR901483 by palladium-catalysed cyclization of vinyl bromides with ketone enolates. *Tetrahedron Letters*, **44**, 8387–8390.
10. Ono, N., Kamiura, A., Miyake, H. *et al.* (1985) New synthetic methods. Conjugate addition of alkyl groups to electron deficient olefins with nitroalkanes as alkyl anion equivalents. *The Journal of Organic Chemistry*, **50**, 3692–3698.
11. Chung, J.Y.L., Grabowski, E.J.J. and Reider, P.J. (1999) Sequential nitromethane conjugate addition/elimination-Pd-catalysed allylation of β-trifloxy acrylates. Application to carbapenem synthesis. *Organic Letters*, **1**, 1783–1785.
12. Mamdani, H.T. and Hartley, R.C. (2000) Phosphazene bases and the anionic oxy-Cope rearrangement. *Tetrahedron Letters*, **41**, 747–749.
13. Corey, E.J. and Kania, R.S. (1996) First enantioselective total synthesis of a naturally occurring dolabellane. Revision of absolute configuration. *Journal of the American Chemical Society*, **118**, 1229–1230.
14. Blid, J. and Somfai, P. (2003) Lewis acid mediated [2,3]-sigmatropic rearrangement of allylic ammonium ylides. *Tetrahedron Letters*, **44**, 3195–3162.

15. Schuster, T., Bauch, M., Dürner, G. and Göbel, M.W. (2000) Axially chiral amidinium ions as inducers of enantioselectivity in Diels-Alder reactions. *Organic Letters*, **2**, 179–181; Schuster, T., Kurz, M. and Göbel, M.W. (2000) Catalysis of a Diels–Alder reaction by amidinium ions. *The Journal of Organic Chemistry*, **65**, 1697–1701.
16. Panek, J.S. and Liu, P. (2000) Total synthesis of the actin-depolymerizing agent (−)-mycalolide A: application of chiral silane-based bond construction methodology. *Journal of the American Chemical Society*, **122**, 11090–11097.
17. Jiang, W.Y., Hong, J. and Burke, S.D. (2004) Stereoselective total synthesis of antitumor macrolide (+)-rhizoxin D. *Organic Letters*, **6**, 1445–1448.
18. Blanchette, M.A., Choy, W., Davis, J.T. et al. (1984) Horner–Wadsworth–Emmons reaction: Use of lithium chloride and an amine for base-sensitive compounds. *Tetrahedron Letters*, **21**, 2183–2186.
19. Nicolaou, K.C., Snyder, S.A., Huang, X. et al. (2004) Studies toward diazonamide A: initial synthetic forays directed toward the originally proposed structure. *Journal of the American Chemical Society*, **126**, 10162–10173.
20. Schmidt, U., Griesser, H., Leitenberger, V. et al. (1992) Amino-acids and peptides. 81. Diastereoselective formation of (Z)-didehydroamino acid-esters. *Synthesis*, 487–490; Burk, M.J., Gross, M.F. and Martinez, J.P. (1995) Asymmetric catalytic synthesis of β-branched amino acids via highly enantioselective hydrogenation of α-enamides. *Journal of the American Chemical Society*, **117**, 9375–9376.
21. Burk, M.J., Feaster, J.E., Nugent, W.A. and Harlow, R.L. (1993) Preparation and use of C_2-symmetric bis(phospholanes): production of α-amino acid derivatives via highly enantioselective hydrogenation reactions. *Journal of the American Chemical Society*, **115**, 10125–10138; Burk, M.J., Allen, J.G. and Kiseman, W.F. (1998) Highly regio- and enantioselective catalytic hydrogenation of enamides in conjugated diene systems: synthesis and application of γ,δ-unsaturated amino acids. *Journal of the American Chemical Society*, **120**, 657–663.
22. Evans, D.A., Katz, J.L., Peterson, G.S. and Hintermann, T. (2001) Total synthesis of teicoplanin aglycon. *Journal of the American Chemical Society*, **123**, 12411–12413.
23. Toumi, M., Couty, F. and Evano, G. (2007) Total synthesis of the cyclopeptide alkaloid abyssenine A. Application of inter- and intramolecular copper-mediated coupling reactions in organic synthesis. *The Journal of Organic Chemistry*, **72**, 9003–9009.
24. Endo, A., Yanagisawa, A., Abe, M. et al. (2002) Total synthesis of ecteinascidin 743. *Journal of the American Chemical Society*, **124**, 6552–6554.
25. Allen, J.R. Harris, C.R. and Danishefsky, S.J. (2001) Pursuit of optimal carbohydrate-based anticancer vaccines: preparation of a multiantigenic unimolecular glycopeptide containing the Tn, MBr1, and Lewis[y] antigens. *Journal of the American Chemical Society*, **123**, 1890–1897.
26. Shoji, M., Akiyama, N., Tsubone, K. et al. (2006) Total synthesis and biological evaluation of neodysiherbaine A and analogues. *The Journal of Organic Chemistry*, **71**, 5208–5220; Sasaki, M., Tsubone, K., Shoji, M. et al. (2006) Design, total synthesis and biological evaluation of neodysiherbaine A derivative as potential probes. *Bioorganic & Medicinal Chemistry Letters*, **16**, 5784–5787.
27. Burk, M.J. (2000) Modular phospholane ligands in asymmetric catalysis. *Accounts of Chemical Research*, **33**, 363–372.
28. Davis, J.R., Kane, P.D., Moody, C.J. and Slawin, A.M.Z. (2005) Control of competing N–H insertion and Wolff rearrangement in dirhodium(II)-catalysed reactions of 3-indolyl diazoketoesters. Synthesis of a potential precursor to the marine 5-(3-indolyl)oxazole martefragin A. *The Journal of Organic Chemistry*, **70**, 5840–5851.
29. Okano, K., Tokuyama, H. and Fukuyama, T. (2006) Total synthesis of (+)-yatakemycin. *Journal of the American Chemical Society*, **128**, 7136–7137.
30. Ellervik, U. and Magnusson, G. (1997) Guanidine/guanidinium nitrate; a mild and selective O-deacetylation reagent that leaves the N-Troc group intact. *Tetrahedron Letters*, **38**, 1627–1628.
31. Flamme, E.M. and Roush, W.R. (2005) Synthesis of the C(1)-C(25) fragment of amphidinol 3: application of the double-allylboration reaction for synthesis of 1,5-diols. *Organic Letters*, **7**, 1411–1414.

32. Allin, S.M., Thomas, C.I., Doyle, K. and Elsegood, M.R.J. (2005) An asymmetric synthesis of both enantiomers of the indole alkaloid deplancheine. *The Journal of Organic Chemistry*, **70**, 357–359.
33. Tanino, K., Onuki, K., Asano, K. *et al.* (2003) Total synthesis of ingenol. *Journal of the American Chemical Society*, **125**, 1498–1500.
34. Trudeau, S. and Morken, J.P. (2005) Short and efficient total synthesis of fraxinellone limonoids using the stereoselective Oshima–Utimoto reaction. *Organic Letters*, **7**, 5465–5468.
35. Futagami, S., Ohashi, Y., Imura, K. *et al.* (2000) Total synthesis of ravidomycin: revision of absolute and relative stereochemistry. *Tetrahedron Letters*, **41**, 1063–1067.
36. Hughes, R.A., Thompson, S.P., Alcaraz, L. and Moody, C.J. (2005) Total synthesis of the thiopeptide antibiotic amythiamicin D. *Journal of the American Chemical Society*, **127**, 15644–15651.
37. Balsamini, C., Bedini, A., Galarini, R. *et al.* (1994) Reactions of 3-carbomethoxy-2-aza-1,3-butadiene derivatives with dienophiles. *Tetrahedron*, **50**, 12375–12394.
38. Williams, D.R., Lowder, P.D., Gu, Y.-G. and Brooks, D.A. (1997) Studies of mild dehydrogenations in heterocyclic systems. *Tetrahedron Letters*, **38**, 331–334.
39. Doi, T., Yoshida, M., Shin-ya, K. and Takahashi, T. (2006) Total synthesis of (*R*)-telomestatin. *Organic Letters*, **8**, 4165–4167.
40. Knapp, S., Madduru, M.R., Lu, Z. *et al.* (2001) Synthesis of griseolic acid B by π-face-dependent radical cyclization. *Organic Letters*, **3**, 3583–3585.
41. Barton, D.H.R., Bashiardes, G. and Fourrey, J.-L. (1983) An improved preparation of vinyl iodides. *Tetrahedron Letters*, **24**, 1605–1608.
42. Paquette, L.A., Hu, Y., Luxenburger, A. and Bishop, R.L. (2007) Studies toward the total synthesis of dumsin. 2. A second generation approach resulting in enantioselective construction of a functionalized ABC subunit of the tetranortriterpenoid insect antifeedant. *The Journal of Organic Chemistry*, **72**, 209–222.
43. Ohgiya, T. and Nishiyama, S. (2004) Total synthesis of (+)-tanikolide, a toxic and antifungal δ-lactone, utilizing bromoalkene intermediates conveniently synthesized from vicinal dibromoalkane by regioselective elimination. *Tetrahedron Letters*, **45**, 8273–8275; Ohgiya, T., Nakamura, K. and Nishiyama, S. (2005) Total synthesis of (+)-tanikolide, using regioselective elimination of a vicinal dibromoalkane. *Bulletin of the Chemical Society of Japan*, **78**, 1549–1554; Kutsumura, N., Yokoyama, T., Ohgiya, T. and Nishiyama, S. (2006) 1,2-Dibromoalkanes into alkynes by elimination reaction under DBU conditions and their application to total synthesis of sapinofuranone B. *Tetrahedron Letters*, **47**, 4133–4136.
44. Wender, P.A., Rice, K.D. and Schnute, M.E. (1997) The first formal asymmetric synthesis of phorbol. *Journal of the American Chemical Society*, **119**, 7897–7898.
45. Tang, B., Bray, C.D. and Pattenden, G. (2006) A biomimetic total synthesis of (+)-intricarene. *Tetrahedron Letters*, **47**, 6401–6404.
46. Knapp, S., Thakur, V.V., Madduru, M.R. *et al.* (2006) Short synthesis of octosyl nucleosides. *Organic Letters*, **8**, 1335–1337.
47. Hanessian, S., Reddy, G.J. and Chahal, N. (2006) Total synthesis and stereochemical confirmation of manassantin A, B, and B$_1$. *Organic Letters*, **8**, 5477–5480.
48. Kusama, H., Hara, R., Kawahara, S. *et al.* (2000) Enantioselective total synthesis of (−)-taxol. *Journal of the American Chemical Society*, **122**, 3811–3820.
49. Wipf, P. and Lynch, S.M. (2003) Synthesis of highly oxygenated dinaphthyl ethers via S$_N$Ar reactions promoted by Barton's base. *Organic Letters*, **5**, 1155–1158.
50. Ebisawa, M., Ueno, M., Oshima, Y. and Kondo, Y. (2007) Synthesis of dictyomedins using phosphazene base catalysed diaryl ether formation. *Tetrahedron Letters*, **48**, 8918–8921.
51. Li, P., Evans, C.D. and Joullié, M.M. (2005) A convergent total synthesis of ustiloxin D via an unprecedented copper-catalysed ethynyl aziridine ring-opening by phenol derivatives. *Organic Letters*, **7**, 5325–5327.
52. Forbeck, E.M., Evans, C.D., Gilleran, J.A. *et al.* (2007) A regio- and stereoselective approach to quaternary centers from chiral trisubstituted aziridines. *Journal of the American Chemical Society*, **129**, 14463–14469.

53. Draper, R.W., Hu, B., Iyer, R.V. et al. (2000) An efficient process for the synthesis of trans-2,3-disubstituted-2,3-dihydro-4H-1-benzopyran-4-ones (Chroman-4-ones). *Tetrahedron*, **56**, 1811–1817.
54. Corey, E.J., Gin, D.Y. and Kania, R.S. (1996) Enantioselective total synthesis of ecteinascidin 743. *Journal of the American Chemical Society*, **118**, 9202–9203.
55. Wood, J.L., Thompson, B.D., Yusuff, N. et al. (2001) Total synthesis of (±)-epoxysorbicillinol. *Journal of the American Chemical Society*, **123**, 2097–2098.
56. Genski, T., Macdonald, G., Wei, X. et al. (1999) Epoxidation of electron deficient alkenes using tert-butyl hydroperoxide and 1,5,7-triazabicyclo[4.4.0]dec-5-ene and its derivatives. *Synlett*, 795–797;Genski, T. and Taylor, R.J.K. (2002) The synthesis of epi-epoxydon utilising the Baylis–Hillman reaction. *Tetrahedron Letters*, **43**, 3573–3576; Quesada, E., Stockley, M. and Taylor, R.J.K. (2004) The first total syntheses of (±)-preussomerins K and L using 2-arylacetal anion technology. *Tetrahedron Letters*, **45**, 4877–4881.
57. Barros, M.T., Maycock, C.D. and Ventura, M.R. (1997) Enantioselective total synthesis of (+)-eutypoxide B. *The Journal of Organic Chemistry*, **62**, 3984–3988. The reaction in the presence of DBU (1 equiv.) was also reported; Takano, S., Moriya, M. and Ogasawara, K. (1993). Concise enantiodivergent synthesis of eutypoxide B. *Journal of the Chemical Society – Chemical Communications*, 614–615.
58. Piers, E., Gilbert, M. and Cook, K.L. (2000) Total synthesis of the cyathane diterpenoid (±)-sarcodonin G. *Organic Letters*, **2**, 1407–1410.
59. Waters, S.P., Tian, Y., Li, Y.-M. and Danishefsky, S.J. (2005) Total synthesis of (−)-scabronine G, an inducer of neurotrophic factor production. *Journal of the American Chemical Society*, **127**, 13514–13515.
60. Corey, E.J., Jardin, P.A.S. and Rohloff, J.C. (1988) Total synthesis of (±)-forskolin. *Journal of the American Chemical Society*, **110**, 3672–3673.
61. Renneberg, D., Pfander, H. and Leumann, C.J. (2000) Total synthesis of coraxeniolide-A. *The Journal of Organic Chemistry*, **65**, 9069–9079.
62. Clark, J.S., Conroy, J. and Blake, A.J. (2007) Rapid synthesis of the A-E fragment of ciguatoxin CTX3C. *Organic Letters*, **9**, 2091–2094.

8
Related Organocatalysts (1): A Proton Sponge

Kazuo Nagasawa
Tokyo University of Agriculture and Technology, 2-24-16 Naka-cho,
Koganei, Tokyo 184-8588, Japan

8.1 Introduction

1,8-Bis(dimethylamino)naphthalene, known as Proton Sponge (**1**), shows unusually high basicity [1]. Its pK_a value was reported to be 12.3, so it is more than seven orders of magnitude more basic than related aromatic amines, such as aniline (pK_a 4.6), *N,N*-dimethylaniline (pK_a 5.1) and 1,8-diaminonaphthalene (pK_a 4.6). The strong basicity of **1** is considered to arise from the formation, upon monoprotonation, of a strong $[N-H \cdots N]^+$ hydrogen bond as shown in **2** [2] (Figure 8.1). The protonated **2** is highly stabilized because of the release of the unfavourable interaction of the nitrogen lone pairs of **1**. Therefore, the deprotonation reaction rate of **2** is extremely low. Owing to its characteristic strong basicity, Proton Sponge (**1**) has found many applications in the field of synthetic organic chemistry, together with other superbases. In this chapter, organic reactions using Proton Sponge (**1**) as a hindered, less nucleophilic and nonionic strong base are described.

Superbases for Organic Synthesis: Guanidines, Amidines, Phosphazenes and Related Organocatalysts
Edited by Tsutomu Ishikawa
© 2009 John Wiley & Sons, Ltd

Proton Sponge™ (1)
(1,8-bis(dimethylamino)naphthalene)

2

Figure 8.1 Structure of Proton Sponge (**1**)

8.2 Alkylation and Hetero Michael Reaction

8.2.1 Amine Synthesis by N-Alkylation

N-Alkylation with an alkylating agent and Proton Sponge (**1**) has been reported (Scheme 8.1). Reaction of a pyrrolidine derivative and 3-butenyl triflate using Proton Sponge (**1**) gave a tertiary amine in 75% yield [3]. N-Methylation of aziridine was conducted using dimethyl sulfate and Proton Sponge (**1**) [4].

Dealkylation of a tertiary amine was firstly reported by Olofson et al. [5]. In this reaction, Ratz et al. reported the effective use of Proton Sponge (**1**) for carbamate formation [6]. Thus, the reaction of the tertiary amine with 2,2,2-trichloroethyl chloroformate (ACE–Cl) and Proton Sponge (**1**), and subsequent hydrolysis of the resulting carbamate with sodium hydroxide (NaOH) gave amine in 75% yield without loss of optical purity (Scheme 8.2).

8.2.2 Ether Synthesis by O-Alkylation

Several mild and efficient alcohol alkylation (alkoxylation) methods are known, including the classic Williamson reaction, alkyl halide and silver oxide in combination, and alkyl triflate and 2,6-di-*tert*-butyl-4-methylpyridine. However, for sterically hindered or optically labile alcohols, alternative powerful and mild methodology is required.

Diem et al. reported a mild alkoxylation reaction using trialkyloxonium tetrafluoroborate (Meerwein's salt) [7]. In this reaction, Proton Sponge (**1**) was found to be an effective base, as well as Hünig's base (diisopropylethylamine) (Table 8.1). The reaction of (R)-(+)-1-phenylethyl alcohol with Meerwein's trimethyloxonium tetrafluoroborate (2.1 equiv.) and Proton Sponge (**1**) (2.1 equiv.) in dichloromethane at room temperature gave methyl ether in 69% yield without loss of optical purity (Table 8.1, run 1). (S)-(−)-2-Methyl-1-butanol was also methylated in 57% yield (run 3).

Evans et al. used the same reaction conditions for ether formation in the synthesis of Ionomycin A [8]. The hindered alcohol in a cyclic polyether was efficiently methylated with trimethyloxonium tetrafluoroborate and Proton Sponge (**1**) (5 equiv. each, 0 °C) to give the desired ether along with 16% recovery of the starting material (Scheme 8.3). In this reaction, other methylation conditions examined were ineffective, and elevated temperature caused the decomposition of the starting polyether. De Brabander et al. reported the selective formation of anisole derivative from phenol using Meerwein's salt and Proton Sponge (**1**) in combination, without formation of isocoumarin by-products [9,10] (Scheme 8.3).

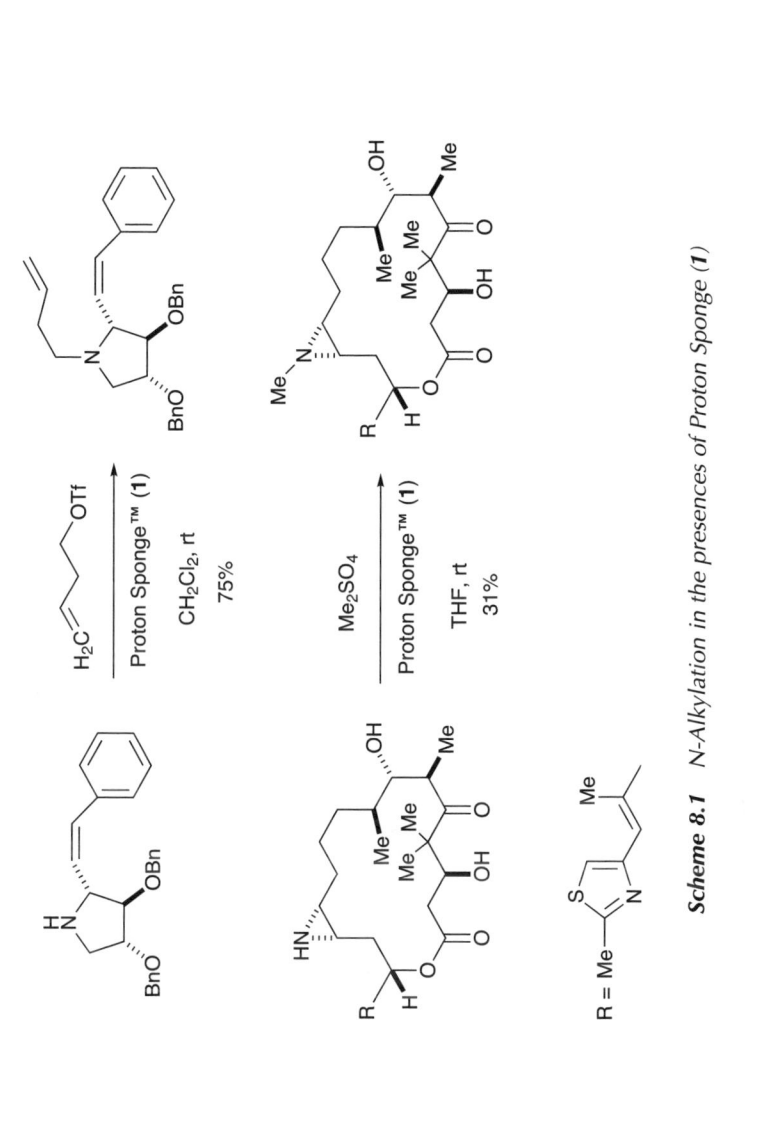

Scheme 8.1 N-Alkylation in the presences of Proton Sponge (**1**)

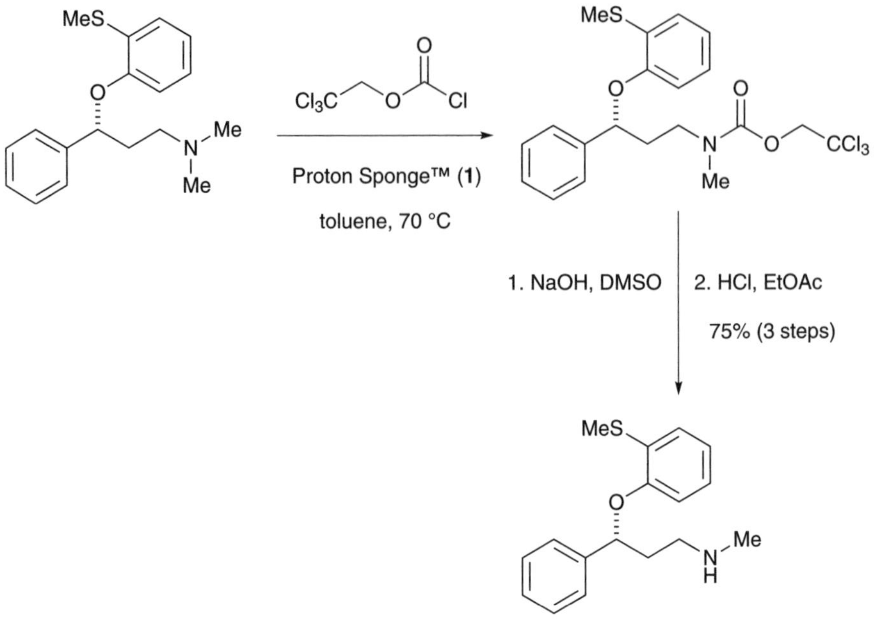

Scheme 8.2 Dealkylation of tertiary amine

The drastic effect of molecular sieves of 4 Å (4 Å MS) was reported by Ireland et al. [11]. Although, methylation of alcohol with Meerwein's salt and Proton Sponge (**1**) in combination took place much faster than in the alternative protocol, the reaction stopped at about 50% conversion in the case of alcohol, and decomposition of the spiroketal moiety was observed

Table 8.1 Alkoxylation of alcohol using Meerwein's salt in the presence of Proton Sponge (**1**)

Run	Alcohol	Oxonium Salt/Amine	Ether Product	Yield (%)
1	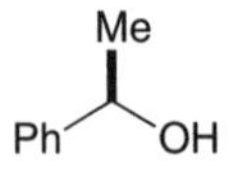 Ph-CH(Me)-OH	Me₃OBF₄ (2.1 equiv) Proton Sponge (**1**) (2.1 equiv.)	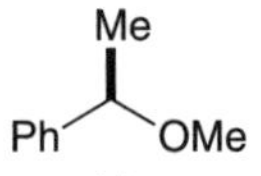 Ph-CH(Me)-OMe	69
2		Et₃OBF₄ (1.5 equiv) ⁱPr₂NEt (3.0 equiv.)	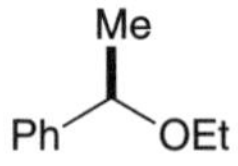 Ph-CH(Me)-OEt	65
3	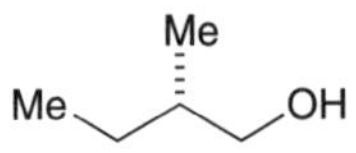 Me-CH(Me)-CH₂-OH	Me₃OBF₄ (1.0 equiv) Proton Sponge (**1**) (1.2 equiv.)	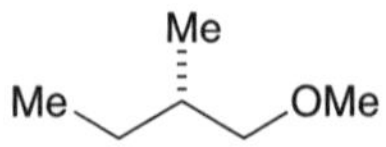 Me-CH(Me)-CH₂-OMe	57
4		Et₃OBF₄ (1.0 equiv) ⁱPr₂NEt (1.2 equiv.)	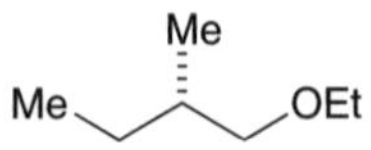 Me-CH(Me)-CH₂-OEt	89

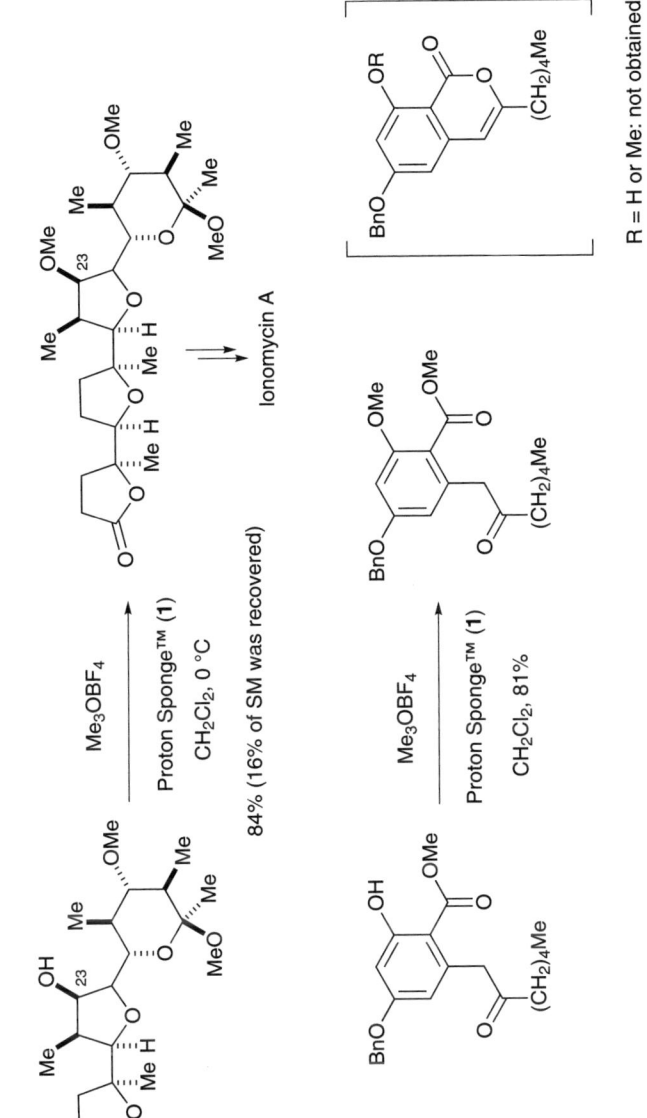

Scheme 8.3 Synthetic application of ether formation using Meerwein's salt

due to the Lewis acidic nature of Meerwein's salt. In this reaction, addition of 4 Å molecular sieves was found to greatly improve the reactivity of methylation with Meerwein's salt and Proton Sponge (**1**) in combination and gave desired ether in 95% yield (Scheme 8.4).

Similar effects of 4 Å molecular sieves were reported by Shea *et al.* and Hansen *et al.*, as depicted in Scheme 8.5 [12,13].

For the synthesis of various ethers from alcohols, the combination of alkyl triflate and Proton Sponge (**1**) is available. The *O*-alkylation reaction of symmetrical alcohol proceeds in high yield with octyl triflate in the presence of Proton Sponge (**1**) as a base (Scheme 8.6a). Low yields were obtained with triethylamine or DMAP instead of Proton Sponge (**1**) [14]. Octadecenyl triflate was also effectively alkylated with the alcohol in the presence of Proton Sponge (**1**) in 66% yield [15] (Scheme 8.6b).

Proton Sponge (**1**) is effective for the ether-type alcohol protection reaction. Reaction of hindered tertiary and secondary alcohols in a modified triol system with benzyl chloromethyl ether (BOM–Cl) and Proton Sponge (**1**) in the presence of sodium iodide (NaI) gave bis-BOM ether in 84% yield [16] (Scheme 8.7a). Perhydroxylated hexane derivative was also protected as the bis-BOM ether without generating the acyl migration product [17] (Scheme 8.7b). Allyl alcohol was protected with (trichloroethoxy)methyl ether using (2,2,2-trichloroethoxy)methyl bromide and Proton Sponge (**1**) in 70% yield [18] (Scheme 8.7c). Triisopropylsilyl (TIPS) ether formation of the secondary alcohol with TIPS triflate and Proton Sponge (**1**) was observed with an excellent yield, although other bases, such as 2,6-lutidine and Hünig's base were not effective [19] (Scheme 8.7d).

8.2.2.1 Thioether Synthesis by Hetero Michael Reaction

Kanemasa *et al.* reported an asymmetric conjugate addition reaction of thiol to *N*-crotonyl oxazolidinone in the presence of Ni(II)-DBFOX/Ph catalyst [20] (Table 8.2). In this reaction, Proton Sponge (**1**) is indispensable for high enantioselectivity, and thioether was obtained in 84–99% yield with 91–94% ee.

8.3 Amide Formation

Sigurdsson *et al.* developed the synthesis of isocyanates from aliphatic amines [21]. Thus, the reaction of an aliphatic amine with trichloromethyl chloroformate (diphosgene) in the presence of Proton Sponge (**1**) (2 equiv.) at 0 °C gave isocyanate in 81% yield. Azide isocyanate was synthesized from azide amine by means of a similar procedure by Keyes *et al.* [22] (Scheme 8.8).

Benzoyl quinidine (BQ) (**3**) catalysed β-lactam synthesis from acid chloride and tosyl imine, as reported by Taggi *et al.* [23]. In this reaction, ketene generated from the acid chloride with BQ (**3**) via dehydrohalogenation reacted with imine to give *cis*-β-lactam selectively with quite high ee (Table 8.3). The BQ (**3**)–HCl was regenerated to BQ (**3**) with Proton Sponge (**1**).

β-Amino acid synthesis was achieved via BQ (**3**) catalysed alcoholysis of β-lactam obtained from the *N*-benzoyl-α-chloroglycine and acid chlorides in the presence of Proton Sponge (**1**) with high selectivity [24] (Table 8.4).

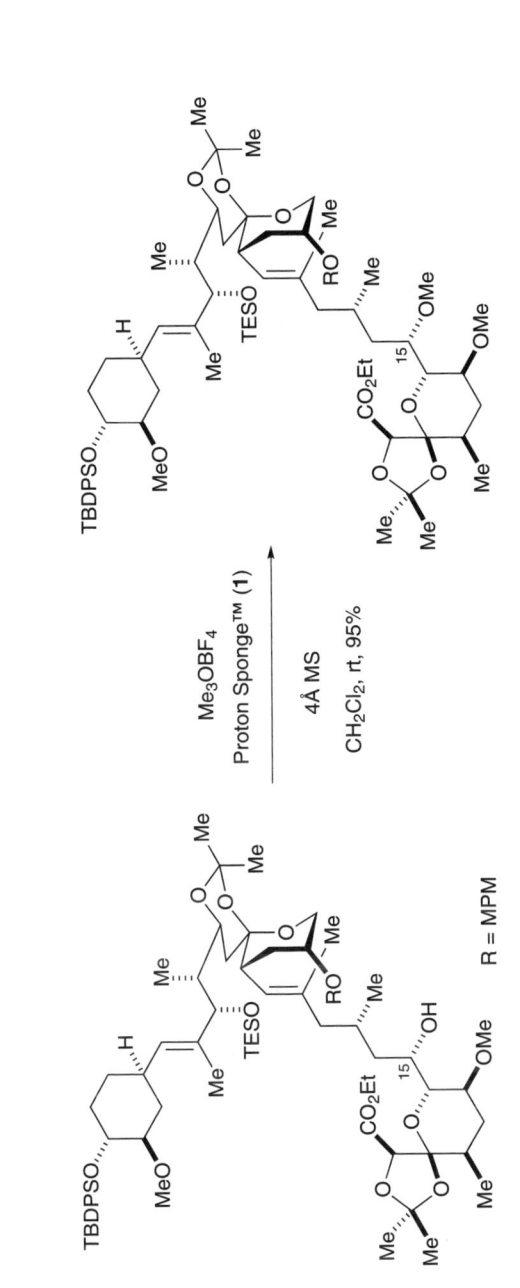

Scheme 8.4 Effect of 4 Å molecular sieves on ether synthesis – I

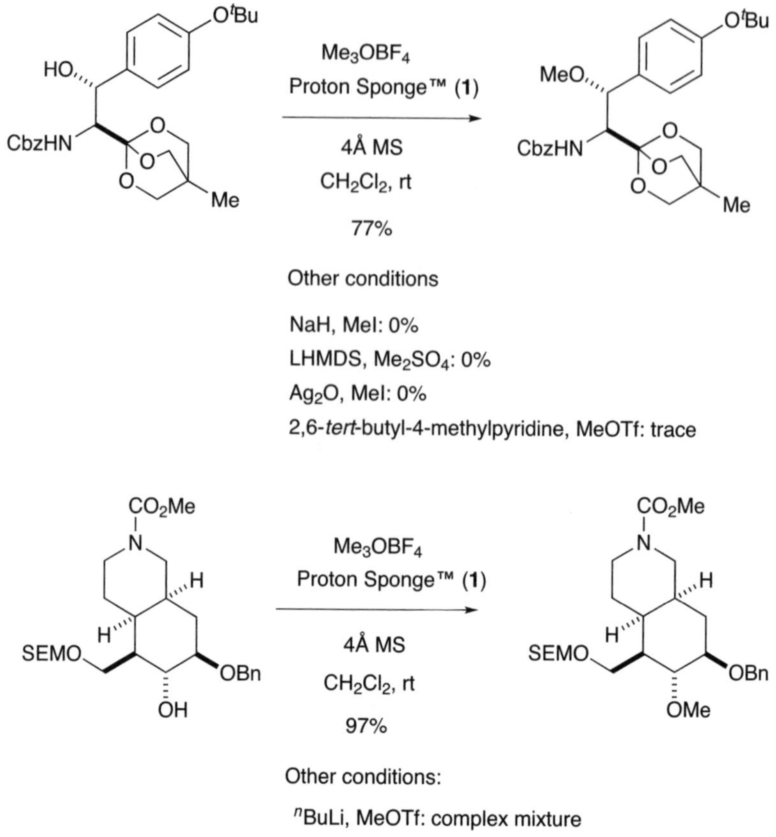

Scheme 8.5 Effect of 4 Å molecular sieves on ether synthesis – II

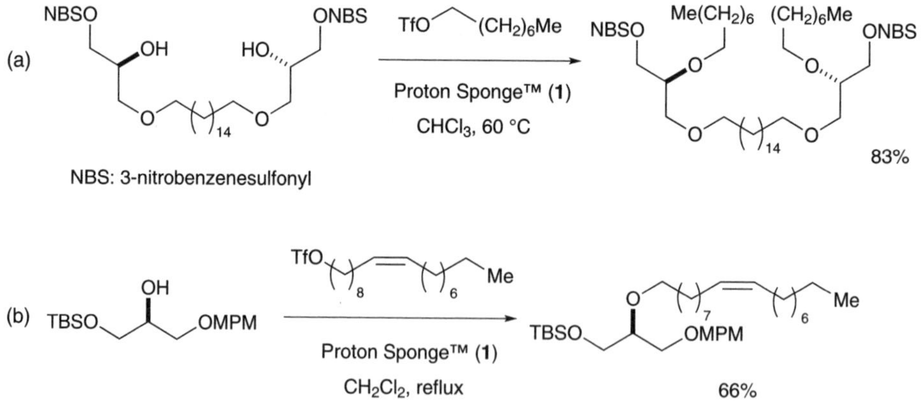

Scheme 8.6 O-Alkylation reaction with alkyl triflate in the presence of Proton Sponge (**1**)

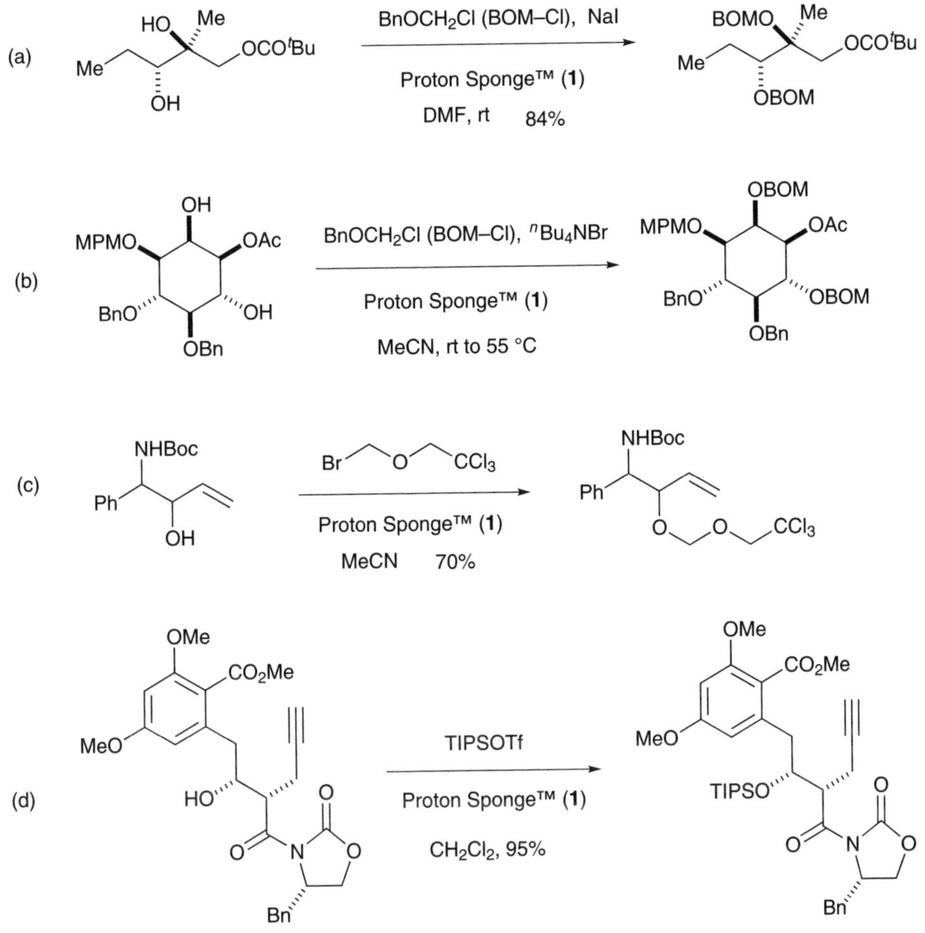

Scheme 8.7 Protection of hydroxyl group with ethers

Recently, Weatherwax et al. reported a highly selective *trans*-disubstituted β-lactam synthesis using an imidazoline catalyst in the presence of Proton Sponge (**1**) as a stoichiometric base [25] (Table 8.5).

8.4 Carbon–Carbon Bond Forming Reaction

8.4.1 Alkylation and Nitro Aldol Reaction

The condensation reaction of β-dicarbonyl compounds with α-haloketones to generate hydroxydihydrofuran is known as an interrupted Feist–Bénary reaction. Calter et al. reported an enantioselective version of this reaction [26]. The aldol reaction of diketone with α-bromo-α-ketoester followed by cyclization proceeded in the presence of dimeric cinchona alkaloid catalyst to give cyclized product in high yield with high ee

Table 8.2 Asymmetric conjugate addition reaction of thiol

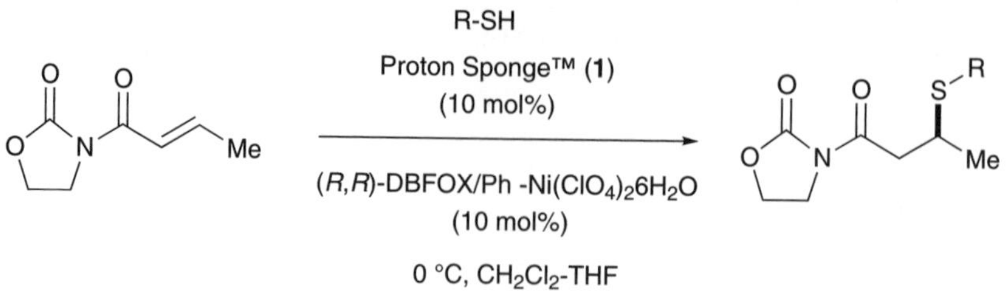

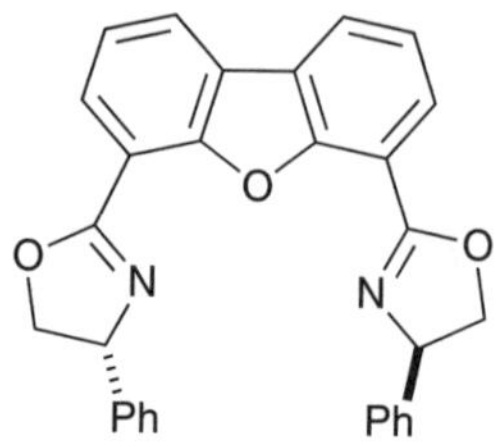

R	yield (%)	ee (%)
phenyl	84	94
o-tolyl	99	95
2-naphthyl	89	91

(Table 8.6). In this synthesis, a slight excess of Proton Sponge (**1**) is indispensable for high selectivity.

Enantioselective nitroaldol reaction (Henry reaction) of simple trifluoromethyl ketone was reported by Tur and Saá [27] (Table 8.7). Reaction of trifluoromethyl ketones with nitromethane in the presence of lanthanum (III) triflate salt complex and Proton Sponge (**1**) (0.25 equiv. each) gave tertiary nitroaldols in 50–93% yields with 67–98% ee.

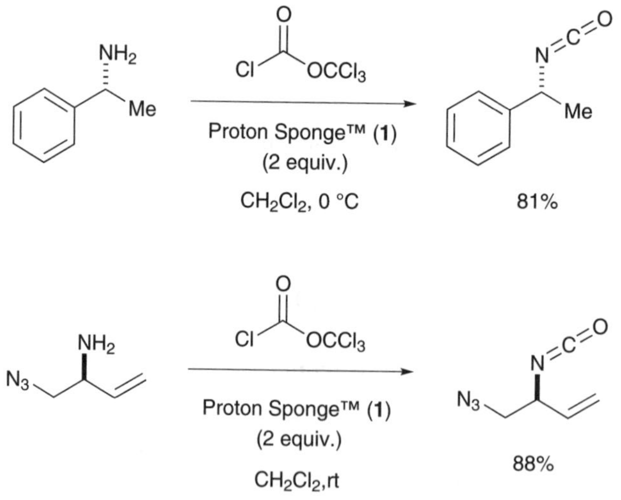

Scheme 8.8 Synthesis of isocyanates

Table 8.3 BQ (3) catalysed β-lactam synthesis

R¹	yield (%)	cis/trans	ee (%)
Ph	65	99/1	96
Bn	60	33/1	96
BnO	65	99/1	96
Et	57	99/1	99
CH=CH$_2$	58	99/1	98
N$_3$	47	25/1	98

8.4.2 Pericyclic Reaction

Some applications of Proton Sponge (1) for pericyclic reactions have been reported. Intramolecular Diels–Alder (IMDA) reaction of triene proceeded under thermal conditions to give bicyclic compound. In this reaction, the yield of the IMDA reaction for the Lewis acid labile substrate was significantly improved to 57% in the presence of a catalytic amount of Proton Sponge (1) (0.3 equiv.) [28] (Scheme 8.9). A similar effect was observed in the IMDA reaction of ester derivative. Whitney et al. conducted the IMDA reaction of acrylate

Table 8.4 BQ (3) catalysed β-amino acid synthesis

R¹	yield (%)	cis/trans	ee (%)
Ph	62	12/1	95
PhO	63	14/1	95
p-MeOC$_6$H$_4$	62	10/1	94
p-ClC$_6$H$_4$	60	12/1	94

Table 8.5 Synthesis of trans-disubstituted β-lactam

R	yield (%)	cis/trans
Ph	50	1/37
p-OMePh	70	1/13
o-OMePh	51	1/50
PhS	35	1/14
2-thiophenyl	69	1/5

in a sealed tube in toluene, obtaining a bicyclic system in 74% yield [29]. The yield of this reaction was also improved by addition of Proton Sponge (**1**) and 2,6-di-*tert*-butyl-4-methylphenol (BHT) (each 0.5 equiv.), and provided the product with complete regiochemical selection.

Table 8.6 Asymmetric interrupted Feist–Bénary reaction

R	yield (%)	a : b	ee (%)
Me	95	98 : 2	94
^{n}pentyl	96	96 : 4	93
^{t}Bu	94	97 : 3	96
Ph	94	96 : 4	93

Table 8.7 Enantioselective nitroaldol reaction

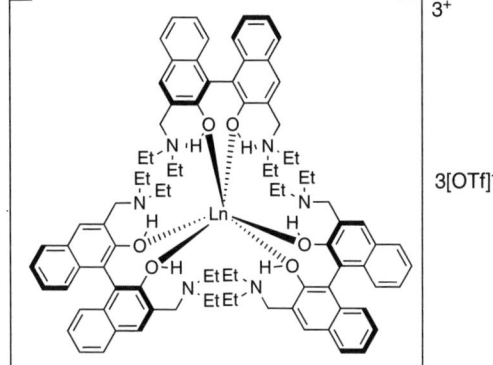

catalyst: [(Δ,S,S,S)-Binolam]₃-Ln(OTf)₃

R	yield (%)	ee (%)
Et	55	85
Bn	93	92
Ph	78	96
3-CF$_3$C$_6$H$_4$	55	67
4-^{t}BuC$_6$H$_4$	50	98

Enantioselective decarboxylative allylic amidation of allylic benzyl imidodicarbamates using iridium (I) catalyst was reported by Singh and Han [30]. Reaction of imidodicarbonates in the presence of iridium (I) and chiral phosphoramidite ligand gave secondary allyl amine in 80–92% yield with high ee. This reaction requires a base for generating the active

Scheme 8.9 Intramolecular Diels–Alder reaction in the presence of Proton Sponge (1)

Table 8.8 Enantioselective decarboxylative allylic amidation

R	yield (%)	ee (%)
Me–CH₂– (Me\~\~\~)	92	94
Me–CH=CH–	80	92
Ph–	80	> 99

catalyst, and both DBU and Proton Sponge (**1**) were found to be vital for high yield and enantioselectivity (Table 8.8).

Reaction of an aniline derivative and ethyl methylthioacetate in the presence of sulfuryl chloride and Proton Sponge (**1**) generated an azasulfonium salt, which showed [2,3]-rearrangement under basic conditions to give oxindole (modified Gassman oxindole synthesis [31]) [32]. In this reaction, Proton Sponge (**1**) plays a role as a hydrogen chloride scavenger. When the reaction was carried out with *tert*-butyl methylthioacetate, *o*-aminophenylacetic acid derivative was obtained [33] (Scheme 8.10).

8.5 Palladium Catalysed Reaction

Proton Sponge (**1**) sometimes drastically affects the enantiomeric excess and yield in the asymmetric palladium catalysed Heck reaction.

Ozawa *et al.* reported the Pd(OAc)₂-BINAP catalysed asymmetric intermolecular Heck reaction of aryl triflate and 2,3-dihydrofuran in the presence of a base [34]. In this reaction, the enantiomeric excess was significantly affected by the base, and Proton Sponge (**1**) gave the best results among the various bases, such as triethylamine, diisopropylethylamine, pyridine derivatives and inorganic bases (Table 8.9).

In the case of intermolecular Heck alkenylation of vinyl triflate and 2,2-disubstituted-2,3-dihydrofuran, Proton Sponge (**1**) gave higher chemical yield and enantioselectivity compared to the trialkylamines [35] (Table 8.10).

Kiely and Guiry reported an asymmetric version of the intramolecular Heck reaction of aryl triflate [36]. In this reaction, Proton Sponge (**1**) also greatly improved the reactivity,

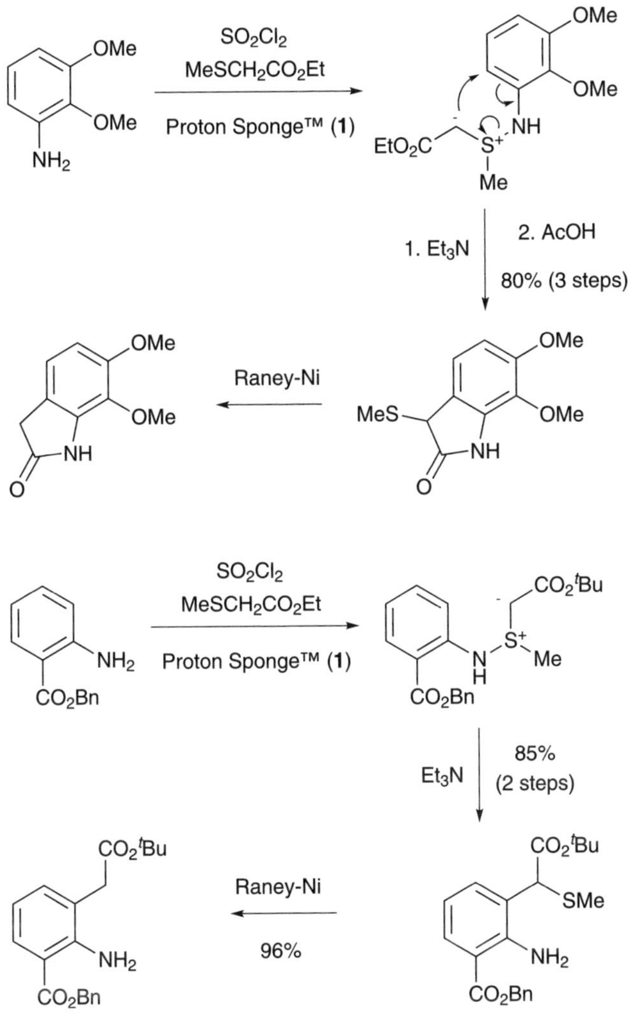

Scheme 8.10 Modified Gassman oxindole synthesis in the presence of Proton Sponge (**1**)

affording cyclized product in 71% yield with 82% ee (regioselectivity; >99:1) (Table 8.11).

Dounary et al. also investigated the intramolecular Heck reaction [37]. In this reaction, Proton Sponge (**1**) minimizes the double bond isomerization of oxindole caused by palladium (II) hydride species and spiro-oxindole is obtained in 88% yield with 60% ee (Scheme 8.11a). The intramolecular Heck reaction onto a tetrasubstituted olefin was reported by Frey et al. [38]. Sterically hindered highly substituted olefin was cyclized in the presence of palladium (0) catalyst with Proton Sponge (**1**) in refluxing toluene to give cyclized product in 57% yield along with recovered starting material (15%) (Scheme 8.11b).

Table 8.9 Asymmetric intermolecular Heck reaction of aryl triflate

ArOTf + (dihydrofuran) →[Pd(OAc)₂, (R)-BINAP / Proton Sponge™ (1) / benzene, 40 °C] product

Ar	Yield (%)	α : β	ee of α (%)
Ph	100[b]	98 : 2	>96
Ph[a]	100[b]	71 : 29	75
m-ClC$_6$H$_4$	88[c]	72 : 28	>96
o-ClC$_6$H$_4$	78[c]	72 : 27	92
p-NCC$_6$H$_4$	78[c]	66 : 34	>96

[a] Et$_3$N was used instead of proton sponge.
[b] Determined by GLC.
[c] Isolated yield.

Table 8.10 Asymmetric intermolecular Heck reaction of vinyl triflate

base	yield (%)	ee (%)
Proton Sponge™ (1)	68	97
^{i}Pr$_2$NEt	60	40
Et$_3$N	26	38

Table 8.11 Asymmetric intramolecular Heck reaction of aryl triflate

base	yield (%)	F$_{2,3}$: F$_{3,4}$	ee (%)
Proton Sponge™ (1)	71	>99 : 1	82
^{i}Pr$_2$NEt	55	98 : 2	57
PMP[a]	71	>99 : 1	53

[a] PMP = 1,2,2,6,6-pentamethylpiperidine.

Scheme 8.11 Intramolecular Heck reaction in the presence of Proton Sponge (**1**)

Arranz and Boons introduced the 2-(allyloxy)phenylacetyl (APAC) group as a protecting group for oligosaccharide synthesis [39] (Table 8.12). This protective group can be removed by the combination of Pd(PPh$_3$)$_4$ with Proton Sponge (**1**) (run 3).

The APAC group in disaccharides could be removed in good yield under optimized conditions without loss of the benzoyl and/or acetyl group (Scheme 8.12).

Table 8.12 Trials for cleavage of APAC protective group

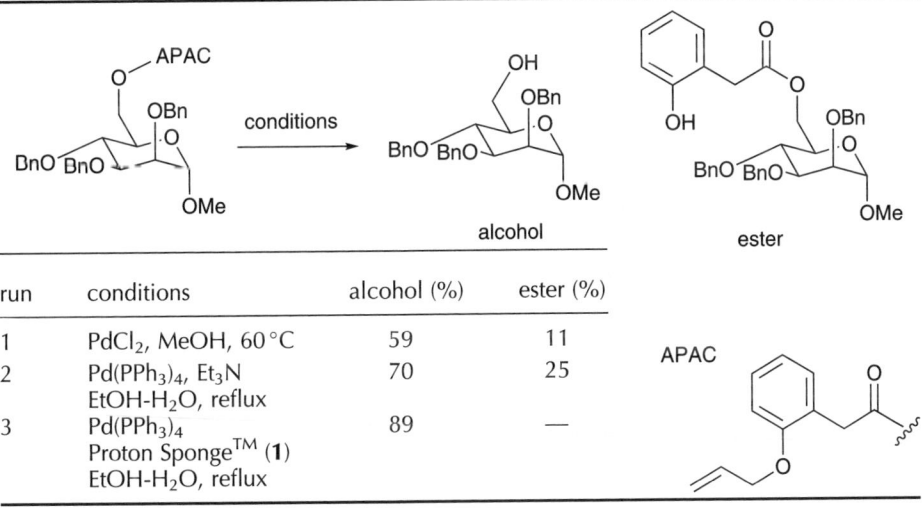

run	conditions	alcohol (%)	ester (%)
1	PdCl$_2$, MeOH, 60 °C	59	11
2	Pd(PPh$_3$)$_4$, Et$_3$N EtOH-H$_2$O, reflux	70	25
3	Pd(PPh$_3$)$_4$ Proton Sponge™ (**1**) EtOH-H$_2$O, reflux	89	—

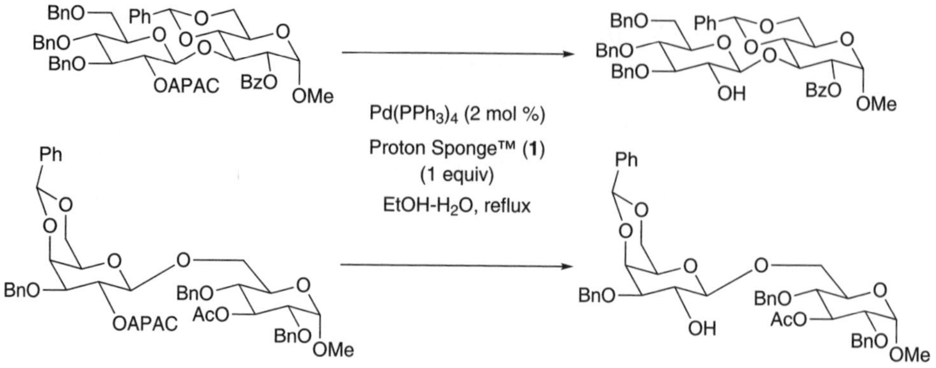

Scheme 8.12 *Palladium catalysed deprotection of APAC group in the presence of Proton Sponge (1)*

8.6 Concluding Remarks

In this chapter, the synthetic utility of Proton Sponge (**1**) was reviewed. This superbase, although not a main player, is indispensable for various mild and selective transformations in organic synthesis. Despite the unique characteristics of superbases, their exploitation is still limited. Recently, various types of proton sponges, including chiral ones, have been developed, and are likely to have a wide range of applications in organic and asymmetric synthesis.

References

1. Alder, R.W., Bowman, P.S., Steele, W.R.S. and Winterman, D.R. (1968) The remarkable basicity of 1,8-bis(dimethylamino)naphthalene. *Chemical Communications*, 723–724.
2. Staab, H.A. and Saupe, T. (1988) 'Proton sponges' and the geometry of hydrogen bonds: aromatic nitrogen bases with exceptional basicities. *Angewandte Chemie – International Edition*, **27**, 865–879; Alder, R.W. (1989) Strain effects on amine basicities. *Chemical Reviews*, **89**, 1215–1223.
3. Kim, I.S., Zee, O.P. and Jung, Y.H. (2006) Regioselective and diastereoselective amination of polybenzyl ethers using chlorosulfonyl isocyanate: total syntheses of 1,4-dideoxy-1,4-imino-D-arabinitol and (−)-lentiginosine. *Organic Letters*, **8**, 4101–4104.
4. Choi, D., Yoo, B., Colson, K.L. et al. (1995) C(10) halogen 10-des(carbamoyloxy)porfiromycins – synthesis, chemistry, and biological activity. *The Journal of Organic Chemistry*, **60**, 3391–3396; Regueiro-Ren, A., Borzilleri, R.M., Zheng, X. et al. (2001) Synthesis and biological activity of novel epothilone aziridines. *Organic Letters*, **3**, 2693–2696.
5. Olofson, R.A., Martz, J.T., Senet J.-P. et al. (1984) A new reagent for the selective, high-yield N-dealkylation of tertiary-amines – improved synthesis of naltrexone and nalbuphine. *The Journal of Organic Chemistry*, **49**, 2081–2082.
6. Magnus, P. and Thurston, L.S. (1991) Synthesis of the vinblastine-like antitumor bis-indole alkaloid navelbine analogue desethyldihydronavelbine. *The Journal of Organic Chemistry*, **56**, 1166–1170; Ratz, A.M. and Weigel, L.O. (1999) S_NAr reactions of 2-haloarylsulfoxides with alkoxides provide a novel synthesis of thiotomoxetine. *Tetrahedron Letters*, **40**, 2239–2242.

7. Diem, M.J., Burow, D.F. and Fry, J.L. (1977) Oxonium salt alkylation of structurally and optically labile alcohols. *The Journal of Organic Chemistry*, **42**, 1801–1802.
8. Evans, D.A., Ratz, A.M., Huff, B.E. and Sheppard, G.S. (1994) Mild alcohol methylation procedure for the synthesis of polyoxygenated natural products. Applications to the synthesis of lonomycin A. *Tetrahedron Letters*, **35**, 7171–7172.
9. G-Fortanet, J., Debergh, J.R. and De Brabander, J.K. (2005) A photochemical run to depsides: synthesis of gustastatin. *Organic Letters*, **7**, 685–688.
10. Paterson, I. and Coster, M.J. (2002) Total synthesis of altohyrtin A (spongistatin 1): an alternative synthesis of the CD-spiroacetal subunit. *Tetrahedron Letters*, **43**, 3285–3289; Paterson, I., Findlay, A.D. and Florence, G.J. (2007) Total synthesis and stereochemical reassignment of (+)-dolastatin 19, a cytotoxic marine macrolide isolated from *Dolabella auricularia*. *Tetrahedron*, **63**, 5806–5819; Jaworski, A.A. and Burch, J.D. (2007) Diastereoselective synthesis of the polyol-containing side chain of the *ent*-bengamides. *Tetrahedron Letters*, **48**, 8787–8789.
11. Ireland, R.E., Liu, L., Roper, T.D. and Gleason, J.L. (1997) Total synthesis of FK-506. Part 2: completion of the synthesis. *Tetrahedron*, **53**, 13257–13284.
12. Sparks, S.M., Gutierrez, A.J. and Shea, K.J. (2003) Preparation of perhydroisoquinolines via the intramolecular Diels–Alder reaction of *N*-3,5-hexadienoyl ethyl acrylimidates: a formal synthesis of (±)-reserpine. *The Journal of Organic Chemistry*, **68**, 5274–5285.
13. Hansen, D.B., Wan, X., Carroll, P.J. and Joullié, M.M. (2005) Stereoselective synthesis of four stereoisomers of β-methoxytyrosine, a component of callipeltin A. *The Journal of Organic Chemistry*, **70**, 3120–3126; Hansen, D.B. and Joullié, M.M. (2005) A stereoselective synthesis of (2*S*,3*R*)-β-methoxyphenylalanine: a component of cyclomarin A. *Tetrahedron: Asymmetry*, **16**, 3963–3969.
14. Thompson, D.H., Svendsen, C.B., Meglio, C.D. and Anderson, V.C. (1994) Synthesis of chiral diether and tetraether phospholipids: regiospecific ring opening of epoxy alcohol intermediates derived from asymmetric epoxidation. *The Journal of Organic Chemistry*, **59**, 2945–2955.
15. Jiang, G.W., Xu, Y. and Prestwich, G.D. (2006) Practical enantiospecific syntheses of lysobisphosphatidic acid and its analogues. *The Journal of Organic Chemistry*, **71**, 934–939.
16. Martin, S.F., Lee, W.-C., Pacofsky, G.J. *et al.* (1994) Strategies for macrolide synthesis. A concise approach to protected seco-acids of erythronolides A and B. *Journal of the American Chemical Society*, **116**, 4674–4688.
17. Chen, J., Feng, L. and Prestwich, G.D. (1998) Asymmetric total synthesis of phosphatidylinositol 3-phosphate and 4-phosphate derivatives. *The Journal of Organic Chemistry*, **63**, 6511–6522.
18. Kanazawa, A.M., Correa, A., Denis, J.-N. *et al.* (1993) A short synthesis of the taxotere side chain through dilithiation of Boc-benzylamine. *The Journal of Organic Chemistry*, **58**, 255–257; Evans, D.A., Kaldor, S.W., Jones, T.K. *et al.* (1990) Total synthesis of the macrolide antibiotic cytovaricin. *Journal of the American Chemical Society*, **112**, 7001–7031.
19. Vintonyak, V.V. and Maier, M.E. (2007) Synthesis of the core structure of cruentaren A. *Organic Letters*, **9**, 655–658; Aggarwal, V.K., Sandrinelli, F. and Charmant, J.P.H. (2002) Synthesis of new, highly hindered C_2-symmetric *trans*-(2*S*,5*S*)-disubstituted pyrrolidines. *Tetrahedron: Asymmetry*, **13**, 87–93.
20. Kanemasa, S., Oderatoshi, Y. and Wada, E. (1999) Asymmetric conjugate addition of thiols to a 3-(2-alkenoyl)-2-oxazolidinone catalysed by the DBFOX/Ph aqua complex of nickel(II) perchlorate. *Journal of the American Chemical Society*, **121**, 8675–8676.
21. Sigurdsson, S.T., Seeger, B., Kutzke, U. and Eckstein, F. (1996) A mild and simple method for the preparation of isocyanates from aliphatic amines using trichloromethyl chloroformate. Synthesis of an isocyanate containing an activated disulfide. *The Journal of Organic Chemistry*, **61**, 3883–3884.
22. Keyes, R.F., Carter, J.J., Zhang, X. and Ma, Z. (2005) Short and efficient synthesis of a vinyl-substituted tricyclic erythromycin derivative. *Organic Letters*, **7**, 847–849.
23. Taggi, A.E., Hafez, A.M., Wack, H. *et al.* (2000) Catalytic, asymmetric synthesis of β-lactams. *Journal of the American Chemical Society*, **122**, 7831–7832; Taggi, A.E., Wack, H., Hafez, A.M. *et al.* (2002) Generation of ketenes from acid chlorides using NaH/crown ether shuttle-deprotonation for use in asymmetric catalysis. *Organic Letters*, **4**, 627–629; France, S.,

Wack, H., Hafez, A.M. *et al.* (2002) Bifunctional asymmetric catalysis: a tandem nucleophile/ Lewis acid promoted synthesis of β-lactams. *Organic Letters*, **4**, 1603–1605; Taggi, A.E., Hafez, A.M., Wack, H. *et al.* (2002) The development of the first catalysed reaction of ketenes and imines: catalytic, asymmetric synthesis of β-lactams. *Journal of the American Chemical Society*, **124**, 6626–6635; Taggi, A.E., Hafez, A.M., Dudding, T. and Lectka, T. (2002) Molecular mechanics calculations as predictors of enantioselectivity for chiral nucleophile catalysed reactions. *Tetrahedron*, **58**, 8351–8356.

24. Dudding, T., Hafez, A.M., Taggi, A.E. *et al.* (2002) A catalyst that plays multiple roles: asymmetric synthesis of β-substituted aspartic acid derivatives through a four-stage, one-pot procedure. *Organic Letters*, **4**, 387–390; Hafez, A.M., Dudding, T., Wagerle, T.R. *et al.* (2003) A multistage, one-pot procedure mediated by a single catalyst: a new approach to the catalytic asymmetric synthesis of β-amino acids. *The Journal of Organic Chemistry*, **68**, 5819–5825.

25. Weatherwax, A., Abraham, C.J. and Lectka, T. (2005) An anionic nucleophilic catalyst system for the diastereoselective synthesis of *trans*-β-lactams. *Organic Letters*, **7**, 3461–3463.

26. Calter, M.A., Phillips, R.M. and Flaschenriem, C. (2005) Catalytic, asymmetric, 'interrupted' Feist–Bénary reactions. *Journal of the American Chemical Society*, **127**, 14566–14567.

27. Tur, F. and Saá, J.M. (2007) Direct, catalytic enantioselective nitroaldol (Henry) reaction of trifluoromethyl ketones: an asymmetric run to α-trifluoromethyl-substituted quaternary carbons. *Organic Letters*, **9**, 5079–5082.

28. Gwaltney, S.L., II, Sakata, S.T. and Shea, K.J. (1996) Bridged to fused ring interchange. Methodology for the construction of fused cycloheptanes and cyclooctanes. Total syntheses of ledol, ledene and compressanolid. *The Journal of Organic Chemistry*, **61**, 7438–7451.

29. Whitney, J.M., Parnes, J.S. and Shea, K.J. (1997) Total synthesis of a *Plocamium* monoterpene marine natural product. Synthetic applications of bridgehead allylsilanes. *The Journal of Organic Chemistry*, **62**, 8962–8963.

30. Singh, O.V. and Han, H. (2007) Iridium(I)-catalysed regio- and enantioselective decarboxylative allylic amidation of substituted allyl benzyl imidodicarbonates. *Journal of the American Chemical Society*, **129**, 774–775; Singh, O.V. and Han, H. (2007) Iridium(I)-catalysed regio- and enantioselective allylic amidation. *Tetrahedron Letters*, **48**, 7094–7098.

31. Gassman, P.G. and van Bergen, T.J. (1974) Oxindoles. New, general method of synthesis. *Journal of the American Chemical Society*, **96**, 5508–5512; Gassman, P.G. and van Bergen, T.J. (1974) Generation of azasulfonium salts from halogen sulfide complexes and anilines. Synthesis of indoles, oxindoles and alkylated aromatic amines bearing cation stabilizing substituents. *Journal of the American Chemical Society*, **96**, 5512–5517; Wierenga, W. (1981) Synthesis of the left-hand segment of the anti-tumor agent CC-1065. *Journal of the American Chemical Society*, **103**, 5621–5623.

32. Warpehoski, M.A. (1986) Total synthesis of U-71,184, a potent new anti-tumor agent modeled on CC-1065. *Tetrahedron Letters*, **27**, 4103–4106.

33. Johnson, P.D. and Aristoff, P.A. (1990) General procedure for the synthesis of *o*-aminophenylacetates by a modification of the Gassman reaction. *The Journal of Organic Chemistry*, **55**, 1374–1375.

34. Ozawa, F., Kubo, A. and Hayashi, T. (1992) Palladium-catalysed asymmetric arylation of 2,3-dihydrofuran: 1,8-bis(dimethylamino)naphthalene as an efficient base. *Tetrahedron Letters*, **33**, 1485–1488.

35. Hennessy, A.J., Malone, Y.M. and Guiry, P.J. (1999) 2,2-Dimethyl-2,3-dihydrofuran, a new substrate for intermolecular asymmetric Heck reactions. *Tetrahedron Letters*, **40**, 9163–9166; Hennessy, A.J., Malone, Y.M. and Guiry, P.J. (2000) The asymmetric cyclohexenylation of 2,2-dimethyl-2,3-dihydrofuran. *Tetrahedron Letters*, **41**, 2261–2264; Hennessy, A.J., Connolly, D.J., Malone, Y.M. and Guiry, P.J. (2000) Intermolecular asymmetric Heck reactions with 2,2-diethyl-2,3-dihydrofuran. *Tetrahedron Letters*, **41**, 7757–7761.

36. Kiely, D. and Guiry, P.J. (2003) Palladium complexes of phosphinamine ligands in the intramolecular asymmetric Heck reaction. *Journal of Organometallic Chemistry*, **687**, 546–561.

37. Dounary, A.B., Hatanaka, K., Kodanko, J.J. *et al.* (2003) Catalytic asymmetric synthesis of quaternary carbons bearing two aryl substituents. Enantioselective synthesis of 3-alkyl-3-aryl

oxindoles by catalytic asymmetric intramolecular Heck reactions. *Journal of the American Chemical Society*, **125**, 6261–6271.
38. Frey, D.A., Duan, C. and Hudlicky, T. (1999) Model study for a general approach to morphine and noroxymorphone via a rare Heck cyclization. *Organic Letters*, **1**, 2085–2087.
39. Arranz, E. and Boons, G.-J. (2001) The 2-(allyloxy) phenylacetylester as a new relay protecting group for oligosaccharide synthesis. *Tetrahedron Letters*, **42**, 6469–6471.

9

Related Organocatalysts (2): Urea Derivatives

Waka Nakanishi
Graduate School of Pharmaceutical Sciences, Chiba University, 1-33 Yayoi, Inage, Chiba 263-8522, Japan

9.1 Introduction

Activity in a living organism could be, in a chemical sense, managed by a variety of catalytic reactions [1]. A typical example is an enzyme-participating reaction, in which a highly controlled hydrogen bonding network plays an important role. In general, an enzyme specifically recognizes a substrate (in some cases, also a reactant) and then catalyses a variety of reactions (Figure 9.1a). A metal catalysed reaction is artificially developed in which a metal ion serves as a Lewis acid catalyst (Figure 9.1b). On the other hand, urea, thiourea and guanidine have been shown to serve as metal-free organocatalysts through double hydrogen bonding [2,3] (Figure 9.1c), in which they could act as either Lewis acids (type I) or proton donors in chelation binding (type II). Phosphate and nitro groups are activated in the latter bidentate fashion. In Figure 9.1c, highly asymmetric induction could be expected when the substrate is fixed under an effective chiral environment after tuning of its side chains.

A vast number of valuable metal catalysed reactions has been developed to date. However, metal-free organocatalysis attracts much attention due not only to green chemical but also to economic demands. An organosuperbase has generally a highly conjugated system stabilized in the protonated form, and the resultant conjugated acid could act as a Brønsted acid and/or a Lewis acid catalyst in organic synthesis. In fact, the Lewis acid type of catalysis activation has been reported in reactions in which guanidine or amidine

(a) Enzymatic reaction

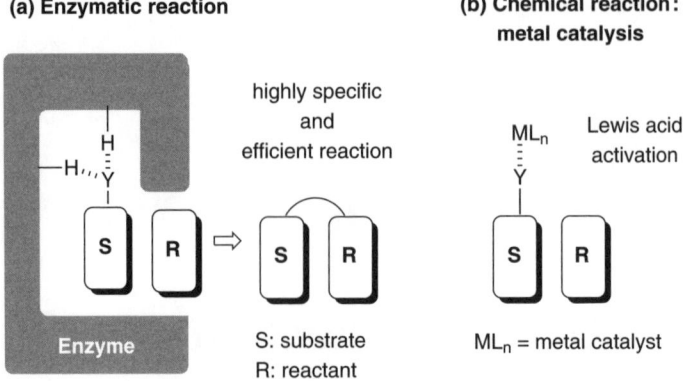

(b) Chemical reaction: metal catalysis

(c) Chemical reaction: metal-free catalysis

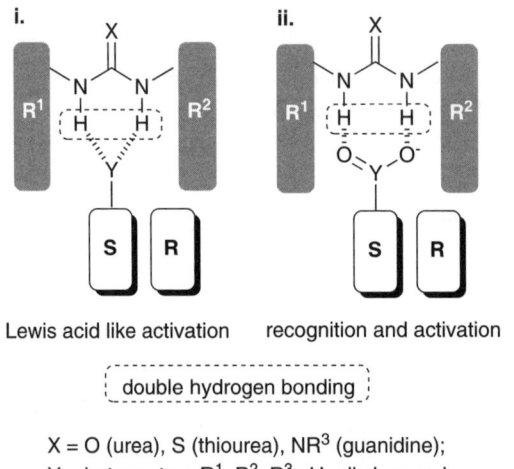

X = O (urea), S (thiourea), NR³ (guanidine);
Y = hetero atom R¹, R², R³ =H, alkyl, or aryl

Figure 9.1 Schematic paths of catalytic reactions

participate. Furthermore, structurally related urea and thiourea compounds are now well developed as alternative organocatalysts [2–5]. In this chapter, reactions catalysed by urea and thiourea will be discussed, mainly focusing on molecular recognition through double hydrogen bonding.

9.2 Bisphenol as an Organoacid Catalyst

9.2.1 Role of Phenol as Hydrogen Donor

Hine *et al.* [6] approached the complex structures of bisphenol (dibenzocyclobutane-1,8-diol) and sp^2 oxygen bearing partners using X-ray crystallographic analysis; the two

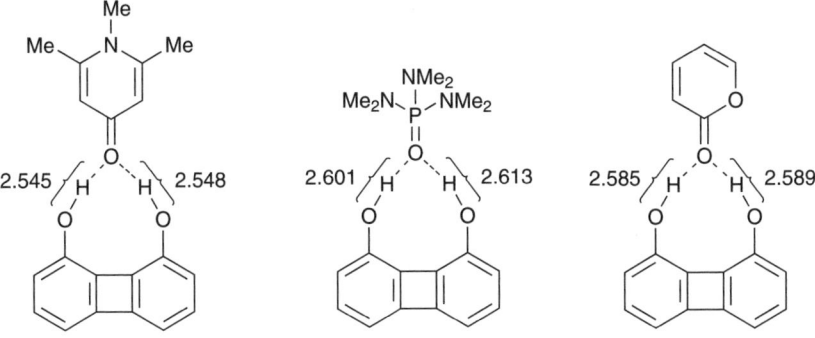

Figure 9.2 Phenol assisted double hydrogen bonding

molecules interact through a double hydrogen bond (Figure 9.2). The two components are almost coplanar, suitable for ideal hydrogen bonding. The O−H−O distances are 2.545 Å and 2.548 Å, and the angles are 177° and 174° for the complex with 1,2,6-trimethyl-4-pyridone (TMP), 2.601 Å and 2.613 Å and 168° and 172° for the complex with hexamethylphosphoric triamide (HMPA), and 2.585 Å and 2.589 Å and 176° and 178° for that with pyrone. In all the three cases, the O−H−O angle is near 180°, which is favourable for hydrogen bonding, and the bond length is smaller than van der Waals radii (3.00 Å for O−H−O) by 0.3 Å, which clearly confirms the existence of strong hydrogen bonding [7].

Later, the same group reported the synthesis of a more acidic 4,6-dinitrodibenzocyclobutane-1,8-diol [8] and examined equilibrium constants for hydrogen bonding with various bases [9] (Table 9.1). Binding affinity of dinitro substituted bisphenol to TMP was 40-fold higher compared to that of unsubstituted one (run 1). Similar but moderate effects were observed with respect to other hydrogen acceptors, HMPA, 2,6-dimethyl-γ-pyrone (DMP), tetramethylenesulfoxide (TMSO), and dimethyl sulfoxide (DMSO) (runs 2–5). Thus, the

Table 9.1 Equilibrium of base and 1,8-dihydroxydibenzocyclobutane

		Equilibrium constants ($10^{-3} M^{-1}$)	
Run	Base	R=H (pK_a=8.31)	R=NO$_2$ (pK_a=5.90)
1	TMP	49	1910
2	HMPA	42	470
3	DMP	1.04	4.3
4	TMSO	1.85	2.3
5	DMSO	1.03	1.60

9.2.2 Bisphenol Catalysed Reaction

Choy el al. [10] reported that asymmetric Diels–Alder reaction between α-hydroxy vinyl ketone and cyclopentadiene proceeded smoothly with excellent diastereoselectivity, in which the formation of a rigid five-membered chelate structure through hydrogen bonding effectively freezes the free rotation, thus making the two diastereotopic faces of the enone system highly distinguishable (Scheme 9.1). This intramolecular hydrogen bonding causes *endo* selectivity eight times higher than *exo* selectivity, resulting in 99% diastereoselectivity of the *endo* product.

Epoxide is effectively opened with diethylamine in the presence of dibenzocyclobutane-1,8-diol, in which the rate is accelerated with more than 10-fold compared with in the use of phenol. The acceleration could be explained by intermolecular double hydrogen bonding between the epoxide and diphenol in the fashion of type I in Figure 9.1c [11] (Scheme 9.2).

Kelly et al. [12] applied these hydrogen bonding systems to control not only *endo/exo* selectivity but also reaction rate in the Diels–Alder reaction of methyl vinyl ketone (MVK) and cyclopentadiene (Scheme 9.3). The presence of dibenzocyclobutane-1,8-diol leads to a 90% yield of the *endo* adduct, whereas the same adduct is given only in 3% yield without the

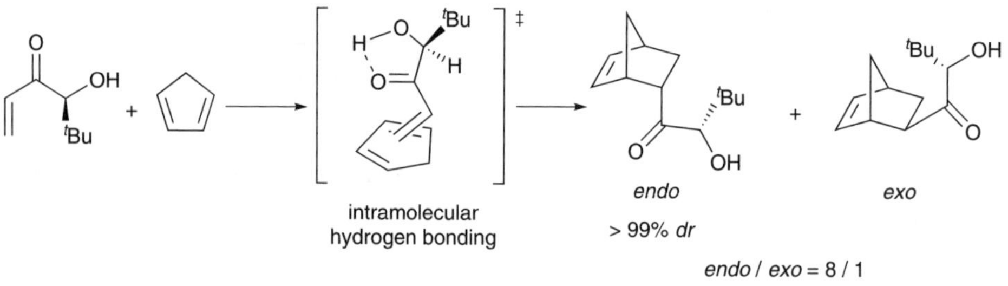

Scheme 9.1 Diels–Alder reaction of α-hydroxy vinyl ketone and cyclopentadiene

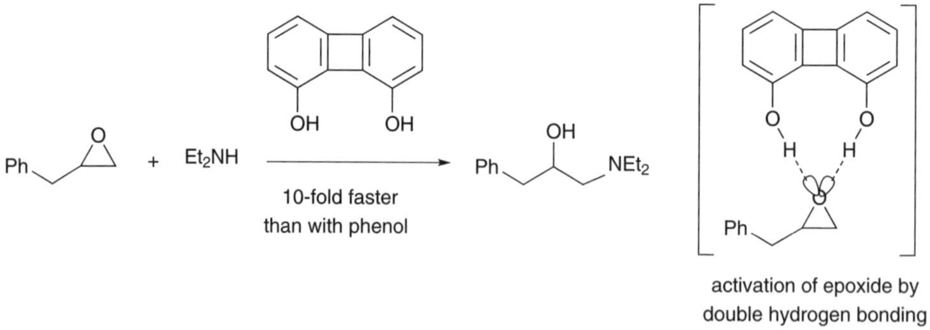

Scheme 9.2 Phenol assisted ring opening reaction of epoxide

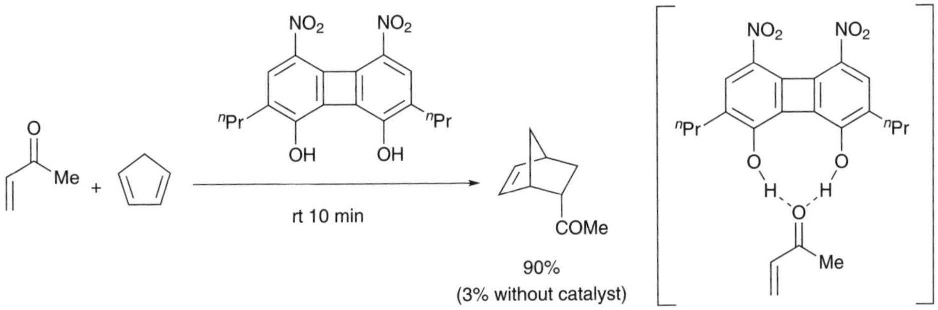

Scheme 9.3 Phenol assisted Diels–Alder reaction

catalyst, suggesting that MVK could be activated through double hydrogen bonding with the bisphenol in the fashion of type I in Figure 9.1c.

9.3 Urea and Thiourea as Achiral Catalysts

9.3.1 Role of Urea and Thiourea as Hydrogen Donors

Three types of urea-participating binding models with heteroatoms through double hydrogen bonding have been recognised by X-ray crystallographic analysis [13]. Introducting a nitro group at the *meta* position of the aryl ring results in effective co-crystallization with weak hydrogen acceptors such as ester and ether. In addition, possible intramolecular CH–O interaction between urea oxygen and acidic hydrogen adjacent to the nitro group may stabilize the complexation, causing reduction of self-complexation (Figure 9.3).

Thiophenol has greater acidity than phenol by 10.7 kcal/mol in dimethyl sulfoxide (DMSO); the difference is explained by unfavourable lone pair–lone pair interactions in the larger thiolate ion (PhS$^-$) ion [14]. Thus, thioureas are, as expected, more acidic than the corresponding urea derivatives (Figure 9.4) [15]. The ability of thiourea as a hydrogen donor was shown by Wilcox *et al.* [16] who systematically analysed substituent effects in the binding of arylurea with a zwitter ion in chloroform (CHCl$_3$) and correlated the acidity of substituted phenols with the binding constants.

It is shown from X-ray crystallography that urea and thiourea can make complexes with various kinds of functionality in the solid state [13]. In solution, the binding affinity of functional groups with alkyl- and arylureas is determined to be in the following order: PhOPO$_3^{2-}$ > PhPO$_3^{2-}$ > PhCO$_2^-$ > PhP(OH)O$_2^-$ > PhOPO$_3^-$ > PhSO$_3^-$ > lactone, and no evidence for hydrogen bonding of nitrobenzene with urea is detected in either CDCl$_3$ or DMSO-d_6 [17] (Table 9.2). The inertness of the nitro group to urea is one of reasons for the wide use of nitro-substituted substrates.

In the complex formation with the carboxylate anion, although 1,3-dimethylthiourea shows a 10-fold increase in stability over urea, the alkylguanidinium ion (pK_a = 14) is found to be the strongest proton donor [18] (Table 9.3).

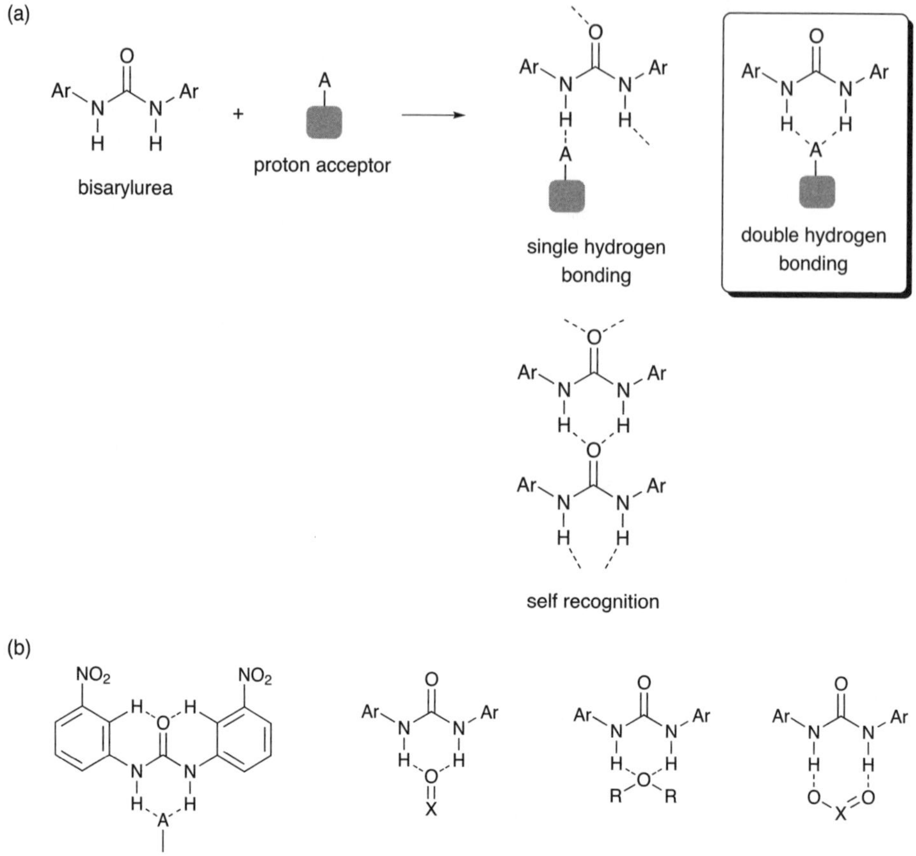

Figure 9.3 (a) Types of coordination of urea through hydrogen bonding; (b) counter part in double hydrogen bonding

On the other hand, inverse affinity between thiourea and urea is reported in the complexation with sulfonate, in which great improvement in the affinity is also observed in the nitro-substituted arylurea derivatives [16] (Table 9.4).

9.3.2 Urea and Thiourea Catalysed Reactions

It was reported by Curran et al. [19] in 1994 that a symmetrical N,N'-diarylurea effectively catalyses the allylation of cyclic sulfinyl radicals with allyltributylstannane (Table 9.5). High *trans* selectivity in the product was observed when a stoichiometric amount of urea catalyst was used (run 4), suggesting activation in the transition state by double hydrogen bonding between the urea hydrogen atoms and the sulfinyl oxygen atom.

Catalytic activity of the same urea and its variations was examined in the Claisen rearrangement of 3-methoxyvinyl vinyl ether [20] (Table 9.6). Great rate acceleration was observed in the use of thiourea, even though completion of reaction could not be estimated due to slow decomposition of the catalyst under the reaction conditions used (run 6).

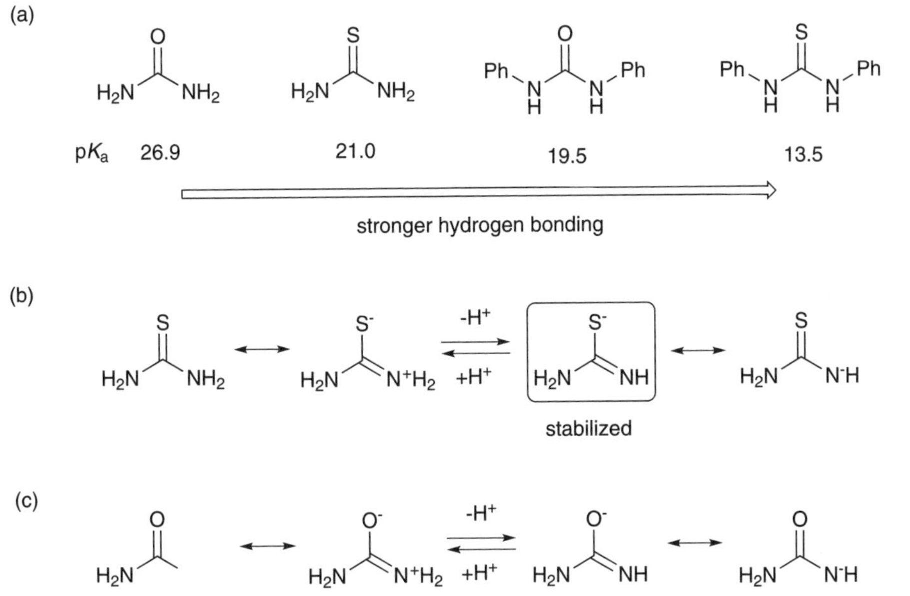

Urea and Thiourea as Achiral Catalysts 279

Figure 9.4 (a) pKa values of urea and thiourea derivatives; (b) resonance forms of thiourea; (c) resonance forms of urea

Table 9.2 Binding affinity of N-octyl-N'-(p-tolyl)urea

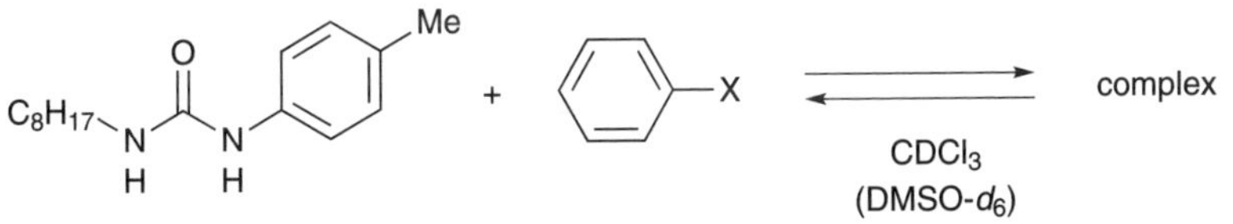

X	NO_2	SO_3^-	OPO_3^-	PO_3H^-	CO_2^-	PO_3^{2-}	OPO_3^{2-}
$K_a(M^{-1})$	ND[a] (ND[a])	NE[b] 13	NE[b] 27	NE[b] 140	1300 150	NE[b] 2500	NE[b] 3600)

[a] Not detected.
[b] Not examined.

Table 9.3 Binding affinity of urea derivatives with acetate

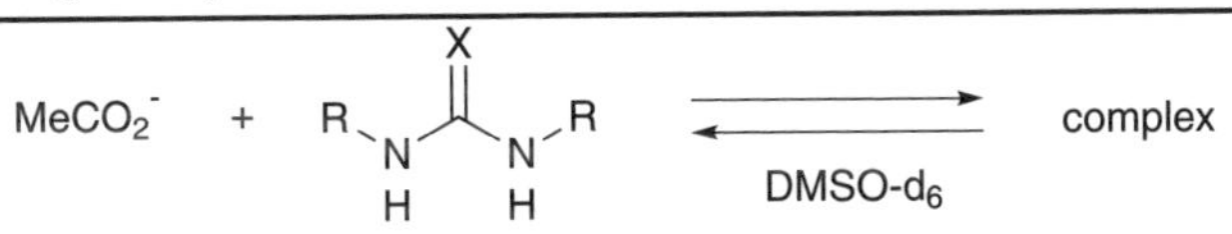

R, X	CH_3, O	CH_3, S	$(CH_2)_2$, $NHCH_2Ph$
$K_a(M^{-1})$	45	340	12 000

Table 9.4 Binding affinity of urea derivatives with N,N,N-tributylammonium butylsulfate

$$Bu_3N^+(CH_2)_4SO_3^- + C_8H_{17}\text{-NH-C(X)-NH-R} \underset{CDCl_3}{\rightleftharpoons} \text{complex}$$

R, X	C_6H_4, S	3-$(NO_2)C_6H_4$, S	3-$(NO_2)C_6H_4$, O	4-$(NO_2)C_6H_4$, O
$K_a(M^{-1})$	45	2100	6300	15 000

Table 9.5 Urea catalysed allylation

run	additive (equiv)	trans/cis	yield (trans)
1	none	5.3/1	59
2	TFE (1.0)	7.1/1	61
3	urea (0.25)	7.1/1	70
4	uera (1.0)	14.1/1	72

TFE: trifluoroethane.

Table 9.6 Urea catalysed Claisen rearrangement of 3-methoxyvinyl allyl ether

(E/Z = 2.6/1)

a: R^1 =CF_3, R^2 =$CO_2C_8H_1$, R^2 =Me, X = O
b: R^1 =R^2 =R^2 =H, X = O
c: R^1 =CF_3, R^2 =$CO_2C_8H_1$, R^2 =H, X = O
d: R^1 =CF_3, R^2 =$CO_2C_8H_1$, R^2 =H, X = S

run	catalyst (equiv)	k (x 10^{-5} s^{-1})	k_{rel}
1	none	0.6	1
2	a (1)	0.6	1.0
3	b (1)	1.0	1.6
4	DMSO (5)	1.2	1.9
5	c / DMSO (1 / 6)	0.8	1.3
6	d (1)	N.D.[a]	3-4

[a] Not determined due to decomposition of the catalyst.

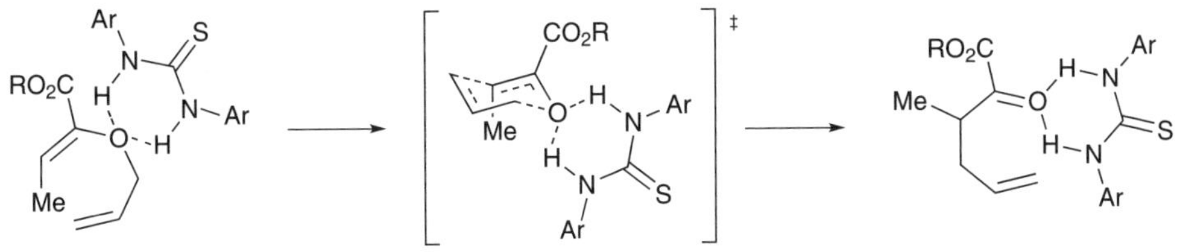

Scheme 9.4 *Supposed mechanism for the thiourea catalysed Claisen rearrangement*

Similar thiourea catalysed Claisen rearrangement was theoretically considered by Kirsten *et al.* [21]. It was suggested that a transition state is significantly stabilized through double hydrogen bonding, whereas the overall effect on the barrier is small due to endergonic conformational changes and complexation (Scheme 9.4).

Urea-type catalyst has to be not only a good hydrogen donor (H-donor) but also a poor hydrogen acceptor (H-acceptor), in order to avoid self-association which disturbs the association of urea and a substance (Figure 9.3). Due to its ability to weakly accept hydrogen, thiourea limits self-association compared to urea. Furthermore, substitution of electron-withdrawing groups at the 3 and 5 positions of the aryl residue not only causes increased acidity, in other word hydrogen is liberated easily, but also lowers the hydrogen accepting ability (Figure 9.5). Amidinium and guanidinium ions, the conjugate acids of organosuperbases, could serve as strong hydrogen donors. However, they are ineffective catalysts due to their too strong ability to form hydrogen bonding, in which a catalyst is trapped by product and/or substrate and cannot be recycled, so-called 'product inhibition'. Thus, thiourea has an advantage in its excellent balance in affinity to product and/or substrate.

Schreiner *et al.* developed thiourea catalyst as a promising hydrogen donor, which has more benefit in solubility, synthesis and catalytic turn over number compared with urea catalyst, in the Diels–Alder reaction of *N*-crotonyloxazolidinone and cyclopentadiene [22,23] (Table 9.7). *N,N'*-Di[3,5-bis(trifluoromethyl)phenyl]thiourea accelerates the reaction and improves stereoselectivity (run 4) similar to a metal catalyst such as aluminium chloride ($AlCl_3$) (run 2) or titanium chloride ($TiCl_3$) (run 3).

Traces of the reaction by NMR and IR spectra together with *ab initio* calculations reveals that the hydrogen bond participated bicyclic structure between the acyloxazolidinone and thiourea mainly controls the reaction course (Figure 9.6).

The thiourea catalysed Diels–Alder reaction of MVK and cyclopentadiene gives useful information for further tuning the structure of the thiourea catalyst: flexible side chains on nitrogen atoms of thiourea are ineffective due to the large entropy in complexation with an hydrogen acceptor [22] (Table 9.8).

The effect of substitution on the aryl group of thiourea has also been examined (Figure 9.7). Substitution at the *ortho* position results in a lowering of catalytic activity due to steric hindrance, and substitution at the *para* position shows activity that is less effective than substitution at the *meta* position.

Trisguanidine is found to recognize phosphate and hydrolyze RNA in an enzyme-like reaction [24] (Scheme 9.5). Although the use of guanidine as a Lewis acid catalyst is limited in organic reaction, it may be a potential catalyst under aqueous conditions. Guanidine catalysed reactions are summarized in Chapter 4.

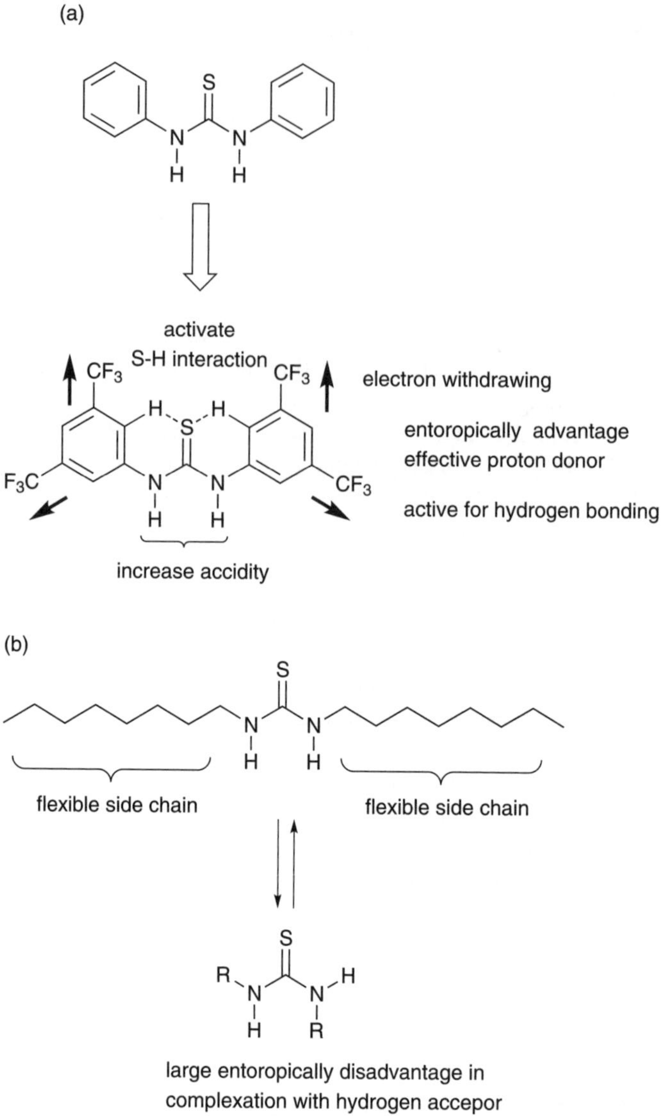

Figure 9.5 (a) Effect on introducing the CF_3 group to aryl thiourea; (b) equilibrium of flexible thiourea

9.4 Urea and Thiourea as Chiral Catalysts

A variety of thiourea (or urea) compounds has been designed as chiral catalysts based on modification of substituent(s) (R^1 and/or R^2) with chiral functionality as shown in type C of Figure 9.1, which could recognize the substrate (and/or reagent) to construct an effective asymmetric environment in transition state. The additional functionality in monothiourea

Table 9.7 *Diels–Alder reaction of N-crotonyloxazolidinone and cyclopentadiene*

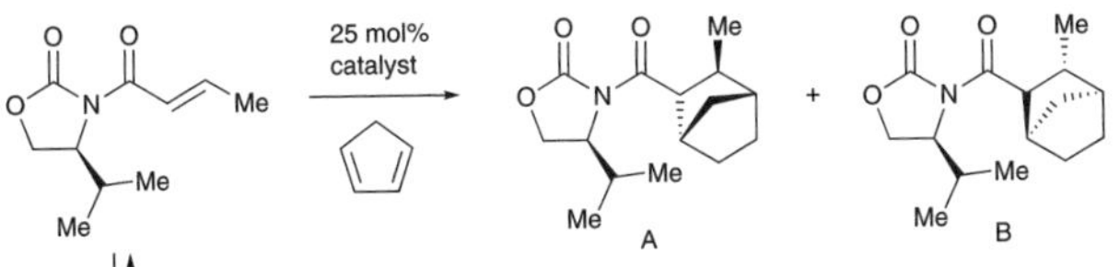
25 mol%
catalyst
Me
A
B

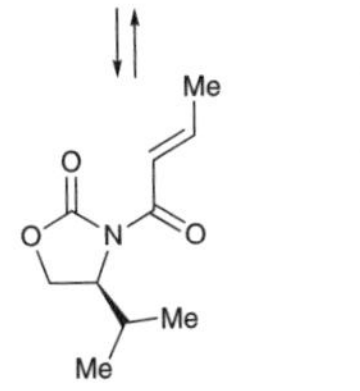
Me

run	catalyst	temp (°C)	time (h)	yield (%)	ratio (A / B)
1	non		96	55	36 / 64
2	$AlCl_3$	-78 °C	1	95	92 / 8
3	$TiCl_3$	-78 °C	1	92	89 / 11
4	thiourea	23 °C	48	78	91 / 19

thiourea

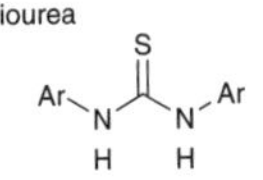
S
Ar
N
H

Ar = 3,5-$(CF_3)_2$-C_6H_3

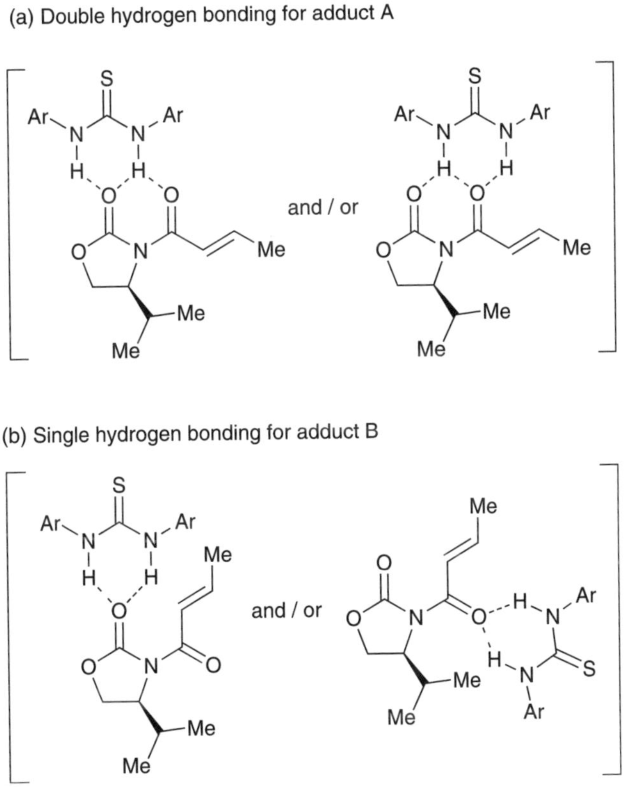
(a) Double hydrogen bonding for adduct A
and / or
(b) Single hydrogen bonding for adduct B
and / or

Figure 9.6 *Possible transition states of the thiourea catalysed reaction for (a) adduct A and (b) adduct B*

Table 9.8 Catalytic activity of thioureas for the Diels–Alder reaction of methyl vinyl ketone and cyclopentadiene

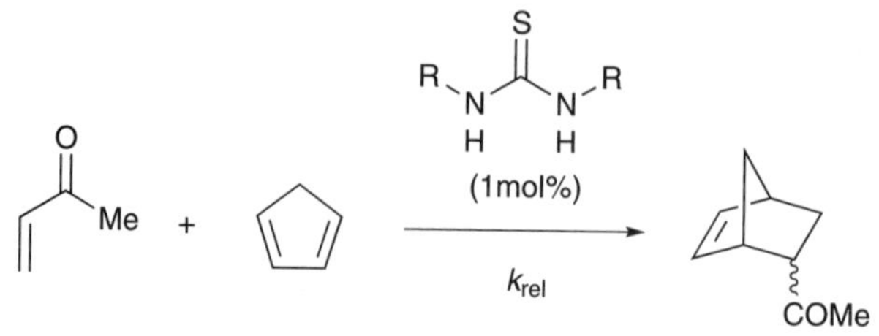

Run	Catalyst	k_{rel}
1	none	1.0
2	R = 3,5-(CF$_3$)$_2$-C$_6$H$_3$	4.8
3	R = C$_6$H$_5$	1.4
4	R = C$_8$H$_{17}$	1.0

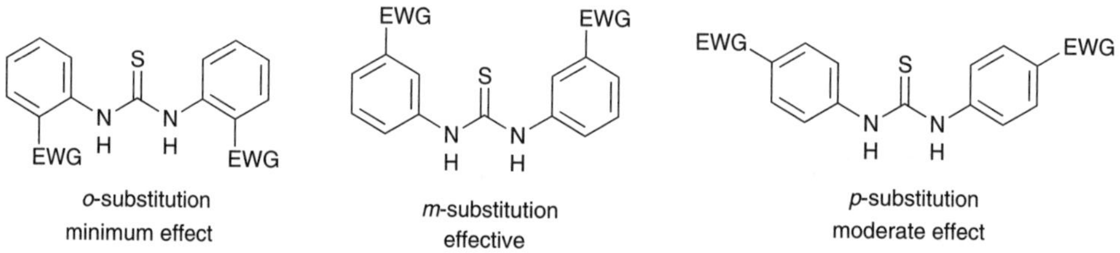

Figure 9.7 Effect of electron-withdrawing groups (EWG) in aryl thioureas

catalyst is historically derived from Schiff base, amine, alcohol and cinchona alkaloid respectively. Bisthioureas are also designed as alternative catalysts.

9.4.1 Monothiourea Catalysts

9.4.1.1 With Schiff Base Function (Jacobsen's Catalyst)

The present establishment of urea-type catalysts as a new category originated from finding by Sigman et al. [25]. They reported the potential catalytic activity of urea-type compounds with Schiff base functionality (Jacobsen's catalyst) in the course of screening their possibility as metal ligand for asymmetric Strecker reaction by parallel synthetic libraries and showed that sterically demanding groups on both urea substituents play important roles for the generation of enantioselectivity (Figure 9.8).

The catalyst (R^1 = PhCH$_2$, R^2 = H, R^3 = OCO^{t}Bu), among more than 70 kinds of urea-type compounds examined, is found to be the most effective one with high catalytic turn over number. Good enantioselectivity is observed in a range of substrates [26] (Table 9.9). Although the size of the substituent in the catalyst is important, both yield and selectivity are unaffected by the geometry of imine substrates.

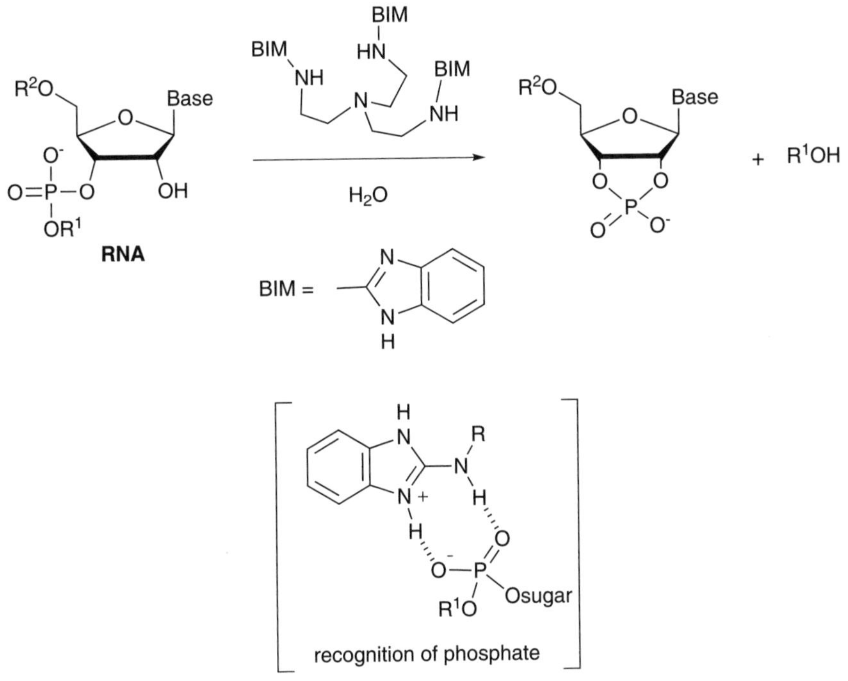

Scheme 9.5 Guanidine catalysed hydrolysis of RNA

Mechanistic approaches to the (thio)urea catalysed Strecker reaction by NMR analysis and theoretical calculations [27] reveal that: the catalyst has a stable 3D structure and isomerises the E-imine substrate to a Z-one; only the hydrogen of the (thio)urea interacts with the Z-imine through double hydrogen bonding; and the double interaction changes to a single one during the reaction, in which the bond strength is weakened and the liberated (thio)urea can be recycled (Scheme 9.6).

Thiourea effectively catalyses hydrophosphorylation of imine [28] (Table 9.10). The products can be deprotected under mild hydrogenolysis conditions to afford the corresponding optically active amino phosphoric acids.

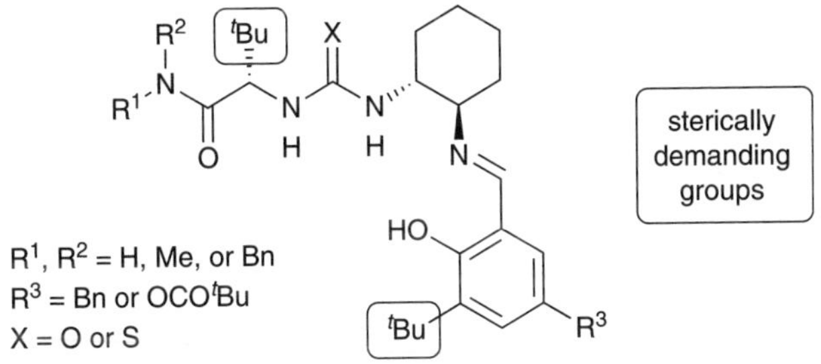

Figure 9.8 Sterically demanding groups of aryl thioureas

Table 9.9 Urea catalysed Strecker reaction

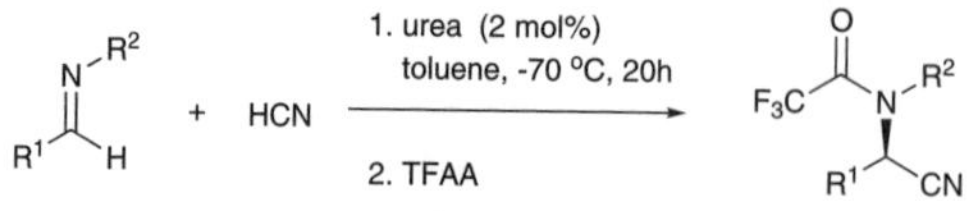

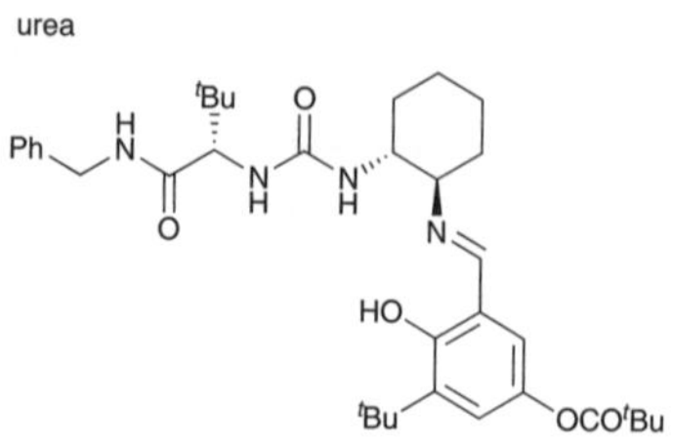

run	R¹	R²	yield (%)	ee (%)
1	aryl[a]	allyl	74-99	77-96
2	^{t}Bu	PhCH$_2$	88	96
3	c-C$_6$H$_{11}$	allyl	88	86
4	n-C$_5$H$_{11}$	PhCH$_2$	69	78

[a] Phenyl, anisyl, tolyl, bromophenyl.

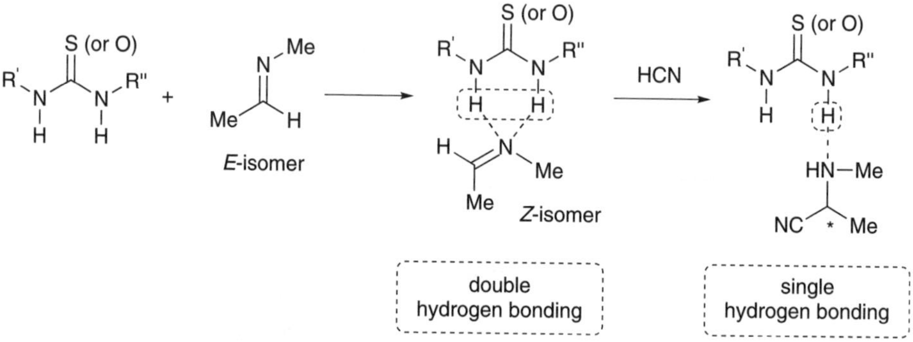

Scheme 9.6 Supposed mechanism of the (thio)urea catalysed Strecker reaction

Table 9.10 Thiourea catalysed hydrophosphorylation of imines

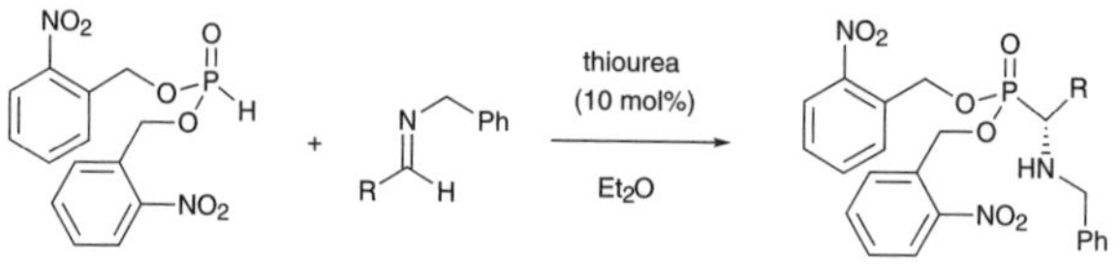

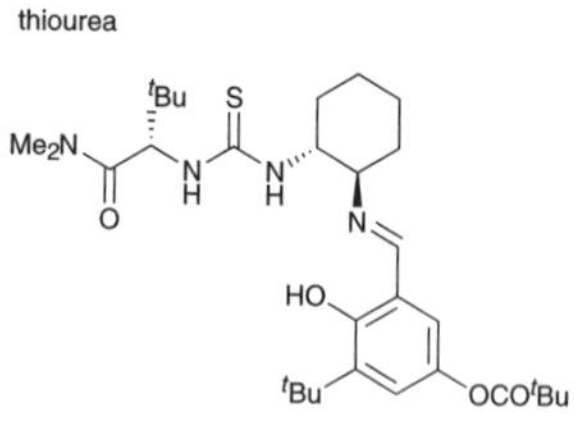

run	R	yield (%)	ee (%)
1	Ph	87	98
2	Et$_2$CH	90	96
3	^{i}Pr	93	90
4	Me$_2$C=CH	91	82

Table 9.11 Thiourea catalysed Michael reaction

Ph–CH=CH–NO$_2$ + CH$_2$(CO$_2$Et)$_2$ →(thiourea catalyst R^{1}NHC(=S)NHR^2, 10 mol%, toluene, rt)→ EtO$_2$C–C(CO$_2$Et)(Ph-CH)–CH$_2$NO$_2$

Run	R^1	R^2	yield (%)	ee (%)
1	3,5-(CF$_3$)$_2$C$_6$H$_3$	(2-NMe$_2$-cyclohexyl)	86	93
2	3,5-(CF$_3$)$_2$C$_6$H$_3$	c-C$_6$H$_{11}$	57[a]	—
3	catalyst[b]		14	35

[a] Copresence of Et$_3$N.
[b] (R,R)-1-acetamindo-2-dimethylaminocyclohexane.

9.4.1.2 With Amine Function

A more simple thiourea catalyst with amino functionality catalyses the asymmetric Michael addition of 1,3-dicarbonyl compound to nitroolefin [29,30]. In the reaction of malonate to nitrostyrene (Table 9.11) the adduct is satisfactorily obtained when N-[3,5-bis(trifluoromethyl)phenyl]-N'-(2-dimethylaminocyclohexyl)thiourea is used as a catalyst (run 1), whereas the reaction proceeds slowly when the 2-amino group is lacking (run2). In addition, chiral amine without a thiourea moiety gives a poor yield and enantioselectivity of the product (run 3). These facts clearly show that both thiourea and amino functionalities are necessary for rate acceleration and asymmetric induction, suggesting that the catalyst simultaneously activates substrate and nucleophile as a bifunctional catalyst.

Furthermore, this catalyst promotes asymmetric Mannich reactions [31] and Michael addition of active methylene compounds to α,β-unsaturated imides [32].

Yoon et al. [33] found that thiourea catalyst with an amine function promotes the stereoselective addition of a range of nitroalkanes to aromatic N-butoxycarbonyl (N-Boc) imines. In the Mannich reaction of nitroethane (Table 9.12) high enantioselectivity, but low yield, is observed when urea is used (run 1), whereas thiourea affords the adduct in >95% yield with 92% ee (run 2). It should be noted that addition of powdered molecular sieves is necessary for reproducible results.

The acyl-Mannich reaction [34], acyl-Pictet-Spengler reaction [35,36] and cyanosilylation [37] catalysed with bifunctional thiourea catalysts have also been developed by fine tuning of the side chain.

9.4.1.3 With Alcohol Function

The alcohol group also works as an alternative functionality in thiourea catalysts. Catalytic enantioselective Friedel–Crafts alkylation of indoles with nitroalkanes using alcohol-

Table 9.12 (Thio)urea catalysed Mannich reaction

Run	X	Yield (%)	syn/anti	ee (%)
1	O	36	11/1	91
2	S	>95	15/1	92

substituted thiourea has been reported [38] (Table 9.13). Reaction in general proceeds in good yield with satisfactory selectivity (runs 1–5); however, the effectiveness is lowered by either protection (run 6) or lack (run 7) of the alcoholic function.

Possible coordination of the thiourea catalyst with not only nitroolefin but also indole through the thiourea hydrogen atom and the hydroxyl oxygen atom, respectively, in the transition state are supposed (Figure 9.9).

Alcohol-substituted thioureas also catalyse the Petasis-type reaction of quinolines [39] and conjugate addition of amine [40].

9.4.1.4 With Cinchona Alkaloid Function

Thiourea catalyst with a modified cinchona alkaloid unit is applied to intramolecular Michael addition of phenol; chiral chromanone is produced in high yield with good

Table 9.13 Thiourea catalysed Michael reaction of nitroolefins with indole

Run	X	R^1	R^2	R^3	Yield (%)	ee (%)
1	OH	H	H	Ph	78	85
2	OH	Me	H	Ph	82	74
3	OH	H	OMe	Ph	86	89
4	OH	H	Cl	Ph	85	71
5	OH	H	H	C_5H_{11}	76	81
6	OTMS	H	H	Ph	18	39
7	H	H	H	Ph	15	0

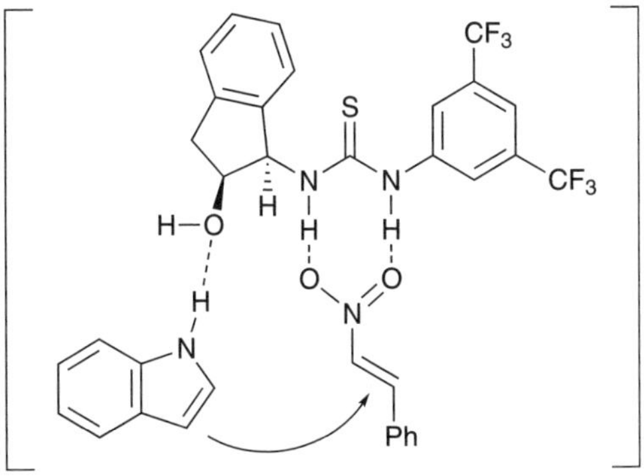

Figure 9.9 *Supposed transition state for thiourea catalysed Michael reaction of nitrostyrene and indole*

stereoselectivity [41] (Scheme 9.7). Incorporation of *tert*-butyl ester at the α-position of α,β-unsaturated ketone function is needed to enhance the reactivity of substrate.

9.4.2 Bisthiourea Catalysts

Thiourea catalyst with additional thoiurea functionality can act as possible bifunctional thiourea catalyst due to the hydrogen bonding ability of the thiourea function. C_2-symmetric bisthiourea has been applied to the Baylis–Hillman reaction of cyclohexenone [42] (Table 9.14). Adduct is obtained in moderate to good yields (runs 1–6), but asymmetric induction is dependent upon the aldehyde electrophile (90% ee in run 6). The use of monothiourea as a catalyst results in low conversion (20%). Thus, it could be reasonably deduced that each thiourea function of bisthiourea independently and effectively interact with cyclohexenone and aldehyde in the transition state (Figure 9.10).

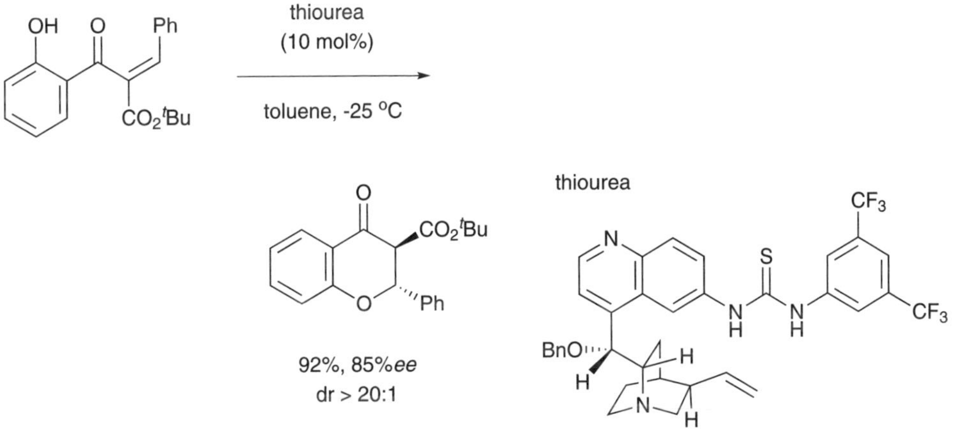

Scheme 9.7 *Thiourea catalysed chromanone synthesis*

Table 9.14 Bisthiourea catalysed Baylis–Hillman reaction

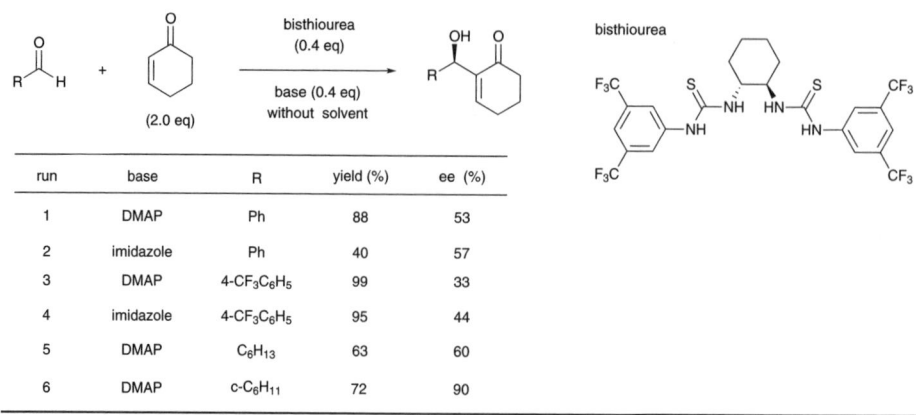

run	base	R	yield (%)	ee (%)
1	DMAP	Ph	88	53
2	imidazole	Ph	40	57
3	DMAP	4-CF$_3$C$_6$H$_5$	99	33
4	imidazole	4-CF$_3$C$_6$H$_5$	95	44
5	DMAP	C$_6$H$_{13}$	63	60
6	DMAP	c-C$_6$H$_{11}$	72	90

DMAP: 4-dimethylaminopyridine.

C_2-symmetric bisthiourea carrying a guanidine function at the molecular centre has been designed as an alternative catalyst and applied to asymmetric Henry reaction; adduct is formed in high yield and high enantioselectivity [43] (Scheme 9.8). Wide generality in substrate is observed. Addition of potassium iodide under heterogeneous conditions is necessary to avoid retro-aldol reaction and self aggregation of the catalyst is suggested by a positive nonlinear effect in yield and enantioselectivity.

9.4.3 Urea-Sulfinimide Hybrid Catalyst

Tan et al. [44] succeeded in developing a new urea-sulfinimide catalyst that promotes the indium mediated allylation of acylhydrazones (Scheme 9.9). Incorporation of Lewis base functionality in proximity to the urea moiety is designed to promote the addition of organometallic reagents to the C=N bond of acylhydrazones through dual activation.

Figure 9.10 Supposed transition state for bisthiourea catalysed Baylis–Hillman reaction

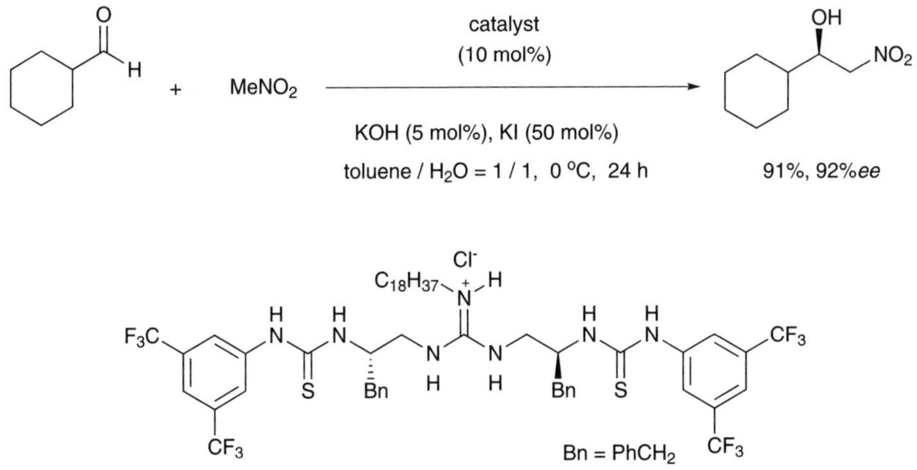

Scheme 9.8 Guanidinobisthiourea catalysed Henry reaction

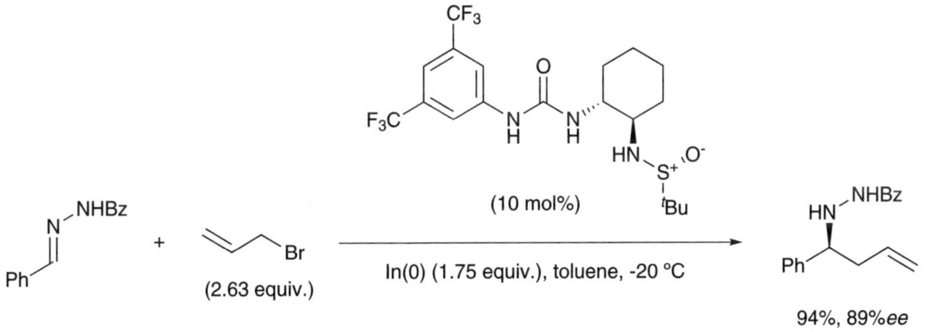

Scheme 9.9 Sulfiminourea catalysed allylation of hydrazone

9.5 Concluding Remarks

Urea (thiourea) catalysts can coordinate and activate hydrogen bonding acceptors such as the carbonyl, nitro and phosphate groups. Introduction of a chiral moiety produces useful urea-based catalysts applicable in several types of asymmetric syntheses under mild conditions. The key aspect in controlling enantioselectivity is an effective hydrogen bonding network, due to additional functionality leading to favourable coordination of catalyst to substrate(s) in the transition state. Schiff base, amine, alcohol, cinchona alkaloid and thiourea itself have been tried as these functionalities. In the future more complex urea systems may be designed for approaches to natural enzyme reaction.

References

1. Ball, P. (2008) Water as an active constituent in cell biology. *Chemical Reviews*, **108**, 74–108.
2. Schreiner, P.R. (2003) Metal-free organocatalysis through explicit hydrogen bonding interactions. *Chemical Society Reviews*, **32**, 289–296.
3. Pihko, P.M. (2004) Activation of carbonyl compounds by double hydrogen bonding: An emerging tool in asymmetric catalysis. *Angewandte Chemie – International Edition*, **43**, 2062–2064.
4. Connon, S.J. (2006) Organocatalysis mediated by (thio)urea derivatives. *Chemistry – A European Journal*, **12**, 5418–5427.
5. Taylor, M.S. and Jacobsen, E.N. (2006) Asymmetric catalysis by chiral hydrogen-bond donors. *Angewandte Chemie – International Edition*, **45**, 1520–1543.
6. Hine, J., Ahn, K., Gallucci, J.C. and Linden, S.M. (1984) 1,8-Biphenylenediol forms two strong hydrogen bonds to the same oxygen atom. *Journal of the American Chemical Society*, **106**, 7980–7981.
7. Emsley, J. (1980) Very strong hydrogen bonding. *Chemical Society Reviews*, **9**, 91–124.
8. Hine, J., Hahn, S., Miles, D.E. and Ahn, K. (1985) The synthesis and ionization constants of some derivatives of 1-biphenylenol. *The Journal of Organic Chemistry*, **50**, 5092–5096.
9. Hine, J. and Ahn, K. (1987) The double hydrogen bonding ability of 4,5-dinitro-1,8-biphenylenediol. *The Journal of Organic Chemistry*, **52**, 2083–2086.
10. Choy, W., Reed, L.A. and Masamune, S. (1983) Asymmetric Diels–Alder reaction: Design of chiral dienophiles. *The Journal of Organic Chemistry*, **48**, 1137–1139.
11. Hine, J., Linden, S.M. and Kanagasabapathy, V.M. (1985) 1,8-Biphenylenediol is a double-hydrogen-bonding catalyst for reaction of an epoxide with a nucleophile. *Journal of the American Chemical Society*, **107**, 1082–1083.
12. Kelly, T.R., Meghani, P. and Ekkundi, V.S. (1990) Diels–Alder reactions: rate acceleration promoted by a biphenylenediol. *Tetrahedron Letters*, **31**, 3381–3384.
13. Etter, M.C., Urbanczyk-Lipkowska, Z., Zia-Ebrahimi, M. and Panunto, T.W. (1990) Hydrogen bond directed cocrystallization and molecular recognition properties of diarylureas. *Journal of the American Chemical Society*, **112**, 8415–8426.
14. Bordwell, F.G., Drucker, G.E., Andersen, N.H. and Denniston, A.D. (1986) Acidities of hydrocarbons and sulfur-containing hydrocarbons in dimethyl sulfoxide solutions. *Journal of the American Chemical Society*, **108**, 7310–7313.
15. Bordwell, F.G., Algrim, D.J. and Harrelson, J.A. (1988) The relative ease of removing a proton, a hydrogen atom, or an electron from carboxamides versus thiocarboxamides. *Journal of the American Chemical Society*, **110**, 5903–5904.
16. Wilcox, C.S., Kim, E.I., Romano, D. *et al.* (1995) Experimental and theoretical studies of substituent effects in hydrogen bond based molecular recognition of a zwitterion by substituted arylureas. *Tetrahedron*, **51**, 621–634.
17. Kelly, T.R. and Kim, M.H. (1994) Relative binding affinity of carboxylate and its isosteres: Nitro, phosphate, phosphonate, sulfonate, and δ-lactone. *Journal of the American Chemical Society*, **116**, 7072–7080.
18. Fan, E., Van Arman, S.A., Kincaid, S. and Hamilton, A.D. (1993) Molecular recognition: hydrogen-bonding receptors that function in highly competitive solvents. *Journal of the American Chemical Society*, **115**, 369–370.
19. Curran, D.P. and Kuo, L.H. (1994) Altering the stereochemistry of allylation reactions of cyclic α-sulfinyl radicals with diarylureas. *The Journal of Organic Chemistry*, **59**, 3259–3261.
20. Curran, D.P. and Lung, H.K. (1995) Acceleration of a dipolar Claisen rearrangement by hydrogen bonding to a soluble diaryl urea. *Tetrahedron Letters*, **36**, 6647–6650.
21. Kirsten, M., Rehbein, J., Hiersemann, M. and Strassner, T. (2007) Organocatalytic Claisen rearrangement: theory and experiment. *The Journal of Organic Chemistry*, **72**, 4001–4011.
22. Wittkopp, A. and Schreiner, P.R. (2003) Metal-free, noncovalent catalysis of Diels–Alder reactions by neutral hydrogen bond donors in organic solvents and in water. *Chemistry – A European Journal*, **9**, 407–414.

23. Schreiner, P.R. and Wittkopp, A. (2002) H-bonding additives act like Lewis acid catalysts. *Organic Letters*, **4**, 217–220.
24. Scheffer, U., Strick, A., Ludwig, V. *et al.* (2005) Metal-free catalysts for the hydrolysis of RNA derived from guanidines, 2-aminopyridines, and 2-aminobenzimidazoles. *Journal of the American Chemical Society*, **127**, 2211–2217.
25. Sigman, M.S. and Jacobsen, E.N. (1998) Schiff base catalysts for the asymmetric Strecker reaction identified and optimized from parallel synthetic libraries. *Journal of the American Chemical Society*, **120**, 4901–4902.
26. Sigman, M.S., Vachal, P. and Jacobsen, E.N. (2000) A general catalyst for the asymmetric Strecker reaction. *Angewandte Chemie – International Edition*, **39**, 1279–1281.
27. Vachal, P. and Jacobsen, E.N. (2002) Structure-based analysis and optimization of a highly enantioselective catalyst for the Strecker reaction. *Journal of the American Chemical Society*, **124**, 10012–10014.
28. Joly, G.D. and Jacobsen, E.N. (2004) Thiourea-catalyzed enantioselective hydrophosphonylation of imines: practical access to enantiomerically enriched α-amino phosphonic acids. *Journal of the American Chemical Society*, **126**, 4102–4103.
29. Okino, T., Hoashi, Y. and Takemoto, Y. (2003) Enantioselective Michael reaction of malonates to nitroolefins catalyzed by bifunctional organocatalysts. *Journal of the American Chemical Society*, **125**, 12672–12673.
30. Okino, T., Hoashi, Y., Furukawa, T. *et al.* (2005) Enantio- and diastereoselective Michael reaction of 1,3-dicarbonyl compounds to nitroolefins catalyzed by a bifunctional thiourea. *Journal of the American Chemical Society*, **127**, 119–125.
31. Okino, T., Nakamura, S., Furukawa, T. and Takemoto, Y. (2004) Enantioselective aza-Henry reaction catalysed by a bifunctional organocatalyst. *Organic Letters*, **6**, 625–627.
32. Inokuma, T., Hoashi, Y. and Takemoto, Y. (2006) Thiourea-catalyzed asymmetric Michael addition of activated methylene compounds to α,β-unsaturated imides: Dual activation of imide by intra- and intermolecular hydrogen bonding. *Journal of the American Chemical Society*, **128**, 9413–9419.
33. Yoon, T.P. and Jacobsen, E.N. (2005) Highly enantioselective thiourea-catalysed nitro-Mannich reactions. *Angewandte Chemie – International Edition*, **44**, 466–468.
34. Taylor, M.S., Tokunaga, N. and Jacobsen, E.N. (2005) Enantioselective thiourea-catalysed acyl-Mannich reactions of isoquinolines. *Angewandte Chemie – International Edition*, **44**, 6700–6704.
35. Raheem, I.T., Thiara, P.S., Peterson, E.A. and Jacobsen, E.N. (2007) Enantioselective Pictet-Spengler-type cyclizations of hydroxylactams: H-bond donor catalysis by anion binding. *Journal of the American Chemical Society*, **129**, 13404–13405.
36. Taylor, M.S. and Jacobsen, E.N. (2004) Highly enantioselective catalytic acyl-Pictet-Spengler reactions. *Journal of the American Chemical Society*, **126**, 10558–10559.
37. Fuerst, D.E. and Jacobsen, E.N. (2005) Thiourea-catalyzed enantioselective cyanosilylation of ketones. *Journal of the American Chemical Society*, **127**, 8964–8965.
38. Herrera, R.P., Sgarzani, V., Bernardi, L. and Ricci, A. (2005) Catalytic enantioselective Friedel–Crafts alkylation of indoles with nitroalkenes by using a simple thiourea organocatalyst. *Angewandte Chemie – International Edition*, **44**, 6576–6579.
39. Yamaoka, Y., Miyabe, H. and Takemoto, Y. (2007) Catalytic enantioselective Petasis-type reaction of quinolines catalyzed by a newly designed thiourea catalyst. *Journal of the American Chemical Society*, **129**, 6686–6687.
40. Sibi, M.P. and Itoh, K. (2007) Organocatalysis in conjugate amine additions. Synthesis of β-amino acid derivatives. *Journal of the American Chemical Society*, **129**, 8064–8065.
41. Biddle, M.M., Lin, M. and Scheidt, K.A. (2007) Catalytic enantioselective synthesis of flavanones and chromanones. *Journal of the American Chemical Society*, **129**, 3830–3831.
42. Sohtome, Y., Tanatani, A., Hashimoto, Y. and Nagasawa, K. (2004) Development of bis-thiourea-type organocatalyst for asymmetric Baylis–Hillman reaction. *Tetrahedron Letters*, **45**, 5589–5592.
43. Sohtome, Y., Takemura, N., Takada, K.*et al.* (2007) Organocatalytic asymmetric nitroaldol reaction: cooperative effects of guanidine and thiourea functional groups. *Chemistry, an Asian Journal*, **2**, 1150–1160.
44. Tan, K.L. and Jacobsen, E.N. (2007) Indium-mediated asymmetric allylation of acylhydrazones using a chiral urea catalyst. *Angewandte Chemie – International Edition*, **46**, 1315–1317.

10

Amidines and Guanidines in Natural Products and Medicines

Takuya Kumamoto

Graduate School of Pharmaceutical Sciences, Chiba University, 1-33 Yayoi, Inage, Chiba 263-8522, Japan

10.1 Introduction

The excellent reviews of Berlinck [1–5] have surveyed a great number of guanidine-type natural products. In addition, some guanidine-derived marine alkaloids have been reviewed by Kobayashi and Ishibashi [6,7]. Also, a recent book gave accounts of marine alkaloids including the phakellins, palau'amines and oroidin-like dimers derived from bromopyrroles and polyketide-derived polycyclic guanidine alkaloids [8].

Some of the medicines that contain amidines and/or guanidines have been reviewed, too [9].

In this chapter, natural products and medicines bearing guanidine and amidine functions are discussed. The topics covered include the amidine and guanidine natural products for which isolation, total synthesis and/or structural revision have recently been reported, as well as medicines, with the focus on activity, stability and mode-of-action.

10.2 Natural Amidine Derivatives

In this section, natural products with the amidine moiety are examined. They have mostly been isolated as fermentation products of actinomycete. For example, isolation of bottromycins [10–12] and coformycin [13,14] has been reported from *Strepromyces* sp. Other amidines were isolated from fungi, marine invertebrates and plants.

10.2.1 Natural Amidines from Microorganisms and Fungi

10.2.1.1 Birnbaumins

Bartsch *et al.* reported the isolation of birnbaumins A and B (**1–2**), unusual 1-hydroxyindole pigments from fungus, the 'flower pot parasol' *Leucocoprinus birnbaumii* (Corda) Singer [15]. The structure of birnbaumin A (**1**) consisted of 1-hydroxyindoleglyoxyl amide with a side chain containing a very rare *N*-hydroxybisamidine moiety, which was determined by the permethylation with diazomethane and the comparison of spectral data with the calculated value of another structural candidate (Figure 10.1).

10.2.1.2 Efrapeptins and Neoefrapeptins

Efrapeptins are a complex mixture of peptide antibiotics produced by the fungus *Tolypocladium niveum* (syn. *Tolypocladium inflatum*, *Beauveria nivea*), a soil hyphomycete. The peptides are inhibitors of mitochondrial oxidative phosphorylation and ATPase activity and photophosphorylation in chloroplasts. The structure of efrapeptin D was first proposed by Bullough [16] as polypeptides composed of 15 amino acid residues including a large number of α-aminoisobutyric acid (Aib), pipecolic acid (Pip), β-alanine and L-isovaline (Iva) with an uncharacterized C-terminal group (X), which consisted of a basic part with leucine residue. Gupta *et al.* [17] reported the full characterization of efrapeptin C (**3**) and the derivatives D (**4**)–G based on 2D-NMR analysis and mass fragmentation. They showed that the C-terminal group contains the bicyclic amidine ring, the same structure as that of the versatile artificial bicyclic amidine 1,5-diazabicyclo[4.3.0]non-5-ene (DBN) (Figure 10.2 for efrapeptin D; Table 10.1). Efrapeptins C (**3**) and E (**5**) showed insect toxicity against beetle *Leptinotarsa decemlineata* (Coleoptera) (LC_{50} at 18.9 and 8.4 ppm, respectively). Efrapeptins C–G showed mitochondrial ATPase inhibitory activity against fungi (*Metarhizium anisopliae* and *T. niveum*) and insect (flight muscles from *Musca domstica*).

Recently, Fredenhagen *et al.* [18] reported the isolation of neoefrapeptins A (**6**) to N from the culture broth of the strain SID 22 780, identified as *Geotrichum candidum* Link:F. The structures were very similar to those of efrapeptins, which contain a bicyclic amidine part at the C-terminus and Aib, Iva and Pip, as well as other unnatural amino acids such as 3-methylproline (3M-Pro) and 1-aminocyclopropanecarboxylic acid (Acc).

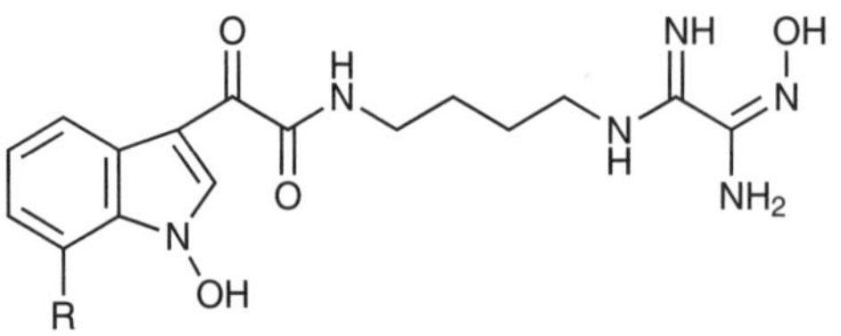

birnbaumin A (**1**): R = H
birnbaumin B (**2**): R = OH

Figure 10.1 *Structures of birnbaumins A-B (**1–2**)*

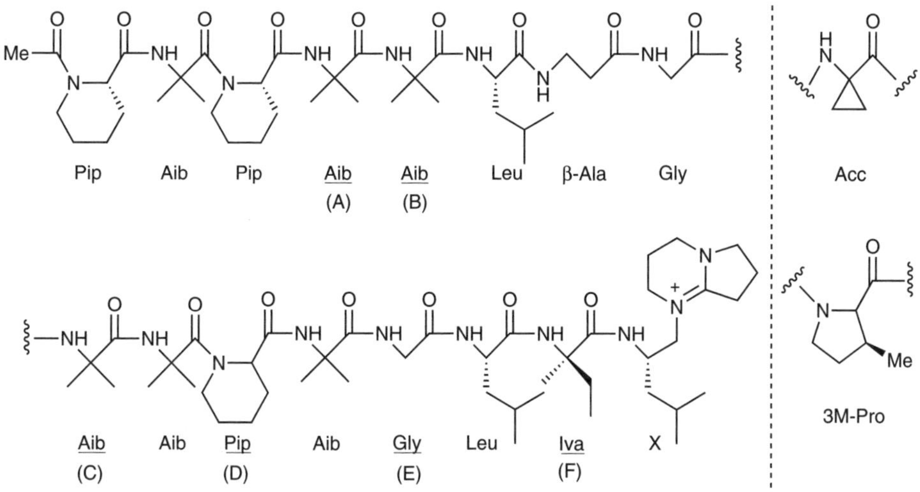

Figure 10.2 Structures of efrapeptin D (**4**) (amino acid with underline shows variable residues)

10.2.1.3 Pyrostatins

Imada *et al.* reported the isolation of pyrostatins A and B from culture broth of strain *Streptomyces* sp. SA-3501 isolated from a marine sediment [19,20]. The structures were determined as 2-iminopyrrolidine carboxylic acid derivatives **7–8**. These compounds showed specific inhibitory activity against *N*-acetylglucosaminidase but no antimicrobial activity. Recently, Castellanos *et al.* [21] reported the isolation of a compound with same structure as pyrostatin B (**8**) from Caribbean marine sponge *Cliona tenuis*; however, the spectral data were not identical with those published for pyrostatin B. Total synthesis of the compound with the structure of **8** was achieved to demonstrate that the synthesized **8** was identical with the compound isolated by Castellanos but not with pyrostatin B previously isolated by Imada *et al.* Further searching of the literature lead to the conclusion that the actual NMR data reported for pyrostatin B matched those of ectoine (**9**) [22,23], another amidine alkaloid isolated from extremely halophilic species of the bacterial genus *Ectothiorhodospira* [24].

Isolations of noformycin (**10**) [25] with antiviral activity [26] and an inhibitory effect towards inducible-nitric oxide synthase [27] and other iminopyrroglutamic acid derivatives **11–13** have also been reported. The latter were inhibitors towards *Erwinia amylovora*, a bacterium responsible for the fire light disease of apple and pear trees [28] (Figure 10.3).

Table 10.1 Amino acid residues in efrapeptins and neoefrapeptins

efrapeptins	(A)	(B)	(C)	(D)	(E)	(F)	neoefra-peptins	(A)	(B)	(C)	(D)	(E)	(F)
C (**3**)	Aib	Aib	Aib	Pip	Gly	Aib	A (**6**)	Iva	Aib	Acc	Pip	Gly	Iva
D (**4**)	Aib	Aib	Aib	Pip	Gly	Iva	B	Iva	Iva	Acc	Pip	Gly	Iva
E (**5**)	Iva	Aib	Aib	Pip	Gly	Iva	C	Iva	Aib	Acc	Pip	Gly	Iva
F	Aib	Aib	Aib	Pip	Ala	Iva	F	Iva	Aib	Acc	3M-Pro	Gly	Iva
G	Iva	Aib	Alb	Pip	Ala	Iva	I	Iva	Iva	Acc	3M-Pro	Gly	Iva

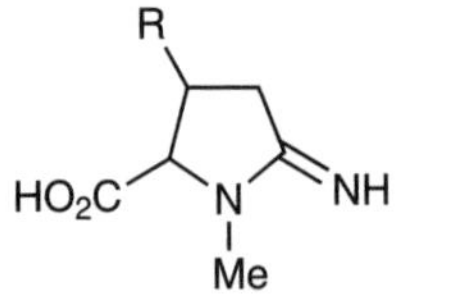

298 *Amidines and Guanidines in Natural Products and Medicines*

pyrostatin A (**7**): R = OH
pyrostatin B (**8**): R = H

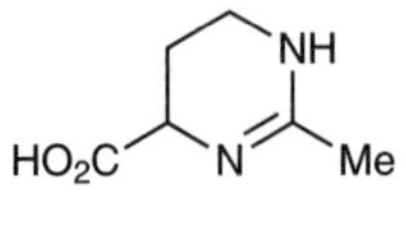

ectoine (**9**)

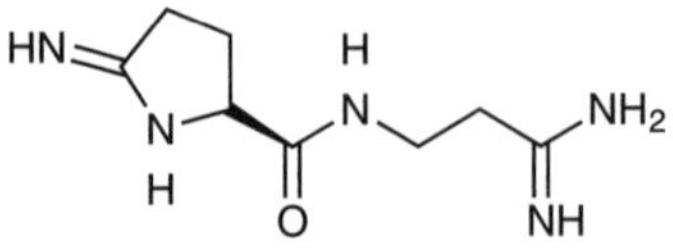

noformycin (**10**)

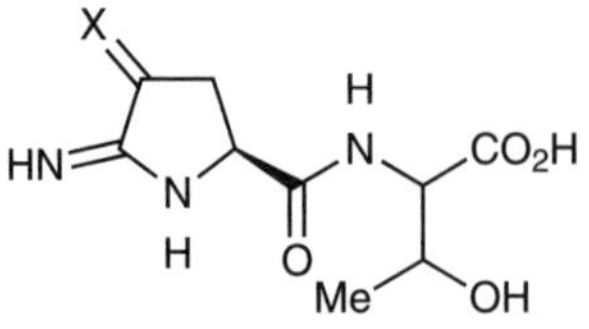

11: X = CH₂; **12**: X = 2H

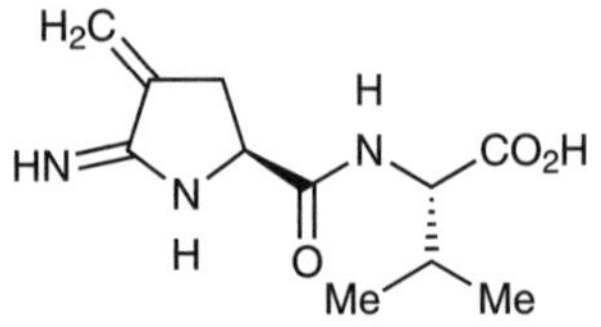

13

Figure 10.3 *Structures of pyrostatins A and B (7–8), ectoine (9) and other iminopyrrolidine natural products*

10.2.2 Natural Amidines from Marine Invertebrates

10.2.2.1 Flustramine C

Flustramine C (**14**) was isolated from bryozoan *Flustra foliacea* (L.) [29,30]. This compound possesses the amidine moiety in brominated pyrroloindole structure with 1,1-dimethylallyl group (Figure 10.4). It has been suggested that the *Flustra* alkaloids are important for the bryozoan by controlling bacterial growth on its surface [31]. This compound is levorotatory [32], however, the absolute configuration has not been determined yet. Total syntheses of racemic one were achieved by three groups [33–35].

10.2.2.2 Perophoramidine

Perophoramidine (**15**), halogenated alkaloid, was isolated from the Philippine ascidian *Perophora namei* Hartmeyer and Michaelson (Perophoridae) [36a]. This compound contains fused hexahydropyrrolopyridine, indole and dihydroquinoline rings. Amidine parts exist at the fusing part of the indole and quinoline rings. The stereochemistry was determined by

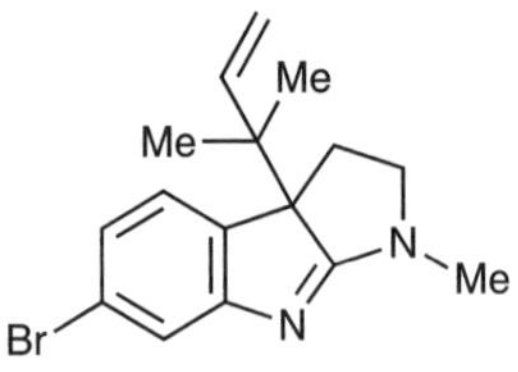

flustramine C (**14**)

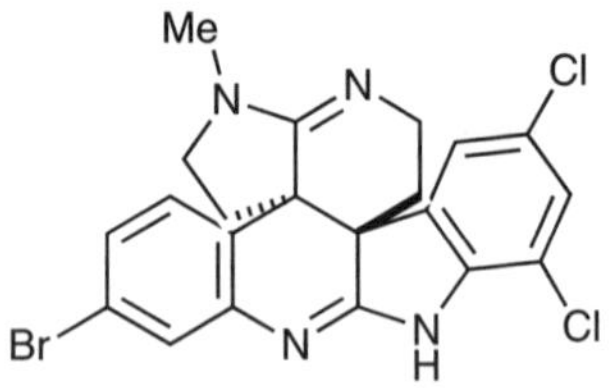

perophoramidine (**15**)

Figure 10.4 *Structures of flustramine C (14) and perophoramidine (15)*

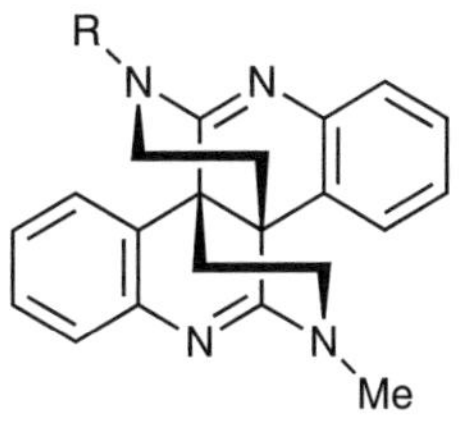

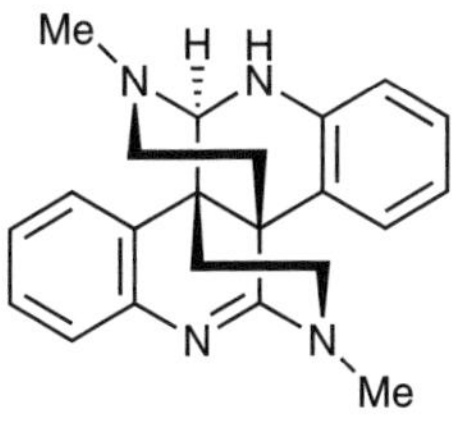

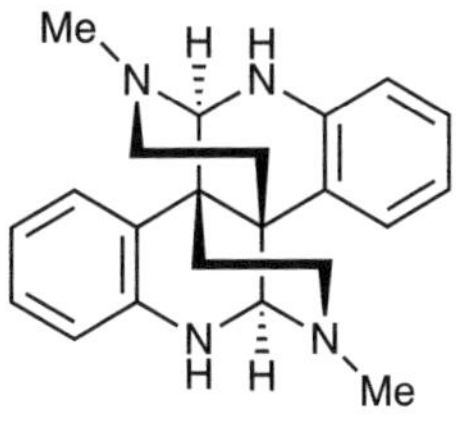

glomerulatine A (16): R = Me
glomerulatine B (17): R = H

glomerulatine C (18)

(-)-calycanthine

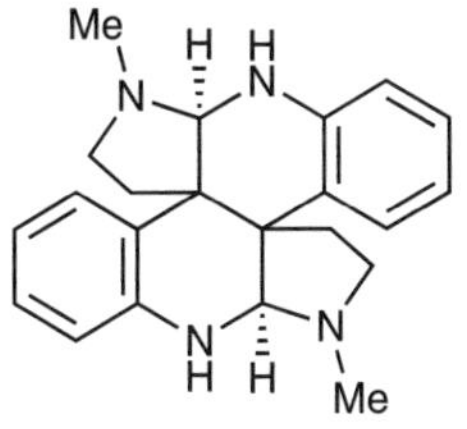

isocalycanthine

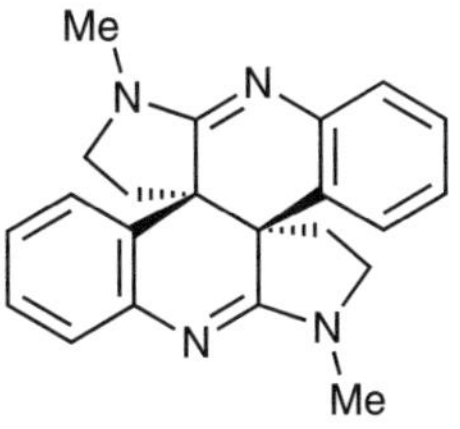

19

Figure 10.5 Structures of glomerulatines A–C (**16–18**) and related compounds

comparison of ^{1}H-NMR data with the computer analysis data (Figure 10.4). Total synthesis of racemic perophoramide has been reported [36b].

10.2.3 Natural Amidines from Higher Plants

10.2.3.1 Glomerulatines

Glomerulatines A–C (**16–18**) were isolated from the aerial part of a shrub, *Psychotria glomerulata* (Don. Smith) Steyermark, previously known as *Cephaelis glomerulata* J.D. Sm [37]. The dimeric structures were determined by spectroscopic means. The absolute configuration of glomerulatine A (**16**) was deduced from those of (−)-calycanthine, which possesses similar optical rotation ([α]$_D$ −466 for **16**, −489 for (−)-calycanthine). On the other hand, isocalycanthine type amidine compound, (8–8a),(8′–8′a)-tetrahydroisocalycanthine (**19**) was isolated from *Psychotria colorata* (Willd. ex R. and S.) Muell. Arg., which belong to the same genus as *P. glomerulata* [38] (Figure 10.5).

10.3 Natural Guanidine Derivatives

In this section, natural products with guanidines are studied. A large group of these products consist of cyclic depsipeptides and polypeptides with arginine as an amino acid residue, produced by mainly actinomycete and cyanobacteria in some cases (ex. microcystin-LR, noduralin) [39]. Isolation of aminoglycosides (streptomycins, streptothricins [40] and their derivatives [41,42]), guanidines connected to cyclic peptides and macrolides through long alkyl chains (fusaricidins [43,44] and azalomycins [45], respectively), and cyclic (capreomycins [46] and tuberactinomycins [47]) and acyclic guanidine derivatives (miraziridine A [48]) were also reported. Another series of natural products was found in higher plants, marine sources such as dinoflagellates (saxitoxins) [7,49], pufferfish (tetrodotoxin) [6,7], sea firefly (*Vargula* luciferin) [50], sponges and so on.

10.3.1 Natural Guanidines from Microorganisms

10.3.1.1 Argifin

Argifin (**20**) was isolated from the cultured broth of a fungal strain *Gliocladium* sp. FTD-0668 by Ōmura *et al.* [51,52] and the structure determined to be a cyclic peptide with arginine residue [53] (Figure 10.6). This compounds has inhibitory activity towards chitinase. Total synthesis [54] and computational analysis of a chitinase-argifin complex [55] have also been reported.

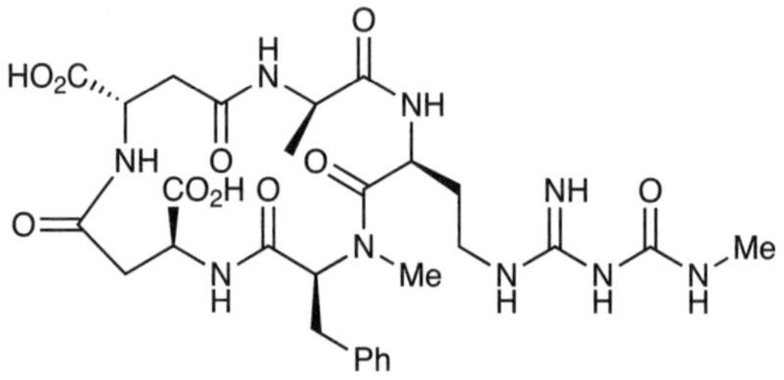

argifin (**20**)

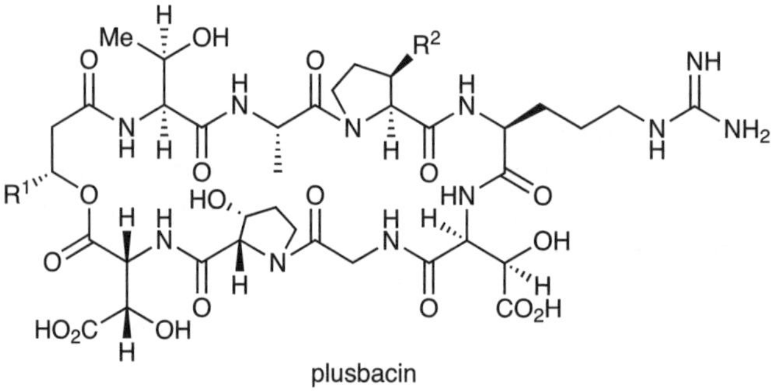

plusbacin

A_1 (**21**): R^1 = $(CH_2)_{10}Me$, R^2 = OH; B_1: R^1 = $(CH_2)_{10}Me$, R^2 = H

A_2: R^1 = $(CH_2)_9CHMe_2$, R^2 = OH; B_2: R^1 = $(CH_2)_9CHMe_2$, R^2 = H

A_3 (**22**): R^1 = $(CH_2)_{10}CHMe_2$, R^2 = OH; B_3: R^1 = $(CH_2)_{10}CHMe_2$, R^2 = H

A_4: R^1 = $(CH_2)_{12}Me$, R^2 = OH; B_4: R^1 = $(CH_2)_{12}Me$, R^2 = H

Figure 10.6 Structures of argifin (**20**) and plusbacins

10.3.1.2 Plusbacins

Plusbacin A$_1$ (**21**) and the derivatives A$_2$–A$_4$ and B$_1$–B$_4$ are lipodepsipeptides isolated from a strain numbered PB-6250 related to the genus *Pseudomonas* obtained from a soil sample collected in the Okinawa Pref., Japan [56]. These compounds contain arginine residue and lactone linkage with characteristic 3-hydroxy fatty acids [57] (Figure 10.6). Plusbacin A$_3$ (**22**) showed inhibitory activity against methicillin resistant *Staphylococcus aureus* [56,58]. Recent total synthesis of this compound was reported and the absolute configuration of the lactone residue was determined as *R* [59].

10.3.1.3 Guadinomines

Guadinomines A and B (**23–24**) were recently isolated from the culture broth of *Streptomyces* sp. K01–0509 FERM BP-08504 strain by Ōmura *et al.* [60]. The structure contained alanine and valine residues, 1,2-diamine and a cyclic guanidine. The absolute configuration of the latter guanidine part was deduced from those of K01-0509B, isolated from the same strain (Figure 10.7) [61]. These compounds possess activity towards a pathogenic Gram negative bacterium in a type III secretion mechanism (IC$_{50}$ **23** = 0.01 μg/mL, **24** = 0.007 μg/ml).

10.3.2 Natural Guanidines from Marine Invertebrates

10.3.2.1 Palau'amine

Palau'amine was isolated from a sponge, *Stylotella agminata*, collected in the Western Caroline Islands. The structure was originally determined as **25** with hexacyclic bisguanidine, in which bicyclo[3.3.0]azaoctane system (D and E) is *cis*-fused (at C11 and C12) [62,63]. Recently the structure of palau'amine was revised from **25** to *trans*-fused **26** based on synthetic studies of the compound with the structure in **25** [64], the computational analysis of coupling constants of tetrabromostyloguanidine (**27**) [65] and further NMR experiments of palau'amine-class metabolites [66,67]; the absolute configuration was revised from (12*S*,17*R*) to (12*R*,17*S*) [68] (Figure 10.8). Palau'amine is less nontoxic (LD$_{50}$ 13 mg/kg; i.p. mice); it is quite active against P-388 and A549 (IC$_{50}$ 0.1 and 0.2 mg/mL, respectively), less so against other cancer cell lines (HT-29 and KB), and possesses antibiotic activity (against *Staphylococcus aureus* and *Bacillus subtillis*) and antifungal activity (against *Penicillium notatum*).

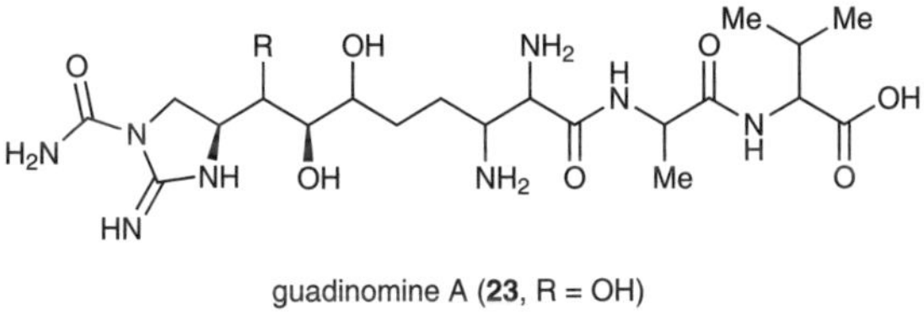

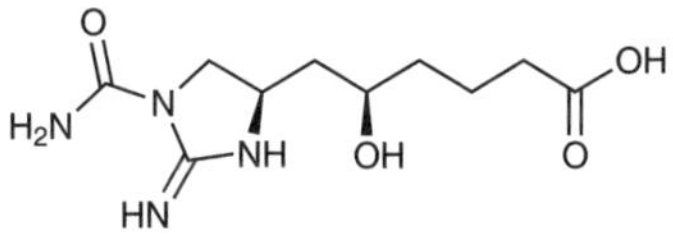

Figure 10.7 Structures of guadinomine A and B (**23–24**) and K01-0509B

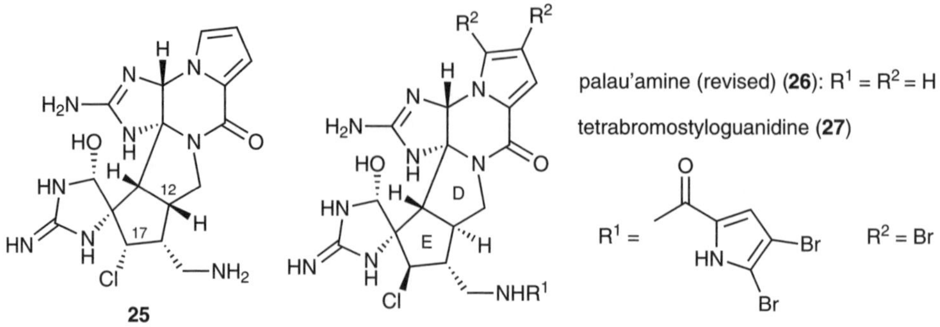

Figure 10.8 Structures of palau'amine (**26**) and tetrabromostyloguanidine (**27**)

10.3.2.2 Crambescins, Ptilomycalins and Batzelladines

Berlinck et al. reported the isolation of polycyclic guanidine alkaloids from the Mediterranean marine sponge *Crambe crambe*. The series of these compounds was at first named as crambins [69], later on renamed as crambescins [70]. Crambescins A (**28**), B (**29**) [69] and C1 (**30**) [70,71] possess a common framework containing the six-membered ring cyclic guanidine, a long alkyl chain and an ester function with terminal guanidine group. Only crambescin B possess a spiro bicyclic ring. The stereochemistry of crambescin B (**29**) was revised by synthetic studies of model compounds [72]. Ptilocaulin (**31**) is a polycyclic guanidine derivative isolated from the Carribean *Batzella* sp, wrongly identified as *Ptilocaulis spiculifer*, and a red sponge *Hemimycale* sp. from the Red Sea [73]. From other marine sponges, ptilomycalins (A, **32**) [74,75] and crambescidin 800 (**33**) [75] have also been isolated as a series of guanidine alkaloids (Figure 10.9).

Batzelladine A (**34**) and the derivatives B–E (**35–38**) are the first natural products of small molecule weight that have been shown to inhibit the gp120-CD4 interaction [76]. On the other hand, batzelladines F–I (**39–42**) induced a dissociation of a p56ck-CD4 binding complex [77]. The structures of batzelladines A (**34**), D (**37**) [78] and F (**39**) [79] have been revised (revised structures are shown in Figure 10.10). Recently, ptilomycalin D (**43**) [80], batzelladines J (**44**) [81] and K–N (**45–48**) [82] have been isolated from sponges *Monanchona dianchora* and *M. unguifera*, respectively (Figure 10.10). Total syntheses of batzelladines were recently reported [79,83,84].

10.3.3 Natural Guanidines from Higher Plant

10.3.3.1 Martinelline and Martinellic Acid

Two pyrroloquinoline alkaloids, martinelline (**49**) and martinellic acid (**50**), have been isolated from an organic extract of root of *Martinella iquitosensis* A. Sampaio (Bignoniaceae) by Witherup et al. (part of Merck's research group) as bradykinin receptor antagonists (Figure 10.11) [85]. The optical rotations of natural **49** and **50** were reported as $[\alpha]_D$ +9.4 and −8.5, respectively, however the synthetic (−)-**50** showed a considerably larger value with the same sense ($[\alpha]_D$ −122.7; [86,87] −164.3 [88]). A compound with the

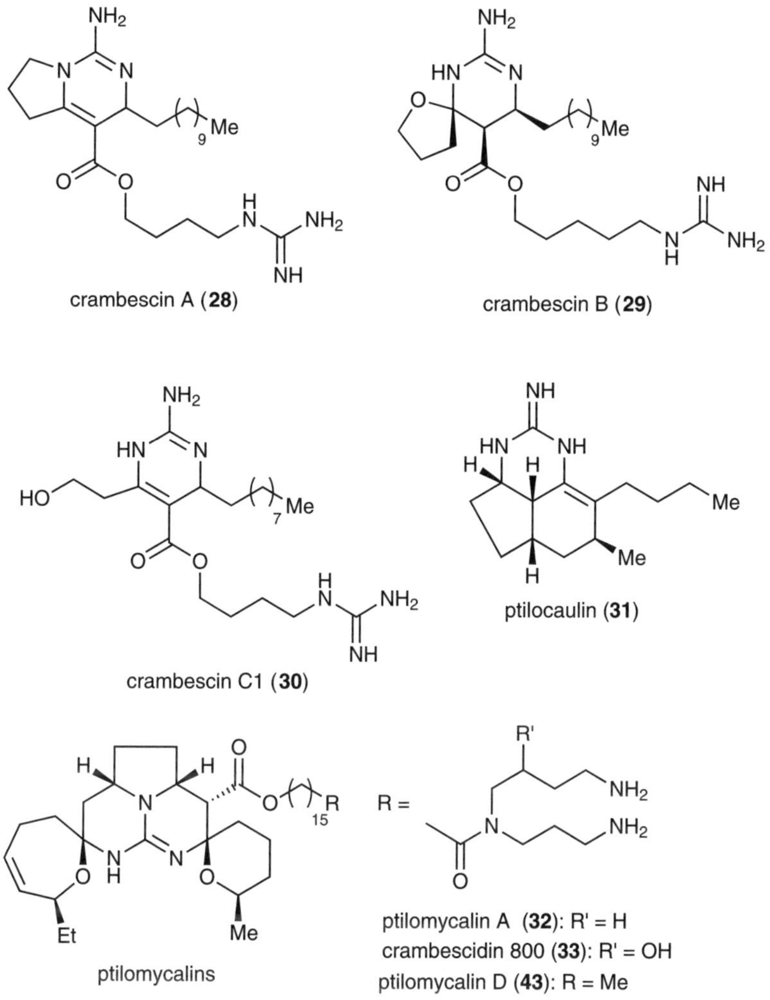

crambescin A (28)

crambescin B (29)

crambescin C1 (30)

ptilocaulin (31)

ptilomycalins

ptilomycalin A (32): R' = H
crambescidin 800 (33): R' = OH
ptilomycalin D (43): R = Me

Figure 10.9 Structures of crambescins and ptilomycalins

structure in **49** was also synthesized, however a larger optical rotation with the opposite sense was observed ($[\alpha]_D^{28}$ −108.0). It was supposed that either the natural martinellic acid (**50**) may be partially racemic or too dilute a solution was used when its specific rotation was measured by the chemists in Merck's group [88].

10.4 Medicinal Amidine and Guanidine Derivatives

In this section, medicinal amidine and guanidine derivatives are detailed. Amidines and guanidines also play important roles in medicinal chemistry in terms of the control of the basicity, high coordination ability and nitric oxide source.

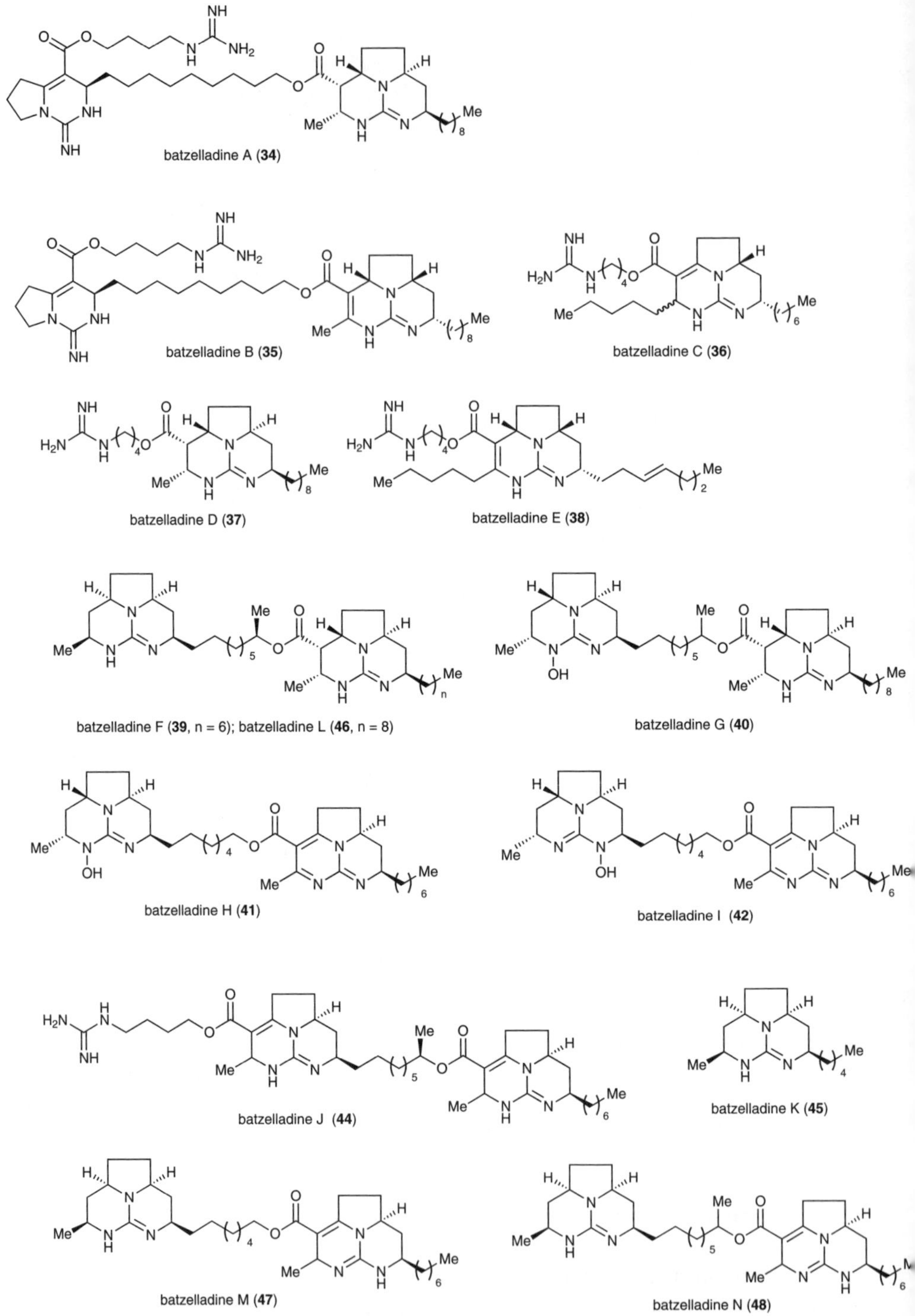

Figure 10.10 Structures of batzelladines

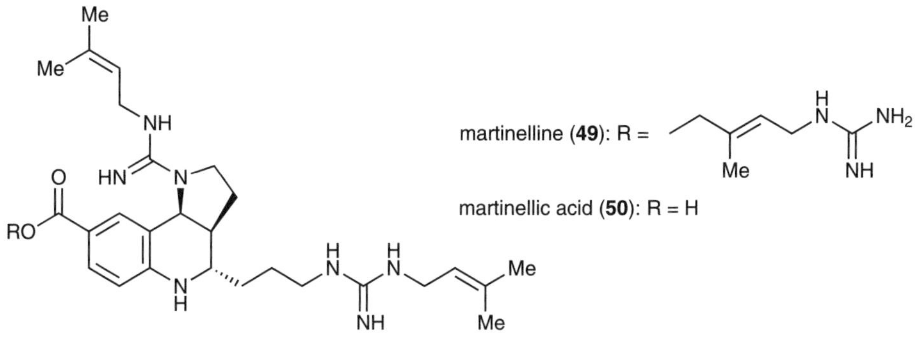

Figure 10.11 Structures of martinelline (49) and martinellic acid (50)

10.4.1 Biguanides

Lowering of blood glucose by the infusion of guanidine [89], biguanides and two linked guanidine moieties has proved to be useful for the treatment of diabetes mellitus. Three compounds became available for diabetes therapy, phenformin (**51**), buformin (**52**) and metformin (**53**) (Figure 10.12). Phenformin (**51**) was withdrawn due to lactic acidosis [90]. Metformin (**53**), a less lipophilic biguanide, was recently approved for use in the USA after 20 years of use in Europe [91].

10.4.2 Cimetidine

Black *et al*. [92] reported the classification and specific blockage of the receptors involved in mepyramine-insensitive, non-H_1 (H_2) histamine responses and the discovery of the selective antagonist brimamide (**54**), which inhibited histamine-induced gastric acid secretion and suppressed some other histamine effects not eliminated by H_1 histamine receptor blockers. Modification of brimamide (**54**) led to the orally active antagonist methiamide (**55**), which proved sufficiently active to allow the exploration of the therapeutic potential of this new type of drug. Side effects of kidney damage and agranulocytosis with methiamide (**55**) might be attributed to the presence of the thiourea group in the drug molecule. Owing to the tendency of guanidinohistamine (**56**) to show weak activity as an H_2-receptor antagonist, derivatives with the guanidine moiety were synthesized and, finally, cimetidine (**57**), with a cyanoguanidine group with protons in similar acidity as those of thiourea derivatives such as **54** and **55**, was found as an effective histamine H_2-receptor antagonist [93]. Famotidine (**58**) is known as another H_2-blocker containing amidine and guanidine parts in the molecule (Figure 10.13).

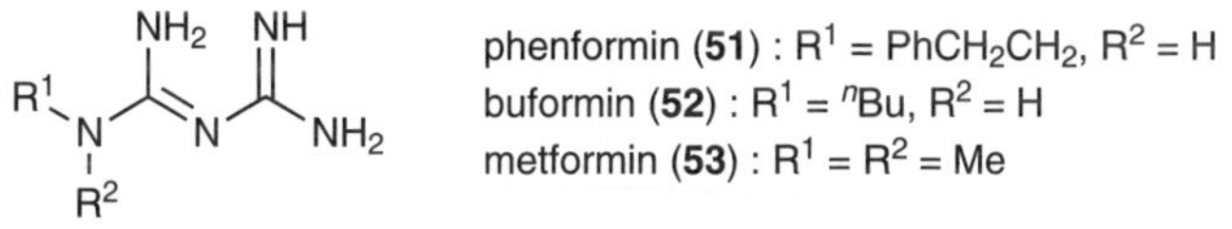

phenformin (**51**) : R^1 = PhCH$_2$CH$_2$, R^2 = H
buformin (**52**) : R^1 = ^{n}Bu, R^2 = H
metformin (**53**) : R^1 = R^2 = Me

Figure 10.12 Structures of biguanides

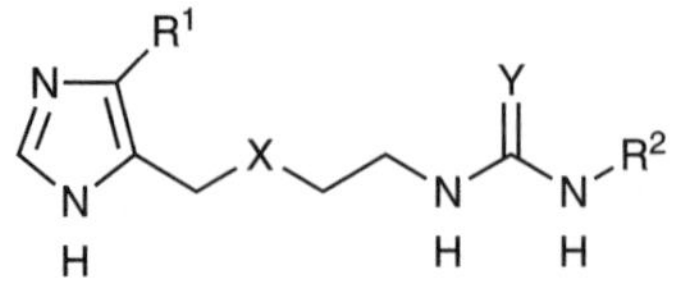

brimamide (**54**): R^1 = H, R^2 = Me, X = CH$_2$, Y = S
methiamide (**55**): R^1 = Me, R^2 = H, X = Y = S
cimetidine (**57**): R^1 = R^2 = Me, X = S, Y = NCN

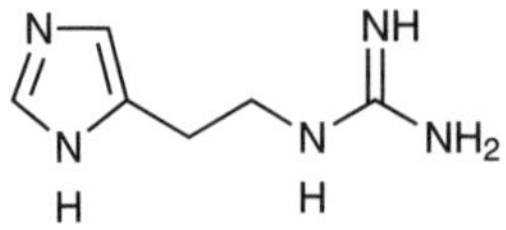

guanidinohistamine (**56**)

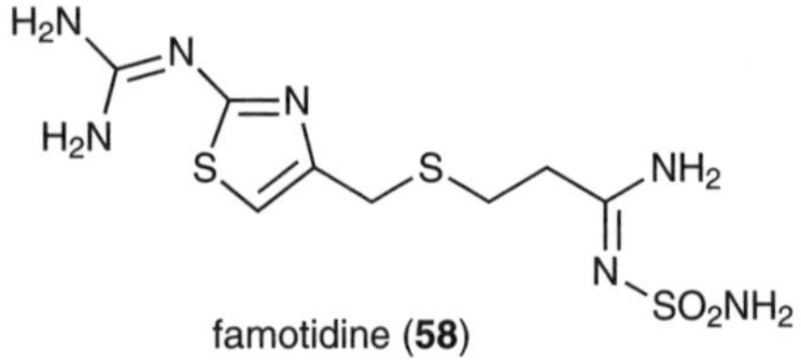

famotidine (**58**)

Figure 10.13 Structures of H$_2$-blockers

10.4.3 Imipenem

Imipenem (**59**) is one of the carbapenem, which possess a broad spectrum of antimicrobial activity against Gram positive as well as Gram negative bacteria, such as *Pseudomonas aeruginosa* [94]. This compound was transformed from thienamycin (**60**) (Figure 10.14) produced by *Streptomyces cattleya*. This antibiotic was known to be rather unstable, for example, less stable than benzylpenicilline at pH 7. Low stability of **60** was also found when the concentration was high, which was deduced from the intermolecular aminolysis of the azeitidinone by the cysteamine side chain in **60**, however, basic functionality at the terminal of carbon chain was found to be necessary because the corresponding *N*-acetyl derivative has lost the activity. Thus, the conversion of the amino group to a more basic one would result in a compound with increased stability by protonation of the basic group in physiological conditions. As expected, **59** with a formylimidoyl group was 5–30 times more stable compared with **60** but kept the same or similar antimicrobial activity [95].

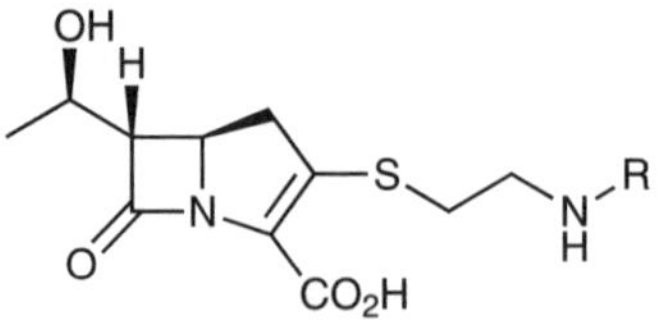

imipenem (**59**): R = CH=NH
thienamycin (**60**): R = H

Figure 10.14 Structures of imipenem (**59**) and thienamycin (**60**) arginine (**61**) *N*-OH-arginine (**62**) citrulline (**63**)

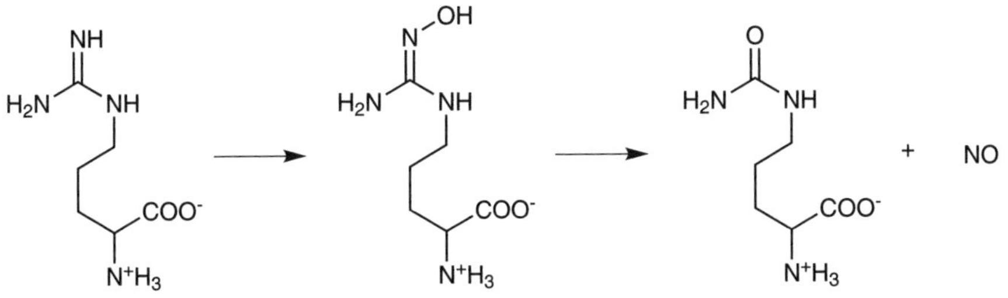

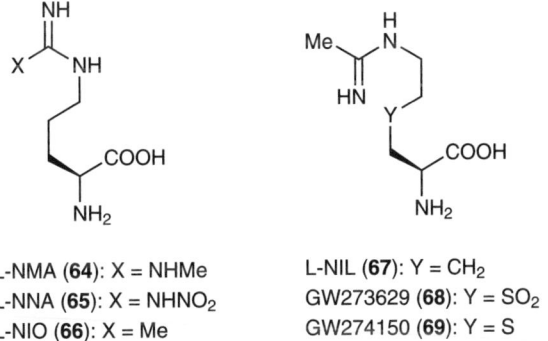

Scheme 10.1 *Generation of nitric oxide from* L*-arginine* (***61***)

10.4.4 NOS Inhibitors

Nitric oxide (NO) is a molecular messenger synthesized by nitric oxide synthase (NOS) enzymes. NOS carries out the oxidation of the guanidine moiety of L-arginine (**61**) to deliver citrulline (**62**) and NO via *N*-OH-arginine (**63**) (Scheme 10.1).

Analogues of L-arginine (**61**), such as N^G-methyl-L-arginine (L-NMA, **64**) [96,97], N^G-nitro-L-arginine (L-NNA, **65**) [98], *N*-iminoethyl-L-ornithine (L-NIO, **66**) [99] and L-N^6-(1-iminoethyl)-L-lysine (L-NIL, **67**) [100] were known as NOS inhibitors. Based on the structure of L-NIL, sulfur containing derivatives such as GW273629 (**68**) and GW274150 (**69**) were designed and found the selectivity towards iNOS, one of the isoforms of NOS [101] (Figure 10.15).

10.4.5 Pentamidine

A series of *p*-alkoxy amidinobenzene derivatives, such as phenamidine (**70**) and pentamidine (**71**), were used in veterinary medicine as an antiprotozoal towards Babesia. Pentamidine (**71**) has been used for the treatment of human protozoan infections [102], and

L-NMA (**64**): X = NHMe
L-NNA (**65**): X = $NHNO_2$
L-NIO (**66**): X = Me

L-NIL (**67**): Y = CH_2
GW273629 (**68**): Y = SO_2
GW274150 (**69**): Y = S

Figure 10.15 *Structures of nitric oxide synthase inhibitors* (***64–69***)

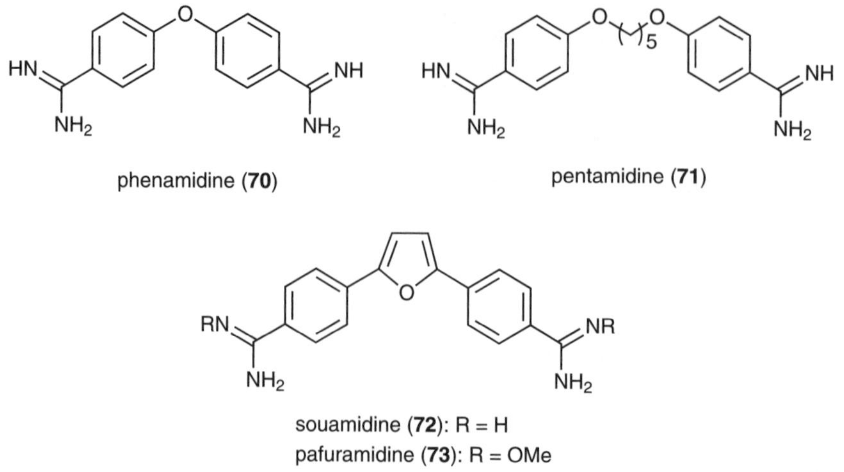

phenamidine (70)

pentamidine (71)

souamidine (72): R = H
pafuramidine (73): R = OMe

Figure 10.16 *Structures of phenamidine (70) and related compounds*

is currently still the clinical drug of choice against African trypanosomiasis, antimony resistant leishmaniais and *Pneumocystis carinii* pneumonia (PCP) [103]. This compound was supposed to show inhibitory activity with insertion into the double helix of DNA [104] and tRNA [105] and was found to inhibit the RNA function [106]. Recently, souamidine (**72**) and the corresponding methoxime pafuramidine (**73**) are currently in clinical trials for treatment of African Sleeping sickness, malaria, PCP and *Pneumocystis jiroveci* pneumonia [107] (Figure 10.16).

References

1. Berlinck, R.G.S. (1995) Some Aspects of Guanidine Secondary Metabolites. *Fortschritte der Chemie Organischer Naturstoffe*, **66**, 119–295.
2. Berlinck, R.G.S. (1996) Natural guanidine derivatives. *Natural Product Reports*, **13**, 377–409.
3. Berlinck, R.G.S. (1999) Natural guanidine derivatives. *Natural Product Reports*, **16**, 339–365.
4. Berlinck, R.G.S. (2002) Natural guanidine derivatives. *Natural Product Reports*, **19**, 617–649.
5. Berlinck, R.G.S. and Kossuga, M.H. (2005) Natural guanidine derivatives. *Natural Product Reports*, **22**, 516–550.
6. Kobayashi, J. and Ishibashi, M. (1992) Marine Alkaloids II, in *The Alkaloids* (eds A. Brossi and G.A. Cordell), **41**, 41–124.
7. Kobayashi, J. and Ishibashi, M. (1999) Marine natural products and marine chemical ecology, in *Comprehensive Natural Products Chemistry* (eds D. Barton and K. Nakanishi), **8**, 415–649.
8. Fattorusso E. and Taglialatela-Scafati O. (eds) (2007) *Modern Alkaloids*, Wiley-VCH Verlag GmbH, Weinheim.
9. Greenhill, J.V. and Lue, P. (1993) Amidines and guanidines in medicinal chemistry. *Progress in Medicinal Chemistry*, **30**, 203–326.
10. Nakamura, S. and Umezawa, H. (1966) The structure of bottromycin A2, a new component of bottromycins. *Chemical & Pharmaceutical Bulletin*, **14**, 981–986.
11. Schipper, D. (1983) The revised structure of bottromycin A2. *The Journal of Antibiotics*, **36**, 1076–1077.

12. Kaneda, M. (2002) Studies on bottromycins. II. Structure elucidation of bottromycins B2 and C2. *The Journal of Antibiotics*, **55**, 924–928.
13. Nakamura, H., Koyama, G., Iitaka, Y. *et al.* (1974) Structure of coformycin, an unusual nucleoside of microbial origin. *Journal of the American Chemical Society*, **96**, 4327–4328.
14. Ohno, M., Yagisawa, N., Shibahara, S. *et al.* (1974) Synthesis of coformycin. *Journal of the American Chemical Society*, **96**, 4326–4327.
15. Bartsch, A., Bross, M., Spiteller, P. *et al.* (2005) Birnbaumin A and B: two unusual 1-hydroxyindole pigments from the 'flower pot parasol' *Leucocoprinus birnbaumii*. *Angewandte Chemie – International Edition*, **44**, 2597–2599.
16. Bullough, D.A., Jackson, C.G., Henderson, P.J.F. *et al.* (1982) The amino acid sequence of efrapeptin D. *Biochemistry International*, **4**, 543–549.
17. Gupta, S., Krasnoff, S.B., Roberts, D.W. *et al.* (1991) Structures of the efrapeptins: Potent inhibitors of mitochondrial ATPase from the fungus *Tolypocladium niveum*. *Journal of the American Chemical Society*, **113**, 707–709.
18. Fredenhagen, A., Molleyres, L., Böhlendorf, B. and Laue, G. (2006) Structure determination of neoefrapeptins A to N: Peptides with insecticidal activity produced by the fungus *Geotrichum candidum*. *The Journal of Antibiotics*, **59**, 267–280.
19. Aoyama, T., Kojima, F., Imada, C. *et al.* (1995) Pyrostatins A and B, new inhibitors of N-acetyl-β-D-glucosaminidase, produced by *Streptomyces* sp. SA-3501. *Journal of Enzyme Inhibition and Medicinal Chemistry*, **8**, 223–232.
20. Imada, C. (2005) Enzyme inhibitors and other bioactive compounds from marine actinomycetes. *Antonie van Leeuwenhoek*, **87**, 59–63.
21. Castellanos, L., Duque, C., Zea, S. *et al.* (2006) Isolation and synthesis of (−)-(5S)-2-imino-1-methylpyrrolidine-5-carboxylic acid from *Cliona tenuis*: Structure revision of pyrostatins. *Organic Letters*, **8**, 4967–4970.
22. Inbar, L. and Lapidot, A. (1988) The structure and biosynthesis of new tetrahydropyrimidine derivatives in actinomycin D producer *Streptomyces parvulus*. Use of ^{13}C- and ^{15}N- labeled L-glutamate and ^{13}C and ^{15}N NMR spectroscopy. *The Journal of Biological Chemistry*, **263**, 16014–16022.
23. Inbar, L., Frolow, F. and Lapidot, A. (1993) The conformation of new tetrahydropyrimidine derivatives in solution and in the crystal. *European Journal of Biochemistry*, **214**, 897–906.
24. Galinski, E.A., Pfeiffer, H.P. and Truper, H.G. (1985) 1,4,5,6-Tetrahydro-2-methyl-4-pyrimidinecarboxylic acid. A novel cyclic amino acid from halophilic phototrophic bacteria of the genus *Ectothiorhodospira*. *European Journal of Biochemistry*, **149**, 135–139.
25. Peck, R.L., Shafer, H.M., Wolf, F.J. and Peck, H.M. (1957) β-(5-Imino-2-pyrrolidinylcarbonylamino)propamidine, U.S. Pat. 2804463.
26. Furusawa, E., Cutting, W., Buckley, P. and Furusawa, S. (1964) Complete cure of DNA virus infection in cultured cells by drug combinations. *Proceedings of the Society for Experimental Biology and Medicine*, **116**, 938–944.
27. Green, B.G., Chabin, R. and Grant, S.K. (1996) The natural product noformycin is an inhibitor of inducible-nitric oxide synthase. *Biochemical and Biophysical Research Communications*, **225**, 621–626.
28. Mitchell, R.E. and Teh, K.L. (2005) Antibacterial iminopyrrolidines from *Burkholderia plantarii*, a bacterial pathogen of rice. *Organic and Biomolecular Chemistry*, **3**, 3540–3543.
29. Carle, J.S. and Christophersen, C. (1979) Bromo-substituted physostigmine alkaloids from a marine bryozoa *Flustra foliacea*. *Journal of the American Chemical Society*, **101**, 4012–4013.
30. Carle, J.S. and Christophersen, C. (1981) Marine alkaloids. 3. Bromo-substituted alkaloids from the marine bryozoan *Flustra foliacea*, flustramine C and flustraminol A and B. *The Journal of Organic Chemistry*, **46**, 3440–3443.
31. Peters, L., Konig, G.M., Wright, A.D. *et al.* (2003) Secondary metabolites of *Flustra foliacea* and their influence on bacteria. *Applied and Environmental Microbiology*, **69**, 3469–3475.
32. Peters, L., Konig, G.M., Terlau, H. and Wright, A.D. (2002) Four new bromotryptamine derivatives from the marine bryozoan *Flustra foliacea*. *Journal of Natural Products*, **65**, 1633–1637.

33. Kawasaki, T., Terashima, R., Sakaguchi, K. et al. (1996) A short route to 'reverse-prenylated' pyrrolo[2,3-b]indoles via tandem olefination and Claisen rearrangement of 2-(3,3-dimethylallyloxy)indol-3-ones: First total synthesis of flustramine C. Tetrahedron Letters, 37, 7525–7528.
34. Fuchs, J.R. and Funk, R.L. (2005) Indol-2-one intermediates: mechanistic evidence and synthetic utility. Total syntheses of (±)-flustramines A and C. Organic Letters, 7, 677–680.
35. Lindel, T., Brauchle, L., Golz, G. and Bohrer, P. (2007) Total synthesis of flustramine C via dimethylallyl rearrangement. Organic Letters, 9, 283–286.
36. Verbitski, S.M., Mayne, C.L., Davis, R.A. et al. (2002) Isolation, structure determination, and biological activity of a novel alkaloid, perophoramidine, from the Philippine ascidian *Perphore namei*. The Journal of Organic Chemistry, 67, 7124–7126; Fuchs, J.R. and Funk, R.L. (2004) Total synthesis of (±)-perophoramidine. Journal of the American Chemical Society, 126, 5068–5069.
37. Solís, P.N., Ravelo, A.G., Palenzuela, J.A. et al. (1997) Quinoline alkaloids from *Psychotria glomerulata*. Phytochemistry, 44, 963–969.
38. Verotta, L., Pilati, T., Tato, M. et al. (1998) Pyrrolidinoindoline alkaloids from *Psychotria colorata*. Journal of Natural Products, 61, 392–396.
39. Harada, K. (2004) Production of secondary metabolites by freshwater cyanobacteria. Chemical & Pharmaceutical Bulletin, 52, 889–899.
40. Waksman, S.A. (1951) Streptomycin, isolation, properties, and utilization. Journal of The History of Medicine and Allied Sciences, 6, 318–347.
41. Hisamoto, M., Inaoka, Y., Sakaida, Y. et al. (1998) A-53930A and B, novel N-type Ca^{2+} channel blockers. The Journal of Antibiotics, 51, 607–617.
42. Kim, B.T., Lee, J.Y., Lee, Y.Y. et al. (1994) *N*-Methylstreptothricin D- a new streptothricin-group antibiotic from a *Streptomyces* spp. The Journal of Antibiotics, 47, 1333–1336.
43. Kajimura, Y. and Kaneda, M. (1996) Fusaricidin A, a new depsipeptide antibiotic produced by *Bacillus polymyxa* KT-8: Taxonomy, fermentation, isolation, structure elucidation and biological activity. The Journal of Antibiotics, 49, 129–135.
44. Kajimura, Y. and Kaneda, M. (1997) Fusaricidins B, C and D, new depsipeptide antibiotics produced by *Bacillus polymyxa* KT-8: Isolation, structure elucidation and biological activity. The Journal of Antibiotics, 50, 220–228.
45. Chandra, A. and Nair, M.G. (1995) Azalomycin F complex from *Streptomyces hygroscopicus*, MSU/MN-4-75B. The Journal of Antibiotics, 48, 896–898.
46. Black, H.R., Griffith, R.S. and Brickler, J.F. (1963) Preliminary laboratory studies with capreomycin. Antimicrobial Agents and Chemotherapy, 161, 522–529.
47. Nagata, A., Ando, T., Izumi, R. et al. (1968) Studies on tuberactinomycin (tuberactin), a new antibiotic. I. Taxonomy of producing strain, isolation and characterization. The Journal of Antibiotics, 21, 681–687.
48. Nakao, Y., Fujita, M., Warabi, K. et al. (2000) Miraziridine A, a novel cysteine protease inhibitor from the marine sponge *Theonella* aff. *mirabilis*. Journal of the American Chemical Society, 122, 10462–10463.
49. Llewellyn, L.E. (2006) Saxitoxin, a toxic marine natural product that targets a multitude of receptors. Natural Product Reports, 23, 200–222.
50. Goto, T., Inoue, S. and Sugiura, S. (1968) *Cypridina* bioluminescence IV. Synthesis and chemiluminescence of 3,7-dihydroimidazo[1,2-a]pyrazin-3-one and its 2-methyl derivative. Tetrahedron Letters, 3873–3876.
51. Omura, S., Arai, N., Yamaguchi, Y. et al. (2000) Argifin, a new chitinase inhibitor, produced by *Gliocladium* sp. FTD-0668. I. Taxonomy, fermentation, and biological activities. The Journal of Antibiotics, 53, 603–608.
52. Arai, N., Shiomi, K., Iwai, Y. and Omura, S. (2000) Argifin, a new chitinase inhibitor, produced by *Gliocladium* sp. FTD-0668. II. Isolation, physico-chemical properties, and structure elucidation. The Journal of Antibiotics, 53, 609–614.
53. Shiomi, K., Arai, N., Iwai, Y. et al. (2000) Structure of argifin, a new chitinase inhibitor produced by *Gliocladium* sp. Tetrahedron Letters, 41, 2141–2143.

54. Dixon, M.J., Andersen, O.A., van Aalten, D.M. and Eggleston, I.M. (2005) An efficient synthesis of argifin: a natural product chitinase inhibitor with chemotherapeutic potential. *Bioorganic & Medicinal Chemistry Letters*, **15**, 4717–4721.
55. Gouda, H., Yanai, Y., Sugawara, A. *et al.* (2008) Computational analysis of the binding affinities of the natural-product cyclopentapeptides argifin and argadin to chitinase B from *Serratia marcescens*. *Bioorganic and Medicinal Chemistry*, **16**, 3565–3579.
56. Shoji, J., Hinoo, H., Katayama, T. *et al.* (1992) Isolation and characterization of new peptide antibiotics, plusbacins A_1-A_4 and B_1-B_4. *The Journal of Antibiotics*, **45**, 817–823.
57. Shoji, J., Hinoo, H., Katayama, T. *et al.* (1992) Structures of new peptide antibiotics, plusbacins A_1-A_4 and B_1-B_4. *The Journal of Antibiotics*, **45**, 824–831.
58. Maki, H., Miura, K. and Yamano, Y. (2001) Katanosin B and plusbacin A_3, inhibitors of peptidoglycan synthesis in methicillin-resistant *Staphylococcus aureus*. *Antimicrobial Agents and Chemotherapy*, **45**, 1823–1827.
59. Wohlrab, A., Lamer, R. and VanNieuwenhze, M.S. (2007) Total synthesis of plusbacin A_3: a depsipeptide antibiotic active against vancomycin-resistant bacteria. *Journal of the American Chemical Society*, **129**, 4175–4177.
60. Omura, S., Tomoda, H., Abe, A. *et al.* (2007) Guadinomine antibiotics manufacture with *Streptomyces*, WO 2007/102229.
61. Tsuchiya, S., Sunazuka, T., Hirose, T. *et al.* (2006) Asymmetric total synthesis of (+)-K01-0509 B: determination of absolute configuration. *Organic Letters*, **8**, 5577–5580.
62. Kinnel, R.B., Gehrken, H.-P. and Scheuer, P.J. (1993) Palau'amine: a cytotoxic and immunosuppressive hexacyclic bisguanidine antibiotic from the sponge *Stylotella agminata*. *Journal of the American Chemical Society*, **115**, 3376–3377.
63. Kinnel, R.B., Gehrken, H.-P., Swali, R. *et al.* (1998) Palau'amine and its congeners: A family of bioactive bisguanidines from the marine sponge *Stylotella aurantium*. *The Journal of Organic Chemistry*, **63**, 3281–3286.
64. Lanman, B.A., Overman, L.E., Paulini, R. and White, N.S. (2007) On the structure of palau'amine: Evidence for the revised relative configuration from chemical synthesis. *Journal of the American Chemical Society*, **129**, 12896–12900.
65. Grube, A. and Köck, M. (2007) Structural assignment of tetrabromostyloguanidine: does the relative configuration of the palau'amines need revision? *Angewandte Chemie – International Edition*, **46**, 2320–2324.
66. Buchanan, M.S., Carroll, A.R., Addepalli, R. *et al.* (2007) Natural products, stylissadines A and B, specific antagonists of the P2X7 receptor, an important inflammatory target. *The Journal of Organic Chemistry*, **72**, 2309–2317.
67. Kobayashi, H., Kitamura, K., Nagai, K. *et al.* (2007) Carteramine A, an inhibitor of neutrophil chemotaxis, from the marine sponge *Stylissa carteri*. *Tetrahedron Letters*, **48**, 2127–2129.
68. Köck, M., Achim, G., Seipe, I.B. and Baran, P.S. (2007) The pursuit of palau'amine. *Angewandte Chemie – International Edition*, **46**, 6586–6594.
69. Berlinck, R.G.S., Braekman, J.C., Daloze, D. *et al.* (1990) Two new guanidine alkaloids from the mediterranean sponge *Crambe crambe*. *Tetrahedron Letters*, **31**, 6531–6534.
70. Jares-Erijman, E.A., Ingrum, A.A., Sun, F. and Rinehart, K.L. (1993) On the structures of crambescins B and C1. *Journal of Natural Products*, **56**, 2186–2188.
71. Gerlinck, R.G.S., Braekman, J.C., Daloze, D. *et al.* (1992) Crambines C1 and C2: Two further ichthyotoxic guanidine alkaloids from the sponge *Crambe crambe*. *Journal of Natural Products*, **55**, 528–532.
72. Snider, B.B. and Shi, Z. (1992) Biomimetic synthesis of the bicyclic guanidine moieties of crambines A and B. *The Journal of Organic Chemistry*, **57**, 2526–2528.
73. Harbour, G.C., Tymiak, A.A., Rinehart, K.L. *et al.* (1981) Ptilocaulin and isoptilocaulin, antimicrobial and cytotoxic cyclic guanidines from the Caribbean sponge *Ptilocaulis* aff. *P. spiculifer* (Lamarck, 1814). *Journal of the American Chemical Society*, **103**, 5604–5606.
74. Kashman, Y., Hirsh, S., McConnell, O.J. *et al.* (1989) Ptilomycalin A: a novel polycyclic guanidine alkaloid of marine origin. *Journal of the American Chemical Society*, **111**, 8925–8926.

75. Palagiano, E., De Marino, S., Minale, L. et al. (1995) Pitlomycalin A, crambescidin 800 and related new highly cytotoxic guanidine alkaloids from the starfishes *Fromia monilis* and *Celerina heffernani*. *Tetrahedron*, **51**, 3675–3682.
76. Patil, A.D., Kumar, N.V., Kokke, W.C. et al. (1995) Novel alkaloids from the sponge *Batzella* sp.: Inhibitors of HIV gp120-human CD4 binding. *The Journal of Organic Chemistry*, **60**, 1182–1188.
77. Patil, A.D., Freyer, A.J., Taylor, P.B. et al. (1997) Batzelladines F-I, novel alkaloids from the sponge *Batzella* sp.: Inducers of p56lck-CD4 dissociation. *The Journal of Organic Chemistry*, **62**, 1814–1819.
78. Snider, B.B. and Chen, J. (1996) Synthesis of the tricyclic portions of batzelladines A, B and D. Revision of the stereochemistry of batzelladines A and D. *Tetrahedron Letters*, **37**, 6977–6980.
79. Cohen, F. and Overman, L.E. (2006) Enantioselective total synthesis of batzelladine F and definition of its structure. *Journal of the American Chemical Society*, **128**, 2604–2608.
80. Bensemhoun, J., Bombarda, I., Aknin, M. et al. (2007) Ptilomycalin D, a polycyclic guanidine alkaloid from the marine sponge *Monanchora dianchora*. *Journal of Natural Products*, **70**, 2033–2035.
81. Gallimore, W.A., Kelly, M. and Scheuer, P.J. (2005) Alkaloids from the sponge *Monanchora unguifera*. *Journal of Natural Products*, **68**, 1420–1423.
82. Hua, H., Peng, J., Dunbar, D.C. et al. (2007) Batzelladine alkaloids from the Caribbean sponge *Monanchora unguifera* and the significant activities against HIV-1 and AIDS opportunistic infectious pathogens. *Tetrahedron*, **63**, 11179–11188.
83. Arnold, M.A., Day, K.A., Duron, S.G. and Gin, D.Y. (2006) Total synthesis of (+)-batzelladine A and (−)-batzelladine D via [4 + 2]-annulation of vinyl carbodiimides with *N*-alkyl imines. *Journal of the American Chemical Society*, **128**, 13255–13260.
84. Shimokawa, J., Ishiwata, T., Shirai, K. et al. (2005) Total synthesis of (+)-batzelladine A and (−)-batzelladine D, and identification of their target protein. *Chemistry – A European Journal*, **11**, 6878–6888.
85. Witherup, K.M., Ransom, R.W., Graham, A.C. et al. (1995) Martinelline and martinellic acid, novel G-protein linked receptor antagonists from the tropical plant *Martinella iquitosensis* (Bignoniaceae). *Journal of the American Chemical Society*, **117**, 6682–6685.
86. Ma, D., Xia, C., Jiang, J. and Zhang, J. (2001) First total synthesis of martinellic acid, a naturally occurring bradykinin receptor antagonist. *Organic Letters*, **3**, 2189–2191.
87. Ma, D., Xia, C., Jiang, J. et al. (2003) Aromatic nucleophilic substitution or CuI-catalyzed coupling route to martinellic acid. *The Journal of Organic Chemistry*, **68**, 442–451.
88. Ikeda, S., Shibuya, M. and Iwabuchi, Y. (2007) Asymmetric total synthesis of martinelline and martinellic acid. *Chemical Communications*, 504–506.
89. Watanabe, C.K. (1918) Studies in the metabolism changes induced by administration of guanidine bases. I. Influence of injected guanidine hydrochloride upon blood sugar content. *The Journal of Biological Chemistry*, **33**, 253–265.
90. Luft, D., Schmülling, R.M. and Eggstein, M. (1978) Lactic acidosis in biguanide-treated diabetics. *Diabetologia*, **14**, 75–87.
91. Witters, L.A. (2001) The blooming of the French lilac. *The Journal of Clinical Investigation*, **108**, 1105–1107.
92. Black, J.W., Duncan, W.A.M., Durant, C.J. et al. (1972) Definition and antagonism of histamine H_2-receptors. *Nature*, **236**, 385–390.
93. Durant, G.J., Emmett, J.C., Ganellin, C.R. et al. (1977) Cyanoguanidine-thiourea equivalence in the development of the histamine H_2-receptor antagonist, cimetidine. *Journal of Medicinal Chemistry*, **20**, 901–906.
94. Rodloff, A.C., Goldstein, E.J. and Torres, A. (2006) Two decades of imipenem therapy. *The Journal of Antimicrobial Chemotherapy*, **58**, 916–929.
95. Leanza, W.J., Wildonger, K.J., Miller, T.W. and Christensen, B.G. (1979) *N*-Acetimidoyl- and *N*-formimidoylthienamycin derivatives: antipseudomonal β-lactam antibiotics. *Journal of Medicinal Chemistry*, **22**, 1435–1436.
96. Olken, N.M. and Marletta, M.A. (1993) N^G-Methyl-L-arginine functions as an alternate substrate and mechanism-based inhibitor of nitric oxide synthase. *Biochemistry*, **32**, 9677–9685.

97. Feldman, P.L., Griffith, O.W., Hong, H. and Stuehr, D.J. (1993) Irreversible inactivation of macrophage and brain nitric oxide synthase by L-N^G-methylarginine requires NADPH-dependent hydroxylation. *Journal of Medicinal Chemistry*, **36**, 491–496.
98. Furfine, E.S., Harmon, M.F., Paith, J.E. and Garvey, E.P. (1993) Selective inhibition of constitutive nitric oxide synthase by L-N^G-nitroarginine. *Biochemistry*, **32**, 8512–8517.
99. Rees, D.D., Palmer, R.M.J., Schulz, R. *et al.* (1990) Characterization of three inhibitors of endothelial nitric oxide synthase *in vitro* and *in vivo*. *British Journal of Pharmacology*, **101**, 746–752.
100. Moore, W.M., Webber, R.K., Jerome, G.M. *et al.* (1994) L-N^6-(1-Iminoethyl)lysine: A selective inhibitor of inducible nitric oxide synthase. *Journal of Medicinal Chemistry*, **37**, 3886–3888.
101. Alderton, W.K., Angell, A.D.R., Craig, C. *et al.* (2005) GW274150 and GW273629 are potent and highly selective inhibitors of inducible nitric oxide synthase *in vitro* and *in vivo*. *British Journal of Pharmacology*, **145**, 301–312.
102. Sands, M., Kron, M.A. and Brown, R.B. (1985) Pentamidine: a review. *Reviews of Infectious Diseases*, **7**, 625–634.
103. Goa, K.L. and Campoli-Richards, D.M. (1987) Pentamidine isethionate. A review of its antiprotozoal activity, pharmacokinetic properties and therapeutic use in *Pneumocystis carinii* pneumonia. *Drugs*, **33**, 242–258.
104. Fox, K.R., Sansom, C.E. and Stevens, M.F.G. (1990) Footprinting studies on the sequence-selective binding of pentamidine to DNA. *FEBS Letters*, **266**, 150–154.
105. Sun, T. and Zhang, Y. (2008) Pentamidine binds to tRNA through non-specific hydrophobic interactions and inhibits aminoacylation and translation. *Nucleic Acids Research*, **36**, 1654–1664.
106. Makulu, D.R. and Waalkes, T.P. (1975) Interaction between aromatic diamidines and nucleic acids: possible implications for chemotherapy. *Journal of the National Cancer Institute*, **54**, 305–309.
107. Chen, D., Marsh, R. and Aberg, J.A. (2007) Pafuramidine for *Pneumocystis jiroveci* pneumonia in HIV-infected individuals. *Expert Review of Anti-Infective Therapy*, **5**, 921–928.

11
Perspectives

Tsutomu Ishikawa[1] *and Davor Margetic*[2]

[1]Graduate School of Pharmaceutical Sciences, Chiba University, 1-33 Yayoi, Inage, Chiba 263-8522, Japan
[2]Rudjer Bošković Institute, Bijenička c. 54, 10001 Zagreb, Croatia

Organic bases attract much attention as environment friendly chemicals due to their easy structural modification, repeated use of recovered materials and simple operation based on the acid–base concept [1]. The functions of organosuperbase catalysts and of the related intelligent molecules in organic synthesis (reaction) are attributable to their affinity to form substrates through their stronger or lesser proton (or nucleophile) affinity. In the cases of a superbase with strong basicity, salt formation resulting from preliminary reaction with the substrate is crucial for the desired reaction course, whereas with an intelligent molecule that is not necessarily strongly basic but is able to form tight hydrogen bonding with the substrate, control of the reaction is by interaction through hydrogen bond network(s). In particular, for effective asymmetric induction it is very important for the active site in the catalyst to selectively (or specifically) recognize target groups in the substrate and then to construct a rigid, but flexible, chiral environment in transition state. Therefore, complexation not only through mono-interaction between each functional group in the catalyst and in the substrate, but also through multi-interaction containing additional functional group(s) in some cases is required for effective molecular recognition. Thus, a lot of intelligent molecules with multi-functions have been designed and prepared by introducing different functional, but mutually noninteractive, groups to the original molecule [2].

Individual reaction in living organisms is strictly controlled by reactant–substrate specificity, as exemplified in enzymatic reactions, even though the total mode of action is systematically controlled by correlation with other reactions. Host–guest interaction in inclusion chemistry using cyclodextrin [3] is a typical example of specific reactions in an artificial field. On the other hand, nonspecific catalysts that are tolerant to various functional

---

Superbases for Organic Synthesis: Guanidines, Amidines, Phosphazenes and Related Organocatalysts
Edited by Tsutomu Ishikawa
© 2009 John Wiley & Sons, Ltd

groups are generally requested as reaction tools with wide ranges of applicability in chemically related organic reactions. Thus, in the design of more intelligent molecules as synthetic tools it is necessary to overcome antipathy between 'specific' for functionality recognition and 'common' for reactivity in organic synthesis. Computer-aided molecular recognition between ligand and pharmacophore has been progressed in drug discovery research [4]. This concept may give a clue for the design of the new generation.

Interestingly, a number of novel nitrogen-containing superbase backbones have been identified through the extensive computational work of Maksić's group; the representatives are shown in Figure 11.1.

Molecules (e.g. **1**) possessing imino structural and electronic motif have been recognized as important building blocks for the construction of potent superbases with extended π-systems [5]. The principles which make systems thermodynamically stable and highly basic are explained by a large increase in the π-delocalization energy of the corresponding conjugate acids, thus leading to appreciable stabilization of protonated species.

Another important structural motif for construction of the strong organosuperbases are cyclopropeneimines (e.g. **2**) [6]. They exhibit high basicity due to the significant aromatic stabilization of the three-membered ring upon protonation and the basicity can be increased by amine substitutions at the double bond. Amino groups stimulate aromatization of the cyclopropene fragment and also release some of their lone pair electron density, thus contributing to a uniform distribution of the positive charge over the entire molecular system. Further increase in the basicity of cyclopropeneimines could be achieved by intramolecular hydrogen bonding (IMHB) such as depicted in **2** [7].

Quinonimine (e.g. **3**) exhibits a very high basicity, which can be ascribed to significant aromatization of the semiquinoid structure upon protonation by resonance [8]. The amine substitutents increase the conjugation of the planar systems thus enhancing the relaxation effect. The amino group is capable of accommodating the positive charge, thus increasing the double bond character in the iminium fragment resulting from protonation. A further increase of the basicity could be achieved by a domino effect in the extended π-system involving two quinoid fragments.

Extended polycyclic π-systems (e.g. **4**) possessing a carbonyl oxygen terminus serve as a basic proton scavenger [9]. Carbonyl polyenes (e.g. **5**) are also calculated to exhibit high basicities, even belonging to the lower part of superbasicity scale [10]. These open chain and *zig-zag* extended π-systems involve polyenes and a carbonyl functional group at the molecular terminus. Structurally related to carbonyl polyenes are iminopolyenes (e.g. **6**), being extended π-systems and a new class of highly basic compounds [11]. The explanation of iminopolyene basicity is the increase in stabilization triggered by protonation, and amino substitution is crucial for their superbasicity by amplifying the resonance effect.

Some of the extended π-systems (e.g. **7**) possessing imino nitrogen atoms as the most basic sites, which are parts of the [3]iminoradialene or quinonimine structure, are neutral organosuperbases [12]. The diaminophosphono [=P(NR$_2$)$_2$] and diaminomethylene [=C(NR$_2$)$_2$] ends (and 1,3-diamino-2-methylene cyclopropene ring) enable efficient cationic resonance across the extended linear π-system, contributing to enlarged basicity. The =P(NR$_2$)$_3$ end gives the largest basicity.

Poly-2,5-dihydropyrroles (e.g. **8**) represent another class of extended π-system with pronounced basicity [13]. The aromatization of their protonated bases amplifies the susceptibility toward the proton attack. Aromatization of the five-membered ring and

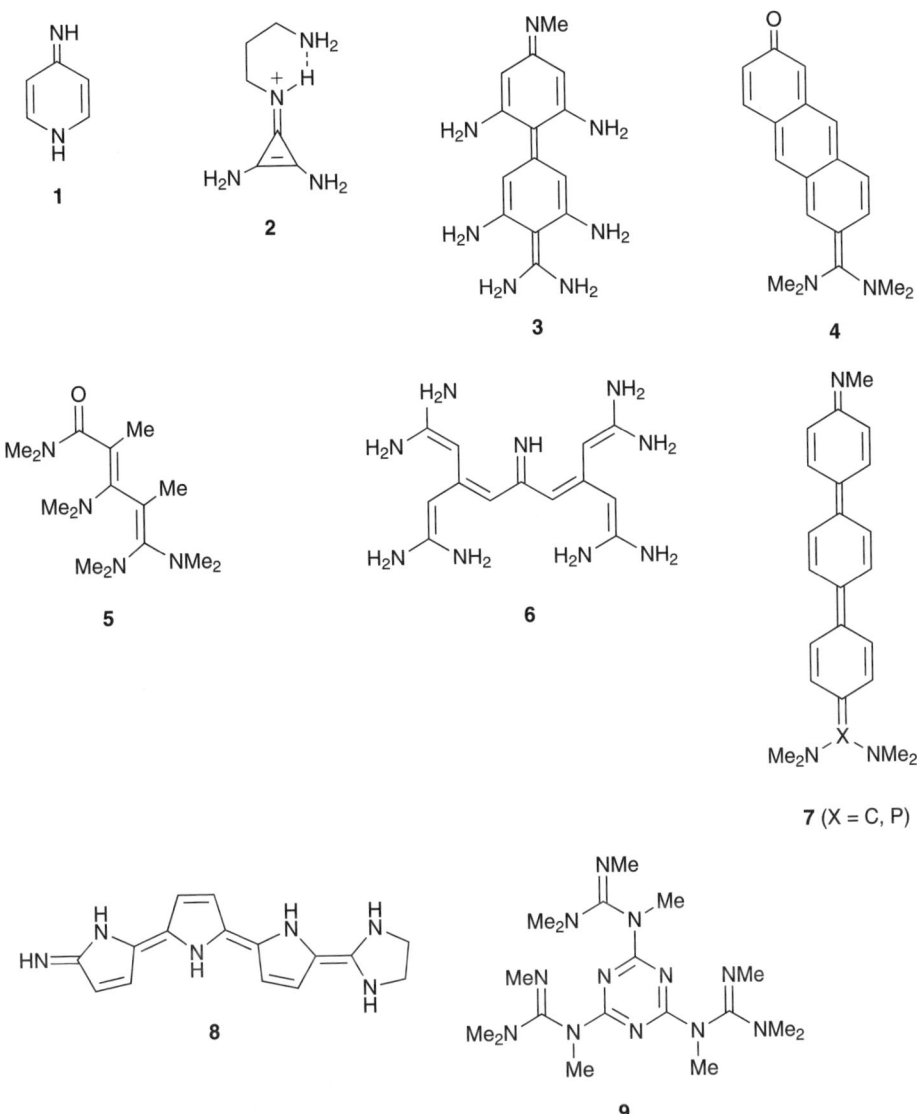

Figure 11.1 Representative structures of computationally designed nitrogen-containing superbases by Maskić's group

accompanying stabilization of the conjugate base, with the important resonance effect, are the main contributions to high basicity.

Triazine (e.g. **9**) can be used as a useful building element for neutral superbases, where incorporation enhances the basicity through thermodynamic stabilization of the molecule [14]. The basicity of triazine is enhanced by dimethylamino substituents such as in **9**. On

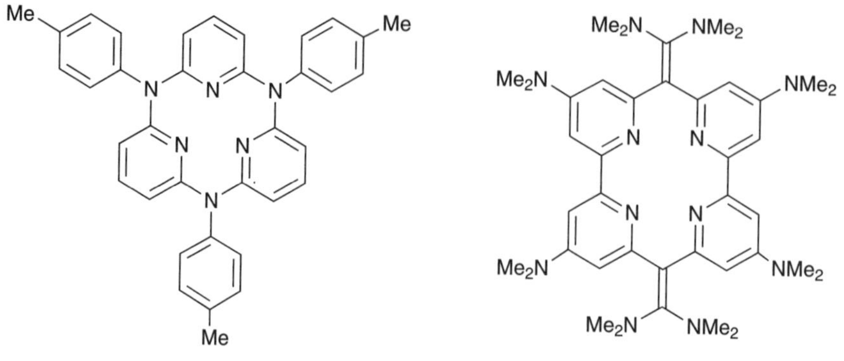

Figure 11.2 Typical structures of N-substituted azacalix[n](2,6)pyridine derivatives

protonation the triazino-guanidine molecule yields an IMHB, which is further stabilized by cationic resonance.

The concept of the IMHB has been proved particularly useful if used in a multiple fashion [15]. Then the cooperative and collective IMHB effects lead to a considerable stabilization of the corresponding conjugate acids. An additional degree of freedom in structures optimizing proton sponge systems is offered by changing the naphthalene moiety for other aromatic spacers, thus varying the N.N distance of the proton pincer. Following these concepts, N-substituted azacalix[n](2,6)pyridines have been synthesized and N,N',N''-tris (p-tolyl)azacalix[3](2,6)pyridine **10** has been found to be the size of the cavity appropriate for a high proton affinity[16] (Figure 11.2). These findings are corroborated by theoretical study by Despotović et al. [17]. A closely related macrocycle **11**, with an additional pyridine ring incorporated, also showed respectable superbasicity [18]. The amplified basicity in both systems is a consequence of strong cationic resonance in conjugate acids supported by stabilization provided by IMHB.

A lot of catalysts incorporating both acidic and basic groups in the molecule have been designed and applied to molecular recognition and asymmetric synthesis, in which intended functionality could be obtained by control of the reaction conditions [19]. In living organisms a variety of α-amino acids with different characters are used as building blocks in the construction of peptides which control the total mode of action. The success in the catalytic use of proline for asymmetric synthesis [20] suggests that alternative enzymatic reactions may propose attractive models in the design of intelligent but simple molecules. Thus, it might be possible to create useful, more sophisticated catalysts from cheap and easily available precursors in the near future.

In the twenty first century it is strongly requested that synthetic chemistry should contribute more extensively to the solution of serious environmental problems for the next generations. One of various chemical approaches is the production of more effective catalysts for the removal of harmful substances, such as arsenic and heavy metals. Basic compounds, in general, show strong affinity to metal salts. In fact, it is known that guanidines can form complexes with a range of metal salts [21], indicating that organo-superbases could serve as effective metal scavengers as well as proton scavengers. This chelation technique may be applied to recycle rare metals from used goods. Organosuperbases with strong basicity could be potential candidates for these purposes.

As discussed in Chapter 4 (Oxazolidinone and Oxazole), guanidines catalyze the insertion of carbon dioxide (CO_2) to alkynic bonds [22]. This behaviour may give a hint to trapping CO_2 by organobase catalyzed chemical reaction, even to necessary modification of reaction conditions. Hopefully, it is the ideal that the CO_2-incorporated products could be used as new energy sources.

References

1. For example: Anestas, P.T. and Warner, J.C. (1998) *Green Chemistry: Theory and Practice*, Oxford University Press, Oxford.
2. For example: Shibuguchi, T., Fukuta, Y., Akachi, Y. *et al.* (2002) Development of new asymmetric two-center catalysts in phase-transfer reactions. *Tetrahedron Letters*, **45**, 9539–9543. Ohshima, T., Shibuguchi, T., Fukuta, Y. and Shibasaki, M. (2004) Catalytic asymmetric phase-transfer reactions using tartarate-derived asymmetric two-center organocatalysts. *Tetrahedron*, **60**, 7743–7754.
3. For example: Atwood, J.L., Davies, J.E. and MacNicol, D.D. (1984) *Inclusion Compounds*, 3, Academic Press, New York.
4. For example: Walters, W.P., Stahl, M.T. and Murcko, M.A. (1998) Virtual screening – an overview. *Drug Discovery Today*, **3**, 160–178.
5. Kovačević, B. and Maksić, Z.B. (1998) Toward engineering of very strong organic bases: pronounced proton affinity of molecules possessing imino structural and electronic motif. *Chemical Physics Letters*, **288**, 289–292.
6. Maksić, Z.B. and Kovačević, B. (1999) Spatial and electronic structure of highly basic organic molecules: cyclopropeneimines and some related systems. *Journal of Physical Chemistry A*, **103**, 6678–6684.
7. Gattin, Z., Kovačević, B. and Maksić, Z.B. (2005) Cooperative intramolecular hydrogen bonding effect and basicity - an *ab initio* and DFT study of the superbasic properties of N-[(dimethylamino)alkyl]-2,3-diaminocycloprop-2-ene-1-imines. *European Journal of Organic Chemistry*, 3206–3213.
8. Maksić, Z.B. and Kovačević, B. (1998) Toward organic superbases: the electronic structure and the absolute proton affinity of quinodiimines and some related compounds. *Journal of Physical Chemistry A*, **102**, 7324–7328.
9. Despotović, I., Maksić, Z.B. and Vianello, R. (2007) Design of Brønsted neutral organic bases and superbases by computational DFT methods: cyclic and polycyclic quinones and [3] carbonylradialenes. *European Journal of Organic Chemistry*, 3402–3413.
10. Despotović, I., Maksić, Z.B. and Vianello, R. (2006) Engineering neutral organic bases and superbases by computational DFT methods – carbonyl polyenes. *European Journal of Organic Chemistry*, 5505–5514.
11. Vianello, R., Kovaević, B. and Maksić, Z.B. (2002) In search of neutral organic superbases – iminopolyenes and their amino derivatives. *New Journal of Chemistry*, **26**, 1324–1328.
12. Despotović, I., Maksić, Z.B. and Vianello, R. (2007) Computational design of Brønsted neutral organic superbases-[3]iminoradialenes and quinoimines are important synthetic targets. *New Journal of Chemistry*, **31**, 52–62.
13. Maksić, Z.B., Glasovac, Z. and Despotović, I. (2002) Predicted high proton affinity of poly-2,5-dihydropyrrolimines – the aromatic domino effect. *Journal of Physical Organic Chemistry*, **15**, 499–508.
14. Despotović, I., Kovačević, B. and Maksić, Z.B. (2007) Pyridine and s-triazine as building blocks of nonionic organic superbases – density functional theory B3LYP study. *New Journal of Chemistry*, **31**, 447–457.

15. Kovacevic, B., Maskic, Z.B., Vianello, R. and Primorac, M. (2002) Computer aided design of organic superbases: the role of intramolecular hydrogen bonding. *New Journal of Chemistry*, **26**, 1329–1334.
16. Kanbara, T., Suzuki, Y. and Yamamoto, T. (2006) New proton-sponge-like macrocyclic compound: synergistic hydrogen bonds of aminopyridine. *European Journal of Organic Chemistry*, 3314–3316.
17. Despotović, I., Kovačević, B. and Maksić, Z.B. (2007) Derivatives of azacalix[3](2,6)pyridine are strong neutral organic superbases: a DFT study. *Organic Letters*, **9**, 1101–1104.
18. Kovačević, B., Despotović, I. and Maksić, Z.B. (2007) In quest of strong neutral organic bases and superbases-supramolecular systems containing four pyridine subunits. *Tetrahedron Letters*, **48**, 261–264.
19. For example: Hatano, M., Miyamoto, T. and Ishihara, K. (2005) Enantioselective addition of organozinc reagents to aldehydes catalyzed by 3,3′-bis(diphenylphosphinoyl)-BINOL. *Advanced Synthesis and Catalysis*, **347** 1561–1568.
20. Göger, H. and Wilker, J. (2001) The application of L-proline as an enzyme mimic and further new asymmetric syntheses using small organic molecules as chiral catalysts. *Angewandte Chemie – International Edition*, **40**, 529–532.

    Cordova, A. (2004) The direct catalytic asymmetric Mannich reaction. *Accounts of Chemical Research*, **37**, 102–112.

    Nota, W., Tanaka, F. and Barbas, C.F. III (2004) Enamine-based organocatalysis with proline and diamines: The development of direct catalytic asymmetric aldol, Mannich, Michael and Diels–Alder reactions. *Accounts of Chemical Research*, **37**, 580–591.

    Marques, M.M.B. (2006) Catalytic enantioselective cross-Mannich reaction of aldehydes. *Angewandte Chemie – International Edition*, **45**, 348–352.
21. For example: Longhi, R. and Drago, R.S. (1965) Transition metal ion complexes of tetramethylguanidine. *Inorganic Chemistry*, **4**, 11–14.
22. Costa, M., Chinsoli, G.P. and Rizzardi, M. (1996) Base-catalyzed direct introduction of carbon dioxide into acetylenic amines. *Chemical Communications*, 1699–1700.

Index

References to figures are given in italic type; references to tables are given in bold type.

acetonitrile 9
acylation reactions, polymer-supported
    188–190
adamanzanes 17, *19*
alcohols, silyation 114, **115, 116**
aldol reactions *126*, 211–214
    copper and amidine cocatalysed 75
    DMAN-catalysed *263*
    guanidine-catalysed 99–101
    phosphazene-catalysed 156
alkenyl sulfones 156
alkylation reactions
    amidine and palladium cocatalysed 75
    amidines 52–53
    DMAN-catalysed 252–256, *258*, 259–261
    guanidine-catalysed 112
    phenols with alkyl halides *196*
    phosphazine-catalysed 151, 152, *155*
        polymer-supported 155
    polymer-supported 194–196
    Michael reactions 190–194
alkylidine phtalates 72
allylation reactions 278–281
2-(allyloxy)phenylacetyl (APAC) group 267
allylsulfone, dimerization 178
amidation reactions
    amidine and iridium cocatalysed 75
    phosphazene-catalysed 153
amides
    in amidine synthesis 54
    synthesis, DMAN-catalysed 256–259
amidines 1, 20–24
    overview 49–52
    applications
        acetoxybromination 65–66
        azidation 67
    as cocatalyst in metal-catalysed
        reactions 74–77
    dehydrohalogenation 70–71
    deprotection 70–71
    deprotonation reactions 71
    displacement reactions 72
    Horner-Wadsworth-Emmons
        reaction 72
    intramolecular cyclization 72, **73**
    isomerization 72
    Michael reaction 77–78
    Nef reaction 78–79
    nucleophilic epoxidation 79
    oxidation 80
    tandem reaction 81–82
    basicity 20–21, **21**
    chiral *64*
    dehydrohalogenation reactions *50*, 51
    ionic, conjugation 2
    medicinal 306–309
    natural derivatives 295–299
    preparation
        by alkylation of acylic amidines 52–53
        by aziridine ring-opening 63–65
        by condensation of 1,2-diamine 53
        by modification of amide derivatives
            54–59
        by multi-component reactions 59–62, *61*
        from thioamides 57–58
        from thioimidates 59
        oxidative 62–63
        via haloiminium salt 56
    rearrangement reactions *50*
amidinium salts 82–86
    synthesis 85–86
amine derivatives 1–2, *2*

Superbases for Organic Synthesis: Guanidines, Amidines, Phosphazenes and Related Organocatalysts
Edited by Tsutomu Ishikawa
© 2009 John Wiley & Sons, Ltd

amines 16–17, 17–20, 53, *54*
    acylation, polymer-supported 189–190
    condensation to amidines 53, *55*
    synthesis using DMAN 252
amino acids 1
amino groups, and gas-phase basicity,
    guanidines 25–26
amphidinol 225, *228*
amythiamycin D 227, *231*
APAC group 267
arginine 1
aryl groups, on amidines 22
arylation reactions
    phosphazene-catalysed 174–175
    phosphazine-catalysed 152, 152–153
    proazaphosphatrane-catalysed 177
arylsilanes 169–170
arylsulfonylbicyclobutanes **133**
aza-Henry reaction 101–102, **101**
azidation reactions 67, **68**, **134**
    alkyl halides **132**
    guanidine-catalysed 113
aziridation reactions, DBU-catalysed 68, **69**
aziridines 125, *129*
    in amidine synthesis 63
    ring-opening reactions **239**

basicity 1
    overview 1–6
    amidines 20–21, **21**
    gas-phase
        amidine derivatives 23
        amidines 22
        definition 41
        guanidines **31**
        phosphazenes **32–33**
    guanidines 25, **25**
        mono- to polyguanidines 28
    guanidinophosphazenes **35**
    measurement 9–10
    phosphazenes 146–148
    proazaphosphatranes **39**
    proton sponges 10–12, *11*
    solution, phosphazenes **32**
    ureas and thioureas **279**
batzelladines *135*, 302, *304*
Bayliss-Hillman reaction **69**
    amidine-catalysed 68
    guanidine-catalysed 102
BEMP 151

benzimidazoles 117, **118**
benzofurans 167–168, 214
birnbaumins 296
bismuth-catalysed reactions 121
bisphenol
    catalytic activity 276–277
    as hydrogen donor 274–276
bispidines 14, *15*
buttressing effect 12
2-*tert*-butylimino-2-diethylamino-1,3,-
    dimethylperhydro-1,3,2,-
    diazaphorine 187
    polymeric 198–199

carbapenems 115, *117*, 217, *218*, 306
carpanone 154–155
chalcone *111*
chiral compounds
    guanidines 94–99, 108–109, *108*
    guanidinium ylides 128
    ureas and thioureas 282–291
ciguatoxin CTX3C 246
cinnamaldehyde 80
Claisen rearrangements 217
    urea and thiourea-catalysed 278–281
click chemistry 59
cobalt-catalysed reactions, amidines 74
conjugate addition 233–236
Cope rearrangements 163
copper-catalysed reactions
    and amidines 75
    with guanidine cocatalyst 117, **118**
coraxeniolide A 240, *245*
crambescins 302, 303
cryptates 17, *19*
CTX3C 246
cyanosilyation 103
cyclization reactions
    amidine-catalyzed 72, *73*
    phosphazene-catalysed 162
cycloaddition reactions 68–69
    TMGA-catalysed **133**
cycloazaphosphines *see* proazaphosphatranes
cyclohexene, oxidative amidination 63
cyclophellitol 211, 211–213, *212*
cyclopropanation, guanidine-catalysed
    114, *127*

DBU *see* 1,5,-diazabicyclo[5.4.0]undec-5-ene
dehydrohalogenation, amidine-catalysed 70

density functional theory (DFT) 41, 146–147, 148–149
  guanidinophosphazenes 36–37
deplacheine 225, *229*
deprotection reactions 70–71, 225
deprotonation reactions 71
diamines 16–17, 53, *54*
1,5-diazabicyclo[4.3.0]non-5-ene (DBN) 20, 21–22, 49–51
  aldol reactions 66–67
  preparation *51*
  synthesis *56*
2,10-diazabicyclo[4.4.0]dec-1-ene 51
1,5,-diazabicyclo[5.4.0]undec-5-ene (DBU) 2, 20, 22, 49–51, 66, 187–188
  aldol-like reaction 66–67
  in azidation reactions 67
  catalysis of aziridination 68
  as cocatalyst in metal-catalysed reactions 75
  in deprotonation reactions 71
  in displacement reactions 72
  in isomerization reactions 72
  in nucleophilic epoxidation 79
  tandem reactions 81–82
diazonamide A 221, *223*
dictyomedins 167–168, *169*, 232, *238*
Diels-Alder reactions
  bisphenol-catalysed 276–277
  catalysed by amidinium salts 82–85, *83*
  guanidine-catalysed 103
  intramolecular 261–264
  urea and thiourea catalysed *283*
diepoxin σ 232, *238*
dimethyl sulfoxide (DMSO) 9
1,8,-bis(dimethylamino)napthalene (DMAN) 5, 9, 10, 37, 251–269, *252*
  alkylation and nitro aldol reactions 259–260
  in amide formation 256–259
  amine synthesis 252
  palladium-catalysed reactions 264–267
  pericyclic reactions 261–264
  related guanidinophosphazene bases *38*
bis(dimethylethyleneguanido)napthalene (DMEGN) 26
2-chloro-1,3,dimethylimidazolium chloride (DMC) 98–99
displacement reactions, in tandem with Michael reaction 125

DMAN *see* 1,8,-bis(dimethylamino)napthalene
DMC 98–99
dollabellatrienone 217, 240, *243*
dumsin 230, *234*

ecteinascidin 743 225, *227*, *241*
efrapeptins 296, **297**
elimination reactions 225–230
enzymes 273–274
epoxidation
  amidine-catalysed 79
  gunidine-catalysed 110–111
epoxysorbicillinol 236, *242*
ertapenem sodium 115, *117*
esterification, gunaidine-catalysed 113
ethers 230–233
  DBAN and DMAN-catalysed synthesis 252–256
  diaryl 163, **166**
eutypoxide L 236, *242*

Feist-Benary reaction **262**
fluorenes 12–13
Flustra foliacea 298
flustramine C 298
folskolin 240, *244*
fraxinellone 225, *229*
furamamide *54*
furan 117, *118*

gelsemine 213, *214*
glomerulatines 299
glycosidation, guanidine-catalysed 114
griseolic acid 230, *233*
guanidines 1, 24, 93–94
  acyclic and monocyclic 95–97
  acyl derivatives, acyclic 30–31
  aminopropyl substitution 29–30
  applications
    addition reactions 99–112
    as cocatalyst in metal-catalysed reactions 121–123
    heterocyclic compound synthesis 117–119
    Horner-Wadsworth-Emmons reactions 119–121
    isomerization 124
    oxidation 123
    reduction 124
    substitution reactions 112–116

guanidines (*Continued*)
  basicity **25**
  bicyclic 97–98
  chiral 94–99
  ionic, conjugation 2
  medicinal 305–309
  natural
    from marine invertebrates 301–302
    from plants 302–303
  natural derivatives, from
      microorganisms 300–301
  overview 24–31
  synthesis, DMC chemistry *100*
  synthesis based on DMC chemistry 98–99
guanidinium salts *107*, 125–136
  ionic liquid 128–131
guanidino-cyclopropenimines 30
guanidinophosphazenes 5–6, 35–37, *38*
gudinomines 301

haloactonization, guanidine-catalysed 104
haloiminium salt 56
Hartree-Fock (HF) modeling, proton affinity
    and gas-phase basicity 41
Heck reaction 123, 264–265, **266**
Henry reaction 74
  guanidine-catalysed 104–105
  nitroalkanes with aldehydes and
      ketones 190
heptane, relative acidity scale 148–149
heterocyclic compounds
  guanidine-catalysed synthesis 117–121
  phosphazene-catalysed synthesis 153
  polymer-supported synthesis 198–200
1,8,Bis-(hexamethyltriaminophosphazenyl)
    napthalene (HMPN) 34–35
histidine 1
HMPN 34–35
Horner-Wadsworth-Emmons reactions 72,
    118–120

IMDA reactions 261–264
IMHB 3
imidates, in amidine synthesis 54–56
imines
  coupling to amidines 53
  cycloaddition reactions 68, *70*
  hydrophosphorylation *286*
iminoamines 24
iminophosphoranes 31–35

imipenem 306
ingenol 221, 225, *229*
intramolecular hydrogen bond (IMHB) 10
  guanidines 30
intricarene 230, *235*
iridium-catalysed reactions, and amidines 75
isoamarine 65–66, *66*
  synthesis 53, *54*
isomerization reactions 72, **74, 237–247**
  guanidine-catalysed 124

Jacobsen's catalyst 284–285
Julia-Kocienski olefination 162, *163*

L-782,392 (anti-MRS agent) 217
β-lactam 262
lactonamycin 215, *216*
Leucocoprinus birnbaumii 296
lysine 1

manassantin A 231, *237*
martefragin A 221, 225
martinelline 302–303, *305*
Meerwein's salt 254, *255*
methyl vinyl ketone (MVK) 107
Michael reaction 77–78, *126*, 215–217
  guanidine-catalysed 106–110, **108**
  polymer-supported 193
  nickel and amidine cocatalysed 75
  phosphazene-catalysed 154
  polymer-supported 191–192, 193, 193–194
  in tandem with Aldol reaction 125
  thiourea-catalysed **287, 288**
molybdenum-catalysed reactions, and
    amidines 75
multicomponent reactions 59–62, *61*
mycalolide A 221, 222

Nef reaction 78
neodysiherbaine A 221, *224*
neoefrapeptins 296, **297**
nickel-catalysed reactions, and amidines 75
nitric oxide synthase inhibitors 307
NMR spectroscopy 148–149
NOS inhibitors 307
nucleophilic epoxidation 79

octosyl acid A 230, *236*
olefination 162, *163*
oxazoles 117, 117–118, *119*, 227

oxazolidinones   117, *119*
oxidation reactions
   amidine-catalysed   80
   guanidine-catalysed   123
oxindole synthesis   265

palau'amine   301
palladium-catalysed reactions   75, 264–268
   with guanidine cocatalyst   117, **118**
palmarumycin CP1   232, *238*
pentamidine   307–308
perhydrohistrionicotoxin   213
pericyclic reactions   217–220
perophoramidine   298–299
Peterson reaction, phosphazene-
   catalysed   170–173
phenols
   catalytic activity   276–277
   glycosidation   114
   as hydrogen donors   274–276
phorbol   230, *235*
phosphazenes   2–3, *4*, 31–35, *147, 150*
   applications
      aryl anion generation   169–170
      arylation of silyated nucleophiles
         165–167
      dictyomedin synthesis   167–168
      halogen-zinc exchange reactions
         173–174
      nucleophile addition to alkynes   164–165
   basicity   **32–33**
   classification   2–3
   overview and classification   145–150
   P1-type   146, *147*
      alkylation reactions   151
      applications
         alkylation reactions   151, 152
         arylation reactions   152, 153
         heterocyclic compound formation   153
   P2-type   156–159
      sigmatropic rearrangments   158
      sulfide formation   159
   P4-type   *148*
      alkylation reactions   159–160
      cyclization, benzofurans   162
      Julia-Kocienski olefination   162, *163*
      sulfinyl carbanion generation   160–162
      synthesis of biaryl ethers   163
   P5-type   164
phosphorus-containing compounds   **40**

piperidone, olefination   **73**
polyamines   16–17, *18, 19*
polymer-supported reactions
   acylations   188–190
   alkylations   190–198
   dehydrohalogenations and
      debrominations   205
   epoxide ring-openings   201
   heterocyclizations   198–200
   overview   187–190
   phosphazenes   154–155
preussomerin L   236, *242*
proazaphosphatranes   5, 37–40
   addition reactions   177
   alkanenitrile arylation reactions   177–178
   allylsilane activation   176–177
   properties   176
   reduction reactions   180
propargyl chloride   *135*
proton affinity (PA)
   definition   41
   proton sponges   10–12
proton sponges
   classical (napthalene-based)   10–12
   other   12–14
   polycyclic   14–20
   see also 1,8,-bis(dimethylamino)napthalene
PS-TBD   196–198
Pseudomonas aeruginosa   306
Psychotria glomerulata   299
ptilomycalins   302, 303
Pudovik-phospha-Brook rearrangement   80
pyrostatins   297
pyrroles   119

quinolenes, as proton sponges   12–13
quinonimine   316
ent-ravidomycin   227, *230*

reduction reactions
   gunaidine-catalysed   124
   proazaphosphatrane-catalysed   180
rhizoxin D   221, *223*
rhodium-catalysed reactions, and
   amidines   75–77
ribonucleic acid (RNA)   281, *285*
RNA   281, *285*

sapinofuranone B   230, *234*
sarcodonin G   240, *243*

scabronine G 240, *244*
Sch 57 050 (estrogen antagonist) 234, *240*
silyation, alcohols, guanidine-catalysed 114, **115**
solvents 9, 22
Stille reactions, proazaphosphatrane-catalysed 179, *180*
Strecker reaction 111, 191, *192*, 286
Streptomyces cattleya 306
sulfides 115
sulfur ylides 39, 157
superbase, definition 6
Suzuki reaction 122–123

taxol 231, *237*, 240, *245*
teicoplanin aglycon 221, *224*
telomestatin 225, *232*
tetrahydrofuran (THF) 9–10
1,1,bis-tetramethylguanido(napthalene) (TMGN) 5, 26
1,1,3,3,-tetramethylguanidine (TMG) 93–94, 99–101, 106–107, *106*, 121, 125, 131–136
tetramethylguanidinium azide (TMGA) 131–135
thermodynamic basicity *see* proton affinity
thienamycin 306
thioamides 57–59
  in amidine synthesis 57–58
thioethers, synthesis, DMAN-catalysed 256, *257, 258*
thioglycosides 153
thioimidates, in amidine synthesis 59, *60*
thioureas
  basicity **279**
  bisthioureas 289–290
  catalytic activity 277–278
    monothioureas
      alcohol-functionalised 287–288
      amine-functionalised 287
      cinchona alkaloid-functionalised 288–289
      Schiff base functionalised 284–285
  role as hydrogen donor 277–278

tin-catalysed reactions 76–77
TMGA 131–135
PS-*p*-toluenesulfonylmethyl isocyanide (TosMIC) 118
triamines 17
trichloromethylcarbinol 66–67
triguanidophosphines 40
tris(2,4,6,-trimethoxyphenyl)phosphine (TTMPP) 80
trimethylsilylcyanation reactions 177
bis(triphenylphosranylidene)ammonium fluoride (PPNF) 74
trisacchardies *202*

ureas
  basicity **279**
  binding affinities **280**
  catalytic activity 277–278
    chiral 278–281, 282–284
  role as hydrogen donor 277–278
  urea-sulfinamide hybrid catalyst 290
ustiloxin D 233, *239*

Verkade's base 37–40, 176–181
Viagra 198–199
vinyl halides 125, *127*
vinyl sulfones 156
vitamin A *50*

Wittig reaction 220–225

yatakemycin 221, *226*
ylides 220–221
  amidine-catalysed synthesis 68, *69*
  ammonium 217–218, *219*
  guanidinium 125–128, **132**
  nitrogen 39
  phosphorus 39
  reactions with aryl aldehydes **132**
  sulfur 39, 157
ytterbium-catalysed reactions 77

zinc-catalysed reactions, and guanidines 123